Promoting Statistical Practice and Collaboration in Developing Countries

Promoting Statistical Practice and Collaboration in Developing Countries

Edited by
O. Olawale Awe, Kim Love, and Eric A. Vance

CRC Press
Taylor & Francis Group
Boca Raton London New York

CRC Press is an imprint of the
Taylor & Francis Group, an **informa** business

A CHAPMAN & HALL BOOK

First edition published 2022
by CRC Press
6000 Broken Sound Parkway NW, Suite 300, Boca Raton, FL 33487-2742

and by CRC Press
4 Park Square, Milton Park, Abingdon, Oxon, OX14 4RN

CRC Press is an imprint of Taylor & Francis Group, LLC

ISBN: 978-1-032-19555-1 (hbk)
ISBN: 978-1-032-19850-7 (pbk)
ISBN: 978-1-003-26114-8 (ebk)

DOI: 10.1201/9781003261148

Typeset in Palatino
by codeMantra

Contents

Part 1 Statistics Collaboration and Practice in Developing Countries: Experiences, Challenges, and Opportunities

## Part 2	Building Capacity in Statistical Consulting and Collaboration Techniques through the Creation of Stat Labs

## Part 3	Statistics Education and Women's Empowerment

Part 4 Statistical Literacy and Methods across Disciplines

Part 5 New Approaches to Statistical Learning in Developing Countries

Part 6 Importance of Statistics in Urban Planning and Development

Part 7 Statistical Literacy in the Wider Society

Foreword

Rarely, but just often enough to rebuild hope, something happens to confound my pessimism about the recent unprecedented happenings in the world. This book is the most recent instance, and I think that all its readers will join me in rejoicing at the good it seeks to do. It is an example of the kind of international comity and collaboration that we could and should undertake to solve various societal problems.

The LISA 2020 program has the potential to be transformational. It has created an infrastructure that connects statisticians in the developing world to each other and to statisticians in the developed world. It enables participants to share software, data, courses, and analytical strategies. And it has created a pipeline that will enable more people to travel, in both directions, between the developed and developing world.

Rich nations consume unfair amounts of resources. Poor nations need to build quickly, before scarcity curtails their development opportunities. Statisticians cannot correct that problem, but we can help to solve it. We know a lot about evidence-based public policy, and we understand optimization. That is insufficient since we still need to gain the trust of the decision-making class, but it is a start. And we are not alone—many others are fellow travelers on this path.

This book is a beautiful example of the power of the possible. Part 1 lays out the need for the LISA 2020 project and describes how the project was created. The next parts explore different facets of the development problem from a statistical perspective. One chapter considers the barriers and opportunities encountered by statisticians working for international development in general, and other chapters consider the particular challenges faced during the establishment of LISA 2020 statistical laboratories in developing countries. There is an important discussion of undergraduate statistics education in developing nations, and the ways in which smart students get deflected from their paths as they move through the system, especially for STEM fields, and especially in statistics.

There is much more. It is a long book, and each chapter adds value and texture.

Upon reflection, my main discovery is that the book tells two stories. One story is about LISA 2020, and what it has done and can do to strengthen the field of statistics in underdeveloped nations. Progress on that front will help us all.

The second story is even more important. The book provides a blueprint for how the LISA 2020 model can be replicated in other fields. Civil engineers, or accountants, or nurses, or any other profession could follow this outline to share expertise and build capacity and promote progress in other countries. It also contain(s) some tutorials for statistical literacy across several fields. The details would change, of course, but ideas are durable and the generalizations seem pretty straightforward. This book shows every other profession where and how to stand in order to move the world. I urge every researcher to get a copy!

Professor David Banks
Duke University, United States

Preface

There have been various reports in the literature on the abysmally low level of development in statistical practice and collaboration in developing countries. To enhance the teaching and learning of statistics in developing countries, the LISA 2020 Program of the University of Colorado Boulder, United States is partnering with universities around the world to create a network of statistical collaboration laboratories (stat labs) in developing countries. This is in a bid to develop statistical practice and collaboration in the developing world. The vision to write this book dates back to almost ten years ago, precisely August 16, 2012 when I expressed this idea to the Global Director of LISA 2020, Professor Eric Vance via an email shortly before my historic visit to his stat lab at Virginia Tech in 2013 as the First LISA Fellow. He quickly obliged but opined that we needed more years to learn, practice, and collaborate with more people before writing such a book.

This book is a collection of thoughts and experiences of eighty erudite scholars from the LISA 2020 Global Network around the world. It is the product of several years of learning the art and practice of statistical collaboration. The task of bringing this book to your table has not only been assiduous but an arduous one that is laden with several sleepless nights of painstakingly ensuring quality contents for the past two years when we brought up the idea again during one of the biweekly zoom meetings of the LISA 2020 Global Network. We are highly grateful to USAID for supporting the idea of funding this book. I thank God that it has finally become a reality! Special thanks also goes to FAPESP and Professor R. Dias for personally supporting me during the course of this project.

The book provides a new insight into the current issues and opportunities in statistics education, statistical consulting, and collaboration. There are 38 chapters in the book and a bonus chapter by one of the most respected scholars in the field of statistical consulting, Emeritus Professor Douglas Zahn. The main objective of the book is to address the topics of individual chapters from the perspectives of the historical context, the present state, and future directions of statistical training and practice, so that readers may fully understand the challenges and opportunities in the field of statistics education and collaboration, especially in developing countries.

In particular, this book will serve as a

- Reference point on statistical practice in developing countries for researchers, scholars, students, and practitioners
- Comprehensive source of present and state-of-the-art knowledge on creating statistical collaboration laboratories within the field of data science and statistics
- Collection of innovative articles for statistical teaching and learning techniques in developing countries

It is expedient for all educational institutions to obtain a copy of this book!

O. Olawale Awe, PhD
First LISA Fellow and LISA 2020 Engagement Ambassador to Africa

Editors

O. Olawale Awe, PhD (Lead Editor) is an elected member of the International Statistical Institute (ISI) and a fellow of the African Scientific Institute, United States. He is the First LISA Fellow and presently the LISA 2020 Engagement Ambassador to Africa in the LISA 2020 Global Network of the University of Colorado, Boulder, United States. He has been a visiting scholar at the following institutions: Department of Statistics, Virginia Tech and BECCA Lab of the University of Pennsylvania, United States (September 2013-September 2014); Department of Statistics, Federal University of Bahia, Brazil (March-September, 2020); Institute of Mathematics, Statistics and Scientific Computing, University of Campinas, Brazil (2021-2023). He has more than fifteen years of experience as a researcher, lecturer, senior lecturer, and professor at various institutions in Nigeria and beyond. His research interests include computational statistics, machine learning, time series econometrics, statistics education, and Bayesian methods. He has facilitated several capacity-building workshops and seminars globally. He has a strong passion for statistics capacity building of researchers in Africa and other developing countries. Awe earned a PhD in Statistics from the University of Ibadan, Nigeria and MBA from Obafemi Awolowo University, Ile-Ife, Nigeria. He is an affiliate member of the African Academy of Sciences (AAS) and an immediate past council member of the International Society for Business and Industrial Statistics (ISBIS) (2017-2021) as well as a country coordinator (Nigeria) of the International Statistical Literacy Program (ISLP) of the ISI, and a Professor at the Global Humanistic University, Curacao.

Kim Love, PhD (Co-Lead Editor) is the LISA 2020 Program Monitoring Specialist. She is also a private statistical collaborator from Athens, Georgia, United States. She is a member of the American Statistical Association (ASA) and is an active leader in the ASA Section on Statistical Consulting, as well as an elected member of the International Statistical Institute (ISI). She earned a PhD in Statistics from Virginia Tech, and a BA in Mathematics from the University of Virginia; she has been an active statistical collaborator for more than 15 years. Her statistical expertise includes regression and linear models, categorical data, generalized linear models, mixed-effects models, nonlinear models, repeated measures, and experimental design; her areas of specialization include nursing and health sciences, as well as forestry, natural resources, and agricultural sciences. Beyond her work with the LISA 2020 Network, her professional mission is to improve researchers' relationships with statistics and data science through education and individual collaboration.

Eric A. Vance, PhD (Co-Lead Editor) is an associate professor of Applied Mathematics and the Director of the Laboratory for Interdisciplinary Statistical Analysis (LISA) at the University of Colorado Boulder, United States. He is the Global Director of the LISA 2020 Network. He researches the micro- and macro-theory of collaboration, i.e., what individual statisticians and data scientists need to know to become effective interdisciplinary collaborators and what institutions can do to promote interdisciplinary collaboration between domain experts, statisticians, and data scientists. He is an elected member of the International Statistical Institute and a member of its Statistical Capacity Building Taskforce. He is a fellow of the American Statistical Association (ASA) and winner of the

2019 ASA Jackie Dietz Award for best paper in the *Journal of Statistics Education* for "The ASCCR Frame for Learning Essential Collaboration Skills." He is also the winner of the 2022 ASA Jackie Dietz Award for best paper in the *Journal of Statistics and Data Science Education* for "Using Team-Based Learning to Teach Data Science." He is a past chair of the ASA Section on Statistical Consulting and the Conference on Statistical Practice. He has traveled through 85 countries and keeps these experiences in mind as he collaborates toward building capacity for sustainable development.

Reviewers

1. Prof. David Banks—Department of Statistical Sciences, Duke University, United States
2. Prof. Jim Cochran—Culverhouse College of Business, The University of Alabama, United States
3. Prof. James L. Rosenberger—Department of Statistics, Pennsylvania State University, United States
4. Prof. Titi O. Obilade—Department of Mathematics, Obafemi Awolowo University, Ile-Ife, Nigeria
5. Professor A. Adedayo Adepoju—Department of Statistics, University of Ibadan, Nigeria
6. Dr. Mathew Druckenmiller—National Snow and Ice Data Center, University of Colorado Boulder, United States
7. Dr. Gordana Popovic—Stats Central, University of New South Wales, Australia

Contributors

Haftom Temesgen Abebe
College of Health Sciences
Mekelle University
Mek'ele, Ethiopia

Albert Ayorinde Abegunde
Department of Urban and Regional
    Planning
Obafemi Awolowo University
Ile Ife, Nigeria

Mumini Idowu Adarabioyo
Department of Mathematical and Physical
    Sciences
Afe Babalola University
Ado-Ekiti, Nigeria

Atinuke Adebanji
Department of Statistics & Actuarial
    Science
Kwame Nkrumah University of Science
    and Technology (KNUST)
Kumasi, Ghana
and
KNUST-Laboratory for Interdisciplinary
    Statistical Analysis (KNUST-LISA)
Kumasi, Ghana

Adetola Adedamola Adediran
Department of Statistics
Federal University of Technology
Akure, Nigeria

Oyelola A. Adegboye
Australian Institute of Tropical Health and
    Medicine
James Cook University
Townsville, Australia
and
Department of Public Health and Tropical
    Medicine
College of Public Health, Medical and
    Veterinary Science
James Cook University
Townsville, Australia

Adeshina I. Adekunle
Australian Institute of Tropical Health and
    Medicine
James Cook University
Townsville, Australia

Monday Osagie Adenomon
Nasarawa State University Laboratory for
    Interdisciplinary Statistics Analysis
    (NSUK-LISA)
Keffi, Nigeria

Adedayo A. Adepoju
University of Ibadan Laboratory for
    Interdisciplinary Statistical Analysis
    (UI-LISA)
Ibadan, Nigeria

Oluwatobi Michael Aduloju
Department of Statistics
Federal University of Technology
Akure, Nigeria

Saheed A. Afolabiand
University of Ibadan Laboratory for
    Interdisciplinary Statistical Analysis
    (UI-LISA)
Ibadan, Nigeria

Oluokun Kasali Agunloye
Department of Mathematics
Obafemi Awolowo University
Ile-Ife, Nigeria

Rafiat Taiwo Agunloye
Department of Educational Management
Obafemi Awolowo University
Ile-Ife, Nigeria

Olawale B. Akanbi
University of Ibadan Laboratory for
    Interdisciplinary Statistical Analysis
    (UI-LISA)
Ibadan, Nigeria

Olalekan J. Akintande
University of Ibadan Laboratory for
    Interdisciplinary Statistical Analysis
    (UI-LISA)
Ibadan, Nigeria

Akanni Akinyemi
Department of Demography and Social
    Statistics
Obafemi Awolowo University
Ile Ife, Nigeria

Oluwayemisi Alaba
University of Ibadan Laboratory for
    Interdisciplinary Statistical Analysis
    [UI-LISA]
Ibadan, Nigeria

Adeyemi Adewale Alade
Department of Urban and Regional
    Planning
University of Lagos
Lagos, Nigeria

Olabimpe B. Aladeniyi
Department of Statistics
Federal University of Technology
Akure, Nigeria

Yakup Ari
Department of Economics
Alanya Alaaddin Keykubat University
Alanya, Turkey

O. Olawale Awe
Institute of Mathematics, Statistics and
    Scientific Computing
University of Campinas
Campinas, Brazil
Nairobi, Kenya
and
Global Humanistic University

Oluwafunmilola Deborah Awe
Department of Medical Statistics
University of Nairobi
Nairobi, Kenya

Olusesan Michael Awoleye
African Institute for Science Policy and
    Innovation
Obafemi Awolowo University
Ile-Ife, Nigeria

Olumide Charles Ayeni
Department of Mathematical Sciences
Anchor University
Lagos, Nigeria

Bayowa Teniola Babalola
Department of Mathematics and Statistics
Kampala International University
Kampala, Uganda

Leonardo César Teonácio Bezerra
Instituto Metrópole Digital
Federal University of Rio Grande do Norte
Natal, Brazil

James Cochran
Culverhouse College of Business
University of Alabama
Tuscaloosa, Alabama, USA

Demisew Gebru Degefu
Department of Statistics
Hawassa University
Hawassa City, Ethiopia

Kelechi Dozie
Imo State University Owerri, Laboratory
    for Interdisciplinary Statistical Analysis
    [IMSU-LISA]
Imo State, Nigeria

Matthew Druckenmiller
National Snow and Ice Data Center
University of Colorado Boulder
Boulder, Colorado, USA

Egwim Evans
Department of Biochemistry
Federal University of Technology
Minna, Niger State, Nigeria

Ifeoma Betty Ezike
Monetary Policy Department
Central Bank of Nigeria
Abuja, Nigeria

Adeniyi Francis Fagbamigbe
Department of Epidemiology and
    Medical Sciences
College of Medicine
University of Ibadan
Ibadan, Nigeria

Ore Fika
Institute of Housing and Urban
    Development Studies (IHUDS)
Erasmus University
Rotterdam, Netherlands

Morufu A. Folorunso
Federal School of Statistics
Ibadan, Oyo State, Nigeria

Serifat A. Folorunso
University of Ibadan Laboratory for
    Interdisciplinary Statistical Analysis
    (UI-LISA)
Ibadan, Nigeria

Ezra Gayawan
Department of Statistics
Federal University of Technology
Akure, Nigeria

David J. Gunderman
Purdue University
College of Engineering
West Lafayette, Indiana
and
Laboratory for Interdisciplinary Statistical
    Analysis [LISA]
University of Colorado Boulder
Boulder, Colorado

Philip Olu Jegede
Institute of Education
Obafemi Awolowo University
Ile-Ife, Nigeria

Asifa Kamal
Lahore College for Women University
Lahore, Pakistan

Kim Love
K. R. Love Quantitative Consulting and
    Collaboration
Athens, Georgia, USA

Vishal Mahajan
Department of Agroforestry, Krishi Vigyan
    Kendra, Kathua
Sher-e-Kashmir University of Agricultural
    Sciences & Technology of Jammu
Jammu and Kashmir, India

Mahnaz Makhdum
Lahore College for Women University
Lahore, Pakistan

Deborah Olufunmilayo Makinde
Department of Mathematics
Obafemi Awolowo University
Ile-Ife, Nigeria

Thiago Valentim Marques
Federal Institute of Rio Grande do Norte
Natal, Brazil

Sandeep K. Maurya
Department of Statistics
Central University of South Bihar
Gaya, India

Thomas Mawora
Department of Statistics and Actuarial
    Science
Maseno University
Kisumu, Kenya

Hope Ifeyinwa Mbachu
Imo State University Owerri, Laboratory
    for Interdisciplinary Statistical Analysis
    [IMSU-LISA]
Imo State, Nigeria

Jacob Wale Mobolaji
Department of Demography and Social
    Statistics
Obafemi Awolowo University
Ile Ife, Nigeria

Lubna Naz
Department of Economics
University of Karachi
Karachi, Pakistan

Marcus Alexandre Nunes
Department of Statistics
Federal University of Rio Grande do Norte
Natal, Brazil

Julius Nwanya
Imo State University Owerri, Laboratory
    for Interdisciplinary Statistical Analysis
    [IMSU-LISA]
Imo State, Nigeria

Tayo Ogundunmade
University of Ibadan Laboratory for
    Interdisciplinary Statistical Analysis
    [UI-LISA]
Ibadan, Nigeria

Oluwadare O. Ojo
Department of Statistics
Federal University of Technology
Akure, Nigeria

Lasun Mykail Olayiwola
Department of Urban and Regional
    Planning
Obafemi Awolowo University
Ile Ife, Nigeria

Abdulhakeem Abayomi Olorukooba
Department of Community Medicine
Ahmadu Bello University
Zaria, Nigeria

Olusanya E. Olubusoye
University of Ibadan Laboratory for
    Interdisciplinary Statistical Analysis
    [UI-LISA]
Ibadan, Nigeria

Adewale P. Onatunji
Department of Statistics
Ladoke Akintola University of Technology
Ogbomoso, Oyo State, Nigeria

Olamide Seyi Orunmoluyi
Department of Statistics
Federal University of Technology
Akure, Nigeria

Tolulope T. Osinubi
Obafemi Awolowo University
Ile-Ife, Nigeria

Oluwaseun A. Otekunrin
University of Ibadan Laboratory for
    Interdisciplinary Statistical Analysis
    (UI-LISA)
Ibadan, Nigeria

Joyce Otieno
Department of Applied Mathematics
University of Colorado Boulder
Boulder, Colorado, USA

Seyifunmi Michael Owoeye
Department of Statistics
Federal University of Technology
Akure, Nigeria

Tomiwa T. Oyelakin
University of Ibadan Laboratory for
    Interdisciplinary Statistical Analysis
    (UI-LISA)
Ibadan, Nigeria

Anton Pak
Australian Institute of Tropical Health and
    Medicine
James Cook University
Townsville, Australia
and
Centre for the Business and Economics of
    Health
The University of Queensland
Brisbane, Australia

Aviral Pandey
A N Sinha Institute of Social Studies
Patna, Bihar, India

Tonya R. Pruitt
NanoEarth
Virginia Tech
Blacksburg, Virginia, USA

Markandey Rai
UN-Habitat
Nairobi, Kenya

Mirian Raymond
Imo State University Owerri, Laboratory
  for Interdisciplinary Statistical Analysis
  [IMSU-LISA]
Imo State, Nigeria

Bbosa Robert
University of Ibadan Laboratory for
  Interdisciplinary Statistical Analysis
  [UI-LISA]
Ibadan, Nigeria

James Rosenberger
Department of Statistics
The Pennsylvania State University
State College, Pennsylvania, USA

Pawan Kumar Sharma
Department of Agricultural Economics,
  Krishi Vigyan Kendra, Kathua
Sher-e-Kashmir University of Agricultural
  Sciences & Technology of Jammu
Jammu and Kashmir, India

Arun Kumar Sinha
Department of Statistics
(Formerly at) Central University of South
  Bihar
Bihar, India

Omodolapo Somo-Aina
Department of Educational Research
  Methodology
University of North Carolina
Greensboro, North Carolina, USA

Salami Suleiman
KNUST-Laboratory for Interdisciplinary
  Statistical Analysis (KNUST-LISA)
Kumasi, Ghana

Muhammad Umair
Department of Economics
University of Karachi
Karachi, Pakistan

Ifeoma Evan Uzoma
Department of Geography and Natural
  Resources Management
University of Uyo
Uyo, Nigeria

Eric A. Vance
Laboratory for Interdisciplinary Statistical
  Analysis, Department of Applied
  Mathematics
University of Colorado Boulder
Boulder, Colorado, USA

Richa Vatsa
Department of Statistics
Central University of South Bihar
Gaya, India

Sara Zahid
Lahore College for Women University
Lahore, Pakistan

Part 1

Statistics Collaboration and Practice in Developing Countries

Experiences, Challenges, and Opportunities

1

Statistics and Data Science Collaboration Laboratories: Engines for Development

Eric A. Vance
University of Colorado Boulder

Tonya R. Pruitt
Virginia Tech

CONTENTS

DOI: 10.1201/9781003261148-2

1.1 Introduction: Why Statistics and Data Science Have Extraordinary Potential for Data-Driven Development

Statistical analysis and data science have enormous potential for enabling and accelerating data-driven development. The discipline of statistics has been around for more than 100 years, and its impacts are widespread and profound. Statistics has been driving development since even before the world's first university statistics department was founded in 1911 at University College London (Hotelling 1988). For example, during the great cholera outbreak in London in 1854, John Snow collected and analyzed data to identify and close off a contaminated water pump (Tulchinsky 2018). In the same year, Florence Nightingale, the founder of modern nursing and a very early member of the International Statistical Institute and the American Statistical Association, used statistics and data visualization to show that simple hygiene measures could drastically reduce infection and death in hospitals (McDonald 1998). As Gunderman and Vance (2021) wrote, the ability to collect and learn from large amounts of data has been a major driver of innovation over recent decades. Everything from health care—including patient analytics (Adebanji et al. 2015; Awe et al. 2021; Kosorok and Laber 2019), wearable devices (Michie et al. 2017), and the COVID-19 response (Aidoo et al. 2021; Gayawan et al. 2020)—to energy (Malakar et al. 2021) to entertainment recommender systems (e.g., Netflix (Bell et al. 2007)), is now driven by data and statistics.

We consider data science to be the science of learning from data (Donoho 2017), which encompasses—at a minimum—statistics, computation, and ways of thinking about data. Clearly, the idea of learning from data is not new. The scientific method is based upon deriving and testing theories from empirical observation, i.e., data. Figure 1.1 shows how the wheel of statistics and data science turns through the scientific process to ultimately apply statistics and data science to solve problems and make decisions for the benefit of society. Starting with an understanding of domain experts' research, business, or policy questions, statisticians and data scientists collaborate with domain experts to design an experiment or study to produce high-quality data, or, alternatively, design an algorithm to collect or scrape already existing data.

Statisticians can furthermore collaborate with data producers to collect high-quality data (Vance and Love 2021). All statisticians and data scientists have the responsibility to understand—and reduce when possible—the sources of variation and potential biases within the data. Statisticians and data scientists visualize the data and analyze data sets with models and algorithms to produce findings relevant to the original research, business, or policy questions. Statistics can help with interpreting the practical and statistical significance of results leading to conclusions about the data and a greater understanding of the underlying questions. This, in turn, may lead to collecting and analyzing more data in another turn of the wheel, in time progressing to solving the problem, making recommendations, and making decisions. The final step for a statistician or data scientist who wants to make a deep contribution is to collaborate with others to implement data-driven solutions, recommendations, and decisions to take action for the benefit of society.

Later in this chapter, Love et al. (2022) ask and answer why we need to develop statistical practice and collaboration in developing countries. In short, local capacity in statistics and data science can enable and accelerate innovative, sustainable solutions to local development challenges. When local statisticians and data scientists collaborate with local domain experts to transform evidence into action—especially when it occurs within a statistics

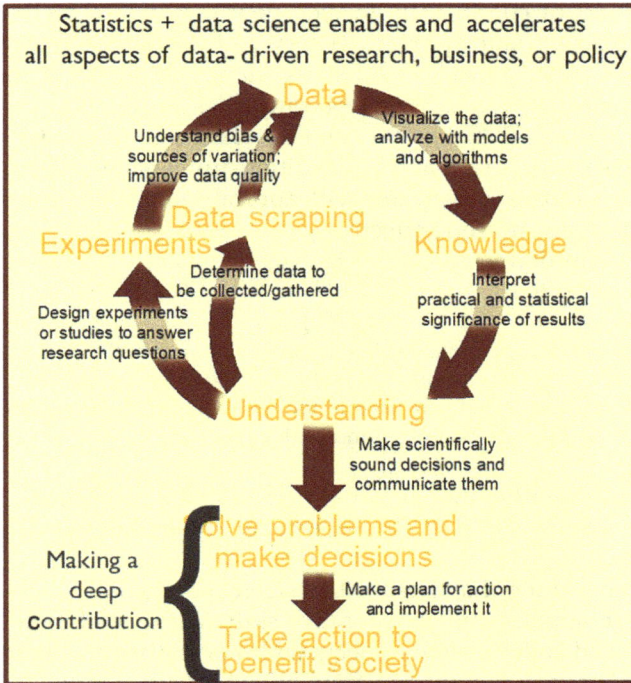

FIGURE 1.1
Statistics and data science enables and accelerates all aspects of data-driven research, business, or policy.

and data science collaboration laboratory—a virtuous cycle of development emerges in which initial collaborations build the capacity of all parties to achieve even more positive development outcomes by applying statistics and data science.

The goal of this chapter is to explain the concept of the statistics and data science collaboration laboratory ("stat lab") and how it can sustainably harness the extraordinary potential of statistics and data science for development. In Section 1.2, we introduce the generator of a stat lab—the collaborative statistician or data scientist. In Section 1.3, we explain what a stat lab is and what it does. We provide three examples of stat labs in Section 1.4. In Section 1.5, we discuss how stat labs can become engines for development and attempt to answer the question: Why should a university create and sustain a stat lab? We walk through the seven steps to create a stat lab in Section 1.6. Based on the lessons learned so far from the LISA 2020 Network, in Section 1.7, we characterize what makes stat labs strong and sustainable. We conclude in Section 1.8.

1.2 Collaborative Statisticians and Data Scientists

A technically trained statistician, who understands the theory and methods of statistics, is well positioned to do many things in academia, business and industry, or the policy domain. One option for the academic statistician is to teach the theory and methods of statistics to others and/or continue to develop new theory and methods. Another path available to all statisticians is to apply the theory and methods of statistics to solve problems that will help benefit society. For a statistician in business, it may be straightforward to identify problems that directly affect companies' products or profits (Oliveira et al. 2016). In academia, on the other hand, it can be a challenge to identify or be exposed to problems that have real-world relevance. In addition, universities, especially in developing countries, may fail to prepare statisticians with appropriate applied statistics skills, including the ability to compute and code effectively (Awe 2012; Love et al. 2022).

Drawing upon a definition proposed by Halvorsen et al. (2020), we define a *collaborative statistician or data scientist* to be

> someone possessing deep technical skills in the theory and methods of statistics or data science and effective collaboration skills who can move between theory and practice to work with domain experts to create solutions to research, business, and policy challenges and achieve research, business, and policy goals.

In other words, a collaborative statistician or data scientist can transform evidence (data) into action, i.e., the solving of research problems or the making of data-driven business and policy decisions.

Fortunately, we have found that those with a strong background in theory and methods of statistics can learn applied and collaborative skills while working on consulting projects and more involved collaborative projects within a statistical collaboration laboratory (Vance et al. 2016). Through mentorship from an experienced individual or via a formal program, a statistician can become a *collaborative statistician or data scientist*; collaborate with domain experts on problems in research, business, or policy (see Figure 1.2); and subsequently have a more positive impact on society (Love et al. 2017; Vance et al. 2017a, 2017b). An individual can also learn collaboration skills through self-study of Vance and Smith's ASCCR (Attitude–Structure–Content–Communication–Relationship) Framework for Collaboration (2019).

Working alone, a collaborative statistician or data scientist can enable and accelerate as many as ten projects or more per year. For example, collaborative statisticians can help understand the policy implications of uneven distribution of costly medical interventions (Vance et al. 2013), impacts of climate change on disease prevalence in Botswana (Alexander et al. 2013), the cost benefits of upgrading domestic water systems in Senegal (Hall et al. 2015a), and the productive uses of piped water in Africa (Hall et al. 2014b).

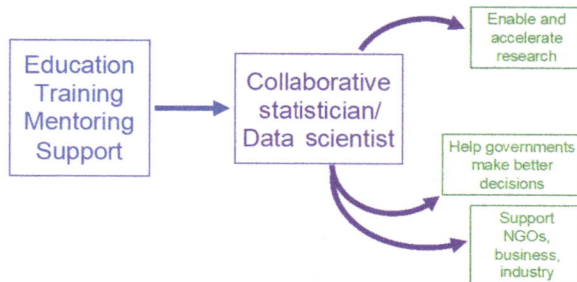

FIGURE 1.2
A well-trained collaborative statistician or data scientist can enable and accelerate many research, business, and policy projects.

Yet another path awaits a collaborative statistician or data scientist, one with orders of magnitude more potential impact. A collaborative statistician or data scientist can create a statistics and data science collaboration laboratory.

1.3 Statistics and Data Science Collaboration Laboratories ("Stat Labs")

A statistics and data science collaboration laboratory, or "stat lab," is a collection of statisticians and data scientists who have been trained to collaborate with domain experts and then do so to enable and accelerate research and make data-driven business and policy decisions. Similar to a statistical consulting center or a university "Center of Excellence," a stat lab is where members of the community can go to access statistics and data science expertise. Stat labs are not "rooms full of computers." Instead, they are more like rooms full of collaborative statisticians and data scientists working with domain experts from multiple disciplines. The term stat lab refers to both the human resources and the physical space they occupy. The statisticians and data scientists who work in stat labs all have different relative strengths—like the comic book and movie superheroes "The Avengers"—and likewise work together to achieve a common purpose: to enable and accelerate data-driven development while training the next generation of collaborative statisticians and data scientists.

1.3.1 The Purpose of Stat Labs

Stat labs can have different missions based on the host institution's statistics and data science needs. Vance and Laga (2017) found in a survey of North American stat labs that their primary mission was (given in order) serving clients and keeping them happy, training students, advancing research, improving clients' statistical skills, generating funds, and advancing the statistical methodology. While a few stat labs' primary mission was to generate funds for their university or department, this potential mission was, overall, the least popular and least relevant for the majority of stat labs, likely because stat labs within universities are not businesses. Rather, their purpose is to educate—not to make money. Successful stat labs that do generate a profit reinvest those funds to support the lab's mission, i.e., building more capacity. If a stat lab were primarily motivated by profit, it could become a private business off campus, or its director/coordinator could become a private statistical consultant.

Stat labs in the LISA 2020 Network (see Part 7) have a threefold mission:

1. Training the next generation of collaborative statisticians and data scientists to move between theory and practice to solve problems for real-world impact
2. Supporting research, innovation, and the transformation of evidence to action through collaboration with domain experts by providing necessary statistical analysis and data science expertise (i.e., create infrastructure for collaboration)
3. Improving statistical skills and data literacy widely by teaching short courses and workshops (i.e., upskill data-capable development actors)

1.3.2 What Stat Labs Do

At a fundamental level, stat labs do four things: they support researchers and data decision-makers, they create new knowledge, they transform knowledge or evidence into action, and they do all of this while building capacity for data-driven development by training the next generation of collaborative statisticians and data scientists.

1.3.2.1 Supporting Domain Experts

Support for the clients or *domain experts* utilizing the stat lab takes several forms. Stat labs teach short courses and workshops to improve statistical analysis capacity and data literacy widely.

Members of stat labs listen to researchers, businesses, and policymakers and provide advice and consultations, and often provide emotional support for their work. When these researchers, businesspeople, and policymakers are not sure about some aspect of their study design, data analysis, or interpretation of data, they can ask an expert in the stat lab. Often that statistician or data scientist may not be able to answer immediately but can find the answer by referencing books or articles, asking a colleague, or searching the internet for a solution.

When the statistician or data scientist goes beyond teaching, advising, or consulting and contributes some aspect to the final result or product of the project, we call this *collaboration* (Love et al. 2017). Examples of collaboration include designing an experiment or study; performing power calculations or determining sample sizes; analyzing data; writing a summary of the statistical methods used; creating plots or graphs to explore and communicate data; interpreting results; communicating findings, conclusions, and recommendations.

One of the three missions of LISA 2020 stat labs is to improve the statistical skills and data literacy of members of their community (e.g., students in other departments, faculty, and staff across the university or institution, local school children, nonuniversity community members). This can occur during the collaboration process. As the domain expert teaches the statistician or data scientist relevant information about their domain, so does the statistician or data scientist teach the domain expert about statistics or data science. Vance et al. (2022a) expatiate on how to create shared understanding between collaborators, both about domain issues and statistics and data science technical issues.

The primary way in which stat labs improve the statistical skills and data literacy of members of their community is by teaching short courses or workshops. These short courses and workshops can be on specific topics the stat lab thinks many people should know, such as how to design an experiment or how to design and analyze a survey. The topics can also be driven by demand from domain experts. For example, if several domain experts come to the stat lab with similar problems or questions about statistics—such as how to conduct an ANOVA—the stat lab could teach a 2-hour short course on that topic so that attendees can learn how to do it themselves. Sometimes a department will request the stat lab to teach a 1- or 2-day workshop such as "Statistical Methods for Biology." For all short courses and workshops, the goal of the stat lab is to improve the statistics and data literacy of the participants. Ideally, this will lead to the participants becoming good users of statistics and data science. At the very least, participants should become more aware of the power of statistics and data science, learn what methods may enable or accelerate their research or decision-making, and recognize when they should collaborate with the stat lab for needed expertise.

1.3.2.2 Creating New Knowledge

Stat labs create new knowledge in at least four ways.

1. **Stat labs help researchers answer questions they could not have answered without expert statistical advice**: Two examples are collaborations in which multiple statisticians from a stat lab helped water and sanitation specialists improve pit latrine management in low-income areas of Malawi (Chirwa et al. 2017) and understand the timing of the infancy-childhood growth transition in rural Gambia (Bernstein et al. 2020). Another example is of a study of agriculturally focused study abroad programs in Asia (Sharma et al. 2014).

2. **Based on an understanding of the domain experts' goals and their data, stat labs may suggest novel questions the data can answer**: In one case, statisticians analyzing data from a water project recognized that the data could answer the question of how willing households were to pay for enhanced toilet systems in Senegal (Hall et al. 2015b). Another example is when statisticians used data from an impact evaluation of a rural water supply project in Mozambique (Hall et al. 2014a) in collaboration with a social scientist to better understand the intracommunity impacts of that project (Van Houweling et al. 2017). In both cases, the research teams deemed the novel questions raised by the statisticians to be very important and impactful, and the statisticians became authors on the papers, leading to the discovery and dissemination of new knowledge.

3. **When stat labs encounter new types of data for which standard statistical methods do not apply, they can create new knowledge by developing novel methods that enable researchers to extract useful information from their data**: In collaboration with a professor of African history, three stat lab statisticians developed a novel integration of spatial statistical models with a Markov decision process to determine the most likely movement paths of people enslaved during the collapse of the Oyo Empire in Africa in the 1820s (Weins et al. 2022). Another example is a novel process of cleaning data during data collection developed by a statistician working on a project in Mozambique (Seiss et al. 2014). Developing new statistical methods to answer research questions arising from stat lab collaborations can be a core component of PhD dissertations; see Vance et al. (2009) and Vance (2008) for an example of this.

4. **Stat labs can generalize their lessons learned to improve statistical practice and collaboration for all**: Based on more than 40 years of combined experience collaborating with domain experts, running stat labs, and training more than 280 collaborative statisticians and data scientists, Vance and Smith (2019) created a framework to learn and teach interdisciplinary collaboration. Additional examples of knowledge creation about collaboration includes an article on goals for statistics and data science collaborations (Vance 2020), how to assess statistical consultations and collaborations (Vance et al. 2020), and—elsewhere in this volume—learning and teaching the ABC of collaboration (Awe and Vance 2022) and using the TEAM framework for collaboration (Adenomon 2022b). With such a rich experience collaborating in different contexts to enable and accelerate data-driven development, more knowledge about best practices in statistical practice and collaboration is sure to be produced by stat labs in the future. In fact, all of the chapters in this volume are examples of knowledge produced by stat labs.

1.3.2.3 Transforming Evidence into Action (TEA)

Stat labs aspire to transform evidence into action (TEA), for what good is knowledge created (i.e., academic evidence) if it leads nowhere and fails to make a positive impact on society? Olubusoye et al. (2021, p. 13) describe the logic of TEA:

> If we can create and sustain stat labs comprised of well-trained statisticians and data scientists who can move between theory and practice to apply statistics to solve problems and make decisions, and if they can collaborate with development actors empowered to produce data and take action on development issues, then the members of the stat labs can transform data into evidence to answer their collaborators' research questions or help them make data-driven policy decisions that will lead to development impacts and benefits to society.

TEA projects are collaborations between stat labs and *data decision-makers*, who are development actors working in a variety of settings: government officials who may be creating or implementing policy, employees of non-governmental organizations who may also be creating or implementing policy, businesses or entrepreneurs making decisions that will ultimately affect how many people they hire or what goods and services they produce, or researchers who drive innovation and discovery based on results of their experiments or studies. The end goal of a TEA project is a positive development impact based on a data-driven innovation.

The following are the reasons described by Vance and Love (2021) why TEA projects are difficult to implement and fully realize:

- Statisticians and data scientists must be skilled collaborators as well as skilled methodologists and analysts. To achieve TEA they must be able to understand the data and projects they are working with on both a deep and broad level. They must also and be able to communicate the results of statistical methods and analytical work in ways that provide actionable evidence to those who can use it to positively impact society.

- Policymakers may not know of the existence of a stat lab; therefore, stat labs must deliberately reach out to them. By reaching out early as skilled collaborators, stat labs can ensure that they are helping to answer questions of interest to local policy or business decision-makers.

- TEA requires a mindset to see a project all the way to its end. Policy change is often a long and complex process; stat labs may feel compelled to move on to a new project before the impacts of the old one are realized. Nevertheless, to transform evidence into action for the benefit of society, statisticians and data scientists must adopt a TEA mindset and have the patience to engage in the complete process.

- Measuring longer term impacts of stat lab activities and describing the contribution of statistics and data science is challenging. So even if a TEA project resulted in a positive development impact, the stat lab may not know about it, or others may not know of the impact the stat lab had on the project's outcome.

However, TEA is worth aspiring toward. Olubusoye et al. (2021, p. 13) write:

> TEA projects have tremendous potential to enable and accelerate data-driven development because they operate in a space occupied by three development actors: data

producers, data decision makers, and data analyzers. Stat labs provide collaborative statistics and data science expertise to bring all of the actors together to produce evidence and transform it into action for development.

1.3.2.4 *Training the Next Generation of Collaborative Statisticians and Data Scientists*

A second terminal goal of TEA projects is to build the capacity of the stat lab to have even more development impact in the future. In fact, the primary purpose of a stat lab is to train students and staff to become effective interdisciplinary collaborators who can move between theory and practice to solve problems for real-world impact.

Vance et al. (2022a) describe an educational paradigm in which statistics and data science students learn collaboration through a five-stage process: preparation, practice, doing, reflecting, and mentoring. As part of their training to work in a stat lab, students prepare by reading articles and observing senior stat lab members collaborate with domain experts, they practice aspects of the ASCCR Framework for collaboration in the classroom (Vance and Smith 2019), they then perform collaboration by working on a stat lab project under the mentorship of senior collaborators, they reflect on what went well and what could be improved, and finally, they reach the stage where they can mentor others in collaboration.

This education and training process requires the involvement of the stat lab to provide students with opportunities to observe (prepare), do, and mentor others on real projects. The other stages can occur within a classroom setting, such as in a capstone course for undergraduates or a statistical consulting and collaboration practicum course (Kolaczyk et al. 2021). Alternatively, the stat lab can conduct weekly staff meetings during which members discuss projects, thereby providing opportunities for students to practice collaboration skills and reflect on how to improve them.

Thus, a stat lab can provide all five stages of the process for education and training in collaboration.

An impactful way to help students progress through the stages of preparation, practice, reflecting, and mentoring is for stat labs to conduct Video Coaching and Feedback Sessions (VCFSs) in which a collaboration project meeting is video recorded (with permission of the domain expert) on a phone, webcam, or video camera and then reviewed in a small group of 4–7 stat lab members. Vance (2014) describes key features of VCFS, which include reviewing only a few (2–4) short clips of the meeting (only 1–5 minutes long), having observers focus on specific behaviors in the short video clips, conducting the VCFS with intention to help the participants improve their collaboration skills, having each participant share one lesson learned from the VCFS, and documenting lessons learned.

Much more is written about educating and training the next generation of collaborative statisticians and data scientists in Vance (2021) and in this edited volume (Gayawan et al. 2022; Nunes 2022). How stat labs can continue to drive such education and training in the practice of statistics will be a focus of continuing research.

1.3.3 Stat Labs Produce Collaborative Statisticians and Data Scientists and Data-Capable Development Actors

Like a laboratory that produces chemicals or invents a new technology, a statistics and data science collaboration laboratory ("stat lab") also produces outputs. We have discussed how stat labs create new knowledge and transform that into action for development. However, the primary output for a stat lab is human capacity in statistical analysis and data science; stat labs produce collaborative statisticians and data scientists and data-capable

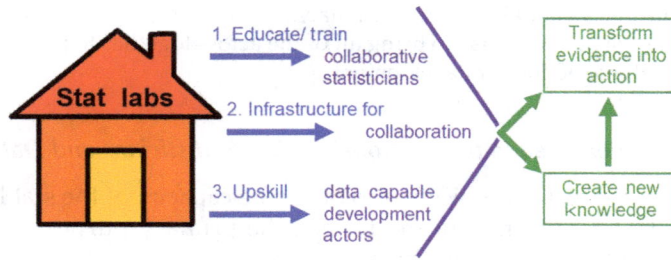

FIGURE 1.3
The activities of stat labs lead to the production of collaborative statisticians and data scientists, data-capable development actors, and space for them to collaborate to transform evidence into action. This leads to data-driven innovations and positive development impacts.

development actors. In sum, stat labs build capacity for data-driven development (Vance and Love 2021).

Figure 1.3 shows a diagram of how stat labs leverage their activities to produce collaborative statisticians and data scientists and data-capable development actors to build capacity for data-driven development. From this capacity, stat labs create knowledge and transform evidence into action, leading to development outcomes and impacts discussed in Section 1.5.

1.4 Exemplar Stat Labs

In an invited talk at the 62nd World Statistics Congress of the International Statistical Institute in 2019, Vance described seven components of a stat lab using an analogy to the human body (see Figure 1.4). Each stat lab has seven "body" parts. The heart is the mission/goals/objectives of the lab. The backbone of the stat lab is its administration. The muscles are the personnel and how they are trained. The head/brain are the stat lab's services. The legs are the lab's communication strategy. Its blood is the support it receives from its university and institution and its budget. And finally, its skin is the (visible) outcomes and impacts of the stat lab. In this section, we describe these seven components for two instantiations of the Laboratory for Interdisciplinary Statistical Analysis (LISA) at Virginia Tech and the University of Colorado Boulder. We also provide an example of a third stat lab, the University of Ibadan's LISA (UI-LISA).

1.4.1 The Laboratory for Interdisciplinary Statistical Analysis (LISA) at Virginia Tech

LISA was created at Virginia Tech in Blacksburg, Virginia, United States in 2008 within the Department of Statistics and lasted for 8 years until 2016 (Vance and Pruitt 2016b). It was built on the foundation of the Statistical Consulting Center (1973–2007) and the Statistical Laboratory (1948–1972), which preceded the establishment of the Department of Statistics in 1949 by 1 year (Arnold et al. 2013). LISA's mission was to train statisticians to become interdisciplinary collaborators, provide research infrastructure to enable and accelerate high-impact research, and engage with the Virginia Tech community in outreach activities to improve statistical skills and literacy (Vance and Pruitt 2016a).

Essential components of stat labs

Heart is *mission/goals/objectives*
Backbone is the *administration*
Muscles are the *personnel/training*
Head/brain are the *services*
Legs are *Communication strategy*
Blood is *University support/budget*
Skin is *impactsandoutcomes*

FIGURE 1.4
Stat labs have seven essential components, analogous to the human body.

LISA provided three main services: collaboration on research projects, Walk-in Consulting to answer quick statistics questions, and short courses to teach new statistical methods. During the 2015–2016 fall and spring semesters, which was LISA's last year of operation at Virginia Tech, the 50 statistical collaborators of LISA met with researchers for individual statistical collaboration meetings on 294 projects. During daily Walk-in Consulting hours, LISA met with 468 faculty, staff, and students to answer quick statistical questions on projects requiring less than 30 minutes of assistance. Fifteen LISA Short Courses were offered to teach 455 graduate students and other university members how to apply statistics in their research. Overall, LISA provided at least 4,701 hours of statistical assistance and education to members of the Virginia Tech community. LISA's statistical collaborators reported 3,750.5 hours for statistical collaboration projects, 650.5 hours for Walk-in Consulting, and 300 hours for Short Courses. The number of hours reported worked in 2015–2016 fall and spring semesters is likely an underestimation of the total hours (Vance and Pruitt 2016b). By counting "Other" responses as half "Yes" and half "No", 91% of LISA clients in the academic year 2014–2015 considered LISA's services "helpful," and 90% were satisfied with their overall LISA experience. See Figure 1.5.

The bulk of LISA's workforce was MS and PhD students from the Department of Statistics. From 2008 to 2016, 223 students and six visiting scholars were educated and trained to move between theory and practice to help researchers apply statistics to answer

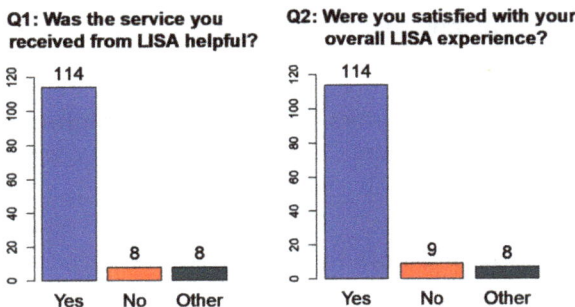

FIGURE 1.5
Summary of quantitative feedback on LISA collaboration meetings in 2014–2015.

their research questions (Vance et al. 2016). During the 2015–2016 fall and spring semesters, 50 students collaborated with researchers on LISA projects. These students had taken a semester-long course in communication and collaboration skills (Vance 2012, 2013) to prepare them for LISA and received continuing training while working in LISA via weekly staff meetings to discuss projects. Additionally, 41 of the students had at least one video recorded meeting with a domain expert reviewed during a VCFS (Vance and Pruitt 2016b). During each semester, four students had a fully funded teaching assistant position (50% time) to work in LISA. Also, a dozen other students each semester had a part-time (25% time) appointment in LISA. The remaining students were volunteers.

LISA was led by a full-time Director, Dr. Eric Vance, throughout its existence at Virginia Tech. From 2010 to 2016, its Assistant Director was Dr. Chris Franck, who was primarily supported financially by external research projects. From 2009 to 2016, Tonya Pruitt was LISA's Administrative Specialist. Her appointment in LISA was for 80% of her time. During the 2015–2016 fall and spring semesters, LISA hired two research associates to collaborate on long-term funded projects. The Director of LISA reported directly to the Statistics Head of Department as well as to the Vice President of Research and the Dean of the College of Science. Having strong, dedicated personnel contributed immensely to the success of LISA.

Statistical assistance from LISA was free for Virginia Tech faculty, staff, and students. LISA was funded jointly by the Office of the Vice President of Research, the College of Science, the Graduate School, the Office of the Provost, and all seven other colleges: Agriculture and Life Sciences, Architecture and Urban Studies, Engineering, Liberal Arts and Human Sciences, Natural Resources and Environment, the Pamplin College of Business, and the Virginia-Maryland Regional College of Veterinary Medicine. The Department of Statistics provided funding for a dozen 25%-time LISA statistical collaborators each semester. It also provided physical space and other infrastructure support such as internet access, accounting, printing, and computers for LISA's activities. The total annual budget for LISA from its joint funders was $300,000.

LISA disseminated its annual reports to college deans and other university administrators. It advertised its services mainly through flyers around campus and word of mouth. However, its short courses were advertised in weekly email newsletters sent by the Graduate School to graduate students, and these short courses always communicated the collaboration and Walk-in Consulting services of LISA. LISA was featured in several national statistics publications (Carzolio 2014; Crandell 2015; Stallings 2014) and was featured in six consecutive invited sessions of the Joint Statistical Meetings (JSM) from 2011 to 2016.

LISA's primary outcomes were the 229 students and visiting fellows it trained to become collaborative statisticians and the thousands of clients who benefited from LISA's services. There were 2,719 total LISA collaboration projects, 2,962 Walk-in Consulting visitors, and 3,748 short course attendees. At least 80 papers stemming from LISA projects were authored by LISA collaborators and published in high-impact journals, with at least 49 of them coauthored by students in LISA. LISA faculty were on 40 grant-funded projects.

1.4.2 LISA at the University of Colorado Boulder

LISA moved to the Department of Applied Mathematics at the University of Colorado Boulder (CU Boulder) in 2016 when its Director, Prof. Eric Vance, took a position there as an Associate Professor. Because the local context of CU Boulder was very different from that of Virginia Tech, LISA transformed itself into a new lab while retaining many of the characteristics that previously made it strong. LISA's current goal is to increase the quantity and quality of statistics and data science applied to advance high-impact research at CU Boulder and expert decision-making in the community. Its threefold mission is to train statisticians and data scientists to become interdisciplinary collaborators, provide research infrastructure to enable and accelerate quantitative research around the campus community, and engage with the community to improve statistical skills and data literacy (Vance 2017).

LISA serves as research infrastructure by collaborating with researchers on campus on big and small projects—both externally funded and unfunded—that lead to a variety of outcomes such as coauthored publications, grant proposals, internal reports, preliminary data for future projects, theses and dissertations, etc. From 2016 to 2021, LISA has averaged around 80 such collaboration projects annually. LISA also hosts a research service called "LISA Statistics and Data Science Walk-in/Zoom-in Hours" and has averaged around 70 walk-in (or Zoom-in during the pandemic) visitors. LISA has taught an average of 15 short courses and workshops per year since 2017 to an average of 200 attendees annually. The quantitative feedback from domain experts collaborating with LISA from 2016 to 2021 was remarkably similar to the feedback for LISA in the academic year 2014–2015: 92% considered LISA's services helpful, and 90% were satisfied with their overall LISA experience.

LISA trains statisticians and data scientists to communicate and collaborate with researchers to solve research problems and work with businesses and government agencies to make data-driven decisions. LISA statistical collaborators can help design experiments, studies, and surveys; collect, analyze, and plot data; run statistical software; interpret results; and develop strategies to communicate statistical concepts and results to nonstatisticians. LISA has trained 111 graduate and undergraduate students, two postdoctoral researchers, and one visiting fellow in statistics and data science collaboration principles through a three-credit course on statistical collaboration (Vance and Smith 2019). Each student collaborates on three LISA projects via this course.

Twenty students have taken an advanced course in statistical collaboration during which they engaged in VCFS and collaborated on an additional five to six LISA projects. Students discuss projects during weekly staff meetings as part of both courses.

LISA's Director remains Prof. Vance. Dr. Judith Law taught the LISA collaboration course for two semesters in 2020–2021. A current grant from the National Science Foundation funds administrative support for tracking the projects and their outcomes. LISA had temporary support from the College of Arts and Sciences to hire two graduate students per semester (50% time) to collaborate with researchers, mentor junior students, host Walk-in Hours, and teach short courses. LISA is currently self-sufficiently funded from grant projects and consulting contracts with a budget of approximately $100,000 per year. CU Boulder and the Department of Applied Mathematics provide physical space for collaboration meetings and Walk-in Hours as well as general information technology and accounting support and internet access.

Because the supply of students who work in LISA comes through a course on statistical collaboration, LISA advertises only as much as it needs to so that each student can work on three projects. Most advertisement of LISA's services occurs through its website (www.colorado.edu/lab/lisa), direct emails to campus departments, and word of mouth. LISA's short courses are advertised on campus by the Center for Research Data and Digital Scholarship. LISA's annual reports are sent to the chair of the Department of Applied Mathematics and deans in the College of Arts and Sciences.

LISA's primary outcomes were the 111 students, two postdocs, and one visiting fellow it trained to become collaborative statisticians and the hundreds of domain experts who have benefited from LISA's services. At least 24 papers stemming from LISA projects were authored by LISA collaborators and published in high-impact journals, with at least 18 of them coauthored by students in LISA. An additional 16 manuscripts are under review in high-impact journals. LISA was written into 11 grant-funded projects.

The mission, goals, objectives, services, training, and communication strategy of LISA at Virginia Tech (2008–2016) were nearly the same as that of LISA at CU Boulder (2016–present). However, when LISA moved to CU Boulder, it had to adapt to its new conditions. A primary difference is that around 90% of the students involved in LISA at Virginia Tech were MS and PhD students from the Department of Statistics, whereas the students of LISA at CU Boulder are much more diverse. More than half of the students of LISA at CU Boulder are undergraduates and around one-quarter of the LISA students have come from departments other than Applied Mathematics. Only 20% of the students have been MS or PhD students from the Department of Applied Mathematics. In fact, the number of MS or PhD students in a statistics or data science program of study at CU Boulder is about one-tenth as large as at Virginia Tech. The Director of the two LISAs has remained the same, but there was considerably more supporting staff and financial support at Virginia Tech. Additionally, at Virginia Tech, the Director was able to devote his efforts full-time to LISA, a luxury few stat labs and their directors can afford. In contrast—and like most other stat lab coordinators or directors—at CU Boulder Dr. Vance has the regular responsibilities of a standard faculty member, i.e., he has additional teaching, research, and service leadership responsibilities on top of his LISA duties. Consequently, LISA at Virginia Tech had more impacts and outcomes.

1.4.3 The University of Ibadan LISA (UI-LISA)

The University of Ibadan LISA (UI-LISA) was established in 2015 as the third member of the LISA 2020 Network through the influence of O. Olawale Awe, the First LISA Fellow. The primary objective of UI-LISA is to train statistics students to be able to employ the power of statistics to solve societal problems, become effective interdisciplinary collaborators, demonstrate the value of statistical thinking, and be excellent statistical communicators. UI-LISA has designed, tested, and implemented several programs, two of which were unique. These innovative programs include: (1) one-hour with a statistician, where basic and popularly used statistical tools and methods are introduced and explained to a diverse audience; (2) walk-in consultation where domain experts are provided with instant solutions to their statistical problems; (3) providing short statistical courses designed to improve statistical literacy and skills of all users of statistics at all levels; (4) a mobile statistical clinic where the UI-LISA team stations itself at public locations such as halls of residence, conference areas, and recreational areas to provide on-the-spot solutions to problems and inquiries related to statistics; and (5) engaging in statistical collaboration with development actors, including non-governmental and governmental agencies. Much more detail about this exemplar stat lab is provided in Olubusoye et al. (2022) and Adepoju et al. (2022).

Examples of additional stat labs are provided elsewhere in this chapter (Mawora et al. 2022), in Chapter 2 (Adenomon 2022a), as well as in Msemo and Vance (2015), Goshu (2016), Amin and Vance (2016), Vivacqua et al. (2018), and Esterhuizen et al. (2021).

1.5 Stat Labs Can Become Engines for Development

1.5.1 Theory of Change

Stat labs dedicated to the threefold mission of training the next generation of collaborative statisticians and data scientists; supporting research, innovation, and the transformation of evidence to action; and improving statistical skills and data literacy widely in their communities will have begun the process of building their capacity for data-driven development. If these stat labs also have an appropriate administrative structure, adequate training of their staff and students to provide services consistent with their mission, an effective communication strategy, and sufficient support from their university or institution, then they will be well positioned to engage in a positive feedback loop in which projects they collaborate on lead to innovations in development and positive development outcomes, the strengthening of the lab's capacity to collaborate on such projects, and consequently more opportunities and projects to make a positive impact on society (see Figure 1.6).

This positive cycle occurs because stat labs that reach out to the local community of researchers, business leaders, government agencies, and non-governmental organizations to provide both training and customized statistical support cultivate data-capable development actors (Vance and Love 2021). The community becomes more aware of the necessity of statistics and data science expertise, more capable in its use, and presents the lab with more requests for collaboration. New projects provide greater opportunities for the lab to train its students and staff and develop its own capacity to support data-driven development.

Stat labs aspire to be agents for change in their university or institution, surrounding community, and country. Stat labs can create a culture of interdisciplinary collaboration to transform evidence into action for development. A key difference between stat labs and more traditional statistical consulting centers is the stat lab's emphasis on education and training, i.e., building capacity to make data-driven decisions (Vance 2015). By focusing on the training of their own personnel and domain experts in the community to transform evidence into action, stat labs can unlock the potential of statisticians, data scientists, and domain experts alike. In short, stat labs can become engines for driving sustainable development.

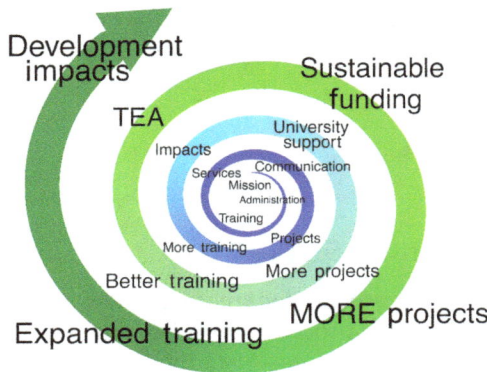

FIGURE 1.6
The virtuous cycle for a stat lab to become strong and sustainable.

1.5.2 Benefits and Impacts of a Stat Lab

We believe that there are many benefits that accrue from creating a stat lab. Stat labs provide *enhanced visibility* for the work of collaborative statisticians and data scientists. Many statisticians already collaborate with domain experts informally. Creating a stat lab elevates this work and provides a mechanism for statisticians to gain formal credit for their collaborations.

Stat labs also make the work of collaborative statisticians and data scientists more *accessible*. Creating a centralized place to go for statistics and data science expertise opens research opportunities to a more diverse set of researchers, businesses, and policymakers.

For university administrators (Vice-Chancellors, Deans, et al.) who aspire to lead their institutions toward a higher profile and a greater positive impact on society, the benefits of creating and sustaining a stat lab are numerous. Stat labs can help foster a culture of interdisciplinary collaboration at the university. Immediately following from this is that stat labs enable more academic staff to conduct high-quality research. A widespread gap in research is an inability to use—or ignorance of—appropriate statistical methods for designing, analyzing, and describing one's research (Vance et al. 2022b). Furthermore, appropriate statistical analyses can enable researchers to publish in high-impact journals. More academic staff publishing research in high-impact journals can lead to an improved reputation for the university, as well as a greater societal impact. A related benefit is for post-graduate students writing theses and dissertations. Assistance with the statistics and data science required for their research improves the quality of their research and shortens time to graduation.

At a lower administrative level, stat labs can help deans and heads of departments accomplish their goals for distinguishing their departments of statistics or mathematics as leaders in statistics and data science education and research. The opportunity for statistics/mathematics students to gain experience applying their statistics and data science skills on real projects positively impacts their education. Students with experience in stat labs report learning statistical theory better by applying it. They also gain practical skills working in a stat lab highly useful and sought after in the job market (Vance et al. 2016). Departments with stat labs can also connect with a global network of stat labs via the LISA 2020 Network (Vance et al. 2022b). These connections can help guide curricular development and global, collaborative research.

Table 1.1 contrasts potential outcomes stemming from creating and sustaining a stat lab with the vision laid out in this paper to what may be the current, status-quo reality at most universities that either have no stat lab or have a statistical center that does not operate like a stat lab.

1.6 Seven Steps for Creating a Stat Lab

We have developed a seven-step process for creating a new stat lab and becoming a member of the LISA 2020 Network. This process is described in Table 1.2. Step 5—opening the lab to train students and provide services—usually occurs after labs have completed the preparatory steps 1–4. However, the steps could be taken out of order. For example, a stat lab could open before they have completed the full stat lab plan/proposal of step 3, though we do not recommend it. The full stat lab plan/proposal available at www.lisa2020.org/

TABLE 1.1

Difference between the Status Quo and the Extraordinary Potential of Stat Labs

The Status Quo: Consulting Centers	What Could Be: Stat Labs
Staff engage in consulting	Staff and students engage in collaboration
Consulting on projects after data are collected	Collaborations with domain experts through all phases of research
Statistics is equations and proofs on a chalkboard (theory only)	Collaborative statisticians use data to solve real-world problems
	Provides research infrastructure
	Statistics drives a laboratory for creating knowledge
Minimal training of students	Students learn by doing
	Students trained by experienced mentors
Guessing what happened with clients	Analyzing video data on what really happened
Only academic staff consult	Students get jobs based on collaboration experience
Dreams of international travel	Students and staff travel internationally to do statistics and data science
Students struggle with statistics	Statistics becomes a strength for the university
Students take too long to complete their MSc and PhD	Collaboration with statisticians speeds time to graduation
Statistics is a barrier to research	Statistics enables and accelerates research at the university
Academic staff do not publish as many articles as they could	Staff publish in high-impact journals
	Staff are promoted
	University gains prestige
	University meets SOCIETAL NEEDS
	Become leading statistical collaboration laboratories
Department/center is isolated on campus	Stat lab connects department to research and educational initiatives across campus
Consulting center is isolated	Stat lab is connected to a global network

about/processes-and-policies guides potential labs through exercises to think about and plan for aligning the stat lab's mission with its institution's mission, identifying the stakeholders in the lab and the services that would make the lab a success to them, determining the lab's personnel and how they will be trained, establishing the lab's administrative structure, considering the budgetary needs of the lab and what support its institution will provide, communicating with the lab's stakeholders, and documenting the lab's outcomes and impacts. When a lab connects with the LISA 2020 Network (usually at step 1 or potentially at step 6 if their mentor is not from the network), the LISA 2020 Secretariat forms a committee to review and provide feedback on the proposed lab's plan/proposal.

Labs that identify a mentor from the LISA 2020 Network and complete steps 1–3 are considered "Proposed Members" of the network. Labs successfully responding to the feedback on the full lab plan/proposal (step 4) become "Transitional Members" of the network. The typical next step is to begin operation of the stat lab (step 5) and stay connected with the LISA 2020 Network (step 6). The final steps required to become a "Full Member" of the LISA 2020 Network are to report a full quarter of metrics and introduce the lab at a semimonthly LISA 2020 Network online Zoom meeting. A two-thirds majority vote of current full members is required to become a "Full Member." Detailed instructions are available at www.osf.io/2gqzqa/.

TABLE 1.2

Process for a Stat Lab to Become a Member of LISA 2020

Stat Lab Stages	Steps Required to Progress to the Next Stage
Potential	1. Identify the director/coordinator 1.1 Identify a mentor from within the LISA 2020 Network 2. Gather and document support 3. Complete the full lab plan/proposal
Proposed	3.1. Submit the full lab plan/proposal 4. Revise the proposal based on reviewer feedback
Transitional	4.1. Review committee approves revisions 5. Open the stat lab 5.1. Train students and staff 5.2. Provide research infrastructure 5.3. Teach short courses/workshops 5.4. Report activities, outcomes, and impacts (metrics) 6. Engage with the network 7. Submit a full quarter (3 months) of metrics 7.1. Give a formal presentation to introduce the stat lab to the network
Full	The Stat Lab's Full Lab Plan/Proposal, submitted metrics, and introduction presentation are sent to the LISA 2020 Network members for a final vote to determine if the transitional stat lab will be admitted as a full member. 2/3 Majority vote required to become a full member.

1.7 What Makes a Stat Lab Strong and Sustainable

In completing the seven steps described above, a new stat lab will have built a strong foundation. They will have secured initial departmental and university support. Their purpose, mission, goals, and objectives will provide a guiding light for lab members. Who will lead and manage the stat lab and participate as collaborators and what their roles and responsibilities will entail will have been established. The stat labs that have completed the process will have considered their unique institutional situation and determined which services they will provide. With this strong foundation—supplemented by feedback and recommendations received from their review panel—these new stat labs are ready to begin operation and build themselves into strong and sustainable engines for development.

The path to a strong and sustainable stat lab begins with training effective collaborative statisticians and data scientists who can move between theory and practice to support research, innovation, and the transformation of evidence into action within and beyond their institutions. This focus on training—especially to impart an evidence to action mindset—creates the conditions needed to transform academic evidence into data-driven innovations that lead to sustainable development impacts. Training, therefore, is the first step in a virtuous cycle where increased collaborations between statisticians and domain experts to develop data-driven innovations to solve local development challenges leads to strengthened stat labs with increased capacity to collect, organize, visualize, analyze, and present development-relevant data (see Figure 1.6). The more projects a stat lab works on, the more opportunities there are for senior statisticians to mentor junior statisticians, and the more quickly junior statisticians strengthen their capacity. These collaborations also lead to enhanced capability of development actors to access and use statistical data for development and greater awareness of the power of statistics and data science to strengthen research and inform decision-making.

Initial successes create a positive feedback loop in which more researchers and development actors want to collaborate with the stat labs and become "data-capable," and more statistics students, faculty, and staff want to work in the stat lab. We have learned that successful projects beget many more requests from domain experts to collaborate with statisticians (Vance and Love 2021). Stat labs that focus on using projects as opportunities to train students rise to the challenge of building capacity quickly to successfully complete a greater number of projects.

When projects come with funding, senior students and faculty can be compensated to both work on projects and mentor junior students. Successful, high-profile projects may also engender support for stat labs from administrators within their institutions—potentially loosening restrictive rules and removing institutional barriers to success—and attract more students to study statistics and data science. This self-reinforcing cycle of using experience on projects to build capacity to work on more projects is a key to the sustainability of strong stat labs.

Effective personnel and a positive reputation are essential for a stat lab, but these alone are not enough to ensure the lab is strong and sustainable. Figure 1.4 is a diagram expressing the analogy of the essential components of a stat lab being like seven parts of a human body—the Heart is the Mission/Goals/Objectives; the Backbone is the Administration; the Muscles are the Personnel/Training; the Head/Brain are the Services; the Legs are Communication Strategy; the Blood is University Support/Budget; and the Skin is Impacts & Outcomes. As the health of each of these body parts is essential for a vigorous human body, the same can be said for a strong and sustainable stat lab. If any one part becomes weak, it impairs the whole system in the short term and can debilitate the stat lab for the long term if the issues are not addressed. Some issues can be addressed internally: ensuring the Mission/Goals/Objectives are integrated with those of the home department and institution, reviewing which services have been successful and which could be improved, exercising the muscles/training the collaborators, accurately keeping track of impacts and outcomes. Other issues may be more challenging and require working with those outside of the stat lab: insufficient operating funds, communicating the value of the stat lab, seeking additional collaboration opportunities. How to address each of these issues depends on the unique situation of the stat lab, but learning from the experiences and best practices of other stat labs—such as those in the LISA 2020 Network—provides direction and guidance. For example, stat labs struggling with statistical collaborator participation can explore incentivization methods that other network stat labs have implemented to determine which might work best at their institution.

In this respect, another fundamental aspect of sustainability is a connection to and being part of an organization larger than itself. Connections with lab mentors and the LISA 2020 Network provide encouragement, support, and guidance as the stat lab encounters both challenges and triumphs. The stat labs can share their experiences, lessons learned, and best practices to strengthen the individual labs and the network as a whole. Regular network meetings, symposia, and staff exchanges encourage communication and engagement between stat labs and enhance opportunities for collaboration. The impetus, development, and growth of the LISA 2020 Network is described in detail later in this chapter (Vance et al. 2022b).

1.8 Conclusion

The goal of this chapter is to explain how a collaborative statistician or data scientist who wants to have more positive impact on society can create a statistical collaboration laboratory, what such a stat lab does, and how stat labs can be created and sustained to become engines for development. We hope this article provides a useful guide for such collaborative statisticians or data scientists that can also be used to help convince academic staff and administrators to support the creation and sustainable activities of stat labs. When operated on the strong and sustainable path, a virtuous cycle of development emerges in which initial collaborations build the capacity of all parties to achieve even more positive development outcomes, enabling the stat lab to achieve its extraordinary potential to apply statistics and data science to positively impact humanity.

References

Adebanji, A., Adeyemi, S., and Gyamfi, M. (2015), "Empirical Analysis of Factors Associated with Neonatal Length of Stay in Sunyani, Ghana," *Journal of Public Health and Epidemiology, Academic Journals*, 7, 59–64. https://doi.org/10.5897/JPHE2014.0679.

Adenomon, M. O. (2022a), "Statistical Consulting and Collaboration Practices: The Experience of NSUK-LISA Stat Lab." *Promoting Statistical Practice and Collaboration in Developing Countries*, eds. O.O. Awe, K. Love, and E.A. Vance, Boca Raton, FL: CRC Press, pp. 131–138. DOI: 10.1201/9781003261148

Adenomon, M. O. (2022b), "Teaching and Learning Statistics in Nigeria with the Aid of Computing and Survey Data Sets from International Organization." *Promoting Statistical Practice and Collaboration in Developing Countries*, eds. O.O. Awe, K. Love, and E.A. Vance, Boca Raton, FL: CRC Press, pp. 371–383. DOI: 10.1201/9781003261148

Adepoju, A. A., Olubusoye, O. E., Ogundunmade, T., Tomiwa, O. T., and Robert, B. (2022), "The Complementary role of UI-LISA in Statistical Training and Capacity Building at the University of Ibadan, Nigeria." *Promoting Statistical Practice and Collaboration in Developing Countries*, eds. O.O. Awe, K. Love, and E.A. Vance, Boca Raton, FL: CRC Press, pp. 111–122. DOI: 10.1201/9781003261148

Aidoo, E. N., Ampofo, R. T., Awashie, G. E., Appiah, S. K., and Adebanji, A. O. (2021), "Modelling COVID-19 incidence in the African sub-region using smooth transition autoregressive model," *Modeling Earth Systems and Environment*. https://doi.org/10.1007/s40808-021-01136-1.

Alexander, K. A., Carzolio, M., Goodin, D., and Vance, E. (2013), "Climate Change is Likely to Worsen the Public Health Threat of Diarrheal Disease in Botswana," *International Journal of Environmental Research and Public Health*, 10, 1202–1230.

Amin, M., and Vance, E. A. (2016), "Inevitability of Interdisciplinary Statistical Laboratories for Food and Agricultural Research in Developing Countries: The case of Pakistan," in *Proceedings of ICAS VII Seventh International Conference on Agricultural Statistics*, Rome, Italy, pp. 1–8. https://doi.org/10.1481/icasVII.2016.h47b.

Arnold, J. C., Hinkelmann, K., Vining, G. G., and Smith, E. P. (2013), "Virginia Tech Department of Statistics," in *Strength in Numbers: The Rising of Academic Statistics Departments in the U. S.*, eds. A. Agresti and X.-L. Meng, New York, NY: Springer, pp. 537–546. https://doi.org/10.1007/978-1-4614-3649-2_39.

Awe, O. O. (2012), "Fostering the Practice and Teaching of Statistical Consulting among Young Statisticians in Africa," *European Journal of Business and Management*, 3, 39–44.

Awe, O. O., Dogbey, D. M., Sewpaul, R., Sekgala, D., and Dukhi, N. (2021), "Anaemia in Children and Adolescents: A Bibliometric Analysis of BRICS Countries (1990–2020)," *International Journal of Environmental Research and Public Health*, 18, 5756. https://doi.org/10.3390/ijerph18115756.

Awe, O. O., and Vance, E. A. (2022), "The ABC of Successful Statistical Collaborations: Adapting the ASCCR Framework in Developing Countries." *Promoting Statistical Practice and Collaboration in Developing Countries*, eds. O.O. Awe, K. Love, and E.A. Vance, Boca Raton, FL: CRC Press, pp. 149–156. DOI: 10.1201/9781003261148

Bell, R. M., Koren, Y., and Volinsky, C. (2007), "The Bellkor Solution to the Netflix Prize," *KorBell Team's Report to Netflix*.

Bernstein, R. M., O'Connor, G. K., Vance, E. A., Affara, N., Drammeh, S., Dunger, D. B., Faal, A., Ong, K. K., Sosseh, F., Prentice, A. M., and Moore, S. E. (2020), "Timing of the Infancy- Childhood Growth Transition in Rural Gambia," *Frontiers in Endocrinology*, 11(142). https://doi.org/10.3389/fendo.2020.00142.

Carzolio, M. (2014), "Masters Without Borders," *Amstat News*, 444, pp. 20–22.

Chirwa, C. F. C., Hall, R. P., Krometis, L.-A. H., Vance, E. A., Edwards, A., Guan, T., and Holm, R. H. (2017), "Pit Latrine Fecal Sludge Resistance Using a Dynamic Cone Penetrometer in Low Income Areas in Mzuzu City, Malawi," *International Journal of Environmental Research and Public Health*, 14, 87. https://doi.org/10.3390/ijerph14020087.

Crandell, I. (2015), "Cultural Values, Statistical Displays," *Amstat News*, 455, pp. 18–19.

Donoho, D. (2017), "50 Years of Data Science," *Journal of Computational and Graphical Statistics*, 26, 745–766. https://doi.org/10.1080/10618600.2017.1384734.

Esterhuizen, T. M., Li, G., Young, T., Zeng, J., Machekano, R., and Thabane, L. (2021), "Advancing Collaborations in Health Research and Clinical Trials in Sub-Saharan Africa: Development and Implementation of a Biostatistical Collaboration Module in the Masters in Biostatistics Program at Stellenbosch University," *Trials*, 22, 478. https://doi.org/10.1186/s13063-021-05427-x.

Gayawan, E., Awe, O. O., Oseni, B. M., Uzochukwu, I. C., Adekunle, A., Samuel, G., Eisen, D. P., and Adegboye, O. A. (2020), "The spatio-temporal epidemic dynamics of COVID-19 outbreak in Africa," *Epidemiology & Infection*, 148. https://doi.org/10.1017/S0950268820001983.

Gayawan, E., Somo-Aina, O., Aladeniyi, O. B., Owoeye, S. M., Aduloju, O. M., Adediran, A. A., and Orunmuluyi, O. S. (2022), "Technology and Multimedia in Statistical Education and Collaboration." *Promoting Statistical Practice and Collaboration in Developing Countries*, eds. O.O. Awe, K. Love, and E.A. Vance, Boca Raton, FL: CRC Press, pp. 515–556. DOI: 10.1201/9781003261148

Goshu, A. T. (2016), "Strengthening Statistics Graduate Programs with Statistical Collaboration— The Case of Hawassa University, Ethiopia," *International Journal of Higher Education*, 5, 217–221.

Gunderman, D., and Vance, E. (2021), "Low- and Middle-Income Countries Lack Access to Big Data Analysis—Here's How to Fill the Gap," *The Conversation*. https://theconversation.com/low-and-middle-income-countries-lack-access-to-big-data-analysis-heres-how-to-fill-the-gap-159412.

Hall, R. P., Davis, J., van Houweling, E., Vance, E. A., Seiss, M., and Russel, K. (2014a), *Impact Evaluation of the Mozambique Rural Water Supply Activity*, Blacksburg, VA: Virginia Tech School of Public and International Affairs.

Hall, R. P., Vance, E. A., and van Houweling, E. (2014b), "The Productive Use of Rural Piped Water in Senegal," *Water Alternatives*, 7, 480–498.

Hall, R. P., Vance, E. A., and van Houweling, E. (2015a), "Upgrading Domestic-Plus Systems in Rural Senegal: An Incremental Income-Cost (IC) Analysis," *Water Alternatives*, 8, 317–336.

Hall, R. P., Vance, E. A., van Houweling, E., and Huang, W. (2015b), "Willingness to Pay for VIP Latrines in Rural Senegal," *Journal of Water, Sanitation and Hygiene for Development*, 5, 586–593. http://dx.doi.org/10.2166/washdev.2015.053.

Halvorsen, K. T., Hanford, K. J., Vance, E. A., Wilson, J., and Zahn, D. (2020), "Transforming Your Stumbling Blocks into Stepping Stones," in *JSM Proceedings*, Alexandria, VA: American Statistical Association: Statistical Consulting Section, pp. 2523–2541.

Hotelling, H. (1988), "Golden Oldies: Classic Articles from the World of Statistics and Probability: The Place of Statistics in the University," *Statistical Science*, 3, 72–83. https://doi.org/10.1214/ss/1177013002.

Kolaczyk, E. D., Wright, H., and Yajima, M. (2021), "Statistics Practicum: Placing 'Practice' at the Center of Data Science Education," *Harvard Data Science Review*. https://doi.org/10.1162/99608f92.2d65fc70.

Kosorok, M. R., and Laber, E. B. (2019), "Precision Medicine," *Annual Review of Statistics and Its Application*, 6, 263–286. https://doi.org/10.1146/annurev-statistics-030718-105251.

Love, K., Awe, O. O., Gunderman, D. J., Druckenmiller, M., and Vance, E. A. (2022), "LISA 2020 Network Survey on Challenges and Opportunities for Statistical Practice and Collaboration in Developing Countries." *Promoting Statistical Practice and Collaboration in Developing Countries*, eds. O.O. Awe, K. Love, and E.A. Vance, Boca Raton, FL: CRC Press, pp. 47–60. DOI: 10.1201/9781003261148

Love, K., Vance, E. A., Harrell, F. E., Johnson, D. E., Kutner, M. H., Snee, R. D., and Zahn, D. (2017), "Developing a Career in the Practice of Statistics: The Mentor's Perspective," *The American Statistician*, 71(1), 38–46. https://doi.org/10.1080/00031305.2016.1255257.

Malakar, S., Goswami, S., Ganguli, B., Chakrabarti, A., Roy, S. S., Boopathi, K., and Rangaraj, A. G. (2021), "Designing a long short-term network for short-term forecasting of global horizontal irradiance," *SN Applied Sciences*, 3, 477. https://doi.org/10.1007/s42452-021-04421-x.

Mawora, T., Otieno, J., and Vance, E. A. (2022), "Exploring the Need for a Statistical Collaboration Laboratory in a Kenyan University: Experiences, Challenges, and Opportunities." *Promoting Statistical Practice and Collaboration in Developing Countries*, eds. O.O. Awe, K. Love, and E.A. Vance, Boca Raton, FL: CRC Press, pp. 61–70. DOI: 10.1201/9781003261148

McDonald, L. (1998), "Florence Nightingale: Passionate Statistician," *Journal of Holistic Nursing*, 16, 267–277. https://doi.org/10.1177/089801019801600215.

Michie, S., Yardley, L., West, R., Patrick, K., and Greaves, F. (2017), "Developing and Evaluating Digital Interventions to Promote Behavior Change in Health and Health Care: Recommendations Resulting from an International Workshop," *Journal of Medical Internet Research*, 19, e7126. https://doi.org/10.2196/jmir.7126.

Msemo, E., and Vance, E. A. (2015), "LISA 2020: Impacting Agricultural Productivity in Tanzania through the Wheels of Statistics," in *Proceedings of the International Statistical Institute's 60th World Statistics Congress*, Rio de Janeiro.

Nunes, M. A. (2022), "Modernizing the Curricula of Statistics Courses through Statistical Learning." *Promoting Statistical Practice and Collaboration in Developing Countries*, eds. O.O. Awe, K. Love, and E.A. Vance, Boca Raton, FL: CRC Press, pp. 351–362. DOI: 10.1201/9781003261148

Oliveira, T., Oliveira, A., Mahmoudvand, R., Ravishankar, N., and Banks, D. (2016), *Book of Abstracts ISBIS 2016: Meeting on Statistics in Business and Industry*, Lisbon: Universidade Aberta.

Olubusoye, O. E., Akintande, O. J., and Vance, E. A. (2021), "Transforming Evidence to Action: The Case of Election Participation in Nigeria," *CHANCE*, 34(3), 13–23.

Olubusoye, O. E., Alaba, O., Vance, E., Folorunso, S., and Akintande, O. J. (2022), "Promoting and Sustaining a Virile Statistical Laboratory in Nigeria's Premier University: Lesson from UI-LISA Experience." *Promoting Statistical Practice and Collaboration in Developing Countries*, eds. O.O. Awe, K. Love, and E.A. Vance, Boca Raton, FL: CRC Press, pp. 95–110. DOI: 10.1201/9781003261148

Seiss, M., Vance, E., and Hall, R. (2014), "The Importance of Cleaning Data During Fieldwork: Evidence from Mozambique," *Survey Practice*, 7(4), 2864.

Sharma, A., Marchant, M. A., Vance, E. A., Smith, E. P., Richardson, W. W., and Hightower, L.S. (2014), "National Survey of Study Abroad Programs Conducted in Asia Using the Food and Agriculture Education Information System (FAEIS) Database," *NACTA Journal*, 58, 142–149.

Stallings, J. (2014), "Type IV Errors: How Collaboration Can Lead to Simpler Analyses," *Amstat News*, 440, 24–25.

Tulchinsky, T. H. (2018), "Chapter 5—John Snow, Cholera, the Broad Street Pump; Waterborne Diseases Then and Now," in *Case Studies in Public Health*, ed. T. H. Tulchinsky, Academic Press, pp. 77–99. https://doi.org/10.1016/B978-0-12-804571-8.00017-2.

Van Houweling, E., Hall, R., Carzolio, M., and Vance, E. (2017), "'My Neighbour Drinks Clean Water, While I Continue To Suffer': An Analysis of the Intra-Community Impacts of a Rural Water Supply Project in Mozambique," *The Journal of Development Studies*, 53, 1147–1162. https://doi.org/10.1080/00220388.2016.1224852.

Vance, E. (2012), "Communication in Statistical Collaborations: Teaching Students How to Be Effective Interdisciplinary Collaborators," in *4th Annual Conference on Higher Education Pedagogy*, Blacksburg, VA: Virginia Tech's Center for Instructional Development and Educational Research, p. 233.

Vance, E. (2013), "Using Team-Based Learning to Teach Effective Communication and Collaboration," in *5th Annual Conference on Higher Education Pedagogy*, Blacksburg, VA: Virginia Tech's Center for Instructional Development and Educational Research, pp. 296–297.

Vance, E. A. (2008), "Statistical Methods for Dynamic Network Data," Ph.D., Durham, North Carolina: Duke University.

Vance, E. A. (2014), "LISA Video Coaching and Feedback Sessions," *Collaboration in a Bag*, p. 1. https://doi.org/10.17605/OSF.IO/XMTCE.

Vance, E. A. (2015), "Recent Developments and Their Implications for the Future of Academic Statistical Consulting Centers," *The American Statistician*, 69(2), 127–137. https://doi.org/10.1080/00031305.2015.1033990.

Vance, E. A. (2017), "LISA: Laboratory for Interdisciplinary Statistical Analysis 2016–17 Annual Report," *OSF*. https://doi.org/10.17605/OSF.IO/ZQEJ2.

Vance, E. A. (2020), "Goals for Statistics and Data Science Collaborations," in *JSM Proceedings*, Alexandria, VA: American Statistical Association: Statistical Consulting Section, pp. 2198–2209.

Vance, E. A. (2021), "Using Team-Based Learning to Teach Data Science," *Journal of Statistics and Data Science Education*, 29(3), 277–296. https://doi.org/10.1080/26939169.2021.1971587.

Vance, E. A., Alzen, J. L., and Seref, M. M. H. (2020), "Assessing Statistical Consultations and Collaborations," in *JSM Proceedings*, Alexandria, VA: American Statistical Association: Statistical Consulting Section, pp. 161–169.

Vance, E. A., Alzen, J. L., and Smith, H. S. (2022a), "Creating Shared Understanding in Statistics and Data Science Collaborations," *Journal of Statistics and Data Science Education*, 30(1), 54–64. https://www.tandfonline.com/doi/full/10.1080/26939169.2022.2035286

Vance, E. A., Archie, E. A., and Moss, C. J. (2009), "Social networks in African elephants," *Computational and Mathematical Organization Theory*, 15(4), 273–293. https://doi.org/10.1007/s10588-008-9045-z.

Vance, E. A., Glimp, D. R., Pieplow, N. D., Garrity, J. M., and Melbourne, B. (in press), "Integrating the Humanities into Data Science Education: Reimagining the Introductory Data Science Course," *Statistics Education Research Journal*.

Vance, E. A., LaLonde, D. E., and Zhang, L. (2017a), "The Big Tent for Statistics: Mentoring Required," *The American Statistician*, 71(1), 15–22. https://doi.org/10.1080/00031305.2016.1247016.

Vance, E. A., and Love, K. (2021), "Building Statistics and Data Science Capacity for Development," *CHANCE*, 34(3), 38–46.

Vance, E. A., Love, K., Awe, O. O., and Pruitt, T. R. (2022b), "LISA 2020: Promoting Statistical Practice and Collaboration in Developing Countries." *Promoting Statistical Practice and Collaboration in Developing Countries*, eds. O.O. Awe, K. Love, and E.A. Vance, Boca Raton, FL: CRC Press. DOI: 10.1201/9781003261148

Vance, E. A., Metzger, T., and Pruitt, T. (2016), "The Educational Impact of Working in a Statistical Collaboration Laboratory," in *Conference on Higher Education Pedagogy*, Blacksburg, VA.

Vance, E. A., and Smith, H. S. (2019), "The ASCCR Frame for Learning Essential Collaboration Skills," *Journal of Statistics Education*, 27(3), 265–274. https://doi.org/10.1080/10691898.2019.1687370.

Vance, E. A., Tanenbaum, E., Kaur, A., Otto, M. C., and Morris, R. (2017b), "An Eight-Step Guide to Creating and Sustaining a Mentoring Program," *The American Statistician*, 71(1), 23–29. https://doi.org/10.1080/00031305.2016.1251493.

Vance, E. A., Xie, X., Henry, A., Wernz, C., and Slonim, A. D. (2013), "Computed Tomography Scan Use Variation: Patient, Hospital, and Geographic Factors," *The American Journal of Managed Care*, 19, e93–e99.

Vance, E., and Laga, I. (2017), "Variations in Statistical Practice between North American Stat Labs," in *ASA Conference on Statistical Practice*, Jacksonville, FL.

Vance, E., and Pruitt, T. (2016a), *LISA: Virginia Tech's Laboratory for Interdisciplinary Statistical Analysis Annual Report 2014–15*, Report, Virginia Tech. Laboratory for Interdisciplinary Statistical Analysis.

Vance, E., and Pruitt, T. (2016b), *Virginia Tech's Laboratory for Interdisciplinary Statistical Analysis Annual Report 2015–16*, Report, Virginia Tech. Laboratory for Interdisciplinary Statistical Analysis.

Vivacqua, C. A., de Pinho, A. L. S., Nunes, M. A., and Vance, E. A. (2018), "Integrating Collaboration, Communication and Problem Solving to Promote Innovation in Statistics Education," in *Looking Back, Looking Forward: Proceedings of the Tenth International Conference on Teaching Statistics (ICOTS10, July, 2018)*, Kyoto, Japan: IASE, pp. 1–5.

Weins, A., Lovejoy, H. B., Mullen, Z., and Vance, E. A. (2022), "A Modelling Strategy to Estimate Conditional Probabilities of African Origins: The Collapse of the Oyo Empire and the Transatlantic Slave Trade, 1817–1836," *Journal of the Royal Statistical Society. Series A (Statistics in Society)*, 1–24. https://rss.onlinelibrary.wiley.com/doi/full/10.1111/rssa.12833. DOI:10.1111/rssa.12833

2

LISA 2020: Promoting Statistical Practice and Collaboration in Developing Countries[1]

Eric A. Vance
University of Colorado Boulder

Kim Love
K. R. Love Quantitative Consulting and Collaboration

O. Olawale Awe
University of Campinas

Tonya R. Pruitt
Virginia Tech

CONTENTS

2.1 Introduction

Statistics and data science collaboration laboratories ("stat labs") are collections of collaborative statisticians and data scientists who collaborate with domain experts to enable and accelerate research and make data-driven business and policy decisions (Vance and

---

[1] Any correspondence should be addressed to Eric.Vance@Colorado.EDU.

DOI: 10.1201/9781003261148-3

Pruitt 2022). In short, stat labs drive sustainable development while also building further capacity for development, which we consider to be sustainable actions that affect society positively (Vance and Love 2021).

Stat labs achieve outstanding results by using projects to train their staff and students to further build their capacity, teaching short courses and workshops to improve statistical skills and data literacy widely, and collaborating with domain experts to transform evidence into action to positively impact development (Olubusoye et al. 2021; Vance 2015b; Vance and Smith 2019). Members of stat labs can learn from each other. When a project requires specialized statistical methodology or data science skills (e.g., spatial statistics, causal analysis, Bayesian analysis, Big Data), an individual working on the project can seek advice from other stat lab members with the needed skills.

Similar to how a stat lab can upscale the impact of an individual collaborative statistician or data scientist by surrounding her with mentors and stat lab members who can complement her technical skills and teach her new ones, a network of stat labs can multiply the impact of its constituent stat labs. Research networks—and the international collaborations they stimulate—have been shown to increase the production of scientific papers and their citations (Jacob and Meek 2013; OECD 2011). Networks facilitate mentorship between network members (Vance et al. 2017a, 2017b), enable exchanges, share best practices, provide purpose and motivation to other members, and yield increased opportunities for all members.

We believe that the LISA 2020 Network is such a network. LISA 2020 is a program to build statistical analysis and data science capacity in developing countries to transform evidence into action. It does this by helping to create, strengthen, and sustain stat labs that have three missions:

1. Train statisticians and data scientists to become effective, interdisciplinary collaborators who can move between theory and practice to solve problems for real-world impact
2. Serve as research infrastructure for researchers and decision-makers to collaborate with statisticians and data scientists to enable and accelerate research and data-based decisions that make a positive impact on society
3. Teach short courses and workshops to improve statistical skills and data literacy widely.

LISA 2020 is centered at the Laboratory for Interdisciplinary Statistical Analysis (LISA) at the University of Colorado Boulder, United States. Since its creation in 2012, the goal of LISA 2020 has been to create a network of at least 20 stat labs in developing countries by 2020 (hence the name "LISA 2020"), specifically by 20 October 2020, which was the third UN-designated World Statistics Day.

The goal of this chapter is to explain the history, current state, and future directions of the LISA 2020 Network. In Section 2.2, we describe the origins of the LISA 2020 Network and key events in its history. Section 2.3, characterizes the current state of the network. In Section 2.4, we discuss the purpose of the LISA 2020 Network, how the network can improve outcomes and impacts of all its members, and how we can sustain this. We preview the future directions for the LISA 2020 Network in Section 2.5, and we conclude in Section 2.6.

2.2 The Origin and History of the LISA 2020 Program

2.2.1 The Beginning of LISA 2020

LISA 2020 was created in 2012 by Dr. Eric Vance, who was the Director of LISA at Virginia Tech in Blacksburg, Virginia, United States. The program came from three international visions for LISA he publicized in 2011 on the LISA website and in the *LISA Annual Report for 2010–2011* (Vance 2012b):

> **2011 Vision 1. ON-THE-GROUND STATISTICIANS**: LISA graduate students will work on large, global, interdisciplinary research projects as "on-the-ground" statisticians (Seiss et al. 2014).
>
> **2011 Vision 2. STATISTICAL EXCHANGES**: LISA students will learn to be internationally aware and culturally competent collaborative statisticians by participating in short-term (6-month) student exchanges within a network of statistical consulting and collaboration centers.
>
> **2011 Vision 3. CAPACITY BUILDING**: Establish statistical consulting and collaboration centers at universities in countries currently lacking the infrastructure and statistical capacity to assist researchers with statistics.

Vision 1, on-the-ground statisticians, was based on lessons learned from two large, international LISA research projects. The first was a study sponsored by the World Bank of water supply and demand in rural Senegal (Hall et al. 2014b). Local enumerators asked more than 300 questions of around 2,000 households about their use of water for various domestic and economically productive purposes (e.g., growing garden crops, irrigating fields, watering livestock). The overall research question was whether investments in constructing water supply infrastructure (e.g., pumps, holding tanks, distribution pipes) would cause enough economic activity to pay for the initial investment (Hall et al. 2015a). LISA was contracted to analyze the data (Hall et al. 2015b) and found that some relevant data was missing (i.e., an important source of water [surface water] was not queried) and that the extant data was riddled with inconsistencies and ridiculous values due to enumerator error. For example, households were asked about their sales of irrigated crops; the data indicated that one farmer had, incredibly, sold one watermelon for US$50,000.

The challenge of understanding, cleaning, and analyzing data about a context foreign to LISA statisticians prompted the idea of embedding a statistician within the team of project designers to ensure that no relevant question was accidentally dropped from the survey and within the team of enumerators to flag suspicious data points, clean the data in real time, and retrain enumerators as necessary. We called this person an "On-the-Ground Statistician" and implemented this idea on a new project sponsored by the Millennium Challenge Corporation to understand the impact of installing water hand pumps in rural Mozambique (Hall et al. 2014a). The scheme worked exactly as intended! Statisticians on the ground helped redesign the study at the last minute, ensured that all relevant data was collected, verified all data points in the field within hours after they had been collected, retrained enumerators when necessary, and provided weekly updates and summary tables to the research team and funding agency. The result was a high-quality dataset that could be easily modeled and analyzed by the same on-the-ground statisticians who now understood the local context of the data production (Seiss et al. 2014; Van Houweling et al. 2017). However, an on-the-ground statistician from the United States is expensive and infeasible for most international projects. A more feasible solution would be to embed a local statistician within the research team who already understood the local context of the research and had the necessary technical skills to guide the research team toward collecting, modeling, and analyzing high-quality data to answer relevant research, business, and policy questions.

Vision 2, statistical exchanges, was motivated by a desire to improve LISA at Virginia Tech by exposing it to perspectives from around the world so as to become *the* premier academic statistical consulting and collaboration laboratory and to provide opportunities for LISA graduate students to travel abroad. At the time, LISA had funding for four, full-time (20 hours/week) graduate students to engage in statistical collaboration. If we could find a match with a foreign university who trained and funded their graduate students to be full-time (20 hours/week) statistical collaborators, then we could make an exchange, with LISA paying its graduate student to work in the foreign stat lab and the foreign university paying their graduate student to work in LISA. As a result, LISA would learn from the experience and ways of collaborating from the foreign exchange student, and the foreign stat lab would learn about LISA's methods. Several LISA students investigated options for exchanges but never found a foreign stat lab that trained and funded graduate students to engage in statistical consulting and collaboration. All of the labs that we found employed permanent staff to engage in *consulting* rather than graduate students to engage in statistical *collaboration*.

Vision 3, capacity building, was motivated by the realization that the LISA model of training students to become effective interdisciplinary collaborators to assist researchers, businesses, and policymakers in statistics was exceptional, had great potential for impact (Vance and Pruitt 2022), and could be adapted and adopted widely for the benefit of society. Furthermore, to train future on-the-ground statisticians who fully understood the local context of projects (Vision 1) and facilitate exchanges of students between LISA and foreign stat labs (Vision 2), more capacity would have to be built. In other words, LISA would need to help establish new stat labs around the world.

In 2012, O. Olawale Awe published a paper calling for all universities in Africa to teach statistical consulting and establish statistical consulting units (Awe 2012). In his paper, Awe summarized the training methods of LISA and concluded, "There is a need for other institutions of higher learning across the continent [of Africa] to follow the same pattern."

While searching for possible exchange opportunities for his students or potential locations for creating new stat labs, Vance read Awe's paper and realized that there was, in fact, an expressed need for and demand to create stat labs in Africa modeled after LISA. A new vision was then established: to build statistical capacity and research infrastructure in developing countries.

Vance explained this new vision to Tim Howland, an officer at Virginia Tech responsible for cultivating charitable donations to the university. On March 2, 2012, Howland and Vance had a conversation that established the goals and the name of the LISA 2020 program:

Howland: "How will you build statistical capacity in developing countries?"
Vance: "By helping to create stat labs in the pattern of LISA."
Howland: "How many stat labs?"
Vance: "Twenty."

Howland: "By when?"
Vance: "Twenty by the year 2020, and we'll call the program 'LISA 2020.'"

Thus was established the program, the goal of creating 20 stat labs in the pattern of LISA by the year 2020, and the name "LISA 2020."

2.2.2 The Origins of LISA 2020

In one sense, the roots of LISA 2020 go as far back as the roots of LISA, which is to 1948 when Dr. Boyd Harshbarger created the Statistical Laboratory at Virginia Polytechnic Institute in Blacksburg, Virginia within the university's Agricultural Experiment Station. Based on the success of the Statistical Laboratory, in 1949 the Department of Statistics—offering an M.S. degree—was approved and housed in the College of Arts and Sciences (Arnold 2000). It is important to emphasize that the Statistical Laboratory predated and paved the way for the Department of Statistics.

In 1973, the Statistical Laboratory was reorganized to become the Statistical Consulting Center under the direction of Dr. Ray Meyers to expand services across the university (now named Virginia Polytechnic Institute and State University) (Arnold 2000). A major change implemented by Myers was to emphasize the training of graduate students in statistical consulting (Arnold et al. 2013). The Statistical Consulting Center was reborn in 2008 as the Laboratory for Interdisciplinary Statistical Analysis (LISA) under the leadership of interim director Laura Freeman and its first (and only) Director Dr. Eric Vance (Arnold et al. 2013). A major focus of LISA was to emphasize collaboration over consultation and to revitalize the training program of its students in interdisciplinary collaboration (Vance 2009, 2015b).

Another transition occurred in 2016 when Vance moved LISA to the University of Colorado Boulder, maintaining LISA's traditions and methods, but operating in the context of a different university and being housed within the Department of Applied Mathematics instead of a statistics department. Concurrently, the LISA 2020 program moved from Virginia Tech to the University of Colorado Boulder while maintaining its original purpose, methods, and personnel (Vance 2017).

LISA 2020 also stems from the travel experiences of Vance as a young man. For 5 years between his undergraduate education and graduate school at Duke University for his M.S. and Ph.D. in Statistical Science, Vance traveled around the world three times through 67 countries in Europe, Australia and New Zealand, South and Southeast Asia, Central and South America, and Africa. In 2002, during his travels through 17 countries in Africa, Vance attempted to hitchhike from the territory of Western Sahara to Mauritania through a part of the Sahara Desert that had no roads (Vance 2012a). He was able to find a ride on the top of a four-wheel-drive vehicle (see Figure 2.1) and rode for five hours bungee-corded to its roof rack to the border of Western Sahara territory and Mauritania, which was laden with mines at the time. The couple who drove Vance to the border felt it would be unsafe to drive through the minefield with him on top of their car.

At the border checkpoint, Vance met two Frenchmen who owned and were driving a bus south from France to sell in West Africa. They agreed to give him a ride in their bus across the minefield after the paperwork to cross the border was completed. At the border checkpoint Vance also met a Ph.D. biologist studying the Saharan desert fox, also known as the fennec fox. When the biologist learned that Vance was to be studying statistics in graduate school, he exclaimed, "A statistician! I need to talk to you." The biologist began to explain how he was struggling to count the foxes and make sense of the data he could collect about them. But before they could get very far into the details of what was Vance's

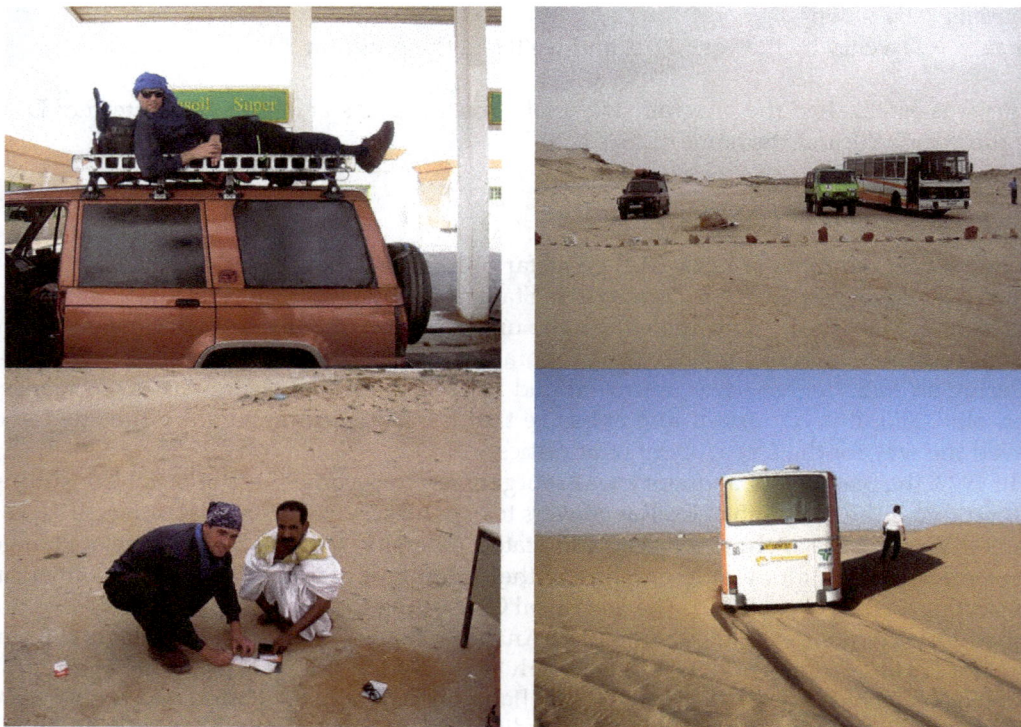

FIGURE 2.1
Photos of Vance hitching a ride on top of the vehicle (top-left), the border checkpoint (top-right), the consultation with the Ph.D.-biologist (bottom-left), and the Frenchmen's bus stuck on a sand dune in the middle of the Sahara Desert (bottom-right).

first-ever statistical consultation, the Frenchmen's bus was ready to depart, and Vance had to abruptly leave the biologist behind to complete the 4-day journey across the desert.

Vance's travels and his abbreviated consultation with the biologist are relevant to LISA 2020 because these experiences informed his motivations. Vance had great experiences traveling around the world and wanted to provide his students in LISA opportunities to travel internationally. He recognized the tremendous benefits of LISA to supply training opportunities for his students to apply statistics to solve real-world problems and for enabling and accelerating research and data-driven decisions within the Virginia Tech community. His experience at the border checkpoint taught him that there are literally Ph.D.-scientists wandering the Sahara Desert looking for statisticians to collaborate with! These experiences combined to help Vance form the idea for the LISA 2020 program because they showed him that it is not just scientists, businesses, and policymakers in the United States who need to collaborate with statisticians and data scientists. Decision makers around the globe need the option to collaborate with statisticians and data scientists, especially in developing countries.

2.2.3 The Timeline of Key Events for LISA 2020

In Table 2.1, we list key events in the history of LISA 2020, which include sources of funding, symposia and conferences, and exchange visits between stat lab personnel.

TABLE 2.1

Timeline of Key Events for LISA 2020

2012

Vance establishes the idea for the LISA 2020 program; travels to the University of Juba in South Sudan

Emanuel Msemo from Tanzania begins pursuing his M.S. in Statistics at Virginia Tech (VT)

2013

Vance is awarded a Google Research Award to establish the LISA Fellows program

O. Olawale Awe from Obafemi Awolowo University (OAU), Ile-Ife, Nigeria selected to serve as the first LISA Fellow, training at Virginia Tech for one year

iAGRI/USAID Award: "Growing Research Capacity at Sokoine University of Agriculture (SUA) by Creating a Statistical Collaboration Laboratory"

Vance visits SUA in Tanzania (also in 2015) and the Federal University of Rio Grande do Norte (also in 2014)

2014

Benedicto Kazuzuru from SUA, Tanzania, visits VT for 6 months

Richard Ngaya from Tanzania begins pursuing his M.S. in Statistics at VT

Jingli Xing from Renmin University in Beijing, China, visits VT for six months

O. Olawale Awe visits Alex Hanlon at the University of Pennsylvania School of Nursing to train with the BECCA lab

Mohammad Djedour from University of Science and Technology Houari Boumediene (Algeria) visits VT for 2 weeks

Ayele Taye Goshu from Hawassa University, Ethiopia, visits VT for 2 months

2015

VT graduate student Ian Crandell spends 6 months in Nigeria at OAU and the University of Ibadan

VT graduate student Adam Edwards travels to Tanzania to support the newly established SUALISA

Vance visits Nigeria, Tanzania, Ethiopia, India, and Brazil

2016

LISA 2020 Advisory Council formed with support from the American Statistical Association (ASA)

Olusanya Olubusoye from UI-LISA visits VT and Purdue University for 3 weeks

Vance moves LISA to the University of Colorado Boulder (CU Boulder)

Imran Khan from Kashmir, India serves as a postdoctoral research fellow to help establish LISA at CU Boulder; visits the UCLA Statistical Consulting Group for 2 weeks in 2017

Demisew Gebru from Hawassa University, Ethiopia, visits Jim Rosenberger at Pennsylvania State University for 3 months and CU Boulder for 1 week

George Uchechukwu spends 6 weeks working at UI-LISA with Olubusoye

2017

First LISA 2020 Symposium in Marrakech, Morocco in conjunction with ISI's 61st WSC with nine participants

Vance presents Keynote at the First International Conference of the Nigerian Statistical Association

2018

LISA 2020 funded by USAID Accelerating Local Potential program

Idowu Adarabioyo spends 6 weeks learning at Anchor University Laboratory for Interdisciplinary Statistical Science & Data Analysis (AULISSDA) from O. Olawale Awe

2019

Second LISA 2020 Symposium in Kuala Lumpur, Malaysia in conjunction with ISI's 62nd WSC with 34 participants

2020–2021

Third LISA 2020 Symposium held virtually with hundreds of attendees; local and regional symposium events held in Ghana (KNUST), Tanzania (Mzumbe), Nigeria (NSUK, MOUAU, OAU), South Africa (Stellenbosch), Pakistan (Lahore), Brazil (UNIOESTE), and Egypt (Alexandria)

Vance's first international travel as Director of LISA 2020 was in June 2012 to the University of Juba in South Sudan. There he met three statisticians eager to start a stat lab (see Figure 2.2), including Loro Gore, who relayed to Vance a story about a pharmacy M.Sc. student who struggled for 9 months to solve a statistics problem in his thesis. The problem was seemingly too difficult for the student to overcome, so he decided to withdraw from his studies. The Registrar at the time, Dr. Martin Baru, asked the student to meet with Gore before dropping out, and 45 minutes later the problem was solved! This anecdote highlights the potential for impact of a collaborative statistician and the need for stat labs in *all* universities so that future researchers, business people, and policymakers do not become so discouraged and frustrated with their lack of capability in statistics and data science.

The first funding support for LISA 2020 was a Google Faculty Research Award to Vance in 2013, facilitated by Dr. Steve Scott. This award created the LISA Fellows program and sponsored one statistician from a developing country to work in LISA at Virginia Tech for

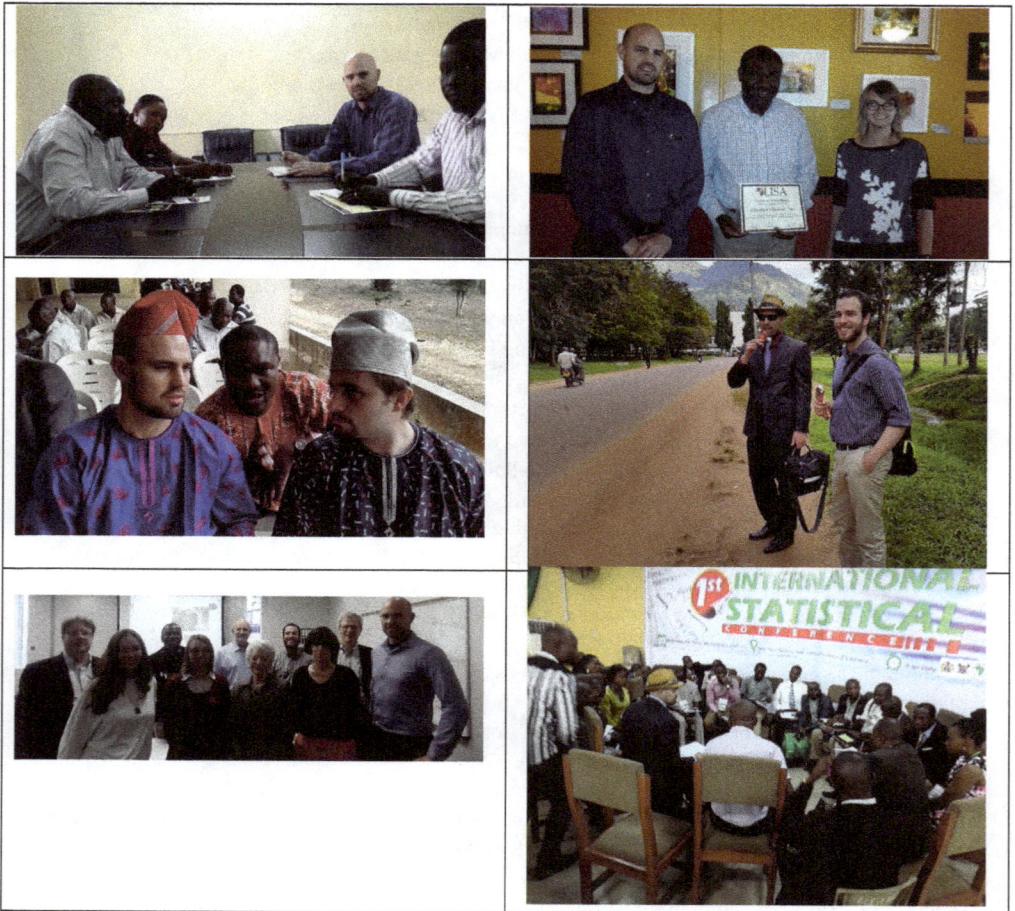

FIGURE 2.2
Photos of Vance with Jaffa, Magdalena, and Gore at University of Juba (top-left); with Awe and Pruitt at VT (top-right); with Awe and Crandell at OAU (middle-left); with Edwards at SUA (middle-right); with Advisory Council (Cochrane, Love, Olubusoye, Altman, Pruitt, Morganstein, Clarke, Edwards, Cule, Rosenberger, and Pollack [not pictured]) in San Diego (bottom-right); at NSA roundtable discussion in Lagos (bottom-right).

1 year. In only 3 weeks, LISA received 108 full fellowship applications from 34 countries. The selection committee named three finalists—Dr. Carla Vivacqua, Genelyn Sarte, and O. Olawale Awe—with Awe selected as the first LISA Fellow. Several other semi-finalists have remained connected to the LISA 2020 Network and created their own stat lab; see, for example, Goshu (2016) and Amin and Vance (2016).

A follow-up Google Faculty Research Award enabled LISA graduate student Ian Crandell to travel to Nigeria for 6 months in 2015 to help Awe build the capacity of Awe's newly created Laboratory for Interdisciplinary Statistical Analysis and Collaboration (LISAC) at Obafemi Awolowo University—the first stat lab established in the LISA 2020 Network. While in Nigeria, Crandell, Vance, and Awe visited the University of Ibadan (UI) to help them establish UI-LISA, the third stat lab in the growing network in June 2015. They also visited Afe Babalola University Ado-Ekiti (ABUAD), which became a full member of the network in 2020.

Also, in 2015, a second LISA graduate student, Adam Edwards, was sponsored by the USAID-funded iAGRI program to visit the newly created Sokoine University of Agriculture Laboratory for LISA (SUALISA) in Tanzania—the second stat lab in the Network—to help build its capacity and sustainability (Msemo and Vance 2015; Vance and Magayane 2014). This visit was facilitated by previous exchanges: an educational visit by Emanuel Msemo from 2012 to 2014 to earn his M.S. in Statistics at Virginia Tech, a visit by Vance to SUA in 2013, a research visit to LISA by Dr. Benedicto Kazuzuru in 2014, and an educational visit by Richard Ngaya from 2014 to 2016 to earn his M.S. in Statistics.

In 2016 the American Statistical Association (ASA) funded Vance's proposal to convene a LISA 2020 Advisory Council. This—along with funding from the Center for Open Science—enabled Prof. Olusanya Olubusoye from UI to join other council members for a meeting at the ASA Conference on Statistical Practice in San Diego, CA. The secretariat of LISA 2020 moved from Virginia Tech to the University of Colorado Boulder in the summer of 2016 when Vance moved LISA.

In 2017, the International Statistical Institute (ISI) provided a room for the first LISA 2020 Symposium held in conjunction with the ISI's 61st World Statistics Congress in Marrakech, Morocco. This was the first opportunity for most of the nine attendees to hear directly from each other about their experience creating and running stat labs. Also, in 2017, Vance delivered a keynote at the First International Conference of the Nigeria Statistical Association in Lagos and, with Awe and Olubusoye, convened a roundtable discussion with several future LISA 2020 members. See Figure 2.2 for photos of this and other key events.

Vance was awarded funding from the United States Agency for International Development (USAID) in 2018 to accelerate local potential at higher education institutions in low- and middle- income countries to transform evidence to action to address development challenges. One of the activities this award funded was the second LISA 2020 Symposium, held in 2019 in Kuala Lumpur, Malaysia in conjunction with the ISI's 62nd World Statistics Congress. Thirty-four participants from 11 countries engaged in person over more than 3 days to share best practices and develop their collaboration and stat lab management skills. See Figure 2.3.

On World Statistics Day, 20 October 2020, during the keynote of the online LISA 2020 Symposium—which highlighted best practices in statistical collaboration and practice in developing countries from stat labs throughout the LISA 2020 Network from October 2020 to May 2021—Vance announced that the LISA 2020 program had achieved its eight-year-old goal of creating a network of at least 20 stat labs and furthermore had surpassed this goal by 40%, welcoming 28 stat labs as full members of the LISA 2020 Network. Table 2.2 shows when the LISA 2020 stat labs were established or joined the network.

FIGURE 2.3
Ten participants from seven countries attended the First LISA 2020 Symposium in Marrakech in 2017. Thirty-four participants from 11 countries attended the Second Symposium in Kuala Lumpur in 2019.

TABLE 2.2

Timeline of the Establishment of Stat Labs in the LISA 2020 Network

2014
1. O. Olawale Awe founds LISAC (Laboratory for Statistical Analysis and Collaboration) at Obafemi Awolowo University, Nigeria
2. Benedicto Kazuzuru and Emanuel Msemo found SUALISA at Sokoine University of Agriculture, Tanzania

2015
3. Olusanya Olubusoye founds UI-LISA at the University of Ibadan, Nigeria
4. Ayele Taye Goshu founds HwU SCC (Hawassa University Statistical Collaboration Center) in Ethiopia
5. Carla Vivacqua joins LEA (Laboratório de Estatística Aplicada) at Federal University of Rio Grande do Norte, Brazil to the LISA 2020 Network

2017
5. O. Olawale Awe founds AULISSDA (Anchor University Laboratory for Interdisciplinary Statistical Science & Data Analysis), Nigeria as LISAC goes dormant
6. Ayele Gebeyehu founds WUSCCC (Wolkite University Statistical Consultancy and Collaboration Center), Ethiopia

(Continued)

TABLE 2.2 (*Continued*)

Timeline of the Establishment of Stat Labs in the LISA 2020 Network

2018

7. Haftom Temesgen founds MULISDA (Mekelle University Laboratory for Interdisciplinary Statistical Data Analysis), Ethiopia
8. M.O. Adenomon founds NSUK-LISA at Nasarawa State University, Keffi, Nigeria
9. Kayode Ayinde and Ezra Gayawan found FUTA-LISA at the Federal University of Technology, Akure, Nigeria

2019

10. Atinuke Adebanji founds KNUST-LISA at Kwame Nkrumah University of Science and Technology, Ghana
11. Bhaswati Ganguli founds data science lab at the University of Calcutta, India
12. Joyce Otieno founds MU-LISA at Maseno University, Kenya; see Mawora et al. (2022)
13. Albert Ayorinde Abegunde, Olubola Babalola and O. Olawale Awe rebranded LISAC to become EDM-LISA OAU.

2020

14. Osama Hussien founds AU-LISA at Alexandria University, Egypt
15. Asifa Kamal founds LISA-LCWU at the Lahore College for Women University, Pakistan
16. George Uchechukwu founds MOUAU-LISA at Michael Okpara University of Agriculture, Umudike, Nigeria
17. Rita Lima founds Statistical Collaboration Laboratory (SCL) at Federal University of Piauí (UFPI), Brazil
18. Tonya Esterhuizen joins SU-Biostatistics at Stellenbosch University, South Africa to the LISA 2020 Network
19. Temidayo Apata founds FUOYE-LISA at the Federal University Oye Ekiti, Nigeria
20. Imran Khan founds SKUAST-K-LISA at Sher-e-Kashmir University of Agricultural Sciences and Technology of Kashmir, India
21. Lubna Naz founds KU-LISA at the University of Karachi, Pakistan
22. Miguel Opazo founds LEE (Laboratório de Estatística Espacial) at the Universidade Estadual do Oeste do Paraná (UNIOESTE), Brazil
23. Adarabioyo Idowu founds ABUAD-CENSAC (Centre for Statistical Analysis and Collaboration) at Afe Babalola University Ado-Ekiti (ABUAD), Nigeria
24. Hope Mbachu founds IMSU-LISA at Imo State University, Nigeria
25. Tadesse Awoke founds UoGStatLab at the University of Gondar, Ethiopia
26. Enobong Udoumoh founds FUAM-LISA at the Federal University of Agriculture, Makurdi, Nigeria
27. Justine Mbukwa founds MULISA at Mzumbe University, Tanzania
28. Daniels Akpan founds CLEDA-LISA at the African Centre for Education Development, Nigeria

2021

29. Bashiru Saeed and Ken Kubuga found TaTU DataScience Lab at Tamale Technical University, Ghana
30. Abdul-Aziz Abdul-Rahaman founds KsTU-DATALINK at Kumasi Technical University, Ghana
31. Happiness Obiora-Ilouno founds NAU-LISA at Nnamdi Azikiwe University, Nigeria
32. Oludare Ariyo founds FUNAAB-LISA at the Federal University of Agriculture, Abeokuta, Nigeria
33. Lamidi-Sarumoh Alaba founds GSU-LISA at Gombe State University, Nigeria
34. Benjamin Odoi and Christiana Nyarko found UMaT-LISA at the University of Mines and Technology, Tarkwa, Ghana
35. Muyiwa Agunbiade founds UNILAG-LISA at the University of Lagos, Nigeria

2.3 The Current State of the LISA 2020 Network

The LISA 2020 Network currently has 35 full members, 6 transitional members, and 7 proposed members (see Figure 2.4). Vance and Pruitt (2022) describe a seven-step procedure for new labs to become full members. Briefly, labs must complete a stat lab plan to become a proposed member, respond to feedback on their plan to become a transitional member,

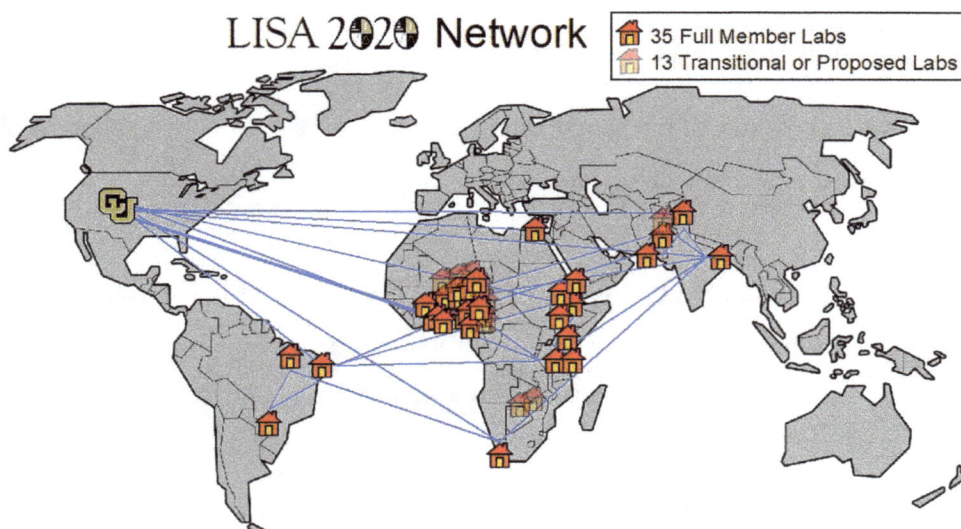

FIGURE 2.4
Map of the LISA 2020 Network's 35 full members and 13 transitional or proposed stat labs indicated by transparency.

and then finalize their plan, report metrics, present their lab to the full network, and be approved by two-thirds of the current full members to become a full member of the LISA 2020 Network.

From 2014 to 2018, the stat labs reported having trained 141 statisticians to communicate and collaborate with more than 1,000 researchers on both short-term consultations and longer term collaborations. They also reported having taught 49 short courses to improve the statistical skills of 1,396 attendees.

In 2019, Dr. Kim Love streamlined the process for stat labs to report their metrics. The network also began to disaggregate metrics referring to numbers of people by gender. This is because gender equality is an important consideration for the LISA 2020 Network. Women bring unique and important perspectives to collaborative settings, development issues, and capacity building, and in many developing countries, they face significantly greater barriers and unrealistic standards relative to their male colleagues. The directors and staff of the LISA 2020 Network stat labs are enthusiastic about gender inclusivity, and by disaggregating these metrics by sex, we remain informed of the state of gender equality in the network's activities and can recommend remedial measures if and where they are necessary.

By 2019 the network had grown to nine full members and would reach 13 full members by the end of the year. One important metric is the number of collaborative statistician trainees produced by the labs. These are individuals (faculty, staff, and students) who have statistics or data science expertise and are trained to use these skills in a collaborative setting. Individuals may be trained in several ways, including (1) participating in stat lab projects, (2) enrolling in a consulting/collaboration course in the department, (3) attending collaborative training specially organized by the lab, (4) participating in collaborative coaching and feedback sessions reviewing video-recorded meetings with domain experts, and/or (5) attending regular meetings where stat lab projects are discussed. Between 2019 and 2021, 36 stat labs have reported training 690 faculty and staff members (32.9% female) in the practice of effective statistics and data science collaboration. Stat labs have reported training 439 graduate (or postgraduate) students (40.8% female) and 912 undergraduate students (41.7% female). From 2014 to December 2021, the total number of LISA 2020 trainees was 2,182.

These 2,182 collaborative statisticians and data scientists have worked on more than 2,274 projects (1,274 projects reported since 2019). Stat lab projects are varied in nature; generally, these projects occur whenever a domain expert collaborates with stat lab staff for statistical advice analysis, or other statistical or data support. These projects may be funded or not; they may require only one meeting, or they may result in a long-term commitment on the part of the stat lab. These projects have directly resulted in a total of at least 134 peer-reviewed publications that include stat lab staff members as authors, as well as 13 non–peer-reviewed publications.

Education and training of non-statisticians take place in several forms. Short courses are brief (less than 1 day); 1-day workshops and multi-day workshops are longer and more involved. All of these have the goal of teaching general statistical and data science principles and techniques to nonstatisticians. Topics could include software applications, specific statistics and data science methods, specialized methods required by domain experts working in specific areas, or anything else in the realm of statistics and data science that benefits the lab's nonstatistician stakeholders. Between 2019 and 2021, 36 stat labs have reported hosting 110 short courses for 2,535 total attendees (31.7% female), 108 1-day workshops for 4,539 total attendees (39.2% female), and 132 multiday workshops for 6,129 total attendees (45.4% female).

2.4 The Purpose of LISA 2020

The original purpose of the LISA 2020 program was to build statistics and data science infrastructure in developing countries by helping to create stat labs to train students and staff to become effective interdisciplinary collaborators; collaborate with researchers, businesses, and policymakers to enable and accelerate research and data-based decisions; and teach short courses and workshops to improve statistical skills and data literacy widely (Vance 2015a). As LISA 2020 has grown and succeeded in achieving its original goal of creating a network of at least 20 stat labs by the year 2020, its purpose has evolved to focus more on the impacts it can facilitate based on the statistics and data science capacity for developing its constituent stat labs have already created.

According to Jacob and Meek, "Leading scientists can play an important bridging role in leveraging global science for local development and serve as conduits between the local and global. But they need support and training, particularly with respect to research management and leadership, if they are to effectively play such bridging roles" (2013, p. 343). The purpose of LISA 2020 is to provide such support and training. Specifically, three intertwined purposes or objectives have emerged for the LISA 2020 program as shown in Figure 2.5:

1. To facilitate and expedite individual stat labs' growth along the path from potential labs to new labs to strong and sustainable labs that serve as mentors to other labs

2. To enable and accelerate stat labs' transforming of evidence into action (TEA) for development

3. To form a community of stat labs to publicize their successes and advocate for their extraordinary potential.

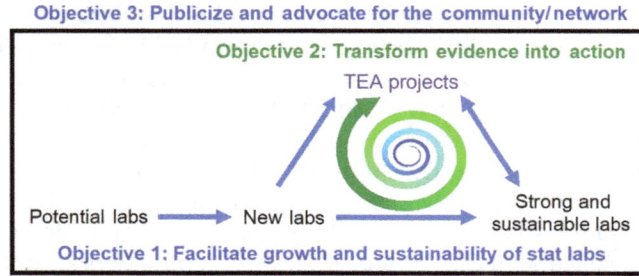

FIGURE 2.5
Three objectives of the LISA 2020 program.

2.4.1 Why LISA 2020 Pursues These Objectives

Stat labs are the foundation of the LISA 2020 program. We believe that the core idea of the stat lab, as described in Vance and Pruitt (2022), is so powerful in advancing the missions of universities for the benefit of society that all universities should have one. But creating a new stat lab is not enough. New stat labs must build their own capacity to become strong and sustainable, and these strong, well-established labs must mentor other labs on their own progression toward sustainability. As a network of stat labs in various stages across this spectrum, LISA 2020 is well positioned to facilitate and expedite individual stat labs' development.

Our vision is for collaborative statisticians and data scientists to collaborate with data producers and data decision-makers to transform evidence (data) into action for development so that more local communities, countries, and regions can benefit from the extraordinary power of statistics and data science to positively impact society. Statistics and data science can be a powerful tool for social impact (Augustin et al. 2021), but statisticians and data scientists must intentionally focus on the outcome of a project, i.e., the *action* for development, rather than on just the production of statistical findings, conclusions, or recommendations. This requires a shift in mindset (Vance and Love 2021) that can be facilitated by the LISA 2020 Network. By intentionally focusing on the *action* for development resulting from the project, stat labs can tailor their statistics and data science work. In many cases, a simpler (though still correct) statistical analysis can be more effective in producing action than a complicated one because it is more understandable and more easily implemented (Stallings 2014).

Since the concept of the stat lab that uses projects to train its students and build its own capacity is still relatively new—having been introduced in the literature by Vance (2015b) not many years ago—a community is needed to support this new concept and to develop and share best practices. Being part of an organization larger than itself can help stat labs become strong and sustainable. As Vance and Pruitt (2022) write, connections with lab mentors and the LISA 2020 Network provide encouragement, support, and guidance to stat labs as they encounter both challenges and triumphs. Stat labs can share their experiences, lessons learned, and best practices to strengthen other labs and the whole network.

Ultimately, LISA 2020 is pursuing these purposes because of the need to build statistics and data science capacity in developing countries (Love et al. 2022). Statistics is widely taught at the university level, and we believe that the application of statistical knowledge has become imperative in virtually all areas of human endeavors because decisions in many areas of modern societies should be based on the collection and analysis of empirical data. Yet, without the proper application of statistical methods, the risk of improper or inefficient data collection increases, as does the risk that the analysis of research data provides suboptimal results, eventually leading to wrong decisions (Awe and Vance 2014; Jerven 2013). It is also necessary for statisticians to possess effective collaboration skills so as to be able to communicate technical knowledge effectively (Vance and Smith 2019). Hence, the importance of stat labs as an aid in teaching and learning statistics in the 21st century cannot be over-emphasized (Vance 2015b).

2.4.2 How LISA 2020 Pursues These Objectives

We believe that many important technical and nontechnical skills are better learned early in one's career as a statistician at the undergraduate and graduate levels in the university (Awe et al. 2015). Collaborative statisticians at LISA 2020 labs in developing countries are being trained to follow Doug Zahn's POWER structure for organizing and facilitating collaboration meetings (Zahn 2022) and create shared understanding (Vance et al. 2022). Two features of this structure are the wanted conversation and the time conversation. The wanted conversation establishes what the participants want to get out of the meeting and the time conversation establishes how long the meeting will last. We have found that when working with American and Brazilian researchers, these techniques add helpful structure to the meetings (Vivacqua et al. 2018).

The LISA 2020 Secretariat, which was funded by USAID from 2018 to 2022, has put in place policies and procedures to help potential labs advance along the path to becoming newly established and operational (see www.lisa2020.org/about/processes-and-policies). These policies and procedures provide a structure that has been successfully effectuated by many stat labs (Vance and Pruitt 2022). They provide a blueprint for stat labs to build a foundation and facilitate and expedite stat labs' growth toward becoming strong and sustainable labs. The primary way, however, that LISA 2020 accomplishes its three purposes is by developing and sharing best practices throughout the network via various communication channels, special network events and committees, and exchanges and site visits.

Communication: LISA 2020 publishes a weekly newsletter to highlight stat lab activities, promote their successes and triumphs, and share best practices to overcome common challenges (see www.lisa2020.org/newsletter). The network also has a WhatsApp group for members to do the same in a less formal manner. Twice per month, the LISA 2020 Network convenes via Zoom to discuss important topics, see presentations of the experiences of other labs, network at a personal level via informal conversation in small breakout rooms, and share best practices. A newly established lab may bring a question to the Zoom meeting that is answered by more experienced labs that had the same challenge and were able to overcome it. Sharing these lessons learned helps to strengthen all labs and form a robust network.

Special network events and committees: LISA 2020 has convened three symposia, the most recent being an online symposium comprised of 12 sessions on Zoom from October 2020 to May 2021 on the overall theme, "Connecting to Build Capacity to Transform Evidence into Action & Celebrate the International Year of Women in Statistics and Data Science." In conjunction with the online plenary symposium events, four stat labs organized local symposium events and multiple stat labs organized five regional events with financial support from the ISI and their World Bank Statistics Capacity Trust Fund. These regional events occurred in Ghana, Northern Nigeria, Tanzania, Southeast Nigeria, and Southwest Nigeria. Both the local and regional events provided opportunities to promote the successes of the stat labs, share best practices, and spur collaborations through in-person networking.

Another notable special activity of LISA 2020 is the development of a certification program by a network committee to both reward and incentivize students, interns, and faculty/staff to work in a stat lab. The program will have local and global levels of certification so that a student, for example, working in one of the labs will first earn the local certificate, and as she gains more experience, can then be nominated to earn the global LISA 2020 Certificate. The certification program is currently in development and will be comprised of trainings in and assessments of technical skills, collaboration skills, and experience collaborating with domain experts to transform evidence into action. The certification program was identified by the network as a major need to help standardize operations across stat labs, improve quality, and overcome the challenge of having too few volunteer statistics and data science collaborators working in the labs.

Exchanges and site visits: We have found that the best way to improve the quality of stat labs, share best practices, and create a robust network is to encourage exchanges between lab staff and site visits to observe other labs in practice. UI-LISA has operated an internship program providing 3 or more months of industrial experience for students to work in the lab since 2015 (Adepoju et al. 2022). Initially, just for its own students, this program has expanded to accept students from other universities who return to their university fully equipped to contribute to their own newly established lab. The COVID-19 pandemic has curtailed much travel and exchange opportunities between stat labs. We hope to resume exchanges and site visits in the coming years and to secure the funding necessary to facilitate them, especially "South-South" exchanges between the stat labs.

2.5 LISA 2020 as a Big Tent for Collaborative Statistics and Data Science

An analogy we have found useful for explaining the purpose of the LISA 2020 program is the "Big Tent." As the ASA's 107th president, Dr. Bob Rodriguez advocated for the ASA to become the "Big Tent" for statistics (Rodriguez 2013). Rodriguez explained, "Big tents do three things. They attract all kinds of people, they serve them on the inside, and they are highly visible on the outside" (2013, p. 2). LISA 2020 can become the Big Tent for collaborative statistics and data science by creating a community of stat labs to share best practices, welcoming statisticians and data scientists to create new stat labs; serving the stat labs within the network through trainings, mentorship, certificates, symposia, exchanges, etc. so that they can become strong and sustainable; enabling and accelerating their transformation of evidence into action (TEA); making the successes of stat labs highly visible to outside observers; and advocating for their extraordinary potential (see Figure 2.6).

By becoming the Big Tent for collaborative statistics and data science, the LISA 2020 program can achieve its purpose in attracting new stat labs, serving to build their capacity to become strong and sustainable labs that transform evidence into action for development, and publicizing their work to make visible the extraordinary impacts of statistics and data science collaboration.

Already, the LISA 2020 program has made the following impacts (Vance and Pruitt 2022):

- Improving statistical literacy among students and researchers in developing countries

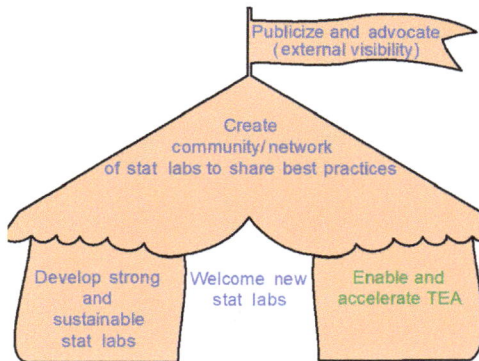

FIGURE 2.6
LISA 2020 can become the Big Tent for statistics and data science collaboration.

- Accelerating research development by offering expert statistical advice to researchers in the institutions
- Aiding graduate students to complete their research appropriately and graduate on time
- Helping to increase the number of Master's and Ph.D. candidates being produced
- Aiding researchers to publish quality papers in high impact journals
- Analyzing data for various research grants, funding proposals, and projects
- Educating graduate students and researchers through regular short courses on modern statistical packages and software
- Causing increased enrollment of students into the various statistics programs in the LISA 2020 Network
- Encouraging collaboration between statisticians and researchers from other disciplines (i.e., domain experts)
- Helping researchers to keeping abreast with current research trends and best practices.

2.6 Conclusion

In this chapter, we explained the history, key events, current state, and future directions of the LISA 2020 program to become the Big Tent for statistics and data science collaboration. With roots as far back as 1948, LISA 2020 began in 2012 and achieved its initial goal of creating a network of at least 20 stat labs in developing countries by the year 2020. Currently comprised of 35 full member stat labs in 10 countries and 13 additional proposed or transitional stat labs, the LISA 2020 Network is well positioned to deliver on its current purposes to help create new stat labs, build their capacity to become strong and sustainable, enable and accelerate their transforming of evidence into action for development, and publicize their inevitable successes in applying statistics and data science to positively impact humanity.

References

Adepoju, A. A., Olubusoye, O. E., Ogundunmade, T., Tomiwa, O. T., and Robert, B. (2022), "The Complementary Role of UI-LISA in Statistical Training and Capacity Building at the University of Ibadan, Nigeria." *Promoting Statistical Practice and Collaboration in Developing Countries*, eds. O.O. Awe, K. Love, and E.A. Vance, Boca Raton, FL: CRC Press, pp. 111–122. DOI: 10.1201/9781003261148

Amin, M., and Vance, E. A. (2016), "Inevitability of Interdisciplinary Statistical Laboratories for Food and Agricultural Research in Developing Countries: The case of Pakistan," in *Proceedings of ICAS VII Seventh International Conference on Agricultural Statistics*, Rome, Italy, pp. 1–8. https://doi.org/10.1481/icasVII.2016.h47b.

Arnold, J. C. (2000), "Virginia Tech Department of Statistics: The First Fifty Years," *Journal of Statistical Computation and Simulation*, 66, 1–17. https://doi.org/10.1080/00949650008812008.

Arnold, J. C., Hinkelmann, K., Vining, G. G., and Smith, E. P. (2013), "Virginia Tech Department of Statistics," in *Strength in Numbers: The Rising of Academic Statistics Departments in the U. S.*, eds. A. Agresti and X.-L. Meng, New York, NY: Springer, pp. 537–546. https://doi.org/10.1007/978-1-4614-3649-2_39.

Augustin, C., Brems, M., and Durgana, D. P. (2021), "Special Issue on Statistics and Data Science for Good," *CHANCE*, 34, 4–5.

Awe, O. O. 2012. "Fostering the Practice and Teaching of Statistical Consulting among Young Statisticians in Africa," *Journal of Education and Practice*, 3(3), 54–59.

Awe, O. O., Crandell, I., and Vance, E. A. (2015), "Building Statistics Capacity in Nigeria Through the LISA 2020 Program," in *Proceedings of the International Statistical Institute's 60th World Statistics Congress*, Rio de Janeiro, Brazil.

Awe, O. O., and Vance, E. A. (2014), "Statistics Education, Collaborative Research, and LISA 2020: A View from Nigeria," *Proceedings of the Ninth International Conference on Teaching Statistics (ICOTS9, July, 2014)*, Flagstaff, AZ.

Goshu, A. T. (2016), "Strengthening Statistics Graduate Programs with Statistical Collaboration—The Case of Hawassa University, Ethiopia," *International Journal of Higher Education*, 5, 217–221.

Hall, R. P., Davis, J., van Houweling, E., Vance, E. A., Seiss, M., and Russel, K. (2014a), *Impact Evaluation of the Mozambique Rural Water Supply Activity*, Blacksburg, VA: Virginia Tech School of Public and International Affairs.

Hall, R. P., Vance, E. A., and van Houweling, E. (2014b), "The Productive Use of Rural Piped Water in Senegal," *Water Alternatives*, 7, 480–498.

Hall, R. P., Vance, E. A., and van Houweling, E. (2015a), "Upgrading Domestic-Plus Systems in Rural Senegal: An Incremental Income-Cost (IC) Analysis," *Water Alternatives*, 8, 317–336.

Hall, R. P., Vance, E. A., van Houweling, E., and Huang, W. (2015b), "Willingness to Pay for VIP Latrines in Rural Senegal," *Journal of Water, Sanitation and Hygiene for Development*, 5, 586–593. http://doi.org/10.2166/washdev.2015.053.

Jacob, M., and Meek, V. L. (2013), "Scientific Mobility and International Research Networks: Trends and Policy Tools for Promoting Research Excellence and Capacity Building," *Studies in Higher Education*, 38, 331–344. https://doi.org/10.1080/03075079.2013.773789.

Jerven, M. (2013), *Poor Numbers: How We Are Misled by African Development Statistics and What to Do about It*, London: Cornell University Press.

Love, K., Gunderman, D. J., Awe, O. O., Druckenmiller, M., and Vance, E. A. (2022), "LISA 2020 Network Survey on Challenges and Opportunities for Statistical Practice and Collaboration in Developing Countries." *Promoting Statistical Practice and Collaboration in Developing Countries*, eds. O.O. Awe, K. Love, and E.A. Vance, Boca Raton, FL: CRC Press, pp. 47–60. DOI: 10.1201/9781003261148

Mawora, T., Otieno, J., and Vance, E. A. (2022), "Exploring the Need for a Statistical Collaboration Laboratory in a Kenyan University: Experiences, Challenges, and Opportunities." *Promoting Statistical Practice and Collaboration in Developing Countries*, eds. O.O. Awe, K. Love, and E.A. Vance, Boca Raton, FL: CRC Press, pp. 61–70. DOI: 10.1201/9781003261148

Msemo, E., and Vance, E. A. (2015), "LISA 2020: Impacting Agricultural Productivity in Tanzania through the Wheels of Statistics," in *Proceedings of the International Statistical Institute's 60th World Statistics Congress*, Rio de Janeiro, Brazil.

OECD (2011), "International mobility," in *OECD Science, Technology and Industry Scoreboard 2011*, Paris: OECD Publishing, pp. 98–99. http://dx.doi.org/10.1787/sti_scoreboard-2011-en

Olubusoye, O. E., Akintande, O. J., and Vance, E. A. (2021), "Transforming Evidence to Action: The Case of Election Participation in Nigeria," *CHANCE*, 34(3), 13–23.

Rodriguez, R. N. (2013), "Building the Big Tent for Statistics," *Journal of the American Statistical Association*, 108, 1–6. https://doi.org/10.1080/01621459.2013.771010.

Seiss, M., Vance, E., and Hall, R. (2014), "The Importance of Cleaning Data During Fieldwork: Evidence from Mozambique," *Survey Practice*, 7(4), 2864.

Stallings, J. (2014), "Type IV Errors: How Collaboration Can Lead to Simpler Analyses," *Amstat News*, 440, 24–25.

Van Houweling, E., Hall, R., Carzolio, M., and Vance, E. (2017), "'My Neighbour Drinks Clean Water, While I Continue To Suffer': An Analysis of the Intra-Community Impacts of a Rural Water Supply Project in Mozambique," *The Journal of Development Studies*, 53, 1147–1162. https://doi.org/10.1080/00220388.2016.1224852.

Vance, E. (2009), *LISA: Laboratory for Interdisciplinary Statistical Analysis Annual Report 2008–2009*, Report. http://hdl.handle.net/10919/50504

Vance, E. (2012a), "International experiences in statistics," *Amstat News*, 420, 17–19.

Vance, E. (2012b), *LISA Annual Report 2010–2011*, Report. https://vtechworks.lib.vt.edu/handle/10919/19012

Vance, E. A. (2015a), "The LISA 2020 Program to Build Statistics Capacity and Research Infrastructure in Developing Countries," in *Proceedings of the International Statistical Institute's 60th World Statistics Congress*, Rio de Janeiro, Brazil.

Vance, E. A. (2015b), "Recent Developments and Their Implications for the Future of Academic Statistical Consulting Centers," *The American Statistician*, 69(2), 127–137. https://doi.org/10.1080/00031305.2015.1033990.

Vance, E. A. (2017), "LISA: Laboratory for Interdisciplinary Statistical Analysis 2016–17 Annual Report," *OSF*. https://doi.org/10.17605/OSF.IO/ZQEJ2.

Vance, E. A., Alzen, J. L., and Smith, H. S. (2022), "Creating Shared Understanding in Statistics and Data Science Collaborations," *Journal of Statistics and Data Science Education*. https://www.tandfonline.com/doi/full/10.1080/26939169.2022.2035286

Vance, E. A., LaLonde, D. E., and Zhang, L. (2017a), "The Big Tent for Statistics: Mentoring Required," *The American Statistician*, 71(1), 15–22. https://doi.org/10.1080/00031305.2016.1247016.

Vance, E. A., and Love, K. (2021), "Building Statistics and Data Science Capacity for Development," *CHANCE*, 34(3), 38–46.

Vance, E. A., and Magayane, F. (2014), "Strengthening the Capacity of Agricultural Researchers through a Network of Statistical Collaboration Laboratories," *Proceedings of the 4th RUFORUM Biennial Conference*, 21st–25th July 2014, Maputo, Mozambique.

Vance, E. A., and Pruitt, T. R. (2022), "Statistics and Data Science Collaboration Laboratories: Engines for Development." *Promoting Statistical Practice and Collaboration in Developing Countries*, eds. O.O. Awe, K. Love, and E.A. Vance, Boca Raton, FL: CRC Press, pp. 3–26. DOI: 10.1201/9781003261148

Vance, E. A., and Smith, H. S. (2019), "The ASCCR Frame for Learning Essential Collaboration Skills," *Journal of Statistics Education*, 27(3), 265–274. https://doi.org/10.1080/10691898.2019.1687370.

Vance, E. A., Tanenbaum, E., Kaur, A., Otto, M. C., and Morris, R. (2017b), "An Eight-Step Guide to Creating and Sustaining a Mentoring Program," *The American Statistician*, 71(1), 23– 29. https://doi.org/10.1080/00031305.2016.1251493.

Vivacqua, C. A., de Pinho, A. L. S., Nunes, M. A., and Vance, E. A. (2018), "Integrating Collaboration, Communication and Problem Solving to Promote Innovation in Statistics Education," in *Looking Back, Looking Forward. Proceedings of the Tenth International Conference on Teaching Statistics (ICOTS10, July, 2018)*, Kyoto, Japan: IASE, pp. 1–5.

Zahn, D. (2022), "Systematically Improving Your Collaboration Practice in the 21st Century." *Promoting Statistical Practice and Collaboration in Developing Countries*, eds. O.O. Awe, K. Love, and E.A. Vance, Boca Raton, FL: CRC Press, pp. 583–599. DOI: 10.1201/9781003261148

3

LISA 2020 Network Survey on Challenges and Opportunities for Statistical Practice and Collaboration in Developing Countries

Kim Love
K. R. Love Quantitative Consulting and Collaboration

O. Olawale Awe
University of Campinas

David J. Gunderman
Purdue University College of Engineering

Matthew Druckenmiller and Eric A. Vance
University of Colorado Boulder

CONTENTS

3.1 Introduction

The need for evidence-based statistical reasoning and capacity in developing countries is growing (Badiee et al. 2004). Statistics and data science collaboration laboratories ("stat labs") have clear advantages for building this statistical capacity as they concurrently benefit local education, public policy-making, economic development, and academic research (Vance and Pruitt 2022). The primary avenue through which stat labs achieve these simultaneous beneficial effects is by providing a mechanism for increasing collaboration between statisticians and researchers, business professionals, and development policy actors (Vance and Love 2021). Of course, these individuals and organizations do not

DOI: 10.1201/9781003261148-4

operate in a vacuum; the environment or setting in which these elements come together plays a crucial role in creating successful outcomes (North et al. 2014).

The creation of academic statistical consulting centers in developed countries has been studied for decades (Gibbons and Freund 1980). These studies posit recommendations regarding best practices for founding stat labs (Does and Zempleni 2001), developing new academic, industry, and government collaborations (Khamis and Mann 1994), advancing educational initiatives (Fletcher 2014), and advancing equity and inclusion for women in statistics (Steffey 2015), among others. While various published surveys have evaluated the challenges of stat labs in developed countries (Windmann and Kauermann 2007, Gullion and Berman 2006, Awe 2012, Vance 2015), it is vital to gain an understanding of the currently existing environment in developing countries, specifically with respect to the background and experiences of the statistical collaborators, along with the current state of statistical practice (and statistical collaboration in particular). In order to develop and orient the network to provide support so that individual stat labs can become independent engines for local development, it is also crucial that this understanding is gained from the perspective of the statistical collaborators themselves, and not from outside individuals in highly developed countries.

Therefore, in August of 2020, the LISA 2020 Network secretariat, which is centered at the Laboratory for Interdisciplinary Statistical Analysis (LISA) at the University of Colorado Boulder (Vance et al. 2022), provided a voluntary survey to all members of the network and its supporters. The purpose of the survey was to learn about the perspective of the lab representatives on the challenges facing the practice of statistics and statistical collaboration in their environments. The survey received 40 responses; 36 of those responses were from individuals representing stat labs, and those 36 responses are included in this article. A copy of the survey can be found online at www.lisa2020.org.

This survey was not given to a random sample of respondents and will not be used for statistical inference purposes. The responses as summarized in this article are only intended to represent the perspectives of the respondents from the LISA 2020 Network. In order to better understand the geographical settings reflected in the responses, Figure 3.1 is a bar chart demonstrating answers to the item "What country is the primary location for your work?" The majority of respondents (55.6%) perform their work in Nigeria, the

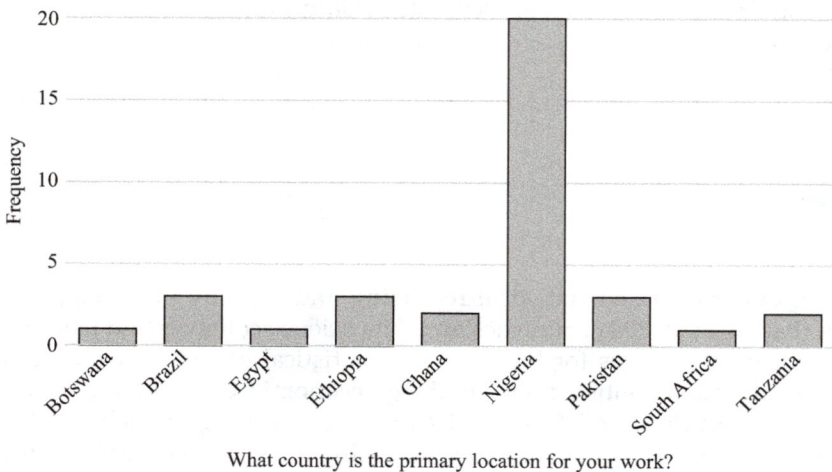

What country is the primary location for your work?

FIGURE 3.1
Primary location of work for survey respondents.

most populous country in Africa; all other countries included in the responses were represented by just 1–3 respondents each. This reflects the makeup of the LISA 2020 Network, which as of September 2020 had 36 full and transitional labs, 17 of which were operating in Nigeria. The responses to the survey will, therefore, be particularly reflective of the perspectives of the network members in Nigeria.

The remainder of this article will focus on the survey responses in five different areas:

- Education, training, and experience of respondents
- Ways to improve statistical practice
- Ways to improve statistical collaboration
- Areas for improvement/development in order to advance statistical practice and collaboration
- Statistical practice within universities.

3.2 Education, Training, and Experience

The survey opened with four items related to education, training, and experience, which were:

- Have you been trained in statistics or a related field?
- Please estimate your level of experience in statistical practice?
- Please estimate your level of experience in statistical collaboration?
- What is the PRIMARY source of your experience in statistical collaboration?

Table 3.1 summarizes the respondents' answers to these items. With regard to education and training, the majority of the respondents are highly educated in the field of statistics or a related field; 86.1% report that they have a graduate degree (Master's or PhD). Only one individual indicated that they have not been trained in statistics or a related field.

Two questions were asked regarding levels of experience. The first question asked about experience with general statistical practice. The majority of respondents (63.9%) report extensive experience; no respondents reported that they have no experience with statistical practice. The second question asked about experience specifically with statistical collaboration. Although many respondents (36.1%) still reported extensive experience with statistical collaboration, levels of experience specific to statistical collaboration are reported to be lower than general experience in statistical practice.

Finally, respondents were also asked about the primary source of their experience in statistical collaboration. The most common answer to this item was that the primary source of experience was "collaborative projects with researchers within your institution/university," with 55.6% of respondents. "Collaborative projects with researchers at other institutions/universities" and "collaboration on student theses" were also common answers. Only one respondent chose the option "direct collaboration with the governmental institutions." There were two other responses available, which were not selected by any of the respondents; those were "direct collaboration with the private (for profit) sector" and "direct collaboration with nongovernmental/nonprofit institutions."

TABLE 3.1

Responses to Four Items on Education, Training, and Experience

Item	Frequency (n=36)	Percent
Have You Been Trained in Statistics or a Related Field?		
No	1	2.8%
Yes—I have educational or professional training in statistics or related field, but DO NOT have an academic degree in statistics or related field	1	2.8%
Yes—I have a Bachelor's degree in statistics or a related field	3	8.3%
Yes—I have a Master's degree or PhD in statistics or a related field	31	86.1%
Please Estimate Your Level of Experience in Statistical Practice?		
Minor experience (<2 years of experience, few or no publications, little teaching and/ or few research projects)	5	13.9%
Moderate experience (2–5 years of experience, several publications, moderate teaching, and/or several research projects)	8	22.2%
Extensive experience (>5 years of experience, many publications, much teaching, and/or many research projects)	23	63.9%
Please Estimate Your Level of Experience in Statistical Collaboration?		
No experience	1	2.8%
Minor experience (<2 years of experience, few collaborations, few or no publications)	8	22.2%
Moderate experience (2–5 years of experience, several collaborations, several publications)	14	38.9%
Extensive experience (>5 years of experience, many collaborations, many publications)	13	36.1%
What Is the PRIMARY Source of Your Experience in Statistical Collaboration?		
Collaboration on student theses	7	19.4%
Collaborative projects with researchers at OTHER institutions/universities	8	22.2%
Collaborative projects with researchers within YOUR institution/university	20	55.6%
Direct collaboration with the governmental institutions	1	2.8%
Direct collaboration with the private (for profit) sector	0	0.0%
Direct collaboration with nongovernmental/nonprofit institutions	0	0.0%

3.3 Ways to Improve Statistical Practice

The survey respondents were given eight items and asked to indicate the level to which they agree or disagree that it will improve statistical practice in their country or region. Those items were:

- Increase capacity within university faculty to practice statistics
- Improve curriculum within university statistics programs
- Strengthen access to computing resources
- Create more job opportunities for statisticians
- Increase pay for statisticians
- Strengthen legal framework for practicing statistics
- Increase overall statistical literacy within society.

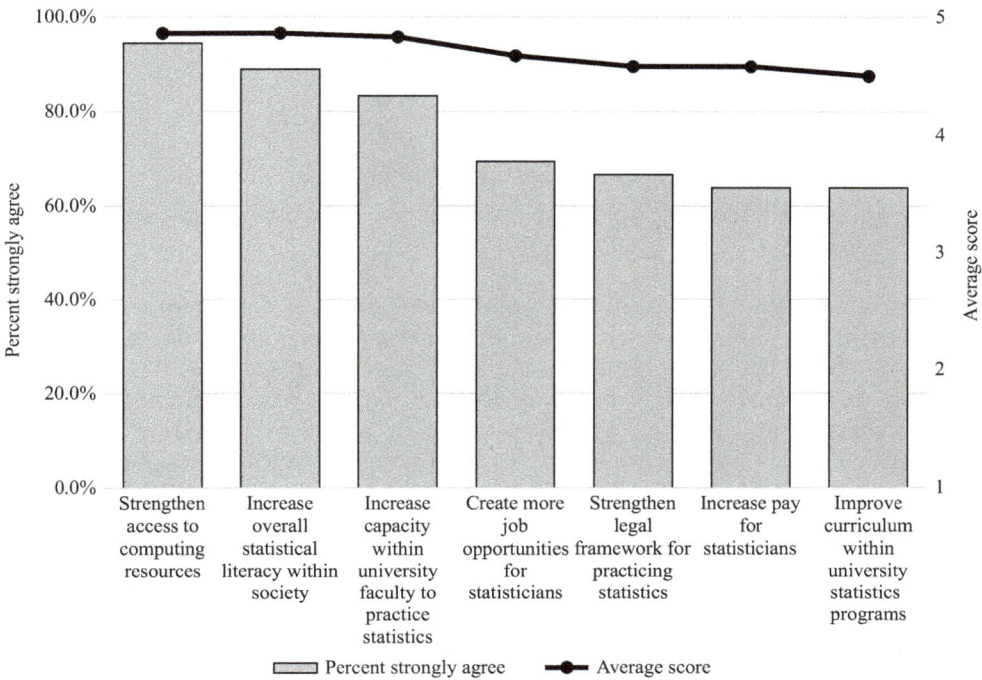

FIGURE 3.2
Percent of respondents answering "strongly agree" (bars) with average score (line) for eight ways to improve statistical practice.

These were Likert-type items, with five possible responses to each item ranging from 1 ("strongly disagree") to 5 ("strongly agree"). There were very few answers on the "disagree" side of the possible responses (just two for "improve curriculum within university statistics programs" and one for "strengthen access to computing resources"); neutral responses were also quite rare. Figure 3.2 is a bar chart of the percent of respondents answering "strongly agree" to each item; scores computed by averaging the numeric Likert-type responses to each item are overlayed.

One thing to note immediately from Figure 3.2 is that for every item, over 60% of respondents (a majority) strongly agreed that it would strengthen statistical practice in their country or setting. All average scores were between 4 (agree) and 5 (strongly agree). The ways of improving statistical practice that were most often rated "strongly agree" were strengthening access to computing resources and increasing overall statistical literacy within society; the ways of improving statistical practice that were least often rated "strongly agree" were increase pay for statisticians and improve curriculum within university statistics programs.

3.4 Ways to Improve Statistical Collaboration

The survey respondents were given six items and asked to indicate the level to which they agree or disagree that it will improve statistical collaboration in their country or setting. Those items were:

- Expand training in statistical collaboration
- Incentivize statistical collaboration within and beyond universities
- Improve access to statisticians by domain experts
- Increase research funding for statistical collaboration
- Increase overall statistical literacy within society
- Increase overall recognition for the role of statistics in decision-making and policy-making.

These were Likert-type items, with five possible responses to each item ranging from 1 ("strongly disagree") to 5 ("strongly agree"). Responses to all items from all participants were either "agree" or "strongly disagree"; there were no answers on the disagree side of the possible responses or neutral responses. Figure 3.3 is a bar chart of the percent of respondents answering "strongly agree" to each item. Scores computed by averaging the numeric Likert-type responses to each item are overlayed. (Note that one of the 36 respondents did not respond to the last item in this section of "increase overall recognition for the role of statistics in decision-making and policy-making" and so the sample size for this item is 35.)

One thing to note immediately from Figure 3.3 is that for every item over 80% (most respondents) strongly agreed that it was a way to improve statistical practice in their country or setting; all average scores were 4.8 or higher (very close to strongly agree). The ways of improving statistical practice that were most often rated "strongly agree" were increasing research funding for statistical collaboration, and expanding training in statistical collaboration; the ways of improving statistical practice that were least often rated "strongly agree" were increase overall recognition for the role of statistics in decision-making and policy-making, and incentivize statistical collaboration within and beyond universities.

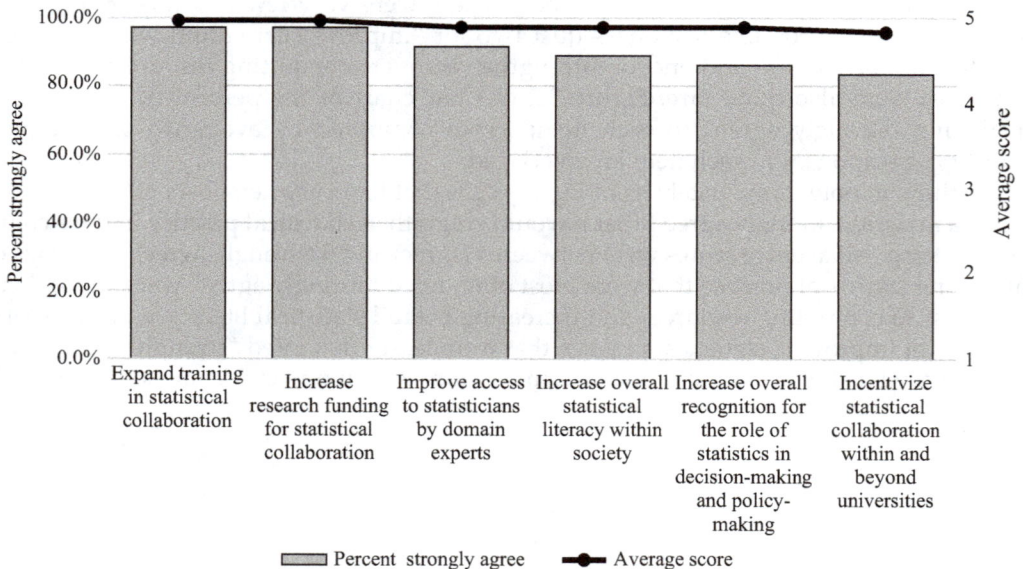

FIGURE 3.3
Percent of respondents answering "strongly agree" (bars) with average score (line) for six ways to improve statistical collaboration.

3.4.1 Areas for Improvement/Development in Order to Advance Statistical Practice and Collaboration

Nine general areas requiring improvement or development in order to advance statistical practice and collaboration were presented to the respondents. Those areas were:

- Math (Education/Training)
- Statistical Theory (Education/Training)
- Statistical Methods (Education/Training)
- Statistical Computing (Education/Training)
- Statistics Applications (Education/Training)
- Communication Skills (Education/Training)
- Computing Resources (Infrastructure)
- Opportunities and Training in Stakeholder Engagement
- Opportunities to Practice Collaborative Statistics.

Respondents were asked to rate whether each had "no/minor need" for improvement (1), "moderate need" for improvement (2), or "significant need" for improvement (3). The percentage responding with "significant need" to each item is presented in Figure 3.4; scores computed by averaging the numeric Likert-type responses to each item are overlayed. (Note that one of the 36 respondents did not complete this section and so the sample size for all items is 35.)

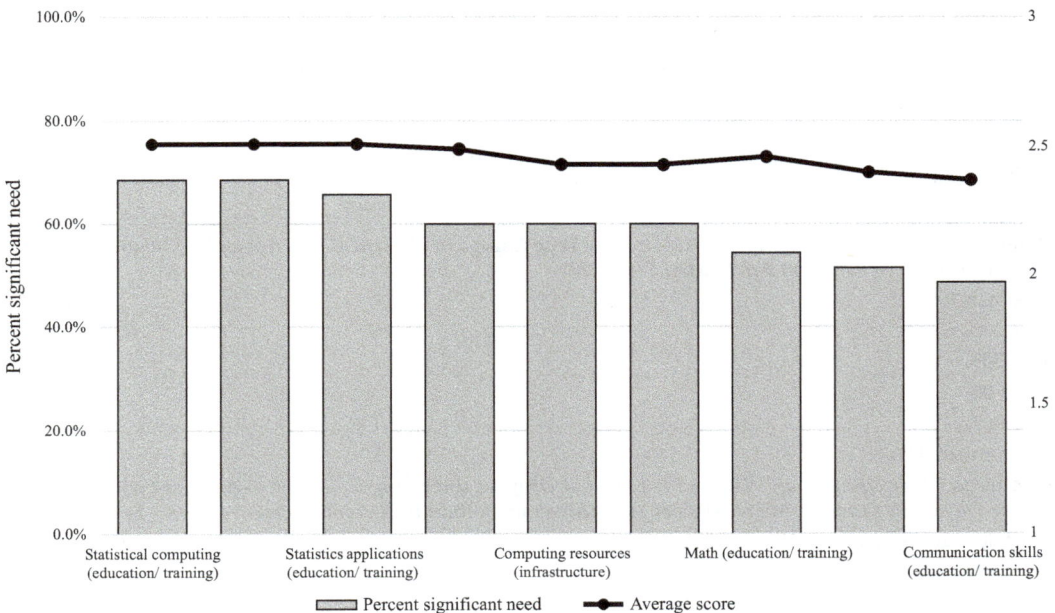

FIGURE 3.4
Percent of respondents answering "significant need" (bars) with average score (line) for nine areas requiring improvement or development in order to advance statistical practice and collaboration.

In Figure 3.4, the percent of respondents who answered "significant need" for each area ranged from 48.6% to 68.6%. The two areas that received the greatest percentage of "significant need" responses were statistical computing (education/training) and opportunities to practice collaborative statistics; the two areas that received the least percentage of responses were communication skills and statistical theory.

3.5 Statistical Practice and Collaboration within Universities

The survey ended with several questions related to statistical practice and collaboration in the respondents' university environment. The first question asked if the respondents worked in a university department or other university unit that engages in statistical practice and/or statistical collaboration. There were three respondents who answered "no" to this item and were not included for the remaining three items.

The remaining three items were:

- What percentage of the statisticians/researchers who engage in statistical practice and collaboration at your university department are MALE?
- What percentage of the statisticians/researchers who engage in statistical practice and collaboration at your university department are FEMALE?
- Relative to other disciplines, what is the extent of dropout (including switching to another field of study) among Bachelor's students in statistics at your university within the last academic year?

The responses to these items are given in Table 3.2. Note that only the question regarding the percentage of the statisticians/researchers who engage in statistical practice and

TABLE 3.2

Responses to Two Items on Statistical Practice and Collaboration within Universities

Item	Frequency	Percent (%)
What Percentage of the Statisticians/Researchers Who Engage in Statistical Practice and Collaboration at Your University Department Are FEMALE? (n = 29)		
Less than 10%	6	20.7%
10%–30%	11	37.9%
30%–50%	7	24.1%
50%–70%	3	10.3%
70%–90%	1	3.4%
Greater than 90%	1	3.4%
Relative to Other Disciplines, What Is the Extent of Dropout (Including Switching to Another Field of Study) among Bachelor's Students in Statistics at Your University within the Last Academic Year? (n = 16[a])		
Much less than other disciplines	2	12.5%
Somewhat less than other disciplines	2	12.5%
Similar to other disciplines	3	18.8%
Somewhat greater than other disciplines	4	25.0%
Much greater than other disciplines	5	31.3%

[a] 12 Responses of "unable to estimate", 5 responses of "not applicable".

collaboration who are female is included in this table; only 29 respondents' answers to this item are included as the remaining responses did not agree with the response to the question about male statisticians and researchers, indicating the question was misunderstood.

From Table 3.2, the most common answer to the percentage of statisticians/researchers engaging in statistical practice and collaboration who are female was "10–30%" (chosen by 37.9% of respondents). Combined with the answer of "less than 10%," we can say that the majority of respondents (58.6%) perceive that 30% or fewer of the statisticians/researchers who engage in statistical practice and collaboration in their university department are female.

With regard to the extent of dropout perceived among bachelor's students in statistics at their university, there were diverse answers among the 16 participants who were able to respond to this item. However, the majority of respondents (56.3%) perceived that dropout rates were either somewhat or much greater than other disciplines. Only 25% of respondents perceived that dropout rates were either somewhat or much less than other disciplines.

3.6 Discussion

There are a number of takeaways from this survey that help us to learn about the current state and future trajectory of stat labs in the LISA 2020 Network.

First, the majority of the LISA stat lab representatives who responded to this survey have extensive training and education; they also have extensive experience in statistical practice. This is a great foundation upon which to build, since it has been noted in the past that statistical expertise is lacking in developing countries (Kiregyera 2010, Awe and Vance 2014, World Bank 2021). Although a number of them also have extensive statistical collaboration experience, in general the proportion who reports this is lower than those who report high levels of education and experience in statistical practice. Some survey participants provided open-ended responses that shed light on this difference. It is clear from some responses that the idea of statistical collaboration at the stat lab level is relatively new, and both potential statistical collaborators and the domain experts they might collaborate with are not aware of best practices. Some difficulties include collaborations that start too late (i.e., after the design stage, when options to answer questions of interest are limited), a lack of funding reserved for statistical collaboration, and a lack of statisticians who are trained or motivated to collaborate. Several respondents recognize the importance of training students to be statistical collaborators while still in school and mention particular difficulties with this that currently exist. For example, two respondents answered: "Majority of the students of Statistics doesn't know how to solve data related problems in other fields not relating to theirs," and "Domain experts do not like engagement of students of Statistics in analyzing data, so it is hurdle in training of future statisticians."

The majority of the primary experience in statistical collaboration comes from respondents working with researchers within their own institutions or at other institutions; primary collaboration experience from government, nongovernmental organizations (NGOs), and private organizations was rare or nonexistent among the respondents. It is important to note that the question in the survey referred only to "primary collaborative experience," and therefore this does not mean that the respondents have no experience with these types of organizations.

However, this suggests that nonuniversity experience is not as common among the stat lab staff. As discussed in Vance and Love (2021) and Olubusoye et al. (2021), academic statistical practice alone does not result in transforming evidence to action for development and capacity building, which is one of the primary goals of the LISA 2020 Network (Vance et al. 2022). It is, therefore, necessary to provide additional training and opportunities for stat labs to collaborate in the intersections of data-driven development, so that they can work with both data producers and decision-makers. A number of respondents indicated that there is reluctance from domain experts (this includes researchers, government officials, and business interests) to collaborate with statisticians. For example, one respondent wrote: "Only few practitioners find it necessary to consult statisticians; there is a need to build trust so they can have confidence in statisticians"; another indicated there was a "lack of will to engage and too much bureaucracy."

Survey respondents strongly agreed with each of the suggested ways to improve statistical **practice** (over 60% of respondents for each item). However, the two most common were to strengthen access to computing resources and increase overall statistical literacy within society. These survey results reflect a lack of access to computing resources in many developing countries across disciplines (Thompson and Walsham 2010). From open-ended answers, this extends to many areas surrounding computing, and not simply hardware and computing facilities; this also extends to a lack of licensed software, a lack of stable internet connectivity (necessary for some software and also for some collaborative interactions), and the inability to emphasize practice and problem-solving in addition to theory. However, the issue of statistical literacy within society may not be limited to the developing world. One respondent indicated there is a "gap of communication between statisticians and nonstatisticians," while another indicated there is "poor societal awareness of statistical techniques as problem solving tools." These issues could be helped by introducing statistical consulting best practices early on in statistical education, perhaps even before methodology education (Taplin 2003).

Survey respondents also strongly agreed with each of the suggested ways to improve statistical **collaboration** (over 80% of respondents for each item). The two ways with the highest average scores were to increase research funding for statistical collaboration and to expand training in statistical collaboration. Research funding is of constant concern in both developing countries and nondeveloping countries; in the case of stat labs, this is how many of the individual faculty and staff members who work in these labs support themselves as well as the students who participate in lab activities (and provide all other resources required for running a stat lab) (Glickman et al. 2010). One respondent specifically indicated that "we have little financial incentive to provide scholarships so that students can dedicate themselves [to stat lab activities]"; another indicated that there are generally "low salaries for statistics collaborators." With regard to expanding training in statistical collaboration, many of the comments made in the open-ended responses relate specifically to training students of statistics and data science to be collaborators. One respondent indicated that "The curriculum in my institution is not designed in a way for statistics to be an interdisciplinary course"; another said that "professional practices occur only in the last semesters, when the number of students is quite small." Taplin (2003) suggests that lab practices and solving real-world problems should take place earlier in a student's educational process, an important pedagogical modification that education in developed countries could benefit from as well. The LISA 2020 Network is currently working on a certification process to improve the quality of statistical collaboration training in its member stat labs (Vance et al. 2022).

With regard to what is required in order for statistical collaboration and practice to be improved, the answer that had the highest average response and the greatest proportion of respondents who said there was a "significant need" for it was improved education/ training in statistical computing.

The next highest rated answer was that there need to be greater opportunities to practice statistical collaboration. Both of these answers can be connected to answers from previous areas on the survey. A need for greater education and training with regard to statistical computing follows from the lack of access to computing resources (Thompson and Walsham 2010); if resources are scarce, it is more difficult to train people to use those resources. A need for increased opportunities to practice statistical collaboration follows from the more limited opportunities the respondents report when it comes to practicing collaboration, particularly when compared to their high levels of education and experience with general statistical practice. Interestingly, the answer that had the lowest average response and percent describing it as a significant need was communication skills; this answer came behind even math and statistical theory. There are multiple ways to interpret this; however, one possible interpretation is that the respondents believe that the communication skills of statisticians are already strong enough. This is a common conception among beginning collaborators in the United States; beginners generally want to focus on technical skills, but experienced collaborators progress to a belief that their weaknesses are not technical skills, but in communication and related collaboration skills (Vance and Smith 2019). As the opportunities to practice statistical collaboration in developing countries increase, it may be the case that the perception of the importance of communication skills also increases.

Finally, with regard to female statisticians and researchers, the survey shows that among the universities of the respondents, there is the perception that gender equality has not been achieved, conforming with the situation in developed countries (Steffey 2015). This, therefore, needs to remain an area of emphasis. With regard to dropout rates, although there are a variety of perceptions here, respondents were more likely to say that the rates are higher than in other disciplines. Babalola (2022) explores the issue of dropout in great detail, emphasizing that there are a number of factors that lead to dropout rates in developing counties that include the potential for academic success, job-related factors, family-related factors, and community-related factors.

3.7 Conclusion

In this paper, we have presented the results of the first broad survey of stat labs in developing countries that the authors know of. The survey respondents were all members of the LISA 2020 Network, but as a first approximation, the results can be used to begin to understand the process of stat lab creation and development in developing nations. The results of this survey suggest some priorities for the LISA 2020 Network as we continue to move forward:

- Continue to promote the value of statistical practice and statistical collaboration among domain experts so that the practice of working together with (and funding) statistical collaborators becomes normalized, and so that training statistics students to be interdisciplinary collaborators becomes an expected part of their education
- Continue to promote the benefits of statistical collaboration within statistics and data science departments and related institutions within universities, so that

training students to become collaborators continues to gain value in the eyes of administrators, faculty, and staff

- Forge relationships between statisticians and domain experts, data producers, and data decision-makers; train statisticians to collaborate in a way that is beneficial and shows the value of the collaboration to the domain experts
- Find ways to connect with general society in developing countries to increase their interest in data-driven solutions to issues they are personally experiencing
- Continue to encourage the inclusion and empowerment of women as statistical collaborators and practitioners
- Encourage LISA stat labs to continue offering workshops to nonstatisticians to help them become more familiar with concepts surrounding data and problem-solving and become more generally statistically literate.

References

Awe, O. O. (2012). Fostering the practice and teaching of statistical consulting among young statisticians in Africa. *Journal of Education and Practice.* 3(3), 54–59. ISSN 2222-1735.

Awe, O. O., & Vance, E. A. (2014). Statistics education, collaborative research, and LISA 2020: A view from Nigeria. In *Sustainability in Statistics Education. Proceedings of the Ninth International Conference on Teaching Statistics (ICOTS9),* pp. 3–4, Flagstaff, AZ.

Babalola, T. (2022). Challenges of statistics education that lead to dropout among undergraduate statistics students in developing countries. *Promoting Statistical Practice and Collaboration in Developing Countries,* edited by Awe, O.O., Love, K, and Vance, E.A., CRC Press, Boca Raton, FL. DOI: 10.1201/9781003261148.

Badiee, S., Belkindas, M., Dupriez, O., Fantom, N., & Lee, H. (2004). Developing countries' statistical challenges in the global economy. In *9th National Convention on Statistics (NCS),* pp. 4–6, Mandaluyong City, Philippines.

Does, R. J. M. M., & Zempleni, A. (2001). Establishing a statistical consulting unit at universities. *Kwantitatieve Methoden, 67,* 51–63.

Fletcher, L. (2014). Statistical consultation as part of statistics education. In *Sustainability in Statistics Education. Proceedings of the Ninth International Conference on Teaching Statistics,* pp. 1–4, Flagstaff, AZ.

Gibbons, J. D., & Freund, R. J. (1980). Organizations for statistical consulting at colleges and universities. *The American Statistician, 34(3),* 140–145.

Glickman, M., Ittenbach, R., Nick, T. G., O'Brien, R., Ratcliffe, S. J., & Shults, J. (2010). Statistical consulting with limited resources: Applications to practice. *CHANCE, 23(4),* 35–42.

Gullion, C. M., & Berman, N. (2006). What statistical consultants do: Report of a survey. *The American Statistician, 60(2),* 130–138.

Khamis, H. J., & Mann, B. L. (1994). Outreach at a university statistical consulting center. *The American Statistician, 48(3),* 204–207.

Kiregyera, B. (2010). The evidence gap and its impact on public policy and decision-makign in developing countries. In *Data and context in statistics education: Towards an evidence-based society. Proceedings of the Eighth International Conference on Teaching Statistics,* pp. 1–6, Ljubljana, Slovenia.

North, D., Gal, I., & Zewotir, T. (2014). Building capacity for developing statistical literacy in a developing country: Lessons learned from an intervention. *Statistics Education Research Journal, 13(2),* 15–27.

Olubusoye, O. E., Akintande, O. J., & Vance, E. A. (2021). Transforming evidence to action: The case of election participation in Nigeria. *CHANCE*, 34(3), 13–23.

Steffey, D. L. (2015). Leadership in statistical consulting. In A. L. Golbeck, I. Olkin, and Y. R. Gel (Eds), *Leadership and Women in Statistics* (p. 145). Boca Raton, FL: CRC Press.

Taplin, R. H. (2003). Teaching statistical consulting before statistical methodology. *Australian & New Zealand Journal of Statistics*, 45(2), 141–152.

Thompson, M., & Walsham, G. (2010). ICT research in Africa: Need for a strategic developmental focus. *Information Technology for Development*, 16(2), 112–127.

Vance, E. A. (2015). Recent developments and their implications for the future of academic statistical consulting centers. *The American Statistician*, 69(2), 127–137.

Vance, E. A., & Love, K. (2021). Building statistics and data science capacity for development. *CHANCE*, 34(3), 38–46.

Vance, E. A., Love, K., Awe, O. O., & Pruitt, T. R. (2022). LISA 2020: Developing statistical practice and collaboration in developing countries. *Promoting Statistical Practice and Collaboration in Developing Countries*, eds. O.O. Awe, K. Love, and E.A. Vance, Boca Raton, FL: CRC Press. DOI: 10.1201/9781003261148

Vance, E. A., & Pruitt, T. R. (2022). Statistics and data science collaboration laboratories: Engines for development. *Promoting Statistical Practice and Collaboration in Developing Countries*, eds. O.O. Awe, K. Love, and E.A. Vance, Boca Raton, FL: CRC Press. DOI: 10.1201/9781003261148

Vance, E. A., & Smith, H. S. (2019). The ASCCR frame for learning essential collaboration skills. *Journal of Statistics Education*, 27(3), 265–274.

Windmann, M., & Kauermann, G. (2007). Statistical consulting at German universities: Results of a survey. *AStA Advances in Statistical Analysis*, 91(4), 367–378.

World Bank (2021). Data on statistical capacity. *Statistical Capacity Indicator*. https://datatopics.worldbank.org/statisticalcapacity/SCIdashboard.aspx.

4

Exploring the Need for a Statistical Collaboration Laboratory in a Kenyan University: Experiences, Challenges, and Opportunities

Thomas Mawora and Joyce Otieno
Maseno University

Eric A. Vance
University of Colorado Boulder

CONTENTS

4.1 Introduction

This paper explores the need for a statistical collaboration laboratory or "stat lab" (Vance and Pruitt 2022) at Maseno University in Kenya. It describes the experiences, challenges, and opportunities for statistics lecturers who established a statistical collaboration laboratory or "stat lab" called the Maseno University Laboratory for Interdisciplinary Statistical Analysis

DOI: 10.1201/9781003261148-5

(MU-LISA) and currently operate it. Section 4.1 introduces the Applied Statistics program at Maseno University and what statistical consultancy services are required in Kenya. Section 4.2 discusses nine challenges to operate a stat lab. Section 4.3 describes MU-LISA and how those involved in the stat lab are overcoming the challenges previously described. Section 4.4 concludes with lessons learned from creating and operating a stat lab in Kenya.

4.1.1 Context of Maseno University

The undergraduate degree program in statistics has been offered in Kenyan institutions of higher learning for more than 30 years. At the turn of the century, Maseno University pioneered the Applied Statistics degree program in the Department of Mathematics. As of now, more than five universities offer a similar program. Maseno further pioneered in introducing Information Technology (IT) to all its programs. This stretched the creativity of teaching at Maseno University and resulted in the then School of Mathematics opting to tailor their IT units to teach the skills their graduates should be expected to have (Stern et al. 2010). Maseno further pioneered being among the first universities in Kenya to offer fully online degree programs through the use of a Learning Management System (Musyoka et al. 2012).

The above activities pushed Maseno University's School of Mathematics, Statistics and Actuarial Science (SMSAS) to be in the limelight for innovations in student instruction. The innovations exposed the lecturers to several processes including working in projects through collaborations with members from the University of Reading's Statistical Service Centre (SSC). The SSC provided statistical consultancy services and had clients within the United Kingdom and internationally. In addition, they provided statistical support services to projects, some of them being in Africa. Apart from the statistical consultancy services, SSC was interested in modernization of teaching statistics. The SSC's collaboration with SMSAS resulted in the introduction of a variety of electronic resources to teach statistics, including Computer-Assisted Statistics Textbooks (CAST) (https://cast.idems.international/), and tailoring the IT units to be more relevant to the skills needed by students in the job market (Stern et al. 2014).

The IT courses were tailored to improve our undergraduate statistics curriculum. They were designed to enhance students' computing skills, needed for data analysis. The courses include basic concepts of IT, which introduces students to computers; web browsing and communication, which introduces students to the internet how to get information from the internet, and communication via the internet; descriptive data analysis and presentation; data management; statistical computing; and problem-based statistical analysis, among others. However, these courses typically run separately from the statistics courses. Only a fraction of students understand that they require these IT skills for statistics and data analysis. Most realize the importance of IT only after being exposed to the work environment. Usually, this is after leaving college and results in them posting on social media entreaties for current students to pay more attention to the IT courses. The few students who are engaged by the IT courses tend to be more interested in coding and not data management, visualization, or analytics. A current opportunity is to incorporate IT components into all statistics courses so that students will understand the importance of computing in statistics.

4.1.2 What Consultancies Are Required in Kenya?

The range of statistical consultancy services needed in a typical Kenyan university is vast. For instance, almost all schools in Kenyan universities offer postgraduate programs that have a project/research component. Many students pay others to do their statistical analysis, many times after data has been collected. The analyst goes ahead to produce the graphs and tables the person wants for topic 4 of their thesis, which is usually included in the results and discussions sections. This is a thriving business in Kenya for statistics graduates out of school and still looking for jobs. They usually come to consult some of their lecturers and in several cases, the data has serious quality issues. These could include (1) the data has not responded to the objectives of the study, (2) the data is in a wrong format, hence the tests they want cannot be done—in some cases the researcher wants to replicate the report of another person, and (3) the data does not yield statistically significant results, much to the dismay of the researcher. As such, the students conducting their studies may need consultancy services not just after data collection but during proposal writing so they are adequately guided on what data to collect and how to collect the data to be of high quality (Seiss et al. 2014).

A good number of lecturers from many of the schools in the university have successfully applied for funded projects. All the projects contain some elements of qualitative and quantitative data collection and analysis, including monitoring and evaluation. Despite the projects being multi-disciplinary, the persons concerned get external collaborators who perform most of the analysis needs. However, sometimes they seek consultancy services from lecturers within the university to help with the analysis, though rarely from the SMSAS. The exclusion of lecturers particularly from SMSAS may be due to the impression the Principal Investigators have concerning the teaching methods in SMSAS being theoretical rather than applicable and suited to current needs in statistical computing.

Conducting short courses on statistical literacy is another area of skills development that has high demand in Kenya. The first online course that Maseno University offered was *e-Statistics Made Simple* (eSMS), offered in 2010 that was developed by the Statistical Services Centre. This was an 8-week course on basic statistical literacy. The course attracted 40 students despite short notice, most of whom were nonstatisticians (Stern et al. 2014). There is more need for face-to-face short courses on the same topic and on more advanced statistical computing skills.

Finally, statistical consultancy services are needed by the local government. The policymakers make many decisions touching on multiple areas without being well informed through data in the region. In addition, development partners are interested in developing local skills and capacity to help respond to the ever-needed interventions they sponsor. Statisticians working with development partners and local governments in the intersection of data production, data analysis, and data decision-making can help make positive impacts on society (Vance and Love 2021).

4.2 The Setbacks to Having a Stat Lab and Human Resources for It

Having worked with the SSC before, one might expect that the university's statistics lecturers should be ready to operate their own statistical collaboration laboratory or "stat lab." However, despite imparting some skills in consultation and collaboration and experience working on consultancy projects, working with the SSC did not prepare lecturers in SMSAS to have their own lab. Below are nine challenges for setting up a stat lab in the Kenyan context.

4.2.1 Minimal Time for Lecturers for Consultations and Collaboration

Lecturing is a prestigious job in Kenya. With the explosion of universities and satellite campuses after the turn of the millennium, many of the lectures are sought after to (1) take more administrative roles and (2) teach courses for both undergraduate and postgraduate students part-time at other universities. These commitments fill up 40 hours-a-week and more, leaving the lecturers with far much less available time for other ventures like offering consultancy services. This is especially so with lecturers who teach mathematics and statistics, because the demand for their teaching expertise is high.

4.2.2 Overly Theoretical Nature of Our Programs

The curriculum provided for the postgraduate students in statistics assumes they already have statistical computing skills, and hence the emphasis is on the theoretical and mathematical aspects of the discipline. The students are required to apply the theory to real-life problems to earn an Applied Statistics degree. The challenge the students experience is primarily related to the big data they collect during the research process. Students are not proficient enough to wrangle data using statistical software to achieve the objectives of the project because education in the theory of statistics does not prepare one to analyze a real dataset.

4.2.3 Limited Statistical Computing Skills in Lectures

In the current realm of statistics, one has to be up-to-speed with the ever-changing statistical computing software and packages, such as the *tidyverse* in R (Wickham et al. 2019). In addition to computing skills, another challenge is that a number of universities do not have the relevant equipment and software to support the students' learning and to provision classrooms for the teaching of statistical computing. For example, only a few universities have subscribed to the limited version of the Statistical Package for Social Sciences (SPSS). This makes it difficult for stat lab members to use the software for training and other consultation services.

4.2.4 University Setup Challenges

Setting up a stat lab can have multiple other administrative challenges. Currently, at Maseno University, there is no defined procedure for setting up a statistical collaboration laboratory and provisioning it with needed office space and an independent physical lab of computers for teaching statistical computing short courses and workshops.

4.2.5 The Challenge of Running Costs and Consistent Funding

Currently, the stat lab at Maseno University strictly runs on volunteers, hence there are no financial implications for the lab or university. However, for long-term sustainability, some essential services for the proper functionality of the stat lab, including accounting and secretarial services, require consistent funds. The lack of such services limits the effectiveness of the lab. Consistent funding would go a long way to attract persons who can offer specialized skills, including writing winning proposals and specialized statistical skills to expand the number and types of projects on which the stat lab can engage.

4.2.6 The Challenge of Collaboration

The perception of other scientists and data users is that statisticians are required only at the data analysis stage of a project, rather than during the design, data collection, interpretation, and transforming evidence into action stages (Vance and Pruitt 2022). In our experience, statistics graduates from Maseno University also tend to consider themselves as just data analysts rather than collaborators in all stages of the research process who have a stake in the outcome of the project and work toward creating shared understanding with domain experts (Vance et al. 2022). Further, the local government and many other partners use external collaborations to help with statistical analysis and other projects. We have observed that there is usually a tendency for most leaders in the African context to overlook their own who have similar skills and opt instead for expatriates. This, coupled with the indecision of the statistics lecturers in engaging in outreach to create working relationships with their leaders and their lack of collaboration skills, inhibits opportunities for consultancies to trickle to the lecturers.

4.2.7 Data Quality

For a long time, there has been a challenge of unavailable and untidy data. This is slowly changing with many organizations embracing the modern methods of data collection. However, many still lack the capacity and skills for ensuring the quality of data is of high standards. As already mentioned, most lecturers have used small and tidy datasets in their classes. In current times, we have big data, with a lot of it being unstructured. There is a need for the statistics lecturers who will work in stat labs to be conversant on how to mine and clean data before they conduct an analysis.

4.2.8 University Process for Accrediting a Course

All university programs in Kenya are regulated by the Commission for Higher Education. Before the programs can be submitted, there are many internal processes and hurdles to clear before they can finally be approved. One main function for a stat lab includes offering short courses that the university can credit. This will boost confidence among learners since many would use the certificates for career progression. Overcoming the hurdles of accreditation for short courses is a major challenge for any stat lab in Kenya.

4.2.9 Pandemic

The whole world felt the pang of the COVID-19 pandemic. This particularly affected physical meetings, which were canceled for over 6 months in 2020 and resulted in squeezed timelines for semesters in Kenyan universities. The pandemic also affected the regular scheduling for in-person stat lab training.

The stat lab is a relatively new concept for lecturers at Maseno University who have been used to conducting classes and training, but not running a stat lab. The skills required include administrative skills of entrepreneurship, management, and leadership and technical skills in statistics and data science. At the same time, the director/coordinator of the lab should be learning new skills to meet the demands in the market, including networking skills to help advertise the lab and its services and connect it to more researchers, local government officials, and nongovernmental organizations. Being pioneers, this is still a challenge which we are rapidly overcoming through the activities mentioned in the next section.

4.3 MU-LISA's Strategies for Success

The MU-LISA was initiated in August 2019 and became the 12th full member of the LISA 2020 Network in March 2020 (Vance et al. 2022). MU-LISA's mission is "To empower learners to use statistical tools and methods to conduct sound analysis for multi-disciplinary research projects." This is accomplished through cooperation within Maseno University and with other universities, corporate, and government institutions for research and informed policy development. One of the goals of MU-LISA is to emphasize statistical thinking in all of our students and all of our projects so that the analysis of data relies less on "recipes" and becomes an activity to foster active learning (Kasturiarachi 2000). The stat lab teaches statistical coding skills, data management, data visualization, and data analytics in its training. In the lab training, we aspire not to be restricted by the semester dates and therefore provide our trainees with a cyclic learning process that develops in four stages. First, the learner encounters concrete experiences, followed by reflections and observations, which direct the learner to abstract conceptualization, and finally ends with active experimentation (Kolb 1983).

Since its inception, MU-LISA has offered over ten training sessions for statistical students, focusing on statistical computing skills such as introduction to R, tidyverse (e.g., dplyr (Wickham et al. 2021), ggplot (Wickham 2016)), managing data in Microsoft Excel, and introduction to python. Despite the multiple challenges mentioned previously, MU-LISA is learning and adapting to overcome these challenges to become successful using several strategies described below. Some are yet to be actualized but are currently ongoing.

4.3.1 Administrative Strategies

First, the process of initiating MU-LISA involved fostering university support by seeking consent from the Vice Chancellor, who agreed for us to initiate the stat lab. Further, the stat lab opened a file with the Directorate of Research and Innovation, hence we submit periodic reports to them. This has caused MU-LISA to be acknowledged as an official entity in Maseno University. As a result of this, we have received some support from the university and SMSAS, including using the physical Maths Lab for MU-LISA training sessions.

MU-LISA members divided roles so that apart from training, we also manage administrative tasks collectively. Members include the coordinator (Dr. Edgar O. Otumba) and trainers (Joyce A. Otieno, Thomas M. Mawora, and James K. Musyoka). All of the time spent is voluntary because MU-LISA does not yet generate funds to pay members to do administrative tasks. MU-LISA members meet regularly to plan and follow up on matters arising from previous meetings. This helps members to keep track of what needs to be done and allocate time for it. Currently, MU-LISA members spend approximately 10 hours every week planning, strategizing, and conducting MU-LISA activities. For the office space, the trainers retained their personal offices as the lab offices and they used existing lecture halls and computer labs for training.

4.3.2 Attending Workshops

MU-LISA members resolved to continual improvement of their general statistical knowledge and statistical computing skills. As such, members have continuously attended workshops, mostly online, to help hone their skills. The workshops have included training on Epidemiology, Online teaching, Tidyverse, Machine Learning, Data Science for

Consultancy, and Statistics Education Software (e.g., Tinkerplots (Brodesky et al. 2008) and Fathom). One organization MU-LISA members have partnered with is the Pak Institute for Statistical Training and Research (PISTAR), which hosted the latter four training sessions. Members who attended the training formed group sessions where they read related books through weekly meetings and practice. The group is now focusing on writing joint publications on subjects of mutual interest.

4.3.3 Continuous Collaboration

When MU-LISA joined the LISA community, members were encouraged to submit proposals for a 1-year project through a Transforming Evidence to Action Capacity in Higher Education (TEACH) fund (Olubusoye et al. 2021). MU-LISA's proposal to build the capacity of staff at Lake Region Economic Bloc (LREB) was selected and awarded $11,000 from the United States Agency for International Development. The LREB Bloc (see https://lreb. or.ke/about/) is a consortium of 14 counties in Western Kenya that share common social interests. The interests are divided into 10 pillars, namely, financial services; infrastructure; information communication technology; education; health; youth, gender and people with disabilities (PWDs); water, environment and climate change; tourism; agriculture; and trade and industrialization. Despite all these interests, the LREB does not have a central repository for the data on these pillars from its 14 counties. MU-LISA's project sought to empower LREB staff so they can set up a data repository center to serve the bloc.

Being awarded the TEACH fund gave MU-LISA members visibility within the university. As a result, they have been involved in multiple collaborative project proposal writing trainings and activities. MU-LISA has received a lot of support and encouragement from the university's Directorate of Research and Innovation. Writing strong proposals requires engaging in a learning process, and even though MU-LISA has not yet secured additional funding, we are optimistic that a funding proposal will be successful as we continuously improve on our project proposals.

The TEACH fund collaboration between MU-LISA and LREB attracted the attention of County Governments. A second concept note was drafted on the need to create a data repository for LREB. This note was created during the regions' governors meeting (summit) in 2020. Currently, LREB and Maseno University are working on a Memorandum of Understanding to help jumpstart the process of designing a repository.

In addition to collaborations within the university and LREB, MU-LISA has partnered with the Society of Applied Statistics Maseno University (SASMU). This partnership helps advertise and market MU-LISA and its activities among students in SMSAS. As a result, MU-LISA offers regular training sessions to members of SASMU and others who come for the training. The practice of collaboration consists of skills that can be learned and improved with practice, especially by using the ASCCR Framework for collaboration of Vance and Smith (2019).

4.3.4 Attachment and Internship Opportunities

Most courses in Maseno University require students to go for a 2-month attachment, which is similar to an internship. This attachment is a requirement for graduation. One negative effect of the Covid-19 pandemic was that many people were losing jobs, hence companies were not in a position to take students for attachment. This need necessitated MU-LISA to take a total of 60 students for attachment in November–December 2020 and January–March 2021.

The attachment program was geared toward providing students with real-life job experiences. As such, MU-LISA created tasks to help students learn practical skills while applying what they learned in their classes. MU-LISA benefitted by tasking the students to create resources that MU-LISA would have otherwise paid for. The students worked on developing "cheat sheets" or study notes for different subjects that can be given to students in order to help them study or review for the courses. In addition, they were given tasks to download data, read about it, develop new study questions, analyze it, and produce reports. At the end of the attachment period, they were provided with letters of recommendation to help them as they source for greener pastures.

Through the attachment, we worked with a group of five students who participated in the International Statistical Literacy Program (ISLP) International Poster Competition. This was a first for Maseno University and Kenya.

MU-LISA also benefited from the African Institute of Mathematical Sciences (AIMS) Alumni Technical Assistance Sponsorship Program. This is a support program to enhance the transition of AIMS graduates to a professional career life (https://wil.nexteinstein.org/wilprograms.htm). In the program, an AIMS alumnus identifies an institution where he would like to go and work for a short period as an intern. The institution provides projects for the AIMS alumnus and reports to AIMS on a monthly basis. AIMS provides the student with a modest stipend. Through this, MU-LISA got its first intern who will work on the ongoing TEACH fund project and other MU-LISA training activities.

4.3.5 Online Career Talks

Upon discussing more with students, MU-LISA members learned that there was still a need to help students understand their place in the job market. However, due to unavailability of time, MU-LISA opted to initiate Saturday Night Career Talks. These 40-minute sessions are recorded via zoom and discuss a variety of topics relating to careers in statistics. Example topics include specializations in statistics, handling data, both structured and unstructured, social network analysis, geospatial analysis, qualitative data analysis, and the professional skills on how to market oneself. The sessions are recorded and uploaded to YouTube (https://www.youtube.com/channel/UCcDKadTekzSYheKrcQL0q7A) so anyone can watch them at their convenience.

The talks focus on professional skills rather than technical programming skills. This was informed by the fact that most of the attendees usually use their phones to view the sessions. Hence, coding might be difficult to follow. We have found that focusing on these important skills makes the career talks more meaningful to them.

4.4 Lessons Learned

Over the course of the 2 years in which MU-LISA has been in existence, we have learned multiple lessons. One lesson we have learned and have come to appreciate is the importance of having this initiative led locally. We believe it is as a result of our local efforts that we have gained massive support from Maseno University in terms of recognizing our accomplishments and supporting us by linking us to many other researchers in the university. The LISA 2020 Network and its mentors provided an impetus for creating the stat lab, but it was the local members who carried the vision to fruition.

Another important lesson we have learned is the need for being part of a consortium or network. The process of joining the LISA 2020 Network helped us understand our goals better, hence we are ever driven to their attainment. Regular meetings in the consortium help rejuvenate us every fortnight as we see more members join and learn from others about how they are overcoming the challenges of their stat labs.

We have learned that even though finance is really important, it is not the very first thing that one needs in order to start a stat lab. Rather, what is required is a few dedicated members and learning how to find different methods of getting in-kind solutions to some of the financial needs. This could include office space, internet, communication, and marketing oneself.

MU-LISA provides an opportunity to train statistical consultants and collaborators. The MU-LISA members intend to collaborate on more projects in the lab and encourage the engagement of statisticians at the problem statement stage, rather than after the data have already been collected. MU-LISA members are expected to collaborate to make decisions on the objectives of the study; help decide how and what data should be collected; assist in how the data is managed, analyzed, stored, and archived; analyze the data; help interpret the results; and finally to help transform the evidence collected and analyzed into action (Vance and Love 2021).

This chapter has shared opportunities that are there for a stat lab to exist in a university in Kenya. There are contextualized hurdles that the lab will have to overcome. However, the stat lab will become more solidly established and sustainable through patience, effort, and focus by its members.

References

Brodesky, A., Doherty, A., and Stoddard, J. (2008), *Digging into Data with TinkerPlots*, Key Curriculum Press, 225 pp. ISBN 978-1-55953-885-5.

Kasturiarachi, A. B. (2000), *Interactive Labs for Introductory Statistics Courses*, Kent State University, Stark Campus, OH.

Kolb, D. A. (1983), *Experiential Learning: Experience as the Source of Learning*, Prentice Hall, Englewood Cliffs, NJ.

Musyoka, J., Otieno, J., and Stern, D. (2012), Using eLearning to Engage Mathematics and Statistics Students in a Kenyan University. *12th International Congress on Mathematical Education*, Seoul, Korea.

Olubusoye, O. E., Akintande, O. J., and Vance, E. A. (2021), "Transforming Evidence to Action: The Case of Election Participation in Nigeria," *CHANCE*, 34(3), 13–23.

Seiss, M., Vance, E., and Hall, R. (2014), "The Importance of Cleaning Data During Fieldwork: Evidence from Mozambique," *Survey Practice*, 7(4), 2864.

Stern, D., Ongati, O. N., Agure, J. O., and Ogange, B. (2010), Incremental Modernisation of Statistics Teaching and Curriculum at Maseno University. *International Conference on Teaching Statistics (ICOTS) 8*, Kenya.

Stern, D., Musyoka, J, and Stern, R. (2014), Building Strength from Compromise: A Case Study of Five Year Collaboration between the Statistical Services Centre of the University of Reading, UK and Maseno University. *International Conference on Teaching Statistics (ICOTS) 9*, Kenya.

Vance, E. A., Alzen, J. L., and Smith, H. S. (2022), "Creating Shared Understanding in Statistics and Data Science Collaborations," *Journal of Statistics and Data Science Education*. https://www.tandfonline.com/doi/full/10.1080/26939169.2022.2035286

Vance, E. A., Love, K., Awe, O. O., and Pruitt, T. R. (2022), "LISA 2020: Promoting Statistical Practice and Collaboration in Developing Countries." *Promoting Statistical Practice and Collaboration in Developing Countries*, eds. O.O. Awe, K. Love, and E.A. Vance, Boca Raton, FL: CRC Press. DOI: 10.1201/9781003261148

Vance, E. A., and Love, K. (2021), "Building Statistics and Data Science Capacity for Development," *CHANCE*, 34(3), 38–46.

Vance, E. A., and Pruitt, T. R. (2022), "Statistics and Data Science Collaboration Laboratories: Engines for Development." *Promoting Statistical Practice and Collaboration in Developing Countries*, eds. O.O. Awe, K. Love, and E.A. Vance, Boca Raton, FL: CRC Press. DOI: 10.1201/9781003261148

Vance, E. A., and Smith, H. S. (2019), "The ASCCR Frame for Learning Essential Collaboration Skills," *Journal of Statistics Education*, 27(3), 265–274. https://doi.org/10.1080/10691898.2019.1687370.

Wickham, H. (2016), *ggplot2: Elegant Graphics for Data Analysis*, Springer-Verlag, New York.

Wickham, H., Averick, M., Bryan, J., Chang, W., McGowan, L. D., François, R., Grolemund, G., Hayes, A., Henry, L., Hester, J., Kuhn, M., Pedersen, T. L., Miller, E., Bache, S. M., Müller, K., Ooms, J., Robinson, D., Seidel, D. P., Spinu, V., Takahashi, K., Vaughan, D., Wilke, C., Woo, K., and Yutani, H. (2019), "Welcome to the Tidyverse," *Journal of Open Source Software*, 4, 1686. https://doi.org/10.21105/joss.01686.

Wickham, H., François, R., Henry, L., and Müller, K. (2021). *dplyr: A Grammar of Data Manipulation*. R package version 1.0.7. https://CRAN.R-project.org/package=dplyr.

5

Barriers, Challenges and Opportunities to Statistical Collaboration

Demisew Gebru Degefu
Hawassa University

James L. Rosenberger
Pennsylvania State University

CONTENTS

5.1 Barriers in Data Quality in Developing Countries

Data collection in developing countries is usually performed by the state institutions since the private sector lacks the resources to be able to collect data. In addition, in most of these countries, the statistical infrastructure is primarily federal (Elahi, 2008). Developing countries encounter challenges in the entire data collection stage as compared to industrialized countries. These challenges are twofold: (1) the challenges may be due to weaknesses other than the Statistical Offices, e.g. lack of appreciation of the importance of data collection, socio-cultural reservations to give personal information, business and industries wanting to avoid sharing data with concerns about paying taxes, infrastructural deficiencies (i.e. inadequate transport and Internet facilities), and lack of reliability and transparency of data (i.e. the government's use of statistics for window-dressing its policy failures) or (2) it may be weaknesses internal to the Statistical Offices such as their inability to attract quality human resources, the perception that official statistics is a government task of minor rank, data entry is centralized and not close to the source of the data, inadequate financial resources allocated for statistics organizations, and statistical organizations lacking a modern statistical structure (Elahi, 2008).

A survey done by the staff members of the Department of Statistics at Hawassa University, Ethiopia, on 200 respondents in Hawassa government organizations, shows that 86% of respondents agreed that the availability of office materials was low or medium, and 75.5% of the respondents agreed that the quality of the training related to data management systems was low or medium. Only about 37.5% of respondents agreed that the

DOI: 10.1201/9781003261148-6

available data represent the reality well (good or very good), and only 17.5% of the respondents agreed that the overall data management system in their organization was good. Key informant interviews also show there is a problem in the reliability of statistical data, especially when the researchers used secondary data: the graduate students in other fields usually lack statistical supervision, teaching methodology focuses more on theoretical concepts rather than on practical applications, and in many organizations, there is no position designed to employ statisticians and selections are usually based on intimacy to the politics even if there are recruitment criteria.

5.2 Opportunities and Challenges Encountered by Stat Labs: The Case of Department of Statistics, Hawassa University, Ethiopia

5.2.1 Opportunities

Before launching the MSc in Applied Statistics in 2008, the Department of Statistics, Hawassa University, Ethiopia was running the BSc program in Statistics. After 2 years, in 2010, the department launched the MSc program in Mathematical & Statistical Modelling (MASTMO) in collaboration with Norwegian Universities, particularly the Norwegian University of Science and Technology and the University of Oslo. Later in 2013, the department enrolled PhD students in Applied Statistics with the assistance of these universities through NORHED PhD-Math-Stat-Sci Project (2014–2018) (Goshu, 2016). Female students also got funds from the MASTMO project to complete their MSc Degree. As of 2019/2020, the department had 7 PhD, 36 MSc and 213 BSc students, among which 93 were female. Two of the PhD students have main supervisors from the University of Cambridge and Lancaster University, United Kingdom; three PhD students have primary supervisors from the University of Oslo, Norway; one student has a primary supervisor from the University of Geneva, Switzerland and the rest have primary supervisors from Ethiopia. The Hawassa University (HwU) staff also benefit from the project through PhD Research leave and conference grants.

The HwU Statistical Collaboration Laboratory (Stat Lab) was established in the department in May 2015, with the rationale that there was a gap between the demands of the researchers and availability of statistical services, and with the vision to help researchers produce the best quality research for solving societal problems (Goshu, 2016). This Stat Lab was established under the LISA 2020 Network, and under this program two faculty members were trained at Virginia Polytechnic Institute and State University and Pennsylvania State University in the USA. Two reports were presented on the special program of the World Statistics Congress held in Marrakech, Morocco (2017) and Kuala Lumpur, Malaysia (2019) about the activities completed so far in the lab.

5.2.2 Challenges

Most literature that compares teacher training or capacity building in statistics originates mainly from developed countries. Thus, there is little attention to the situation in developing countries in this respect (North et al., 2014). In this regard, the undergraduate statistics students (students whose major is statistics) in the Department of Statistics, Hawassa University, Ethiopia face challenges due to the unsuitable curriculum and teaching methodology.

These students are only required to take three mathematics subjects, namely, Algebra I and II, Calculus I and II, and Numerical Analysis. The small number of mathematical courses has a negative influence on students' future careers; for example, this makes it difficult to specialize in postgraduate programs that require advanced mathematical knowledge.

Undergraduate students take fundamental Statistics courses such us Statistical Methods, Introduction to Probability, Sampling Theory and Research Methods, Analysis of Variance and Design of Experiments, Regression Analysis, Time Series Analysis, Categorical Data Analysis, Multivariate Analysis, Statistical Theory of Distribution and Statistical Inference and Biostatistics and Epidemiology. However, these courses focus more on statistical theory than they do on practice and application. The students only use statistical software (SPSS and R software) and Computer Course (Introduction to Computer Sciences, Fundamentals of Programming and Database System) on their Statistical Research (Proposal Writing and Senior Research Project) session. The lack of applied statistical courses and practical training opportunities in the curriculum means that students are often unprepared for practical applications of statistics following their education.

In addition to these challenges with the curriculum itself, undergraduate students face other challenges in the statistics program. Undergraduate students are unable to access texts and reference books for most courses, since the hard copies are limited in number and many are not the most recent editions. Students also have no computer or smartphone access to use soft copies. Undergraduate students usually lack internet access to search for additional materials and to develop basic programming skills that can be used for statistical computing courses. The other challenge for the undergraduate students in recent years is lack of employment opportunities following graduation, as the most vacancies announced by government or NGOs do not typically invite statisticians to apply; positions that statisticians could fill are often occupied by other professionals who do not have an educational background in statistics. This is because these organizations don't realize the importance of statisticians or the nuanced training received by students in a statistics program.

Other challenges emerge beyond the undergraduate education level. Postgraduate students face data quality issues when they use secondary data, and financial constraints when they plan to collect primary data. Despite this, most of the MSc theses focus on applied research requiring data, instead of developing new statistical methodologies. The final challenge to educating postgraduate students in statistics at HwU is that the department lacks financing and advisor time to admit more PhD students. A similar study done by Michael & O'Connell (2014) showed AAU and other universities in Ethiopia remain confronted with several challenges regarding statistics training and statistics education.

The urgent priorities to improve the educational situation among statistics students in the Department of Statistics, Hawassa University, are the following:

- Since the undergraduate curriculum is harmonized nationally, the department should begin an initiative to revise it to alleviate the problems mentioned above.
- The curriculum review committee should investigate the experiences of the best universities in similar developing countries in this aspect.
- Attempts should be made to subscribe journals and book publishers, e.g. (Elsevier, 2022), (ESPA, 2022), and other organizations including Wikipedia, that make resources available for students through digital and/or an online library. Of course, this requires an increase in internet access and more computers, tablets, or

mobile devices and preparing a schedule based on the number of students to use them properly.

- Use the postgraduate students to assist undergraduate students. This may be done through the statistical collaboration laboratory, where teams work together on real science projects.

- The department should implement a teaching methodology that focuses more on practical applications than theory and use statistical software to provide examples with real data.

- To develop closer connections with government, NGOs, and industry, the HwU Statistics department could invite government, NGO, or industry staff to present seminars to describe a practical problem they are trying to solve and invite the students to suggest solutions.

- Increase international collaboration to get more financial support and enroll more postgraduate students, especially PhD students.

References

Elahi, A. (2008). Challenges of Data Collection: With Special Regard to Developing Countries. *Statistics, Knowledge and Policy: Measuring and Fostering the Progress of Societies.* 294, 304. ISBN-978-92-64-04323-7.

Elsevier, (2022). Open Access, 2/16/2022, <https://www.elsevier.com/open-access>

ESPA (2022). Ecosystem Services for Poverty Alleviation, 2/16/2022, <www.espa.ac.uk>

Goshu, A. T. (2016). Strengthening Statistics Graduate Programs with Statistical Collaboration. The Case of Hawassa University. *International Journal Higher Education*, 5(3), 217.

Michael, K. & O'Connell, A. (2014). Statistics Education in Ethiopia: Successes, Challenges and Opportunities. *Sustainability in statistics education.* Proceedings of the Ninth International Conference on Teaching Statistics (ICOTS9, July, 2014), Flagstaff, Arizona, USA.

North, D., Gal, I. and Zewotir, T. (2014). Building Capacity for Developing Statistical Literacy in a Developing Country: Lessons Learned from an Intervention. *Statistics Education Research Journal*, 13(2), 15–27.

6

Challenges of Statistics Education That Leads to High Dropout Rate among Undergraduate Statistics Students in Developing Countries

Bayowa Teniola Babalola
Kampala International University

O. Olawale Awe
University of Campinas

Mumini Idowu Adarabioyo
Afe Babalola University

CONTENTS

6.1 Introduction

Statistics is the science of learning from data, and of measuring, controlling and communicating uncertainties. It provides the navigation essential for controlling the course of scientific and societal advances as well as policy implementations (Awe et al., 2015). Closely related to Statistics is Data Science which is an interdisciplinary field that utilizes scientific methods, processes, algorithms and systems for knowledge extraction and insights from many structured and unstructured data (Dhar, 2013). Education is undoubtedly an important instrument in developing any nation. Its importance cannot be overemphasized. And of course, the importance of Statistics as a course of study among undergraduate students is of massive importance, which is worthy of discussion. Therefore, the challenges related to dropout among students who enrolled for Statistics as their major, and not just students taking a few courses in Statistics alongside their main courses, are highlighted in this chapter.

Dropout, simply put, is an untimely withdrawal from a school without completing the initial program enrolled for. Therefore, in this context, we refer to dropout as withdrawal from studying Statistics (sometimes, withdrawal from the department where the course domiciles). In the higher education context, there are two levels of dropout. The first level in dropout is that which occurs inside the module or subject, whereby the student moves on to another subject or course; the second is complete withdrawal from the course of

DOI: 10.1201/9781003261148-7

study (Cohen, 2017; Burgos et al., 2018). School dropout is becoming a menace in the educational section across all levels of education in the world (Moshin et al., 2004; Bridgeland et al., 2006; Oghuvbu, 2008).

It is interesting to note that dropout rates differ by various demographic factors, including gender, race and ethnicity, immigration status and geographic location in developed countries. The Global Education Digest once revealed that Africa has the world's highest dropout rate. About 42% of African school children will leave school early, with about one in six leaving before Grade 2 (Cos, 2005).

It has been established in literature that the reason students drop out of their studies has been classified into four clusters (Cos, 2005), which are:

1. School-related factors, i.e., limited opportunities for academic success.
   * It has also been established that students who often get low grades or fail in subjects have high odds of discontinuing such course of study prior to graduation. Also, positive relationships between lecturers and students in a department enhance longevity of studentship.
2. Job-related factors.
3. Family and student-related factors, including gender, racial and ethnic minority status, low socioeconomic status, poor school performance, low self-esteem, delinquency, substance abuse, expulsion and pregnancy among others.
4. Community-related factors, i.e., whether schooling is perceived as relevant to the current or future lives of students.
5. Support for schooling in general or for the continued enrollment of students through graduation can vary from community to community and society to society. Each of these can be explained in further detail and be subdivided into smaller scales (Oghuvbu, 2008).

Several research works relating to dropout among students across all levels of the educational systems in the world have been conducted. The predictors of dropout and graduation for urban students among primary and secondary school students were considered in Robison et al. (2017). Their findings suggested that interventions, practices and policies in schools may prevent negative behavior and consequences. Also, "dropout early warning systems for high school students using machine learning" was reported in the work of Chung and Lee (2019). They developed a predictive model with an excellent performance in predicting students' dropouts in terms of various performance metrics for binary classification. The educational factors influencing female students' dropout from high schools in Nepal were reported by Dahal et al. (2019). An educational factor found to influence female students' dropout from high schools in Nepal was weakness in policy implementation, which resulted in low motivation in teachers and students, poor learning achievement and student dropout.

The effect of social media on dropout among first-year university students was the focus of the work of Masserini and Bini (2020). The researchers concluded that participation in social media groups is effective for lowering the dropout rate. Research on peer-relation and dropout behavior among junior school students was conducted with the result showing that both "push out" and "pull out" factors are strongly associated with student dropout (Gao et al., 2019). Student–teacher relationship as a factor for dropout was researched by Lessard et al. (2010). The study analyzed the relationships between the students' commitment, satisfaction, perceived achievement level, attitudes toward teachers, the perceived support and structure provided by teachers and the dropout risk. Their results indicated that for boys, satisfaction and achievement contributed to explaining 18% of the variance, whereas, for girls, commitment, satisfaction and achievement explained 23% of the variance. The effect of parents' loss of job on dropout of students was reported in the work of Maio and Nisticò (2019). The study showed that loss of parents' job increases the risk of dropping out of school. Bradley and Migali (2019) discussed the effect of hikes in school fees on dropout behavior. They concluded that hikes in school fees naturally have a negative impact on most students, especially the indigent ones.

It is a known fact that learning can be face-to-face or online these days. Its common knowledge that there are restrictions (gathering or/and movement restrictions) in some countries due to the COVID-19 pandemic; therefore, online teaching of Statistics has become inevitable. Several authors have been involved in modeling dropout among students studying Statistics online. For instance, predicting early dropout in online Statistics course was investigated by Figueroa-Cañas and Sancho-Vinuesa (2019, 2020). The latter article offered a simple and interpretable procedure for identifying dropout-prone students and fail-prone students of Statistics with the aid of tree-based classification models. It also shows that, in Statistics, regular engagement of students by the teachers is paramount in order to sustain studentship in the online environment. Adult dropout in online degree programs was the focus of Choi and Park (2018).

In Nigeria, like in other developing countries, Statistics is gradually gaining popularity and its relevance is becoming obvious by the day (Awe and Vance, 2014). There are well over 40 universities offering Statistics as a course in Nigeria and about 90 polytechnics who offer Statistics as well. Many universities across the developing countries of the world can boast of a large number of tertiary institutions that present students with the opportunity of taking this course, even up to postgraduate degree levels.

Only very few institutions in developing countries record 100% graduation of students admitted into Statistics, as there is always high rate of drop out among undergraduate students of Statistics. Therefore, the numerous challenges leading to dropout of students are discussed in the next section. It is believed, if tackled, this may help find a lasting solution to this lingering problem faced by both students and staff members associated with the course. On this note, it is important to discuss subject-specific challenges associated with the study of Statistics in tertiary institutions based on experiences collated from Statistics lecturers over the years.

6.2 Challenges Peculiar to Statistics Education

Apart from the aforementioned factors in literature that influence students' dropout, there are other practical factors responsible for dropout from Statistics education, some of which will be discussed in the subsequent paragraphs. It is very important to reemphasize that this section was written based on shared experiences received from Statistics Professors who had been teaching Statistics in different universities/polytechnics for some years in Nigeria and some other African countries. The next paragraph presents a brief background about Ekiti State, which serves as the case study for this study.

Ekiti State is a 25-year-old state that has four universities, out of which two have students taking Statistics as their main course of study (Federal University Oye-Ekiti, at Bachelor of Science level and Ekiti State University up to Doctoral level). We also have the Afe Babalola

University, Ado-Ekiti where students from different colleges and programs take up to six units of Statistics courses before graduation. All of these institutions have students taking Statistics modules as a compulsory subject at one level or the other across almost all the departments. Two polytechnics in the state also have Statistics as a major course of study for students up to the Higher National Diploma level. Based on the findings of the authors, it is estimated that the total number of students taking Statistics as their major course of study in the state as at the time of writing this chapter is about 970, while over 4000 take few courses in Statistics before graduation. With this figure, the lecturers have a reasonable number of students to interact with, which will most certainly be helpful in understanding some basic factors associated with dropout among Statistics students in this part of the world.

In countries with high rates of unemployment and underemployment, like Nigeria, findings showed that caregivers and guardians believe that it is not worth it for their children or wards to study so hard in school and end up being jobless or underemployed. It is therefore their duty to make sure those they are responsible to must, as a matter of necessity, study very lucrative courses, which may, in turn, allow them to establish themselves and become employers of labor. This is what gives birth to students always wanting to study professional courses like Medicine, Engineering and Law-related courses. This is all for the fear of joblessness in the future. There is therefore a perceived lack of societal value associated with Statistics as a course. A very misleading misconception raging among secondary school students is that after graduation, a Statistics graduate has high odds of being jobless when compared to graduates of the aforementioned courses. Therefore, Statistics is never a part of the topmost desired courses among young secondary school leavers for the fear of job prospects after graduation. The truth is that this is not only applicable to Statistics graduates but almost every graduate of any discipline, due to high rates of underemployment and unemployment stated earlier. Even some well-trained Medical doctors and Engineers are jobless, while some are underemployed in developing countries.

One of the topmost reasons for dropout among undergraduate students of Statistics is the poor university admission processes. Most students, especially in developing countries, would rather not choose Statistics as one of their first two choices while attempting to secure university admission. If a student is mathematically inclined, he/she would naturally opt for courses like Engineering or Computer Science. It is a pretty rare occurrence for students to choose Statistics as a first choice while seeking university admission in developing countries. Therefore, the volume of applications received by universities for those preferred courses (compared with the limited slots or quota available) pushes the university management into offering the spill-over applicants (who could not be accommodated in those courses) admission for courses they do not really desire from the onset. Often, Statistics is one of these courses. This in turn affects the psyche of students, making them perceive the course to be more difficult than it actually is. This gives birth to the "I didn't want it in the first place" syndrome. It is believed that if universities only offer Statistics courses to students who themselves chose Statistics as there will be a level of willingness that can serve as self-motivation for them during their stay in the university. But if this is lacking, there could be catastrophe ahead. Imposing a course of study on students is almost always counterproductive.

Moreover, the foundation of any structure is very important to the structure. A chain is as strong as the strength it has in its weakest point. Mathematics forms the basis of all the manipulations and the technicalities in Statistics. Calculus and algebra form the basis of most theories and derivations in Statistics. Poor background in any of these will have a negative impact on the quest for learning Statistics. Although there are some exceptions, a vast portion of Statistics demands a good understanding of Mathematics which has been seen to always overshadow Statistics in most developing countries (Awe and Vance, 2014). While reading a Statistics textbook or an article, if a line or step of solving a question (for example, a proof or a derivation) is skipped, there could be more than one or more mathematical principles that might have been applied in solving the problem before arriving at the next line. Hence, if a student is not well grounded in that regard, the student may drop what he or she is reading and move on to something else without learning what was intended. This will, at the end of the semester, be reflected in the grade of such a student. The deficiency in Mathematics might be due to "laziness" on the part of some students, or it can be due to inadequate study materials for proper mathematical prerequisites, especially for indigent students.

The difficulty of the course is also a very prominent factor that needs to be discussed. The course is perceived as a very difficult course among undergraduate students, because there are quite numerous formulae (and/or analysis methods) that they are introduced to during their course of study. For example, following a 2-hour lecture on Regression analysis or Sampling, a student can go home with 10–15 never seen before formulae or cumbersome lines of statistical expressions. Incidentally, a mastery of all of these, one after the other, is required, as the answer gotten with the first formula will be needed in the second formula up to the last one. No one can just gamble and get a formula he/she never learned (and learned well before) in a quiz or an examination. This alone makes students feel unsafe, as an incomplete or a wrong formula will never yield the desired result, and this could be quite discouraging.

It is common knowledge that some students have phobia for working with data. This is a real big issue among such students and is rampant among students who are not numerically or quantitatively sound. They prefer working with words rather than with numerical figures. It becomes an insurmountable problem when confronted with large numeric measurements. Such students frequently prefer to drop such lecture notes, handouts or textbooks. You hear such students saying "I do not like all the ugly figures, I prefer verbal notes".

The teaching styles or methods of some Statistics Professors should be queried too. Statistics lecturers need to understand the peculiarity of the course and try their best to incorporate bait into their lectures that will be helpful in enticing unwilling hearts to listen to them. It is often said that "if you do not like a teacher, you may end up not liking what the teacher is teaching". It is necessary to make students feel comfortable with a course from the onset by telling about the opportunities and benefits that abound in the course from the first day of lecture.

The introduction to the course should be warm and subsequent lectures should always be as lucid as possible, captivating and very interesting. This would most certainly keep the students going all through the course. There are indeed some challenging Statistics courses, but these courses are made more difficult by some lecturers' teaching methods. Concentrating on letting students know how challenging a Statistics course is one factor that can demoralize students, and of course, failure begins there. Consequently, dropping out may become inevitable even from the outset. Based on this fact, it can be safely inferred that some lecturers have not been trained well to be educators and particularly have not been trained well to be educators in Statistics. The basic principles of educating students are lacking in some Statistics lecturers.

Due to the inadequacy of basic amenities in the developing countries of the world, most especially electricity, it takes a longer time for Statistics students to see the beauty of Statistics hidden in statistical computing. The use of computer systems and statistical software in exploring the teaching of Statistics to make the course very fascinating to students becomes a very difficult problem. Statistical computing should naturally generate interest in students if every institution has access to the necessary equipment (stable electricity, access to uninterrupted and fast internet facilities, state-of-the-art computer systems and licensed computer software, among others). These amenities are lacking in most developing countries (Awe and Vance, 2014).

No one can stop a determined mind. Another very important challenge of Statistics is the fact that students, who are the major stakeholders in this context, do not seem to understand the underlying importance or application of the course in real-life problem-solving. Most of the time, it is not until a student proceeds to a postgraduate degree that such a student begins to understand the important applications of Statistics in several fields, including Science, Medicine, Engineering, Agriculture, Business, Commerce and Geography (to mention a few). This is a key factor that undermines the potential and potency of Statistics in problem-solving.

Poor database in most developing countries is also a cause of discouragement for Statistics students. Most organizations whose data should be useful to students for analysis and learning are never available to students, and if available, they are never in good shape. Poor data management causes a lot of problems for researchers, not excluding undergraduate students. Students of Statistics in developing countries face this challenge when carrying out their final year project during their final year, leaving them frustrated and confused. Since this does not lead the senior students to tell good stories about Statistics, it, therefore, has a ripple effect on the morale of their junior colleagues.

These are some of the challenges that make Statistics students tend to drop out from Statistics to courses that are perceived as easier, more lucrative and more interesting. They believe this will save the rigors associated with the formulae, the calculations and the theories of Statistics at large.

6.3 Opportunities for Statisticians

Contrary to some of the challenges and wrong perceptions discussed earlier, the opportunities available to Statistics graduates are limitless. Statistics is a professional course that is of great relevance in developing countries, where data systems are the backbone of the administration. Data Science is gradually gaining recognition and prominence in developing countries, and, therefore, this automatically gives Statisticians and Data Scientists an edge in involvement in the policy and decision-making of most countries. Their inputs are needed in developing meaningful policies for the betterment of citizens, as these policies will be based on existing figures. Their inputs are never faulted by anyone, not even the government. In the past, data collection was costly and time-consuming; now, recent developments in Information Technology have created a new dimension to data collection by introducing ease to it with minimized cost. This is being explored by both industry and the government. This implies that demand for data analysis is on the increase as new areas such as data analytics (which interconnects Information Technology and Data Science) evolve.

Statistics is a multidisciplinary course with applications in virtually all professions and disciplines. There is Statistics in Medicine (called Medical Statistics/Biostatistics), Finance and Insurance industries (called Actuarial Science), Chemical Systems (called

Chemometrics), Population (called Demography), Economics (called Econometrics), Environmental Science (called Environmental Statistics or Envirometrics), Geography (called Geostatistics), Applied Mathematics (called Operations Research), Psychology (called Psychometrics), Manufacturing industries (called Quality Control) and Mechanics (called Statistical Mechanics), to mention a few fields. This, therefore, suggests a very wide range of potential collaborators waiting for Statistics graduates. Other researchers depend on Statisticians for better outputs and correct interpretation of their analyzed datasets.

Statistics and Data Science is also the backbone of some emerging Information Technology professions such as Business and Data Analytics, which is currently one of the leading and best selling professions in Information Technology around the world now. This is also a field that is craving the expertise of Statisticians. Those skills learnt in schools are needed for world-changing solutions. This will therefore bring about recognition and fulfillment.

One beautiful thing about being a trained Statistician is that the training is incorporated and embedded in one's way of life as the person becomes very calculative and reasons fantastically. The knowledge of numbers and their working become an integral part of one's daily life. This cannot be bought; it is earned because it is a product of diligence and perseverance.

The opportunities are enormous. A trained Statistician can easily diversify into any other field with very little stress. After successful completion of the "rigour" as a Statistician, there is no field that one would not be able to cope with because Statistics teaches stoicism, persistence, consistency and self-motivation. No hard training at the moment seems truly pleasant, especially for inexperienced students; however, later on, it produces good fruits for those who have undertaken difficult training.

All these points nullify the most basic fear of joblessness after graduation. There are numerous opportunities available for Statistics graduates, even in developing countries. Statisticians are needed in every phase of the economy of a developing country, as well as the lives of citizens. This last point has been demonstrated very recently, as the knowledge of Statistics has proven useful in modeling epidemic/pandemic outcomes, particularly in curbing the spread of COVID-19 pandemic with the slogan "let's flatten the curves".

The following are some recommendations that would end dropout among Statistics students in developing countries, or at least reduce it to the barest minimum.

1. Statistics-related jobs should be given to trained Statisticians directly, since they have the required technical-know-how to handle such perfectly.

2. Universities and other tertiary institutions should admit only students who are willing to take Statistics into Undergraduate Statistics programs.

3. Universities and other tertiary institutions should make and enforce policies mandating that only students with good Mathematics background are offered admission to study Statistics.

4. The problem of the perceived volatility of Statistics and the resulting bad grades can be tackled collectively by both the students and their educators. The students should work harder, and the lecturers should be more professional while grading students.

5. Universities and other tertiary institutions should ensure that only students who have demonstrated flair for data and numbers are offered admission to study Statistics.

6. Students should be constantly encouraged by the school, the private sector and the government through workshops and seminars for students, as well as quizzes and the likes with enticing awards (e.g., cash gifts for the young scholars will most certainly do a great job of keeping students encouraged to work hard).

7. Some lecturers are not well trained to be lecturers, and students do not enjoy their classes. This can be solved by constantly organizing training and retraining for lecturers. It will help them discover the best approach to engage in order to record a higher level of Statistics student's retention.

8. Every Statistics Departments should establish a statistical collaboration laboratory (e.g., LISA 2020 Stat lab) and ensure that Statistics students pass through the lab before graduation. More technological and computing tools should be supplied by the school, government, private sector and well-meaning individuals so that students have enough practical tools to work with.

9. Students should be exposed to both local and international conferences, workshops and seminars as early as possible. This will expose them to fascinating discoveries about the application of Statistics, and this will consequently reduce dropout rate among the students.

10. Proper data management should be encouraged. The governments of developing countries should make policies that will make it mandatory for organizations to keep records of all the happenings in the organizations.

If these recommendations are carefully implemented, there will be fewer dropouts among Statistics students in developing countries. As a matter of fact, studying Statistics will be more attractive and more students will enroll in Statistics. Therefore, society will have more Statisticians capable of recommending good policies that will improve the well-being of the citizens. This will also bring about desired developments in the developing countries.

References

Awe, O. O., Crandell, I., and Vance, E. A. (2015). Building statistics capacity in Nigeria through the LISA 2020 Program. In *Proceedings of the International Statistical Institute's 60th World Statistics Congress*, Rio de Janeiro, Brazil.

Awe, O. O., and Vance, E. A. (2014). Statistics education, collaborative research, and LISA 2020: A view from Nigeria. In *Sustainability in Statistics Education. Proceedings of the Ninth International Conference on Teaching Statistics (ICOTS9)* (pp. 3–4), Flagstaf, AZ.

Bradley, S., and Migali, G. (2019). The effects of the 2006 tuition fee reform and the great recession on university student dropout behaviour in the UK. *Journal of Economic Behavior and Organization* 164, 331–356.

Bridgeland, J.M., Dilulio J.J. and Morison K.B. (2006). *The Silent Epidemic*. New York: Civil Enterprises, LLC.

Burgos, C., Campanario, M.L., de la Peña, D., Lara, J.A., Lizcano, D., and Martínez, M.A. (2018). Data mining for modeling students' performance: A tutoring action plan to prevent academic dropout. *Computers and Electrical Engineering* 66, 541–556.

Choi, H.J., and Park, J.H. (2018). Testing a path-analytic model of adult dropout in online degree programs. *Computers & Education* 116, 130–138.

Chung, J.Y., and Lee, S. (2019). Dropout early warning systems for high school students using machine learning. *Children and Youth Services Review* 96, 346–353.

Cohen, A. (2017). Analysis of student activity in web-supported courses as a tool for predicting dropout. *Educational Technology Research and Development* 65, 1285–1304.

Cos, D. (2005). *High School Dropout, Enrollment, and Graduation Rates in California*. Sacramento: California Research Bureau. California State Library.

Dahal, T., Topping, K., and Levy, S. (2019). Educational factors influencing female students' dropout from high schools in Nepal. *International Journal of Educational Research* 98, 67–76.

Dhar, V. (2013). Data science and prediction. *Communications of the ACM* 56 (12): 64–73.

Figueroa-Cañas, J., and Sancho-Vinuesa, T. (2019). Predicting early dropout students is a matter of checking completed quizzes: The case of an online statistics module. *CEUR Workshop* 2415, 100–111.

Figueroa-Cañas, J., and Sancho-Vinuesa, T. (2020). Predicting early dropout and final exam performance in an online Statistics course. *IEEE Revista Iberoamericana De Tecnologias Del Aprendizaje* 15 (2), 100–111.

Gao, S., Yang, M., Wang, X., Min, W., and Rozelle, S. (2019). Peer relations and dropout behavior: Evidence from junior high school students in northwest rural China. *International Journal of Educational Development* 65, 134–143.

Lessard, A., Poirier, M., and Fortin, L. (2010). Student-teacher relationship: A protective factor against school dropout? *Procedia Social and Behavioral Sciences* 2, 1636–1643.

Maio, M.D., and Nisticò, R. (2019). The effect of parental job loss on child school dropout: Evidence from the occupied Palestinian Territories. *Journal of Development Economics* 141, 102375.

Masserini, L., and Bini, M. (2020). Does joining social media groups help students' dropout within the first university year? *Socio-Economic Planning Sciences*, 73, 1–9.

Moshin, A.O., Aslam, M., and Bashir, F. (2004). Causes of dropouts at the secondary level in the Barani area of Punjab. *Journal of Applied Sciences* 4 (1), 155–158.

Oghuvbu, E.P. (2008). The perceived home and school factors responsible for dropout in primary schools and its impact on national development. *Ekpoma Journal of Behavioural Sciences* 1, 234–235.

Robison, S., Jaggers, J., Rhodes, J., Blackmon, B.J., and Church, W. (2017). Correlates of educational success: Predictors of school dropout and graduation for urban students in the Deep South. *Children and Youth Services Review* 73, 37–46.

7

Statistics Education and Practice in Secondary Schools in Nigeria: Experiences, Challenges and Opportunities

Oluokun Kasali Agunloye, Deborah Olufunmilayo Makinde, and Rafiat Taiwo Agunloye

Obafemi Awolowo University

CONTENTS

7.1 Introduction

In Nigeria, statistics is taught as an integral part of the mathematics curriculum at both junior and senior secondary school levels. Despite the huge applicability of statistics in everyday life, statistics is not accorded the same recognition given to other subjects such as mathematics, English language, and Civic Education, which are classified as core subjects in Nigerian secondary schools (Awe and Vance, 2014; Adelodun and Awe, 2013). Students at the secondary school level are generally oblivious of abundant career opportunities that are open to them in the field of statistics. The challenges confronting the development and growth of statistical education and practice in Nigerian secondary schools cannot be said to be unconnected with non-recognition of statistics as a distinct branch of knowledge by policy-makers in the education sector. Most students at the secondary school level are unaware that they can make a big career out of statistics until when such students gain admission into tertiary institutions. At the senior secondary school level, most students would have garnered enough exposure to make decision on their future career choice which in most cases includes but not limited to the following professions medicine, pharmacy, nursing, engineering, law, accountancy, banking and finance, business administration among others. It is very uncommon to see students wishing to study statistics because they lack requisite exposure and information about abundant opportunities that statistics has to offer. Students' exposure in secondary school is primarily determined by two factors, namely home and school factors. The home factor is primarily concerned with parental and family influence on student choice of future career, while the school factor is essentially concerned with the influence of peer-group, personal academic ability and personal interest devoted to a particular extra-curricular activity on the choice of students' future career. The extra-curricular activities in public secondary school system in Nigeria are primarily designed to strengthen students' understanding of concepts taught during classes and at the same time the platform serves as an avenue that could guide students in making an informed decision on future career choices based on students' personal interest. These extra-curricular activities parade students' clubs such as Junior Engineering Technicians (JET) club whose membership is mainly populated by science students, Literary and Debating Society (LDS) whose membership is mostly populated by art students and Press Club whose membership cut across all divides. It is pertinent to emphasize here that none of the above-listed students associations seeks to promote statistical education and practice in the Nigerian secondary school system.

A number of authors have investigated challenges and prospects of statistics education in secondary schools in African countries and the common finding has been that statistics education at that level of education is still evolving due to a number of problems such as dearth of statistics educators, inadequacy of statistics content of mathematics curriculum, inadequacy of instructional materials and teachings aids, to mention just a few. To the best of our knowledge, little attention has been given to the study of statistics education and practice in Nigerian secondary schools in the literature. Adichie (1986) examined statistical education in developing countries of Africa using Nigeria as a case study. This study focused on statistical education at post-secondary level in Nigeria and it was argued that statistical education is fairly well-established. Ogum (1998) examined the problems and prospects of statistical education in Nigeria at primary, postprimary, intermediate and tertiary levels. The author identified four major factors that constitute bane of statistical education in Nigeria as low literacy level, cultural restrictions of nationalities that make up Nigeria, dearth of statistics educators and lack of political will on the part of government to promote statistical education. Steffens (1998) examined the private experience in statistics education in South Africa and Namibia. This study attributed the problem in teaching statistics to rural students in southern Africa to the poor level of school mathematics. Oyesola (2000) opined that statistics education in secondary schools and technical colleges in Nigeria needs to be repackaged to prepare students for quantitative thinking. Odhiambo (2002) reviewed statistics education in Kenyan secondary schools and argued that the statistics content is limited to basic concepts of descriptive statistics and probability. Polaki (2006) investigated statistics education in secondary schools in Lesotho and contended that the statistics content of the two high school mathematics textbooks used by secondary school students in Lesotho is grossly inadequate. Zewotir (2006) examined statistics education at the university level in Ethiopia and argued that statistics as a discipline is not popular among secondary school students as most students prefer other courses to statistics. Zewotir and North (2011) investigated challenges and opportunities of statistics education across all levels of education in South Africa, and it was reported that South Africa has restructured her school curriculum with vast statistics content.

7.2 Sample and Sampling Techniques

Stratified random sampling technique was used for this study. The population for this study consists of the entire senior secondary school students, as well as the mathematics and further mathematics teachers in the ten secondary schools selected for this study. A sample of one hundred and twenty (120) students was drawn from the population through proportionate stratified random sampling technique which was intentionally adopted to ensure that the chosen respondents include students from science, commercial and art classes. From the 10 schools under study, we also considered all the 20 mathematics teachers and 10 further mathematics teachers who serve as statistics educators in the chosen schools to make our sample size to be 150 respondents altogether.

7.3 Instrumentation

This study used a questionnaire as a research instrument. The questionnaire is divided into four sections: A, B, C and D. Section A captured the biodata information of the respondents with section B seeking to assess the respondents' opinion about the adequacy of statistics content of mathematics curriculum in Nigerian secondary schools while section C dealt with respondents' view about the manpower and material resources available for the teaching and learning of statistics in Nigerian secondary schools and lastly section D focused on respondents' view on how professional statistical associations can contribute to the development and growth of statistics education in Nigerian secondary schools.

7.4 Data Analysis

Descriptive statistics were used to analyze the research questions using frequency counts and percentages. The researchers administered questionnaires, the responses of the respondents were analyzed, and the results are presented and interpreted in Table 7.1.

Out of 150 respondents considered for this study, 20 (representing 13.3% of the sample size) were science students, 40 (representing 26.7% of the sample size) were commercial students, 60 (representing 40% of the sample size) were arts students, 20 (representing 13.3% of the sample size) were mathematics teachers and 10 (representing 6.7% of the sample size) were further mathematics teachers (see Table 7.1).

Research question 1: What is the respondents' view about the adequacy of statistics content of the current mathematics curriculum in Nigerian secondary schools?

TABLE 7.1

Distribution of Respondents

Respondents	Frequency ($n=150$)	Percentage
Science students	20	13.3%
Commercial students	40	26.7%
Art students	60	40%
Mathematics teachers (statistics educators)	20	13.3%
Further mathematics teachers (statistics educators)	10	6.7%
Total	150	

TABLE 7.2

Summary of Respondents' View about Adequacy of Statistics Content of the Current Mathematics Curriculum in Nigerian Secondary Schools

Research Question 1	Agreed	Disagreed
Statistics content of junior secondary school mathematics curriculum is adequate	30 (20%)	120 (80%)
Statistics content of senior secondary school mathematics curriculum is adequate	25 (16.7%)	125 (83.3%)

Table 7.2 provides respondents' view about the adequacy of statistics content of the current mathematics curriculum in Nigerian secondary schools.

Table 7.2 summarizes the opinion of students and teachers on the adequacy of statistics content in the mathematics curriculum in Nigerian secondary schools. From this table, 30 respondents (representing 20% of the sample size) agreed that the statistics content of junior secondary school mathematics curriculum is adequate, while 120 respondents (representing 80% of the sample size) disagreed. Similarly, 25 respondents (representing 16.7% of the sample size) agreed that statistics content of senior secondary school mathematics curriculum is adequate, while 125 respondents (representing 83.3% of the sample size) disagreed. This finding shows that majority of the respondents are of the opinion that statistics content is inadequate for both junior and senior secondary classes.

Research question 2: What is the respondents' view on the availability of trained statistics educators in Nigerian secondary schools?

Table 7.3 provides respondents' view on the availability of trained statistics educators in Nigerian secondary schools.

Table 7.3 summarizes the opinion of students on the availability of trained statistics educators in Nigerian secondary schools. Only three students representing 2.5% of the sample size agreed that there is no dearth of statistics educators in Nigerian secondary schools, while 117 students representing 97.5% of the sample size disagreed. This finding might not be unconnected with the fact that none of the 30 teachers who serve as statistics educators in all the ten selected schools is a professional statistics educator by training.

Research question 3: What is the respondents' view on the availability and adequacy of instructional materials and teaching aids on statistics in Nigerian secondary schools?

Table 7.4 provides respondents' view on the availability and adequacy of instructional materials and teaching aids on statistics in Nigerian secondary schools.

Table 7.4 summarizes the opinion of students and teachers on the availability and adequacy of instructional materials and teaching aids required to facilitate teaching–learning process. The table shows that 15 respondents (representing 10% of the sample size) agreed that instructional materials on statistics are available and adequate, while 135 respondents (representing 90% of the sample size) disagreed. Moreover, eight respondents (representing 5.3% of the sample size) agreed that teaching aids on statistics are available and adequate, while 142 respondents (representing 94.7%) disagreed. Furthermore, two respondents (representing 1.3% of the sample size) agreed that statistical laboratory is

TABLE 7.3

Summary of Respondents' View on Availability of Trained Statistics Educators in Nigerian Secondary Schools

Research Question 2	Agreed	Disagreed
There is no dearth of trained statistics educators in Nigerian secondary schools	3 (2.5%)	117 (97.5%)

TABLE 7.4

Summary of Respondents' View on Availability and Adequacy of Instructional Materials and Teaching Aids on Statistics in Nigerian Secondary Schools

Research Question 3	Agreed	Disagreed
Instructional materials on statistics are available and adequate	15 (10%)	135 (90%)
Teaching aids on statistics are available and adequate	8 (5.3%)	142 (94.7%)
Statistical laboratories equipped with computers and statistical software are available and adequate	2 (1.3%)	148 (98.7%)

available, while 148 respondents (representing 98.7% of the sample size) disagreed. Both students and teachers admitted that instructional materials and teaching aids are available but not adequate. Majority of the respondents also agreed that a statistical laboratory is not available.

Research question 4: What is the respondents' view on awareness about the existence of professional statistical bodies in Nigeria?

Table 7.5 provides a concise summary of respondents' views on awareness about the existence of professional statistical bodies in Nigeria.

Table 7.5 summarizes the opinion of students and teachers on awareness about the existence of professional statistical bodies in Nigeria. Five respondents (representing 3.3% of the sample size) admitted they were aware of the existence of Nigerian Statistical Association (NSA), while 145 respondents (representing 96.7% of the sample size) claimed they were unaware. Moreover, three respondents (representing 0.7% of the sample size) admitted they were aware of the existence of Professional Statisticians Society of Nigeria (PSSN), while 147 respondents (representing 99.3%) claimed they were unaware. Furthermore, two respondents (representing 1.3% of the sample size) admitted they were aware of the existence of the International Biometric Society (IBS)-Nigeria Local Group, while 148 respondents (representing 98.7% of the sample size) claimed they were unaware. Similarly, two respondents (representing 1.3% of the sample size) admitted they were aware of the existence of the Royal Statistical Society (RSS)-Nigeria Local Group, while 148 respondents (representing 98.7%) claimed they were unaware. Our finding shows that majority of respondents are unaware of the existence of these professional statistical associations. Hence, these professional associations need to make their impact felt among Nigerian secondary school students through an awareness campaign and also by organizing annual statistics quiz competitions for students.

TABLE 7.5

Summary of Respondents' View on Awareness about Existence of Professional Statistical Bodies in Nigeria

Research Question 4	Agreed	Disagreed
Aware of the existence of the Nigerian Statistical Society (NSA)	5 (3.3%)	145 (96.7%)
Aware of the existence of the Professional Statisticians Society of Nigeria (PSSN)	3 (2%)	147 (98%)
Aware of the existence of the International Biometric Society (IBS), Nigeria Local Group	2 (1.3%)	148 (98.7%)
Aware of the existence of the Royal Statistical Society (RSS), Nigeria Local Group	2 (1.3%)	148 (98.7%)

7.5 Discussion and Conclusion

There are three major important findings of this research work. Of all the 30 teachers involved in this study, none of them is a trained statistics educator. This clearly signifies a major challenge for the growth and development of statistics education at postprimary level in Nigeria. Statistical laboratory is practically nonexistent in all the ten public secondary schools considered in this study, and this also explains the unimpressive status of statistics education in postprimary schools in Nigeria. Last, but not least, there are currently four professional statistical bodies operating in the country but both teachers and students in postprimary schools in Nigeria are yet to feel the impact of these professional statistical bodies. Given our findings as highlighted above, we conclude that statistical education and practice in Nigerian secondary schools is still evolving and it requires urgent attention of all stakeholders such as government at all levels, professional statistical bodies and nongovernmental organizations (NGOs) that are committed to the advancement of statistical education and practice. Given the global trend on statistics education, it is obvious that there is an urgent necessity to introduce statistics as a core subject in postprimary schools in Nigeria. Both the state and the federal government should consult with all stakeholders in the education sector to facilitate the immediate establishment of statistics education programs in all colleges of education and universities nationwide in a bid to produce trained statistics educators that will be employed to teach statistics education at primary and postprimary levels.

7.6 Recommendations

Based on our findings highlighted above, we make the following recommendations:

1. Government should reintroduce statistics as a distinct core subject in postprimary schools in Nigeria.
2. Government at all levels should employ trained statistics educators to teach statistics in postprimary schools in Nigeria.
3. Government should provide functional and well-equipped statistical laboratories in all postprimary schools in Nigeria for the use of statistics educators and students.
4. Government should provide adequate funding to organize capacity-building workshops/refresher courses for statistics educators on regular basis to keep them abreast of the latest happening in statistical education and practice.
5. Government should mobilize school administrators through the prompt release of running grants required for the provision of the much-needed instructional materials and teaching aids to facilitate teaching and learning of statistics in all postprimary schools in Nigeria.
6. Professional statistical bodies like NSA, PSSN, RSS, Nigeria Local Group, IBS, and Nigerian Local Group should collaborate with government in promoting statistical education in Nigerian secondary schools especially in the areas of writing statistics textbooks that capture peculiarities of our local environment for the use of

statistics educators and students. Professional bodies should also organize annual academic competitions, seminars, workshops and career talks aimed at promoting and advancing the cause of statistics education in Nigeria.

7. As part of their corporate responsibilities to our society, organized private sectors and NGO should be encouraged to contribute their quota toward promoting and propagating statistical education and practice in Nigeria through the provision of infrastructural facilities for the statistical laboratories in all public postprimary schools in Nigeria as well as sponsorship of academic competitions on statistics education and practice among students of postprimary schools in Nigeria.

8. The Federal Government of Nigeria should also consider the establishment of the National Statistical Centre that will be responsible for the promotion and advancement of statistics education at all levels of education in Nigeria.

9. The National Commission for Colleges of Education in Nigeria should encourage and facilitate the establishment of the department of statistics in all colleges of education in Nigeria where professional statistics educators will be trained.

References

Adichie, J. N. (1986). Statistical Education in the Developing Countries of Africa: The Nigerian Experience. Paper presented at the *Second International Conference on Teaching Statistics*, Singapore.

Adelodun, O. A., & Awe, O. O. (2013). Statistics Education in Nigeria: A Recent Survey. *Journal of Education and Practice*, 4(11): 214–220.

Awe, O. O., & Vance, E. A. (2014). Statistics education, Collaborative Research, and LISA 2020: A View from Nigeria. In *Sustainability in Statistics Education. Proceedings of the Ninth International Conference on Teaching Statistics (ICOTS9)* (pp. 3–4), Flagstaf AZ.

Odhiambo, J.W. (2002). Teaching of Statistics in Kenya. Paper presented at *the Sixth International Conference on Teaching Statistics*, Cape Town, South Africa.

Ogum, G.E.O. (1998). Statistical education in Nigeria: Problem and Prospect. Paper presented at the *Fifth International Conference on Teaching Statistics*, Singapore.

Oyesola, G.O. (2000) Introducing statistics Education in Secondary Schools in Nigeria. *The Nigerian Journal of Guidance and Counseling*, 7(1): 24–31.

Polaki, M.V. (2006). Looking at the Mathematics Curriculum and Mathematics Textbooks to Identify Statistical Concepts that Lesotho's High School Students Experience. Paper presented at the *Seventh International Conference on Teaching Statistics*, Singapore.

Steffens, F. E. (1998). Statistical Education in the African Region: Private Experiences in South Africa and Namibia. Paper presented at the *Fifth International Conference on Teaching Statistics*, Singapore.

Zewotir, T. (2006). Status of Statistics in Africa: The Case of Ethiopia. Paper presented at the *Seventh International Conference on Teaching Statistics*, Singapore.

Zewotir, T., & North, D. (2011). Opportunities and Challenges for Statistics Education in South Africa. *Pythagoras*, 32(2): 5, Art. #28. http://doi.org/10.4102/pythagoras.v32i2.28.

Part 2

Building Capacity in Statistical Consulting and Collaboration Techniques through the Creation of Stat Labs

8

Promoting and Sustaining a Virile Statistical Laboratory in Nigeria's Premier University: Lesson from UI-LISA Experience

Olusanya Elisa Olubusoye, Oluwayemisi Alaba,
Serifat Folorunso, and Olalekan J. Akintande
University of Ibadan Laboratory for Interdisciplinary Statistical Analysis [UI-LISA]

Eric A. Vance
University of Colorado Boulder

CONTENTS

8.1 Introduction

Education is one of the earliest social services embraced in Nigeria as an instrument of stability, change, and development. Since national independence in 1960, the education system has undergone several phases of transformation. Presently, it is divided into six years of primary, three years of junior secondary, three years of senior secondary, and four years of tertiary education. This system is popularly known as 6-3-3-4. Under this arrangement, statistics is not being taught as a subject at the primary school level. Before 1960, statistics was virtually unheard of even at the secondary school level. Thereafter, it was taught in secondary schools under the subject of mathematics up to 1973. It was introduced as one of the subjects to be taken in the West African School Certificate Examination (WASCE) in 1974 both at the Ordinary Level (O/L) and Advanced (A/L) (Olubusoye and Shittu, 1998). According to Adamu (1986), in 1979, the ratio of the number of candidates taking

DOI: 10.1201/9781003261148-10

statistics to that of mathematics, for example, was 1 to 94 and these candidates were from 138 of about 2,258 schools that registered for the examination. In 1982, the ratio was 1 to 170. With the introduction of the 6-3-3-4 education system in 1990, statistics was removed from the junior secondary school and senior secondary curricula, and subsequently from the WASCE syllabus.

The University of Ibadan (UI), fondly called the premier university in Nigeria, was founded in 1948 as University College, Ibadan (UCI) with 104 foundation students and three founding faculties (Arts, Science, and Medicine). Initially, statistics was taught in Departments of Mathematics and Economics and it was possible to graduate with B.Sc. statistics if the two other subsidiary subjects were in Science, say Chemistry, Physics, and Mathematics or B.A. statistics if the two other subsidiary subjects were in Arts, say English, Latin or History. The Department of Statistics was statutorily established in 1965 but operated in the Department of Economics. It became fully autonomous and started functioning as an independent department in 1973 (Olubusoye and Shittu, 1998) in the Faculty of Science. In addition to the B.Sc. degree programme in statistics, the department also ran the Professional Diploma in statistics (PDS) programme. Higher degree programmes were introduced in 1977. An unresolved debate that continues today is whether the department should remain a department within the Faculty of Science or be made into an autonomous institute.

Following the brief historical discourse enunciated above, statistical training in the department was confronted with several challenges. These problems underline the attraction of the LISA 2020 initiative in the Department of Statistics. For this chapter, three critical problems are identified. The first is the challenge of maintaining a judicious balance between theory and practice. At the inception of the department, much emphasis was placed on traditional statistics with overwhelming importance placed on theory as the driver of statistical concepts without the essential practical skills needed for using statistics to solve real-life problems. The excitement and pride of both staff and students were in glorifying statistics as a hard discipline and for belonging to a noble field of study. Much attention was given to abstract mathematical concepts such as real analysis, ordinary differential equations, etc., to the extent that students were fraught with being pessimistic about the possible future career and the application areas of the subject. The design of the curriculum failed to show where mathematics ends and where statistics begins.

The second problem is connected with the concept, setup, and operation of a statistical laboratory (stat lab). Even though statistical computing labs existed, it was never clear what role stat labs should play in statistical training. The need for a functional stat lab to support the teaching and learning of statistics has long been advocated. According to Kanji (1974), it is an emerging facility in statistical education. Kanji posits that the size of stat lab should depend on the amount and level of statistics being taught, and the utilisation should be widened to include facilities such as a "Statistical Library," a "Data Bank," the publication of a "Statistical Bulletin," etc. The existence of such a stat lab would provide proper statistical training and a better understanding of the subject, as well as facilitate the correct uses of statistics for users with varied interests. However, despite the existence of the Department of Statistics for almost five decades, the need for a statistical laboratory was not contemplated. Prior to the emergence of University of Ibadan Laboratory for Interdisciplinary Statistical Analysis (UI-LISA), it was a mirage to imagine statistics as a discipline that needs a laboratory in a fashion similar to chemical or biological disciplines. One area of friction was the manner of sharing bench fees among the departments with specialist laboratory or fieldwork. Bench fees are payments made by the university to departments to maintain facilities in the laboratory and to cover consumables. While several departments in the Faculty of Science automatically benefit from these payments, the statistics department was painfully exempted due to lack of a functional stat lab.

The third but very notable problem was the poor motivation or rather lack of incentive for interdisciplinary or collaborative research in the university system. The issue of single, co-authorship and multiple-authorship have been debated variously in the literature (Vafeas, 2010, Woods, 1998, Woods et al., 2010, Lei et al., 2016, Awe and Vance, 2014). The university promotion guidelines clearly reward single authors compared with multiple authors. For instance, a journal article attracts a maximum of five points in the promotion guideline. A single-authored article could get 100% of the maximum points but only the first author can obtain about 50% of the points where there are more than five authors. Consequently, four single-authored journal articles can get a candidate seeking promotion to a cadre requiring a minimum of 20 points, while twice that number is required if each of the journal articles has two authors. Faculty members are mostly interested in solo rather than interdisciplinary research which is the strength of any research. Modern-day realities put collaborative research at the epicentre of societal problem-solving.

Following the above, this chapter plans to discuss the emergence of a LISA stat lab in Nigeria's Premier University, University of Ibadan, popularly called UI. The discourse is intended to cover the activities directed at addressing the critical challenges relating to teaching, research and collaboration in the Department of Statistics, UI. The rest of the chapter is organised as follows: Section 8.2 discusses the birth of the UI-LISA; Section 8.3 discusses the lessons drawn from two renowned and well- established stat labs in the United States which helped to build a solid foundation for UI-LISA; Section 8.4 elaborates on UI-LISA advocacy programmes, training activities and collaborations; and, finally, Section 8.5 presents concluding remarks for emerging labs.

8.2 The Birth of UI-LISA

The vision to create twenty stat labs in developing countries by 2020 has been vigorously pursued by Prof. Eric Vance since 2012. The concept and *modus operandi* of LISA 2020 is uniquely and carefully constructed to build the capacity of statisticians and non-statisticians on essential skills needed to enhance effective collaboration. The programmes of LISA 2020 are specially tailored to build statistics and data science capacity and research infrastructure in developing countries to help statisticians collaborate with scientists, government officials, businesses, and NGOs to use data to solve problems and make decisions for real-world impact. The programme was designed to pursue three missions which include (1) training statistics and/or mathematics students and staff to become effective, interdisciplinary collaborators who can move between theory and practice to solve problems for real-world impact; (2) serve as a research infrastructure for researchers and decision-makers at the university or in the surrounding community to collaborate with statisticians to enable and accelerate research and data-based decision-making that will have a positive impact on society; and (3) to teach short courses and workshops to improve statistical skills and literacy widely. All these were highlighted on January 16, 2015, in a paper titled "LISA 2020: Creating a Network of Statistical Collaboration Laboratories" presented by Professor Eric Vance during his first visit to Nigeria to create a network of stat labs that will connect with LISA 2020.

The historic journey to the establishment of LISA in Nigeria started in May 2013 with the selection of Dr O. Olawale Awe, then a doctoral student in the Department of Statistics, University of Ibadan, as the first LISA Fellow to the Department of Statistics, Virginia Tech (VT). Visiting LISA Fellows are made to undergo intensive practical training under the supervision of the LISA 2020 Director and thereafter they return to their home countries to start a statistical collaboration laboratory.

Upon his return in 2014, the first LISA Fellow introduced the LISA 2020 concept to UI but opted to establish the Laboratory for Interdisciplinary Statistical Analysis and Collaboration (LISAC) in October 2014 at Obafemi Awolowo University, where he was employed as a lecturer. By December 2014, the second stat lab was created at Sekoine University of Agriculture (SUALISA), Morogoro, Tanzania (Vance and Magayane, 2014). The birth of UI-LISA followed suit in March 2015 and became officially recognised as a member of the LISA 2020 Network in June 2015.

Constant monitoring and follow-up by the founder of LISA 2020 have helped to sustain UI-LISA and position it to continuously serve as an engine for development through training and collaboration. There were two commissioned visits to UI-LISA to help strengthen its operations and to conduct an assessment of its activities. There was also the exchange visit extended to the lab coordinator to visit two prestigious universities in the United States, VT and Purdue University, to see first-hand how LISA 2020 trains students to communicate and collaborate with non-statisticians and how to manage labs to help clients apply statistics in their research projects. The unforgettable visit by Mr. Ian Crandell, a doctoral student from VT to UI-LISA took place between March 28 and April 17, 2015. Mr. Crandell helped UI-LISA develop a programme similar to LISA at VT and in line with the LISA 2020 programme. During his visit, he made a significant contribution to the setting up of UI-LISA as a fully functional statistical collaboration centre. He spent most of his time developing short courses for basic statistics and statistical software packages. He was instrumental in organising a series of collaboration meetings with graduate students and statisticians in UI. He actively participated in programme design & orientation and provided a series of lectures and meetings at the Faculty and Centre for Petroleum Energy Economics and Law (CPEEL). The memory of his visit and contributions still lingers in the minds of UI-LISA collaborators.

The second visit commissioned by the founder of LISA 2020 during January 2018 was by Ms. Monica Johnston; an expert in statistics education and assessment with expertise and experience working with women statisticians and small business owners in the United States. She is a prominent and active member of the ASA Section on Statistical Consulting. She had the mandate of the founder to conduct an assessment of UI-LISA and to identify some strengths of UI-LISA including potential opportunities to make UI-LISA even stronger and more sustainable. She also shared some ideas that may be useful in future grant proposals to be able to strengthen UI-LISA and the other LISA 2020 stat labs in the future. Finally, she documented some best practices that could be valuable to share with other stat labs in the LISA 2020 Network. Apart from her in-depth assessment and evaluation of UI-LISA activities since inception, she engaged in several other capacity-building and statistical education activities. She had exciting moments with the women in statistics and inaugurated the Women in Statistics Wing of the Department of Statistics, University of Ibadan. She made a public seminar presentation titled "Improving Student Learning; within a Small Business Setting". The lecture was indeed an eye-opener to young graduates of statistics that they could start their statistical practice upon graduation without having to take a government job. She encouraged the early career statisticians regarding the importance of joining a professional body and spreading tentacles across the globe. The hallmark of her visit was the facilitation of one of the flagships of the UI-LISA programme tagged "One hour with a statistician" with a topic *Statistical Method: Communication*, which attracted 87 attendees from three faculties across the university community.

An extract from the executive summary of the report is presented in the textbox below. The report underlines the importance of programme metrics and programme evaluation metrics which regrettably were found missing at UI-LISA. The UI-LISA programme evaluation is presented in Table 8.1. The report assesses UI-LISA programmes and how consistent they are with the stated mission and goals.

ACCORDING TO MS. MONICA JOHNSTON:

Overall, UI-LISA is doing the work of a statistical lab, but it has not demonstrated that it can sustain itself because it has not begun to develop the necessary infrastructure to quantify its success. The key to obtaining funding lies in its ability to quantify and reports its success.

However, the UI-LISA team was highly receptive to constructive feedback and began to make changes immediately in using software that would more readily allow them to quantify the success that they're experiencing. There is a culture of eagerness to learn and systematise their processes in accordance with the LISA Network reporting standards.

Recommendations are that UI-LISA be provided with

1. Clear and direct guidance on building the infrastructure necessary to sustain a statistical laboratory, including advising on software selection, metric development and reporting. Furthermore, that advice should be consistent among the LISA Network members.

2. Financial assistance to motivate students to participate as student collaborators. Students and faculty are deprived of steady electricity, school-based Wi-Fi and basic materials and, given the sustained national economic inequality and deprivation, the spirit of volunteerism has not flourished.

3. A stronger incentive for lab coordinators, one that guarantees the role is independent of the faculty appointment, by re-examining the MOU process the LISA Network uses with universities.

The overall mission of UI-LISA is "Building statistics and Data Science Capacity in Nigeria." The goals are to (1) complement statistical training in the department by developing the non-technical skills of our students to be able to relate with domain experts in other fields; (2) make statistics graduates to be employable in industry or be self-employed through practical experience; (3) promote statistical literacy within and outside the university community; and, (4) use statistics to transform our society by identifying and solving local problems. The UI-LISA programme evaluation is presented in Table 8.1. The Drop-In and Mobile Statistical Clinic (MSC) programmes are found not to be consistent with the goals in terms of metrics.

TABLE 8.1

Summary of Monica Johnston's Assessment of UI-LISA Programmes[a] in January 2017

	Drop-In	Collaborative Training Workshops	Short Courses	Student Training	IT Training	One Hour w/a Statistician	Mobile Statistical Clinic
Consistent w/ mission	Yes	Yes	Yes	Yes	Yes	Yes	Yes
Consistent with goals	No, only 2 of 450 participated	Yes	Yes	Yes	Yes	Yes	No
Programme metrics are in the analysis-ready form	No (paper)	No (Word)	No	No (Word)	No	No (Paper)	No
Evaluation metrics are available	No	Yes, Survey Monkey	No	No	No	No	No
Total # of attendees	84	I: 9 II: 22 III: 15	Unknown	Formal training: Not available Informal: 80+ students tutored (drop-in)	13 trained; upcoming training: 15	Unknown	Unknown

[a] These programmes are discussed in Section 8.4.

The drop-in clients are coming from just 1 to 2 departments, thus creating the need to expand and reach out to other departments. The MSC is least successful probably because it requires more preparation and is of lower priority to the lab. Nearly all the programmes failed in programme metrics and evaluation metrics because the data about them were available mainly in "Word" or "Paper" formats rather than in format such as spreadsheet which could be readily analysed. No programmes, except for the Collaborative Training Workshops, have been formally evaluated using modern survey and analysis software. Emerging start labs must realise very early that programme evaluation is key to lab sustainability and quality service delivery. This has been a very hard lesson for UI-LISA.

The hallmark of the efforts to put UI-LISA on a foothold and strengthen its operations was the exchange visit by the UI-LISA Coordinator in February 2016, to the then headquarters of LISA 2020 at VT in the United States. The essence of the visit was to gain inspiration and ideas needed to continue to stir and manage the UI-LISA stat lab to fulfil its mission of training collaborative statisticians and helping the entire UI researchers (staff and students) apply statistics in their research. Also facilitated was a visit to Purdue University located in Lafayette, Indiana, United States. The abridged version of the report of the visit as presented to the Vice-Chancellor of the University of Ibadan is reproduced in the following section for the benefit of emerging laboratories.

The hallmark of the efforts to put UI-LISA on a foothold and strengthen its operations was the exchange visit by UI-LISA Coordinator to the then headquarters of LISA 2020 at VT in the United States between February 12 and March 16, 2016. The essence of the visit was to gain inspiration and ideas needed to continue to stir and manage UI-LISA stat lab to fulfil its mission of training collaborative statisticians and helping the entire UI researchers (staff and students) apply statistics in their research. Also facilitated was a visit to Purdue University located at Lafayette, Indiana, United States. The abridged version of the report of the visit as presented to the Vice-Chancellor of the University of Ibadan is reproduced in the following section for the benefit of emerging laboratories.

8.3 UI-LISA Setup and Programmes

The overriding objective of the LISA 2020 Programme is to build statistics and data science capacity in developing countries by creating a robust network of statistics and data science collaboration laboratories. It is expected that these stat labs will serve as engines for development by training the next generation of collaborative statisticians and data scientists; serving as infrastructure to support local researchers, businesses, governments, and NGOs; and teaching short courses and workshops to improve statistical skills and data literacy widely. The LISA 2020 Network encourages stat labs to improve their operations by adopting the best practices learned from one another. Each stat lab is allowed to operate based on the distinctive characters of its environment and the established culture of its institution. The freedom enjoyed by network members allows each stat lab to creatively formulate its vision and mission statements, including developing programmes suitable to its host institution.

Consistent with the LISA Network mission, UI-LISA puts UI students at the centre of its mission statement with the sole aim of building their capacity in data science and extending the same to statistics students in higher education institutions in Nigeria. Consequently, the lab programmes and activities are designed to support this mission and goals enumerated. Thus, UI-LISA has seven (7) mission-driven programmes, namely LISA drop-in assistance, Collaborative Training, Short Courses, Student Training (formal and 1-1 tutoring), Industrial Training (IT) for students completing their academic degree, One Hour with a Statistician and the Mobile Statistical Clinic. Two programmes serve the students while the other five serve the entire university community, including statistics students and faculty, students and faculty in other departments, non-academic staff and neighbouring universities.

The UI-LISA lab has been operating since its inception in a standard office space provided by the Department of Statistics, which can accommodate about twenty people at a time. Laboratory equipment, such as a laptop, projector and printer, were provided by the department at take-off. The lab enjoys limited access to departmental facilities, such as computing labs, lecture rooms, and administrative assistance. Several other lab infrastructures are lacking or inadequate which include a camera, statistical kits, whiteboard, flip chart board, furniture, and internet access.

UI-LISA administrative structure is as follows:

- **Lab Coordinator** who oversees the lab and implements the LISA 2020 agenda. The Lab Coordinator decides who works on a project based on interest and areas of expertise.

- The **Deputy Coordinator** assists the Lab Coordinator in implementing UI-LISA Programmes.

- The **Senior Lab Collaborators** are in charge of training and design training modules. Usually, the Senior Collaborators are assigned to work with the expert faculty member to work on a project. All the Senior Collaborators are PhD students in the department. Due to lack of remuneration or compensation to lab collaborators, the level of commitment is very low.

- The **Lab Administrative Officer** keeps lab records, compiles programme metrics and conducts programme evaluation.

- The **Graduate Collaborators** assists the Senior Collaborators and carries out other duties assigned by the Lab Coordinator.

- The **Faculty Members** provide consulting services to clients in their areas of expertise.

In particular, the Administrative Officer and the Collaborators are motivated by funds that are generated by the lab.

UI-LISA has two categories of student collaborators. The first category comprises students in the industrial training (IT) programme (the second article in this chapter addresses this). Usually, statistics students (from UI and other neighbouring institutions) who are still undergoing their degree programmes apply to the lab as trainees. The second category comprises graduates (of statistics and other disciplines) who want to gain practical experience to enhance their job opportunities. All UI-LISA collaborators must go through software training and the rigorous application of statistical methods to real-life problems. The primary programming software for all collaborators is R, and they must be very proficient in its use for data visualisation, descriptive and inferential analysis. The undergraduate and postgraduate trainees are fully involved in the weekly statistics practical sessions facilitated by the Senior Collaborators. All IT students and graduate collaborators are assigned alongside Senior Collaborators to work on clients' projects.

In the following sub-sections, the programmes designed for the lab trainees and for promoting statistical literacy university-wide are discussed.

8.3.1 One Hour with a Statistician

This One hour with a Statistician programme is aimed at improving the level of statistical literacy among students and users of statistics at the university. During the one-hour meeting, an invited statistical expert briefly gives a discourse on a popular statistical concept including its uses and misapplications. This is followed by a constructive engagement in the form of questions and answers from the audience. The one hour with a statistician provides a platform for enlightening the university researchers and for correcting abuses and misuses of statistical techniques and their interpretations. The programme is now very popular with growing participation among staff and students from departments within and outside the Faculty of Science. Some of the subjects covered by eminent facilitators in the previous editions of the programme include

- Concepts and interpretation of the p-value
- Regression analysis: strength and abuse
- Multivariate methods in statistics
- Hypothesis testing
- Sampling and sample size determination
- Exploratory data analysis
- Statistical modelling: its strength and abuse
- Understanding statistical design and analysis of experiments
- Time series analysis: its misconceptions, abuses and strength in scientific research
- Questionnaire design
- Regression analysis and its interpretation

8.3.2 Mobile Statistical Clinic

The Mobile Statistical Clinic programme is aimed at bringing statistical education to the doorstep of university researchers and also publicising the activities of the laboratory to potential beneficiaries. Mobile clinic means that UI-LISA team move out and station at public locations such as halls of residence, conference/meeting areas, lecture theatre, parks, event centres and even recreational areas to provide on-the-spot solutions to statistical problems and enquiries. At the moment, the Mobile Statistical Clinic team consists of five lab collaborators led by the Senior Lab Collaborators. For problems that cannot be solved on-the-spot, clients are referred to the laboratory for further help.

8.4 Short Courses and Collaborative Training Workshops

Training workshops at UI-LISA are generally classified into two. The first is the workshop that is targeted at sundry users of statistics, specifically researchers in any discipline who are interested in the theme of the workshop. In recognition of the fact that disciplines using statistical analysis in their research have peculiar training needs, the UI-LISA team conducts specialised training workshops. Thus, the second is collaborative training organised in partnership with non-statistics–based departments, centres, and faculties focusing on statistical methodologies tailored to the needs of their researchers and students. In the past, Collaborative Training Workshops have been organised with the Faculty of Science, Faculty of Veterinary Medicine, CPEEL, Department of Crop Protection and Environmental Biology, and Department of Zoology.

8.4.1 Collaborative Projects

UI-LISA has provided statistical advice to 65 research projects to date including 18 PhD theses, 46 masters' dissertations, and 1 NGO project. In collaboration with CPEEL, UI-LISA participated in a university-wide energy audit project commissioned by the Vice-Chancellor of UI in 2018. Indeed, the most productive collaboration UI-LISA has had is with CPEEL (https://cpeel.ui.edu.ng/). The collaboration has so far yielded two journal articles on renewable energy in Africa (see Olanrewaju et al., 2019 and Akintande et al., 2020). The lab has just completed a project on *Enhancing Election Participation in Nigeria* in collaboration with The Electoral Institute (TEI), a research and documentation unit of the Independent National Election Commission (INEC). The funding is provided by the LISA at the University of Colorado Boulder in cooperation with the US Agency for International Development (USAID) Accelerating Local Potential (ALP) Programme (Cooperative Agreement Number: 7200AA18CA00022). The project has produced three policy briefs and three research working papers. Additionally, the lab is currently engaged in a sanitation project funded by the USAID TEACH Fund in collaboration with the Ministry of Environment and Natural Resources, Oyo State, Nigeria.

8.5 Conclusion

The laboratory presently enjoys a good reputation and patronage among the university community. The environment of operation is becoming friendlier and more cooperative than when the laboratory was launched in 2015. In recent times, academic staff in the Faculty of Science (comprising 10 departments) have been directing their postgraduate students to the laboratory for statistical advice. Several applications are now received from those who have completed their degrees for industrial training experience in the laboratory. All these developments have created a huge window of opportunities for statistical capacity building, collaborations, statistical education and outreach efforts.

Overall, UI-LISA is doing the work of a statistical lab, and despite many challenges (such as limited funding) UI-LISA has demonstrated remarkable resilience in striving to achieve its vision and mission. The lab is developing the necessary infrastructure to quantify its success which is key to obtaining funding to sustain its activities. Emerging stat labs need to have clear and direct guidance on building the infrastructure necessary to sustain a statistical laboratory, including advising on software selection, metric development, and reporting. Also, it is pertinent to consider giving financial assistance to motivate students to participate as student collaborators and the spirit of volunteerism must equally be encouraged to ensure lab sustainability.

Appendix

Lessons from the Fountainhead

The visits to VT and Purdue Universities were arranged to come after the Conference on Statistical Practice (CSP) organised by the American Statistical Association (ASA) held in San Diego, CA from February 17 to 21, 2016. The conference was indeed a great learning experience for UI-LISA and it provided a good foundation for the sustainability of the lab.

Lessons from the 2016 Conference on Statistical Practice (CSP) in San Diego

This was the 5th ASA Conference on Statistical Practice. It was held at San Diego Westin Hotel. The conference brought together hundreds of applied statisticians and data scientists to focus and engage in discussions on innovations and best practices for the applied statistician. The presentations covered various issues in statistical applications including learning about new statistical techniques and best practices, how to better communicate with clients and colleagues, and how to have a positive impact on our organisations. The conference had the following four concurrent sessions:

i. Communication Impact, and Career Development

ii. Data Modelling and Analysis

iii. Big Data Prediction and Analytics

iv. Programming and Graphic

For obvious reasons, UI-LISA Coordinator attended all the presentations in Communication Impact and Career Development session. The presentations under this theme helped the participants to develop new skills and perspectives towards being effective as a statistician in performing one's role as a leader, strategist, consultant, and collaborator. Precisely, this conference provided new ideas, techniques, and strategies to communicate effectively in a way to make the lab record a greater impact on the university. Some of the lessons learnt include

i. Business ideas for any statistics student to start his/her career as an independent statistical consultant after graduation;

ii. Strategy to move from being a statistical consultant to a trusted adviser who can communicate better with clients and customers to have a positive impact on their projects and research;

iii. The seven habits of a highly effective statistical consultant;

iv. How to use video to improve statistical practise;

v. Common pitfalls and misconceptions in statistics with suggested solutions;

vi. Techniques based on ways to "Just say No!" to ethically, scientific or legally unjustified procedures and analyses;

vii. Mentoring and influencing using motivational interviewing; and

viii. Recognising acumen as a critical skill for all statisticians.

Lessons from Exchange Visit to Virginia Tech (VT) LISA

The visit to VT provided a lot of insights into the style of operation, funding, level of staff and students' participation in VT LISA. The programme of activities during the visit included:

i. Attending Path Meetings, Collaborators' Meetings, and Video Review Meetings with members of VT LISA;

ii. Giving a Departmental Colloquium Seminar presentation titled: *Sustaining a Virile Statistical Laboratory in Nigeria's Premier University*;

iii. Attending the classes on Statistical Consulting course taught by Prof. Eric Vance;

iv. Having private meetings with some academic staff and postgraduate students in the department; and

v. Creating a website for UI-LISA with the assistance of VT LISA's Administrative Staff.

In VT LISA, statistical assistance was provided free of charge to faculty members, staff, and students. The reason for this is that VT LISA was funded jointly by the Office of the Vice President of Research, the College of Science, the Graduate School, the Office of the Provost and all seven other colleges (Agriculture and Life sciences, Architecture and Urban Studies, Engineering, Liberal Arts and Human Sciences, Natural Resources and Environment, the Pamplin College of Business, and the Virginia-Maryland Regional College of Veterinary Medicine) in the institution. The Department of Statistics also provided funding for many of the LISA statistical collaborators and provides other support for LISA's activities.

Users of LISA engaged in sponsored research benefitted from in-depth help and were encouraged to include statistical collaboration in grant proposals. Usually, this took the form of a full or partial graduate research assistantship, partial funding of a faculty member's salary, or a direct-cost line item. VT LISA provided occasional statistical consultation and collaboration on projects outside of VT for a fee. Through statistics in the Community (StatCom), students in the Department of Statistics also provided pro-bono statistical consultation and collaboration for researchers studying topics of local interest and for local community non-profits, schools, and government organisations. In summary, the following lessons were learnt from VT LISA.

i. UI-LISA can emulate VT LISA by supporting UI researchers to benefit from the use of statistics in designing experiments, analysing data, interpreting results, writing grant proposals and using statistical software.

ii. UI-LISA should explore the possibility of securing direct funding support from the UI or units such as Postgraduate College so that statistical services including short courses can be provided free of charge to faculty members and student researchers.

iii. UI-LISA should devise a means of motivating and attracting staff and students to seek statistical collaboration in their projects and studies to accelerate high-quality research outputs and publications in top-rated journals.

iv. UI-LISA should be given the university's recognition and be empowered to provide out-of-classroom training in statistics to our students at all levels and prepare them for a professional career in statistical consulting.

Connecting UI-LISA with the Center for Open Science (COS)

Coincidentally, the visit came up during the Research Week (Monday–Wednesday, February 22–24, 2016) at VT. During this period, April Clyburne-Sherin, the Reproducible Research Evangelist from the Center for Open Science (COS) also visited VT. The Center is a non-profit organisation working to improve the inclusiveness, transparency, and reproducibility of scientific research globally. The Center provides free services to researchers and research institutions, including free infrastructure, free training and new incentives. UI-LISA engagement with the Center emphasised providing free infrastructure for implementing reproducible research and training to the researchers in UI. Some resources including CDs to train and educate researchers about reproducibility research were provided to UI-LISA by the COS Evangelist.

Lessons from the Exchange Visit to Purdue University

The visit to the Statistical Consulting Service (SCS) unit in the Department of Statistics, Purdue University, under the able leadership of Prof. Bruce A. Craig contributed in no small measure to the UI-LISA success story. The programme of activities during the 1-week visit included

i. Participating in the SCS Weekly meeting;

ii. Having lunch meetings with key members of the department including Head of Department, Prof. Hao Zhang; Prof. George McCabe, Associate Dean of the College of Science; Assistant Prof. Arman Sabbaghi, Associate Director of the SCS; Prof. Tom Kuczek, past Associate Director of the SCS; Prof. Kiseop Lee, a Visiting Professor of statistics and Ce-Ce Furtner, the Manager of the SCS;

iii. Participating as an observer in seven Initial Meetings (IMs) of SCS with clients;

iv. Attending Prof. Craig's Statistical Consulting (STAT 582) classes;

v. Giving a colloquium on *Sustaining a Virile Statistical Laboratory in Nigeria's Premier University*; and

vi. Attending a summit on Service-Learning and Engagement.

The Statistical Consulting Service (SCS) provides the university community with free advice on problems involving the use of statistics. The service consists of two parts. For individuals needing assistance with computer software, a drop-in service is provided during normal working hours. Problems involving the design of experiments, the statistical analysis of data, and the interpretation of results are handled via the booking of appointments. SCS staff includes faculty members from the statistics department, graduate students, and support staff. SCS staff members always strive to develop a working relationship with the clients and become involved in the process whereby research ideas are formulated, translated into the framework of a statistical model and investigated in the context of such a model. The emphases of the university are discovery, learning and engagement. The three functions are addressed by SCS staff by providing statistical consulting services. The service provided has a direct impact on the quality of research performed while also giving the graduate students who work as consultants a very valuable learning experience by working with real problems.

The following lessons were learnt from the structure and operations of Purdue's SCS:

i. A line budget from the university provides a steady source of funding for the SCS, which enables it to offer its services free of charge to the members of the university community.

ii. Staff involved in SCS are motivated with a reduction in their teaching workload to have sufficient time to attend to clients.

iii. If the client is a student, the major professor or a member of the client's degree committee must attend the initial meeting and any formal follow-up meetings.

iv. Researchers seem to have cultivated the habit of involving the statistical design consultants from the formulation and design stages of the projects.

v. Consultants work directly with the clients to determine the appropriate analysis, recommend software procedures, and help to interpret results.

The lessons and the connections arising from these visits have been helpful in the following ways:

- Restructuring of UI-LISA for excellent service delivery similar to those visited;

- Mobilising university staff and postgraduate students to seek statistical advice and counselling at UI-LISA and gradually build their confidence and trust;

- Promoting statistical collaboration with non-statisticians by following them through the various stages of their research from study design to reporting writing;

- Training UI-LISA student collaborators to acquire skills and know-how to communicate effectively and prepare them for a career as statistical consultants as part of the effort to make them job creators rather than job seekers; and

- Continuing to pursue and securing stable funding support and infrastructure needed to deliver free consulting services to UI researchers.

The lessons drawn from both the commissioned and the exchange visits gave impetus to UI-LISA to design unique programmes for its operations.

References

Adamu, S. O. (1986). Training of statisticians in the last three decades. *An Invited Paper Presented at the 10th Anniversary of the Conference of the Nigerian Statistical Association*, September 15th–18th, 1986, Conference Centre, University of Ibadan, Ibadan: Evans Brothers (Nigeria Publishers) Ltd.

Akintande, O. J., Olubusoye, O. E., Adenikinju, A. F., & Olanrewaju, B. T. (2020). Modelling the determinants of renewable energy consumption: Evidence from the five most populous nations in Africa. *Energy*:117992.

Awe, O. O. & Vance, E. A. (2014). Statistics education, collaborative research, and LISA 2020: A view from Nigeria. *Proceedings of the Ninth International Conference on Teaching Statistics* (ICOTS9, July 2014), Flagstaff, Arizona, USA.

Kanji, G. K. (1974). The role of the statistical laboratory in the teaching of statistics. *International Journal of Mathematical Education in Science and Technology*, 5(1):53–57, DOI:10.1080/0020739740050107.

Leiva, M., Carrera, E., Bottai, H., Contini, L., & Vaira, S. (1999). *Estadística en la formación de grado y postgrado en las Ciencias Biomédicas*. Mendoza, Argentina: CLATSE V (V Congreso Latinoamericano de Sociedades de Estadística).

Olanrewaju, B. T., Olubusoye, O. E., Adenikinju, A., & Akintande, O. J. (2019). A panel data analysis of renewable energy consumption in Africa. *Renewable Energy*, 140:668–679.

Olubusoye, O. E. and Shittu, O. I. (1998). A diagnostic review of statistics education in Nigeria. *Proceedings of Annual Conference of the Nigerian Statistical Association (NSA)*, pp. 25–33.

Vafeas, N. (2010). Determinants of single authorship. *EuroMed Journal of Business*, 5(3):332–344.

Vance, E. A. & Magayane, F. (2014). Strengthening the capacity of agricultural researchers through a network of statistical collaboration laboratories. *Proceedings of the 4th RUFORUM Biennial Conference*, 21st–25th July 2014, Maputo, Mozambique.

Woods, R. H. (1998). Single vs. co-authored and multiple-authored articles: The views of CHRIE educators on their value. *Journal of Hospitality and Tourism Education*, 10(1):53–56.

Woods, R. H., Youn, H., & Johanson, M. M. (2010). Single vs. co-authored and multi- authored research articles: Evaluating the views and opinions of ICHRIE Scholars. *International CHRIE Conference-Refereed Track*. 11. https://scholarworks.umass.edu/refereed/CHRIE_2010/Saturday/11.

9

The Complementary Role of UI-LISA in Statistical Training and Capacity Building at the University of Ibadan, Nigeria

Adedayo A. Adepoju, Olusanya E. Olubusoye, Tayo P. Ogundunmade, Tomiwa. T. Oyelakin, and Bbosa Robert
University of Ibadan Laboratory for Interdisciplinary Statistical Analysis [UI-LISA]

CONTENTS

9.1 Introduction

Statistical education in the 21st century has continued to receive increasing attention by researchers and vigorous advocacy by professional statistical associations (Tishkovskaya and Lancaster, 2012). Evidence from Bjornsdottir and Garfield (2009) shows over 150 articles and book chapters that were published on statistics education in 2007 alone. Leading the vanguard at promoting statistical literacy, improving teaching and learning of statistics, developing innovative pedagogical instructions, educational technologies, and producing abundant web resources are the International Association of Statistics Education (IASE) and the International Statistical Institute (ISI). The former is the publisher of one of the frontline peer-reviewed electronic journals called *Statistics Education Research Journal* (SERJ). While these efforts are going on at the global level, the advent of the LISA 2020 Network has brought another dimension to accelerating and transforming statistical education. With a clear focus directed mainly on developing countries, the network has intensely promoted the training of statisticians and data scientist toward solving real- life problems.

What is the status of statistical education in Nigeria? This is a fundamental question which should interest the statistical community. Hitherto, statistical education appears to have been given inadequate attention and support in developing countries, particularly Nigeria. Before Nigeria got independence in 1960, statistics education was relatively unknown in both secondary and tertiary institutions. Statistics was only included as a topic in the mathematics syllabus and thus taught as part of mathematics in secondary

DOI: 10.1201/9781003261148-11

schools up to 1973. Statistical education appeared to have gained some prospect in 1974 with the introduction of statistics as a subject in the West African School Certificate (WASC) Examination both at the Ordinary Level (O/L) and Advanced Level (A/L). Regrettably, in 1982, a decline in the number of candidates offering statistics led to the removal of statistics from Junior Secondary School and Senior Secondary School curricula and subsequently from the WASC syllabus in 1990 (Adamu, 1986). Since then, the earliest major encounter with statistical learning in Nigeria is at the tertiary institution level.

In a strategic effort towards meeting its middle-level statistical manpower needs, the Federal Office of Statistics (FOS) established the Federal Office of Statistics Training School in 1961 under the United Nations Technical Assistance programme. The school was designed to impart knowledge of elementary statistical methods data collection, processing, presentation and analysis of official socio-economic statistics. Two additional campuses of the Federal Office of Statistics Training School were built in Kaduna in 1978 for the Northern zone and another in Enugu in 1991 for the Eastern zone. Consequently, the name of the school was changed from Federal Office of Statistics Training School to Federal School of Statistics (FSS) in 1996. The school awards a National Diploma certificate in statistics to its graduates. Thus, it became the first post-secondary training for statistics in Nigeria.

In Nigeria today at the tertiary level, a rapid growth has been recorded in the number of schools (both private and public) offering statistics as a course. Apart from the FSS, statistical training exists in different institutions like Colleges of Education, Polytechnics and Universities. The foundation of statistical education in Nigeria at the university level began with the creation of the Department of Statistics, University of Ibadan (UI) in 1965. The department did not start functioning until 1973 thus becoming the oldest Department of Statistics in Nigeria. The department runs several programmes such as Postgraduate Diploma in statistics, Bachelor, Master, and Doctorate Degrees, and Distance Learning programmes. Amongst its numerous achievements, the department produced the first "made in Nigeria" PhD in statistics in 1984 and its pioneering staff published the first two textbooks on statistics ever to be written by Africans, Afonja (1975) and Adamu and Johnson (1983). The department is recognised by the ECA (Economic Commission for Africa) as one of the Centres of the Statistical Training Programme in Africa (STPA).

Higher education in Nigeria is under strict regulatory supervision of the National Universities Commission (NUC), a parastatal under the Federal Ministry of Education (FME). Amongst other functions, the Commission is responsible for granting approval for all academic programmes, establishment of higher educational institutions offering degree programmes run in Nigerian universities and creating a channel for external support to Nigerian universities: It follows that all universities running statistics degree programmes in Nigeria as well as the curriculum used must comply with NUC guidelines. In exercising its regulatory roles, the Commission developed the Minimum Academic Standard (MAS) for all the programmes taught in Nigerian universities in 1989. After more than a decade of using the MAS, the Commission in 2014 produced a revised version of the document called the Benchmark Minimum Academic Standard (BMAS). The BMAS is designed to make graduates from Nigerian universities in various disciplines including statistics acquire appropriate skills and competencies that will make them globally competitive and capable of contributing meaningfully to Nigeria's socio-economic development.

Sequel to the above, this chapter presents the efforts of the UI Laboratory for Interdisciplinary Statistical Analysis (UI-LISA) to fill a major gap in the implementation of the BMAS for statistics programme in Nigerian universities. The rest of this chapter is divided into the following four parts. Section 9.2 presents the philosophy and summary of the content of the BMAS for statistics programme and compares it with the curriculum being implemented for Bachelor in Statistics at the UI. In Section 9.3, the Students Industrial Work Experience Scheme (SIWES) and the challenges faced by the students are discussed. Section 9.4 discusses the internship training at UI-LISA and the complementary roles being played. Finally, Section 9.5 gives the feedback from UI-LISA.

9.2 Benchmark Minimum Academic Standard (BMAS) for Statistics Programme

This new version referred to as the BMAS is what is being used in all the Nigerian institutions till today. The BMAS document is expected to serve as a guide to the universities in the design of curricula for their programmes in terms of the minimum acceptable standards of input, process and measurable benchmark of knowledge, skills and competences expected to be acquired by an average graduate of each of the academic programmes. The B.Sc. Statistics is a full-time course at the university. A full-time student is required to register for a minimum of 20 units in a given semester, as stipulated by the university. Basic courses in the first year comprise of Statistics, Computer Science, Physics, Economics and General Studies. The BMAS learning outcomes for statistics programme are

- **Regime of subject knowledge**: To develop cognitive abilities and skills relating to statistics,
- **Competencies and skills**: To demonstrate practical skills relating to the solution of statistical problems and their applications,
- **Behavioural attitudes**: To demonstrate general skills relating to non-subject-specific competencies, ICT capability, communication skills, interpersonal skills and organisation skills.

The BMAS course structure covered Probability and Statistics, Applied Statistics, Statistical Inference, Mathematics – Calculus, Algebra and Real Analysis. The attainment level for BMAS stipulates that graduates of statistics are expected to have the ability to apply knowledge and skills to solving theoretical and practical problems in statistics and other related areas in relation to national and societal needs.

Though the NUC BMAS ought to serve as a guide to developing a curriculum for the statistics programme, in most cases, the benchmark minimum standard is barely surpassed. Thus, NUC periodically embarks on the accreditation of academic programmes in Nigerian universities. In the last 30 years, precisely between 1990 and 2020, the statistics programme at UI has received full accreditation except between 1990 and 2000 when the accreditation status was interim. The present full accreditation status is expected to last until 2024.

Table 9.1 compares the curriculum for the undergraduate statistics programme at the UI with the approved NUC benchmark.

This comparison reveals that the minimum standards are largely achieved except in the SIWES which represents Industrial Attachment course in UI curriculum. According to BMAS guidelines, a student should be attached to a relevant organisation for a cumulative period of 24 weeks to be taken equally at the end of the second and third years and participation should earn the student 6 credit units. However, at UI, the student is attached to a relevant organisation for only 12 weeks. Two problems are associated with

TABLE 9.1

UI Undergraduate Statistics Curriculum and BMAS for Statistics Programme

S/N	Requirements	BMAS		UI Curriculum	
		3-year Programme	4-year Programme	3-year Programme	4-year Programme
1	Admission requirements	Candidates with at least two A-level passes (graded A–E) at the GCE/IJMB Advanced Level in relevant subjects (Mathematics, Physics and Further-Mathematics) may be admitted into 200-Level.	At least credit level passes in five subjects including English Language, Mathematics and Physics to form the core subjects and any other two relevant science subjects at the Senior Secondary School Certificate or its equivalent. In addition, an acceptable pass in the Unified Tertiary Matriculation Examination (UTME) with the appropriate subject combination is required for admission into 100-Level.	Candidates with at least two A-Level passes (graded A–E) subjects to include Mathematics or University of Ibadan Professional Diploma in Statistics (PDS) plus five 'O' Level credit passes including English Language and Mathematics	Candidates with five credit passes including English Language, Mathematics and any other two science subjects. The approved science subjects are Agriculture Science, Biology, Chemistry, Geography, Geology, Mathematics, (Pure and Applied), Further Mathematics, Physics, Statistics and Zoology.
2	Minimum Duration	3 years	4 years	3 years	4 years
3	Graduation Requirements	90 units	120 units	90 units	120 units
4	SIWES	• **Minimum Credit**: 6 units • **Duration:** 6 months (24 Weeks) • **Level:** Year 2 and 3 • **Status:** Compulsory		• 3 units • 3 months (12 weeks) • Year 3 only • Compulsory	

5	Resource Requirement:
	• **Personnel**
	• **Physical facilities**
	• **Library and Information Resources**
		• Ratio 1:20 of academic staff to students,
		  • at least six academic staff,
		  • 70% of them must have Ph.D.
		• At least two lecture rooms capable of sitting at least 60 students
		• Library for at least 25% of total students
				• Ratio 1:15 of academic staff to students,
				  • Twenty-one academic staff,
				  • 86% have Ph.D.
				• Three lecture rooms capable of sitting at least 60 students
				• Library facility available for at least 25% of total students	

BMAS guideline on SIWES. First, the concept of "relevant organisation" is very ambiguous. Second, each university is responsible for searching for the "relevant organisation" and the placement of students. In the following section, the process of posting students and the associated challenges as well as the intervention from UI-LISA are discussed.

9.3 Internship Training at the University of Ibadan

An integral part of the training of statistics students in the university is to expose them to some elements of industrial art under the SIWES. The process is managed, administered and fully controlled by the Industrial Training Coordinating Centre (ITCC) established by the UI. ITCC manages the three modules of industrial training (IT) in the university. The first module is for 2 months and this is taken by all 200-level Engineering and Food Technology students in the university. This module of IT is designed to expose the students to engineering and technology operations at the shop floor level. The second module is for 3 months. This is for the 300-level students of Engineering, Food Technology, Geography, Biochemistry, Nursing, Pharmacy, Geology, Chemistry, Physics, Statistics, and Library Science and Veterinary Medicine. The third module is however for 6 months and it is taken by 400-level students of Engineering, Food Technology, Botany, Microbiology, Computer Science, Zoology, Agriculture and Physiotherapy.

In managing the IT of the students, the ITCC is responsible for performing the following functions: soliciting for IT jobs in business, industry, government and service agencies depending upon the needs and qualifications of the students; carrying out the placement of students in industries; supervising the students while on IT; conducting follow-up activities regarding all students' placements by checking regularly each student's job performance through company visits and individual student interviews; organising seminars for students after the IT programme; evaluating feedback from the industry regarding the students' IT; and liaising between the UI and the Industrial Training Fund (ITF), the NUC and other relevant Government Agencies in the operation of SIWES. During the course of statistics students' placement in particular, the ITCC takes cognizance of the following set objectives:

- To provide real-world experience that will help the students to put what they learned in class into practice;
- To gain skills that can be applied to future work;
- To make statistics graduates more competitive in the job market; and
- To make the students see the statistical profession as a choice with high potential.

While placement is the primary responsibility of the ITCC coordinator, most students get an internship position by directly applying to the host organisations themselves and getting it endorsed by the centre. This is because most private organisations (see Table 9.2) are unwilling to accept students for internship for reasons such as inability to pay compensation and the fear of compromising sensitive data and losing classified information to competitors. While on internship, each student is assigned a supervisor who is required to pay regular visits to the student at the place of internship. The visits are to ensure that students are engaged in work that is related to their academic specialisation. "Log books" are provided to the students to serve as monitoring tools to be completed weekly by every intern to indicate the work done in any week. The host organisation supervisor then signs and stamps the "logbook" to indicate agreement with what is written. At the end of the internship period, the host organisation supervisor evaluates each student's performance by completing a student evaluation form. The student is then required to write an internship report and make a seminar presentation at both the department and the IT centre. The seminar presentations, IT report, "logbook" and the evaluation report all form the basis of evaluating each student's performance in the internship. Since IT is a compulsory course, the grade given in the assessment contributes to the student's cumulative grade points average (CGPA). Table 9.2 shows a list of some popular organisations that have accepted statistics undergraduate students in the past for IT. The majority of the organisations are government-owned, thus reflecting the Nigerian reality in which government is the largest employer of labour.

TABLE 9.2

List of Some Organisations Accepting Statistics Undergraduate Students for IT

S/N	Name of Organisation	Ownership Type
1	Transmission Company of Nigeria	Government
2	Ministry of Establishment and Training, Oyo State Secretariat	Government
3	National Bureau of Statistics	Government
4	Ibadan Local Governments Properties Co. Ltd	Government
5	Federal School of Surveys	Government
6	Hope of Motherhood Empowerment Initiative of Nigeria	Private
7	University Health Service	Government
8	National Population Commission	Government
9	Nigerian Meteorological Agency	Government
10	St. Raphael Divine Mercy Specialist Hospital	Private
11	Zincere Koncept (Accounting Solutions)	Private
12	Oyo South-East Local Council Development Area	Government
13	Federal Airports Authority of Nigeria	Government
14	Bursary Department, UI	Government
15	Zoological Garden, UI	Government
16	Institute of Agricultural Research & Training, Moor Plantation	Government
17	Nigerian Institute of Social and Economic Research (NISER)	Government

The following critical challenges have been identified by some of the statistics under-graduate students who have participated in the IT scheme in the department.

1. Little or no practical skills are acquired in most government-owned organisations. In most cases, the placement is not based on the need for the services of the interns but to satisfy the requirements for the award of the bachelor's degree as stipulated in the curriculum.

2. Limited supervision by both the employer and the supervisor due to poor administrative logistics for monitoring the students and the inadequate funding to cover the associated costs including travel allowance(s) and other incentives.

3. Converting IT students to office servants or allotting them trivial jobs. While the students are expected to have a sense of belonging and be an important part of the organisation, they are often made to carry out errands and given unimportant assistance jobs.

4. Unwillingness to permit the interns to handle data management, processing or analysis due to internal data protection and confidentiality rules.

9.4 Internship Training and UI-LISA Complementary Role

With growing unemployment in Nigeria and the popular slogan of "unemployable gradu-ates" from employers of labour, universities in Nigeria have been under serious pressure to help policymakers redress the trend. The appreciation of this national problem forms the basis for setting one of the goals of UI-LISA to ensure that statistics graduates are not only employable but are also able to become employers of labour by practising as consulting statisticians after completing their studies. Hence, the IT programme in the stat lab is designed to be "self-employment-focused" and to train students in a host of skills required for employment.

Over the past years, UI-LISA has been collaborating with the ITCC in UI to build the capacity of the IT students through hands-on and practical skills beyond the formal training provided for the undergraduate statistics programme in the NUC BMAS document. The initiative commenced in 2015 with the posting of two male students to UI-LISA by ITCC and since then the number has been increasing steadily. The reputation of the internship training has spread beyond the shores of UI. Applications for internship training have been received from prospective interns from neighbouring universities, polytechnics and other tertiary institutions. These institutions include the Federal University of Technology, Akure (FUTA) situated in Ondo State; Federal University of Agriculture, Abeokuta (FUNAAB) situated in Ogun State; Obafemi Awolowo University (OAU), Ile-Ife, situated in Osun State; and Ibadan Polytechnic (Ibadan Poly) and FSS, both in Ibadan, Oyo State. Also, graduate students who felt deficient in practical statistical skills from both quantitative and non-quantitative disciplines have also participated in UI-LISA internship training. Specifically, UI- LISA has trained graduates who have completed the National Youth Service Corp (NYSC) and those running a Postgraduate programme with the Centre for Petroleum Energy Economics and Law (CPEEL) in UI (https://cpeel.ui.edu.ng/).

Table 9.3 shows the distribution of intern trainees (completed and on-going) at UI-LISA. The record shows that 55 interns comprising 37 (67%) male and 18 (33%) female trainees have been enlisted between 2015 and 2020. The interns participate in all lab activities including walk-in statistical clinic sessions where statistical advice is provided to members of the university community. They are directly involved with the day-to-day activities of the lab such as attending to clients, participating in the design of survey tools and survey fieldwork. They undergo specialised hands-on training in statistical tools and software, particularly R, statistical consulting practices, statistical collaboration, statistical communication and professional conduct and ethics for practising statisticians. The interns who undertake their IT programme with UI-LISA are closely monitored and thoroughly supervised by both their academic supervisor and the ITCC supervisor compared to their counterparts posted to government organisations. After their training, they complete an online survey to collect feedback about their experience and skills learnt. The summary of the responses from 36 interns who completed their training before the emergence of COVID-19 lockdown is shown in Figure 9.1.

The maiden edition of UI-LISA's end-of-year party for interns and collaborators took place at the UI Botanical Garden on 19 December 2019. The programme had 28 participants in attendance, including UI-LISA coordination team members, the current and past interns from UI and neighbouring institutions, namely FSS, The Polytechnic Ibadan, FUNAAB and FUTA. Apart from being fun-packed, the event was also used to receive testimonies

TABLE 9.3

Number of Industrial Trainees at UI-LISA from 2015 to 2020

Gender/Year	2015	2016	2017	2018	2019	2020	Total (%)
Male	2	4	6	8	6	11	37 (67)
Female	0	3	1	4	1	9	18 (33)
Total	2	7	7	12	7	20	55 (100)

Gender

36 responses

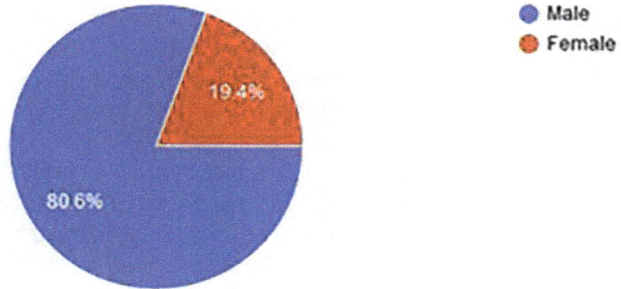

What is the name of your institution?

36 responses

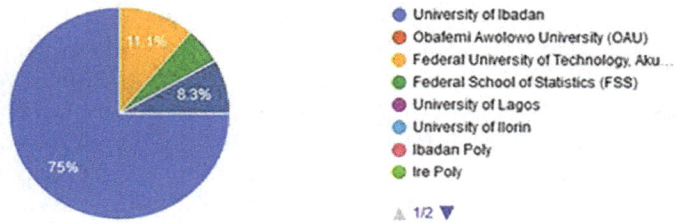

What is/was your main course of study?

36 responses

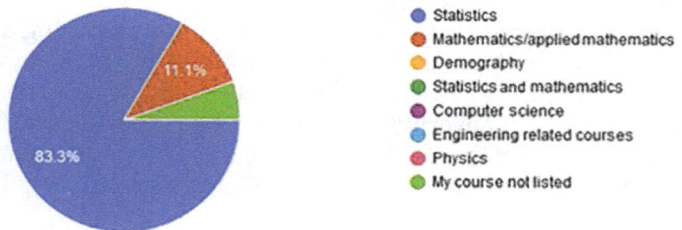

What type of degree programme are you running?

36 responses

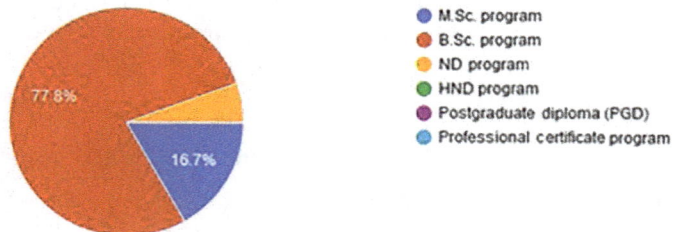

FIGURE 9.1
Summary of feedbacks from UI-LISA interns.

(*Continued*)

What was the duration of your internship programme?

36 responses

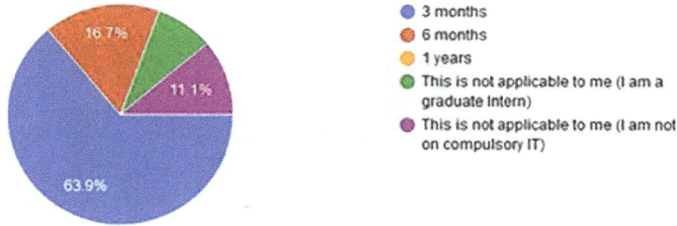

- 3 months
- 6 months
- 1 years
- This is not applicable to me (I am a graduate Intern)
- This is not applicable to me (I am not on compulsory IT)

16.7%
11.1%
63.9%

At the time of your internship at UI-LISA, which category did you belong to?

36 responses

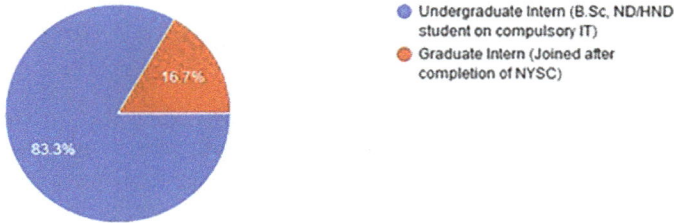

- Undergraduate Intern (B.Sc, ND/HND student on compulsory IT)
- Graduate Intern (Joined after completion of NYSC)

16.7%
83.3%

Which of the following services did you enjoy/participate in during your internship period at UI-LISA?

36 responses

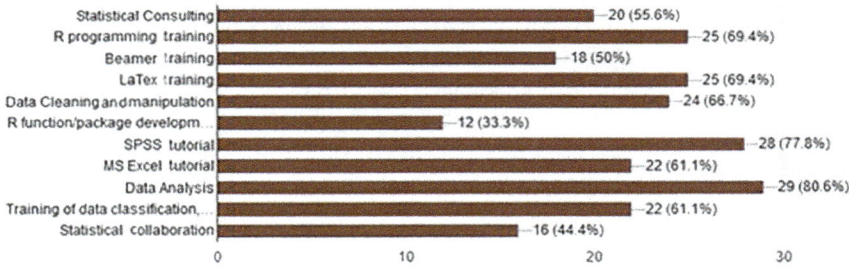

Service	Value
Statistical Consulting	20 (55.6%)
R programming training	25 (69.4%)
Beamer training	18 (50%)
LaTex training	25 (69.4%)
Data Cleaning and manipulation	24 (66.7%)
R function/package developm...	12 (33.3%)
SPSS tutorial	28 (77.8%)
MS Excel tutorial	22 (61.1%)
Data Analysis	29 (80.6%)
Training of data classification,...	22 (61.1%)
Statistical collaboration	16 (44.4%)

What can you say about your internship experience at UI-LISA? Please rate us!

36 responses

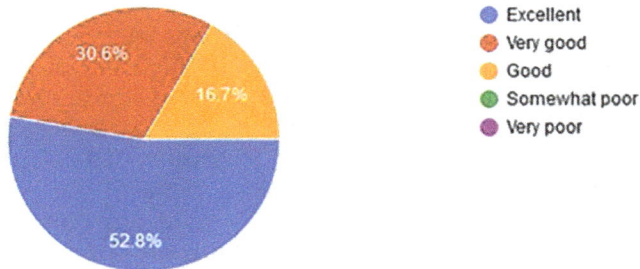

- Excellent
- Very good
- Good
- Somewhat poor
- Very poor

30.6%
16.7%
52.8%

FIGURE 9.1 (*Continued*)
Summary of feedbacks from UI-LISA interns.

FIGURE 9.2
Interns showcasing their certificate of completion.

and feedback on UI-LISA training activities towards building statistical and data science capacity in Nigeria. A testimony of special interest came from Mr. Robert Bbosa who was undertaking a Master's programme in Oil and Gas economics at the CPEEL, UI sponsored by the ACADEMY project. He participated in the internship graduate training at UI-LISA for 3 months where he acquired skills in R-Software, grant proposal writing and peer review journal article writing. He attested to how the training boosted his data analytical skills with R and data visualisation insights. Being an auditor, he discovered the relevance of data science in the understanding of the client's business environment. He graduated with a Bachelor of Commerce from Makerere University, Kampala-Uganda in January 2014 and has since worked for Direct Access Ltd as an accountant; KSK Associates Certified Public Accountants as a senior audit associate; and BRAC Uganda as a senior audit officer. Currently, Robert is the Lead lab collaborator in UI-LISA (Figure 9.2).

9.5 Conclusion

Statistics education has been expanding rapidly in Nigeria since independence in 1960. A good number of tertiary institutions including polytechnics and universities now offer statistics programme in Nigeria compared to the pre-independence situation. A foremost and leading statistics training institution in this pack is the Department of Statistics, UI. The department joined the LISA 2020 Global Network in 2015 by creating the UI-LISA. The emergence of the stat lab has given statistical education a major lift by filling a major gap that exists in the implementation of the BMAS for regulating the quality of statistics graduates produced in Nigeria. The flagship of UI-LISA programme is the specialised training designed for students participating in the SIWES which is a compulsory course in BMAS. The inability of the students to genuinely acquire practical technical skills and competences required for employment after completing their studies made the UI-LISA intervention very timely and necessary.

A host of practical-oriented skills are offered to the interns during their internship training. They participate in all lab activities including walk-in statistical clinic sessions.

They undergo specialised hands-on training in statistical tools and software, particularly R, statistical consulting practices, statistical collaborations, statistical communication and professional conducts and ethics for practising statisticians. They attend collaboration meetings with clients, implement the clients' projects, generate reports and carry out statistical advocacy. In sum, the interns are groomed to have the ability to apply statistical knowledge and skills to solve real-life problems.

Finally, UI-LISA plans to upgrade its current status to become a full-fledged centre meeting the statistical needs of the members of the university community.

References

Adamu, S and Johnson, T. L. (1983). *Statistics for Beginners*. Ibadan: Evans Brothers (Nigeria Publishers) Ltd.

Adamu, S. O. (1986). Training of statisticians in the last three decades. *An Invited Paper Presented at the 10th Anniversary of Conference of the Nigerian Statistical Association*, September 15th–18th, 1986, Conference Centre, University of Ibadan, Ibadan: Evans Brothers (Nigeria Publishers) Ltd.

Afonja, B. (1975). *Introductory Statistics: A Learner's Motivated Approach*. Ibadan: Evans Brothers (Nigeria Publishers).

Bjornsdottir, A., and Garfield, J. (2009). Teaching bits: Statistics education articles from 2008. *Journal of Statistics Education*, 17(1), 1–8.

De Veaux, R. D., Agarwal, M., Averett, M., Baumer, B. S., Bray, A., Bressoud, T. C., ... Kim, A. Y. (2017). Curriculum guidelines for undergraduate programs in data science. *Annual Review of Statistics and Its Application*, 4, 15–30.

Tishkovskaya, S., and Lancaster, G. A. (2012). Statistical education in the 21st century: A review of challenges, teaching innovations and strategies for reform. *Journal of Statistics Education*, 20(2), 1–57.

10

Statistical Consulting and Collaboration Techniques

Hope Ifeyinwa Mbachu, Julius Nwanya, Kelechi Dozie, and Mirian Raymond

Imo State University Owerri, Laboratory for Interdisciplinary Statistical Analysis [IMSU-LISA]

CONTENTS

10.1 Introduction

Statistical consulting is educational, exciting and challenging. Statistical consulting is about communication in any field of study that involves random variation. This means that every problem in every field of study that collects data belongs to this. The purpose of this chapter is to address the skills needed to equip a statistical consultant as an effective problem solver in statistics and data science and also provide strategies for collaboration techniques. For these skills to be fully appreciated, statistical consulting requires a strong interest in science and the art of scientific discovery.

The best source of knowledge is data; it is the symbiosis between data and theory that provides us with scientific knowledge. It is crucial to look at data because theory alone cannot always lead us to valid conclusions. It is the combination of theoretical knowledge and observational study, which is known as the scientific method, that is the procedure followed in scientific discovery. This paradigm, which is worthy to note, was not introduced until the late seventeenth century.

In the recent decade, many areas of the world have been working towards statistical consulting and collaboration techniques, but examples and good practices are still restricted in developing countries in Africa. Experience has shown that the majority of people who should be involved are not specialists in statistical consulting and neither do they have time or motivation to undertake formal study or training when occasionally organized. It is in raising awareness among these individuals and groups that informal capacity building has its role. According to Boen and Zahn (1982), the perspectives of statistics are so diversified that few statisticians are knowledgeable about all designs. Statistical consultants should be knowledgeable in their respective disciplines and familiar with relevant methodologies with the help of mentorship (collaboration). Without the proper statistical knowledge, data collection suffers, the risk of additional cost and effort, the analysis provides suboptimal results, and finally, wrong decisions are obtained.

Ultimately, statistical consulting provides many services which include giving support informing research questions, planning of studies, data collection, data management, statistical programming and scientific writing (interpretation). The purpose of teaching statistical consulting and collaboration among researchers is to make them more proficient in applying statistical techniques and to be good writing communicators of statistical information.

A statistical consultant in data science should know how the data are generated and what measures are employed. Research questions have to be translated to research problems that can be answered with data, determine the appropriate type of statistical analysis and in some cases be the 'honest broker' when the scientific result for validity. In other words, a statistical consultant should know how to deal with data – collect it, clean it, analyse it and interpret the findings. Therefore, working as a statistical consultant in a research field or specific project, there is a need to not only understand but be able to translate the subject matter into a statistical problem and then translate findings back to the client. However, after this introduction the rest of the chapter is subdivided into five sections, namely (1) consulting comes in many colours, (2) collaboration techniques, (3) activities of Laboratory for Interdisciplinary Statistical Analysis (LISA) statistical capacity building, (4) benefits of statistical consulting and collaboration technique and (5) need to improve statistical consulting and collaboration.

10.2 Consulting Comes in Many Colours

Statistical consulting includes, but is not limited to, assisting with study design, data collection, proposing an analysis plan, conducting analysis and writing up the result. Furthermore, it requires estimating the time and cost for your work and possibly invoice for your work. All of these are for the straight project.

10.2.1 Types of Consulting

Different types of consulting as well as different types of consulting jobs are available. There are short-term consultations on one end of the spectrum. This may require just one or two meetings. This is typically for quantitatively skilled researchers who are comfortable doing their analysis but who would prefer an expert giving a touch to their work for proper guidance. On the other side of the spectrum is long-term consulting. In an academic setting, this type of consultation is known as collaboration. It is a situation that requires the consultant for a particular project to be an integral part of the research team from the beginning to the end, which may last for years. However, other projects could be anywhere in between short-term and long-term consulting. This may include being an expert witness in a trial or helping post-graduate students in their thesis. Also, one can work independently (in-house statistician) or as part of a small group of consultants (collaboration) in a consulting firm (Kwasny, 2019).

10.2.2 A Four-Circle Consulting Services

Critical behaviour of successful statistical consulting comes in different models. For on-going expert advice, there are four-circle consulting services models for effective statistical consulting as shown in Figure 10.1. First, identify needs; second, strategic solutions; third, implement solutions and fourth, measure effectiveness (Kirk, 1991). Just like a consultancy collaboration, this relationship of a consulting services model has to be maintained for smooth running and successful consultation.

10.2.3 Important Skills Required for Providing Statistical Consulting

A statistical consultant is a problem solver. The purpose of this work is to address the skills needed to produce effective statistical consulting. One of the skills is that the person should have a strong interest in statistics and data science, and in particular, the art of scientific discovery. The symbiosis between data and theory provides scientific knowledge. In other words, there is a need to look at data, since theory alone cannot give an unbiased conclusion. The scientific method is the combination of observational and theoretical study which leads to scientific discovery (Cabrera & McDougall, 2002, Gile & Liu, 2015).

Good consulting requires a strong technical background in statistics and data science. Strickland (1995) opined that time, patience, understanding and empathy are all essential skills for collaboration between the client and the consultant, but none can substitute for technical competence. The consultant is armed with scientific, statistical computing and communication skills (which includes verbal and writing skills). Communication is to enable you to talk to clients and understand their language, which in effect helps to communicate statistics effectively. Almost every statistician is being called for help with a problem involving statistical data analysis and application of statistical methods. In a consulting relationship, Moolman (2010) identified three ways the clients' questions should be asked and the contents of the information that should be supplied to the consultant. He noted that "the client should; (1) be precise about the objectives of the analysis and give a brief background of the problem. (2) Describe the variables and data that are being studied and possible relationships that might have to be investigated. (3) List some questions/hypotheses to be answered/tested. In addition to the qualities mentioned above, like knowing statistics, learning about the research field you work in and communication skills, one also needs business skills, business acumen, time management skills, networking talents

FIGURE 10.1
Four-circle consulting services.

and a host of other skills. With all these, we can see that statistical consulting is indeed a multifaceted operation which requires sound logic, insightful analytical training and client-centred interpersonal communication skills.

10.3 Collaboration Technique

To collaborate in a working place is when two or more people come together, sharing ideas and thinking of common interest. In other words, the collaboration technique brings together people with different ideas, experiences, knowledge and skills to achieve a common goal or create something. There are three components of a successful team needed in every business or personal relationship to avoid bias: they are collaboration, communication and cooperation. Collaboration technique on its own, to build a good human relationship, has five principles. These five principles of collaboration are applying trust, respect, willingness, empowerment and effective communication. It is worthy to note that when any of these principles is lacking, human relationship is no longer cordial (Moris, 2019).

Some of the strategies required to deepen a student's understanding of collaboration techniques are (1) creation of complex learning activities, for students need a reason to collaborate (2) by preparing students to be part of a team, (3) minimize opportunities for communicating openly and effectively about your team's goal, (4) build in many opportunities for discussions and censuses by giving room for them to interact, exchange ideas to promote a working environment, and finally (5) by focusing on strengthening and stretching expertise, which can be achieved through heritage relationship (Olamide, 2007).

10.4 Activities of LISA Statistical Capacity Building

Some of the higher institutions in developing countries now have a LISA, which serves as a statistical consulting unit. They have statistical consultants who are trained to provide solutions to design experiments, analyse and plot data, run statistical software, interpret results and communicate statistical concepts in the form nonstatisticians can understand. Their areas of expertise include Multivariate Methods, Time Series, Data Science, Quality Control, Biostatistics, Bayesian Statistics, Experimental Design, Publication Preparation, Grant Proposals and various statistical software packages. A statistical consultant is often available at a certain time of the week in the department to attend to clients with research projects. However, many of the higher institutions in Nigeria have similar units operating in their campuses like in University of Ibadan (UI-LISA), Obafemi Awolowo University (OAU-LISA), Michael Okpara University of Agriculture Umudike (MOUAU-LISA), Imo State University (IMSU-LISA), and Federal University Oye Ekiti (FUOYE-LISA), to mention a few. These statistical laboratories are being managed by expert faculty/departmental members who are experienced to be trained in the use of various software like SPSS, MATLAB, SAS, MINITAB, R, and STATA. Some of the activities of LISA statistical laboratories, among others, include

- Regular discussion and training on modern areas of research in statistics and data science and their application;
- Training on current statistical/mathematical software and related packages;
- Conducting various departmental seminars and workshops to train staff and students and young statisticians;
- Engaging in statistical and data science consulting services;
- Posting of research and tutorial materials on the lab's website;
- Crossbreeding of new academic ideas, knowledge and mentorship which leads to collaboration;
- Collaboration among the existing LISA statistical laboratories in Nigeria and beyond for capacity building and
- Statistical collaboration with domain experts in other disciplines like Medicine, Economics, Agriculture and Biological Sciences; some of these activities mentioned serve as short-term services, while others are for medium-term services.

10.5 Benefits of Statistical Consulting and Collaboration Techniques

There is an analogy that statistics and data science are the scaffolding that helps construct or remove a building. Statistical consulting and collaboration techniques provide the means to keep statistical and communication skills sharp. It provides the opportunity to discover different and potentially better ways to explain things. All told, statistical consulting is educational, exciting and challenging.

Statistical consulting and collaboration techniques enable statistical consulting and collaboration among researchers. This is done through individual meetings and support for interdisciplinary research projects. Statistical consulting and collaboration play an important role in scientific studies to seek knowledge and to improve our lives generally. The learning and practice of statistical consulting offer a wide variety of statistical professional opportunities which give solutions to several life problems. This statistical consulting helps to develop graduates with the necessary skills for a job in government, industries or academics (Awe, 2012). Other benefits include enhancing the practical use of statistics and data science by contributing to the body of fundamental statistical science through research; enhancing the quality of decisions and conclusions made on the strength of the statistical approach to research and broader issues; and boosting the self-image of statisticians and ensuring that statistics and data science continues to grow as a field as opined by Thabane et al. (2008).

Furthermore, collaboration between academia and industry is now becoming increasingly beneficial for the development of systematic natural innovation systems. This brings great rewards that are also distinct in developing countries. Some of the benefits are to avoid duplications, stimulate additional investment, exploit synergies complementary of scientific–technological capabilities, expand the relevance of research carried out in public institutions, foster the commercialization of public outcomes and boost the mobility of labour between public and private sectors.

Industries and higher institutions are increasingly recognizing the benefits of collaboration. Institutions are gradually utilizing open innovation strategies to better incorporate other relevant information that can better the collaboration with universities. These universities have begun adopting advanced teaching and research methods which help to improve and rectify the challenges faced by the industry and contribute directly to economic success and data science.

On the other hand, the priorities and scope of university–industry collaboration vary between developed and developing countries. In developing countries like Nigeria, a key consideration is the poor quality of education and lack of university funding which typically leads to the insufficient aptitude to team up with the industry in innovation-related projects. The existing collaboration tends to be more informed and to focus on job employment of graduates rather than research. In most developing countries, university–industry collaboration is hindered by historically based, cultural and institutional barriers, which consequently takes time to prevail.

As universities adapt to globalization, there is an observed increase in the number of prominent universities from developed countries that are operating campuses in developing countries. The universities are now training their students and staff to engage in consulting and collaboration effectively in developing countries, to globally take advantage of their reputation, knowledge base and management practices, which is the mission of LISA. Apart from the statistical, educational and communication opportunities consulting provides, it is also a convenient way to expand your professional network beyond statistics and data science.

10.6 Need to Improve Statistical Consulting and Collaboration

Statistical consulting and collaboration technique as has been stated is a means required for capacity building to do statistical work, be knowledgeable in data processing, and be conversant with development issues and form an effective bridge between users of statistics and statistical information. There is a need to develop key skills for consulting and collaboration and to promote the practice among researchers in higher institutions of learning in developing countries in statistics and data science. These skills we all know are rarely taught in universities. This brings about the missing gap and major hindrance to the development and growth of capacity building in the practice of statistics and data science in Africa. To improve this situation, the advent of the LISA 2020 initiative is bridging the gap with a fast-growing network of statistical laboratories, spread across higher institutions of learning in Nigeria and other developing countries. To achieve this, LISA engages in building capacity to improve informal statistical skills through training and collaborations.

This paper, therefore, suggests the need to improve in other areas to foster practice, teaching and learning of statistical consulting and collaboration in developing countries as follows.

1. There is a need to include statistical consulting as a course in the teaching programme of the training institutions in developing countries.

2. The government and other agencies should make funds available to higher institutions of learning and research institutes, to establish statistical laboratories and equip existing ones, like LISA 2020 is doing presently.

3. Researchers should be sponsored to take short courses in developed countries to learn statistical consulting and collaboration in statistics and data science, for effective mentoring of young statisticians and data scientists.

4. There is a need to organize, promote and embrace conferences, workshops, seminars, symposiums, training sections to enhance effective consulting and collaboration skills among staff, undergraduate and post-graduate students in the universities.

5. Collaborative studies are also necessary among staff and students across disciplines in the university community and beyond.

6. New statistical consultants should not go it alone; find a mentor, colleague or boss who is available for advice. This is especially helpful when you meet one of those difficult clients.

Go! Consult! Collaborate! Have fun.

References

Awe, O. O. (2012). Fostering the practice and teaching of statistical consulting among young statisticians in Africa. *Journal of Education and Practice* 3(3), pp. 54–59.

Boen, J. and Zahn, D. (1982). *The Human Side of Statistical Consulting.* Lifetime Learning: Belmont, CA.

Cabrera, J., McDougall, A. (2002). *Introduction to Statistical Consulting. In: Statistical Consulting.* Springer, © Springer Science+Business Media: New York, NY.

Gile, K. J. and Liu, A. (2015). Statistical Consulting and Collaboration Services. Department of Mathematics and Statistics, College of Natural Sciences. Lederle Grad Research Tower, Box 35415 710 N. Pleasant st, University of Massachusetts Amherst, MA 01003-9305.

Kirk, R. E. (1991). Statistical consulting in a university: Dealing with people and other challenges. *The American Statistician* 45(1), pp. 28–34.

Kwasny, M. (2019). Why Be a Statistical Consultant? American Statistical Association, STATtr@k. A website for new statisticians, professionals, navigating a data-centric world. Stattrak.amsstat. org/2019/09/01/statconsultant/

Moolman, W. H (2010). Communication in Statistical Consultation. ICOTS8 (2010) Invited Paper. National Research Council (1994): Modern Interdisciplinary University Statistics Education, National University Press, Washington, DC.

Morris, S. (2019). Informal Capacity Building. Available from http://www.coastalwiki.org/wiki/Informal_Capacity_Building [accessed on 14-05-2020].

Olamide, A. A (2007). Knowledge and perception of health workers towards tele-medicine application in new teaching hospital in Lagos. *Scientific Research and Essay* 2. http://www.academic-journals.org/SRE.

Strickland, A. J. (1995). *Strategic Management: Concept and Cases* (8th Edition). International Student Edition USA: Irwin. https;//www.cambridge.org.

Thabane, L., et al. (2008). Training young statisticians for the development of statistics in Africa. *African Statistic Journal* 7, pp. 125–148.

11

Statistical Consulting and Collaboration Practises: The Experience of NSUK-LISA Stat Lab

Monday Osagie Adenomon

Nasarawa State University Laboratory for Interdisciplinary Statistics Analysis (NSUK-LISA)

CONTENTS

11.1 General Introduction

In African countries, development partners and statisticians know how useful applying statistics can be to solve problems in business and industry (Sanga et al., 2011). However, the adoption of statistical methods in business and industry can often be slow. This is especially true in developing countries, which lack both demands from industry to hire statisticians and supply of well-trained statisticians from universities who can apply statistics and data science to solve real problems (Schwab-McCoy, 2019). This article presents Nasarawa State University Keffi (NSUK) as a case for a university that is addressing these challenges.

The Nasarawa State University was established under the Nasarawa State Law No. 2 of 2001. It officially took off for effective academic activities in March 2002, following the appointment of a distinguished educationist, Professor Adamu Baikie, as pioneer Vice-Chancellor. The purposes for which the university was established are listed in Sections 4 and 5 of the University's enabling law as follows:

DOI: 10.1201/9781003261148-13

a. To encourage and promote the advancement of learning and hold out all persons without distinction of race, creed or sex.

b. To provide courses of instruction and other facilities and to make available those facilities for the pursuit of learning in all its branches and to make those facilities available on proper terms to such persons(s) as equipped to benefit from them.

c. To serve as both teaching, research and examining body, subject to the provision of the University Law specifying the functions of the University.

The Department of Mathematical Sciences offers three (3) undergraduate programmes leading to the award of the following degrees: BSc (Hons) Mathematics, BSc (Hons) Computer Science and BSc (Hons) Statistics. BSc (Hons) Mathematics was the only option in the department at inception, but BSc (Hons) in Computer Science and BSc (Hons) in Statistics took off during 2003/2004 academic session and in 2018 the Department of Statistics was created as a full department in the faculty of Natural and Applied Sciences. From its inception, BSc Statistics received Full Accreditation in 2013 and again recently in 2019 (Department of Statistics, Handbook, 2019). As of today, the department offers PGD, MSc, MPhil/PhD and PhD programmes in statistics, but unfortunately, there has been a lack both demand from industry to hire statisticians and supply of well-trained statisticians from universities who can apply statistics and data science to solve real problems in North Central Nigeria. The NSUK-LISA stat lab was created to fill this void.

11.2 Challenges of Statistical Development in North Central Nigeria

The present situation in North Central Nigeria regarding the state of statistics can be outlined and summarized as follows:

i. **Low statistical capacity building**

Statistical capacity building in North Central is very low. The region has few professors of statistics and it was not until recently that some universities such as the University of Abuja, Federal University of Technology, Minna and Nasarawa State University, Keffi started their postgraduate studies in statistics. Similarly, grant writing, statistical computing, statistics consulting and collaborative skills are low among postgraduate and tutors of statistics.

ii. **Wrong use of statistical tools**

NSUK started postgraduate studies in statistics in late 2016. Before this time, undergraduate and postgraduate students were unable to seek statistical advice or support on their final year projects. For example, it was common to find students using ordinary least squares to estimate a regression model whose response variable is binary in nature or using ordinary least square to run a regression model whose response variable is count. The incorrect use and interpretation of statistical tools were very pervasive among students and users of statistics in the region.

iii. **Shortage of qualified and experienced statisticians**

As mention earlier, many universities in North Central Nigeria lack adequate and competent manpower in their departments of statistics. For instance, the Department of Statistics, NSUK has eight (8) academic staff with only four (4) PhD holders as of August 2020. Shortage of experts in statistics in the region also affects the research output of the academic staff in statistics as few staff have published papers in journals indexed in Scopus and Web of Science (see publons.com/institution/6073 as an example of Nasarawa State University, Keffi).

iv. **Teaching statistics without real-life applications**

Statistical education in North Central Nigeria is still growing. The teaching of statistics is often done without real-life applications and hands-on experience with statistical software and programming languages. This was because of the low popularity of statistical education in North Central Nigeria. According to Adelodun and Awe (2013), teaching statistics with real-life applications is inadequate and infrequent in Nigeria. This has continued to affect the quality of undergraduate and postgraduate projects in statistics in some of the universities (Uche, 1984).

v. **Low interest by students to pursue statistics as a choice of course**

Courses of study such as medicine, pharmacy, law, accounting, engineering, business administration and economics are very competitive in all the universities in the region because of a high level of awareness about these courses among secondary school students, unlike statistics and mathematics. The low awareness of statistics among secondary school students has had a negative effect on the number of applications for bachelor degree in statistics.

vi. **Poor mentoring culture**

Mentoring is a relationship between two people with the goal of professional and personal development. In some universities in the region, students are transferred to Statistics Department from other departments (especially Computer Science) due to oversubscription by applicants. Poor mentoring culture often discourages the few who completed their bachelor programme from returning for graduate studies in Statistics.

vii. **Lack of scholarship for Statistics students**

Poor funding has been a major setback for the education sector in the region. Worst still is the lack of scholarship for statistics students as well as grants for tutors and researchers in core statistics. These have greatly hindered the rapid growth in the development of statistics in the region.

viii. **Inadequate facilities**

Lack of funding have a direct link to inadequate facilities. Many universities in North Central Nigeria have statistical labs but most of them are not well equipped. Inadequate facilities include lack of computer systems for teaching practicals, access to licensed statistical software such as EViews and STATA and lack of stable power supplies.

11.3 The Birth of NSUK-LISA Stat Lab

In September 2017, the Nigerian Statistical Association hosted its first international conference, during which Prof. Eric Vance introduced the Laboratory for Interdisciplinary Statistical Analysis (LISA) 2020 Network to participants at the conference. The interactive meeting with him exposed us to the three missions of the Network: training students,

FIGURE 11.1
Photo from the First Workshop organised by NSUK-LISA stat lab, 2018.

collaborating to solve problems and teaching statistical skills and literacy widely (Vance et al., 2019; Adenomon, 2019).

NSUK-LISA in Nigeria was created on 31st January 2018 as the 9th member of the LISA 2020 Network. The lab is coordinated by Dr. M. O. Adenomon (Figure 11.1).

The NSUK-LISA was created with the following goals:

1. To support statistical learning in the department through solving real-life problems and demonstrating effective mentoring;
2. To teach short courses; and
3. To improve collaborative skills and interdisciplinary research among staff and students (Vance and Smith, 2019).

11.4 Role and Potentials of NSUK-LISA Stat Lab

NSUK-LISA stat lab has successfully organized and conducted the following short courses since 2018.

11.4.1 Workshop 1: Basic Statistical Methods for Physical Sciences with Examples in R

The workshop was held in 2018. The workshop covered a general introduction to statistics and R software; descriptive and elementary Statistics (including Probability Distributions); correlation and regression analyses including diagnostic testing and Analysis of Variance (ANOVA). The timing was about 2 hours for both undergraduate and postgraduate students.

11.4.2 Workshop 2: Basic Statistical Methods for Physical, Management and Social Sciences with Examples in MINITAB

The workshop was also held in 2018. The workshop covered the general introduction of statistics and MINITAB software; data entry in MINITAB, construction of scatter plot, descriptive and elementary statistics (including Normality Test with MINITAB); simple

TABLE 11.1

Distribution of Participants by Programme, Level and Gender

Undergraduate Classes	Male	Female	Total
400 Level	33	7	40
300 Level	25	5	30
Postgraduate Classes	Male	Female	Total
M.Sc. Level	29	8	37
PGD Level	4	2	6

and multiple regression analyses including diagnostic testing; Chi-square analysis and ANOVA. The duration was about 2 hours for both undergraduate and postgraduate students (Table 11.1).

11.4.3 Workshop 3: Macroeconometric Analysis with EViews

In 2020, a workshop was held on Macroeconometric Analysis using EViews. The workshop covered the following: data entry and graphical analysis, basic statistics and tests of normality, data transformation and generating new variables, regression analysis (simple and multiple regression), diagnostic tests (e.g., serial correlation, heteroscedasticity, multicollinearity, normality tests of residuals, interpretation of macroeconometric analysis) (Table 11.2).

11.4.4 Workshop and Webinar: Capacity Building on Consulting Best Practices Technique for Women in Statistics and Data Science in Nigeria amidst the COVID-19

Capacity Building on Consulting Best Practices Technique for women in Statistics and Data Science in Nigeria amidst the COVID-19 took place on 20th June and 27th June 2020 using Zoom and in-person participation of fewer than 20 persons as prescribed by the Nigerian Government. The highlights of the programme were

Overview of ISI and LISA 2020 Networks, Need for more women in Statistics and Data Science in Nigeria, Proven techniques for Award winning Grant writing, Career opportunities for Women in Statistics and Data Science in Nigeria, Statistical Consulting Practices and Hands-on practical in STATA. The workshop and webinar were hosted by the Department of Statistics, Nasarawa State University, Keffi, Nigeria.

The hands-on covered the following topics: data entry, descriptive statistics, normality testing, regression analysis and diagnostics, correlation analysis such as Pearson and Spearman's correlation analysis, Poisson regression and goodness-of-test fit and negative binomial regression analysis. A total of 20 people participated online and 10 in-person.

TABLE 11.2

Distribution of Participants by Gender

Undergraduate Class	Male	Female	Total
300 Level	17	10	27

11.4.5 Statistical Consulting Practices of NSUK-LISA Stat Lab

i. The NSUK-LISA stat lab carried out her consulting practices through research collaboration.

ii. The NSUK-LISA stat lab also uses attitudinal influence and mentoring. Mentorship and effective collaborative research through interdisciplinary statistical analysis become meaningful if these bring results in addressing real-life problems and enhanced research outputs (Amstat News, 2019). Experience from the NSUK-LISA stat lab has shown that much can be achieved in statistical capacity building in North Central Nigeria and Nigeria as a whole through mentorship and effective collaborative research. The stat lab uses the T.E.A.M. framework to conduct its mentorship and collaborative activities. The T.E.A.M. framework stands for the following: T-Teach with statistical computing; E-Examples with real-life application; A-Attitudinal change (You can do it); M-Mentoring. Through mentorship and effective collaborative research, Dr. Emenogu a collaborator with NSUK-LISA stat lab was able to achieve the following feats.

iii. Successfully completed his PhD in Financial Time Series Analysis to become the first doctorate graduate in Statistics in Nasarawa State University, Keffi, Nigeria.

iv. Developed good academic writing skills and published three journal articles, namely Emenogu et al. (2018), Emenogu et al. (2019) and Emenogu et al. (2020).

11.4.6 Some Progress from NSUK-LISA Stat Lab

The following represents the progress of the NSUK-LISA stat lab:

i. Outreach to Department of Statistics, Federal Polytechnic, Bida, Nigeria on 25 July 2018 to train students on statistical analysis with MINITAB and R.

ii. Outreach to Dorben Polytechnic, Abuja, Nigeria on 17 and 18 May 2019 to conduct a workshop with the theme: "Contemporary Methodology Processes in Research" for staff and students in Social, Management, Physical and Engineering Sciences.

iii. Presented an invited lecture titled "Potentials of LISA 2020 Network and Statistical Capacity Building in North Central Nigeria" at the FUAM-LISA Research Workshop series 1 on 28 February 2020.

The NSUK-LISA stat lab provided statistical advice and support to the following studies.

i. Statistical Analysis of Tuberculosis and HIV Cases in West Africa Using Panel Poisson and Negative Binomial Regression Models.

ii. Modelling Infectious Diseases in Jos, Nigeria using PAR and PEWMA Models.

iii. Modelling Registered Births and Deaths in Nigeria using Generalized Linear Models.

iv. What Are the Determinants of Inflation in Nigeria? Answers from Autoregressive Distributed Lag Model.

v. Poverty Reduction through Increased Literacy Rate and Financing of SMEs in Nigeria: Evidence from Vector Error Corrected Model.

   vi. The Determinants of Unemployment Rate in Nigeria: Evidence from Fully Modified OLS and Error Correction Model.

   vii. The Contribution of Commercial Banks to GDP Growth in Nigeria using Autoregressive Distributed Lag Model.

In summary, before the creation of NSUK-LISA stat lab, the Department of Statistics, Nasarawa State University, Keffi did not have any publications in journals indexed in Scopus and Web of Science but now the department has three papers in Scopus and one in a journal indexed in Web of Science through its collaboration and consulting activities. Lastly, the paper titled "Modeling Registered Births and Deaths in Nigeria using Generalized Linear Models" won the best paper presentation award at the 1st International Conference of the Royal Statistical Society Nigeria Local group held in February 2020.

11.5 Conclusion

The shortage of well-trained statistical consultants in North Central Nigeria constitutes a major hindrance to the growth and development of statistical capacity building in the region. There is a need to teach undergraduate and postgraduate statistics students statistical consulting and collaboration techniques. The activities carried out in the NSUK-LISA stat lab in Nigeria are geared towards bridging the gap between theory and application in the region. The plans include:

- To have more collaborations with other departments in the university and the neighbouring polytechnics;
- To initiate a memorandum of understanding with the Postgraduate School on statistical training to graduate students;
- To reach out to the National Bureau of Statistics (NBS) for collaboration in the production of official statistics;
- To seek for funding and sponsorship opportunities for lab collaborators; and
- To promote statistical literacy through career talk to students in Senior Secondary School classes.

References

Adelodun, O. A., & Awe, O. O. (2013): Statistics education in Nigeria: A recent survey. *Journal of Education and Practice*, 4(11):207–215.

Adenomon, M. O. (2019): Building manpower in the field of statistical sciences for business and industry in North Central Nigeria. *A Paper Presented at the 62nd ISI World Statistics Congress 2019*, 18–23 August 2019, Kuala Lumpur, Malaysia.

Amstat News (2019): Consulting Best Practices: Practice and Mentorship. The Membership Magazine of the American Statistical Association, Issue #507. http://magazine.amstat.org.

Emenogu, N. G., Adenomon, M. O. & Nweze, N. O. (2018): On the performance of Garch family models using the root mean square error and the mean absolute error. *Benin Journal of Statistics*, 1:45–60.

Emenogu, N. G., Adenomon, M. O. & Nweze, N. O. (2019): Modeling and forecasting daily stock returns of Guaranty Trust Bank Nigeria Plc using ARMA-GARCH models, persistence, half-life volatility and backtesting. *Science World Journal*, 14(3):1–22.

Emenogu, N. G., Adenomon, M. O. and Nweze, N. O. (2020): On the volatility of daily stock returns of Total Petroleum Company of Nigeria: Evidence from GARCH models, value-at-risk and backtesting. *Financial Innovation*, 6(1):18, Springer.

Sanga, D., Dosso, B. & Gui-Didy, S. (2011): Tracking progress towards statistical capacity efforts: The African statistical development index. *International Statistical Review*, 79(3):303–335.

Schwab-McCoy, A. (2019): The state of statistics education research in client disciplines: Themes and trends across the university, *Journal of Statistics Education*, DOI: 10.1080/10691898.2019.1687369.

Uche, P. I. (1984): Undergraduate project work in statistics at a Nigerian University. *Journal of Royal Statistical Society Series D (The Statistican)*, 33(3):295–300.

Vance, E.; Awe, O.; Ganguli, B.; Olubosoye, O.; Adenomon, M. O. & Gebru, D. (2019): Statistical practice around the world. *A Paper Presented as the American Statistical Association (ASA) Conference on Statistical Practice*, New Orleans, LA, USA. February 14th 2019.

Vance, E. A. & Smith, H. S. (2019): The ASCCR frame for learning essential collaboration skills. *Journal of Statistics Education*, DOI: 10.1080/10691898.2019.1687370.

12

Statistics: The Practice of Data Surgeon

Olusanya E. Olubusoye and Olalekan J. Akintande

University of Ibadan Laboratory for Interdisciplinary Statistical Analysis (UI-LISA)

CONTENTS

12.1 Introduction

Statistics is a word with different meanings; a subject (taught in schools), a profession (career or occupation) and numerical aggregates (as in data, figures, news, records, journals, etc.) (Afonja et al., 2014, Chp. 1). The latter is interesting to the majority of users, while the background process that produces the aggregates or summaries (such as the design, collection, cleaning, methods, and analysis) is less appealing. Due to the importance of statistics, almost every researcher has assumed the position of a data analyst, statistician or scientist. Consequently, statistics remains widely counterfeited and this has led to many misconceptions, misuses and abuses of its principles and methods. Unfortunately, the process of journal review does not mandate the certification (in most cases) of statistical contents by qualified statisticians. Hence, there are numerous statistical abuses sighted in journal articles including those in high impact journals. The fact that someone took a few elective or compulsory statistics courses should not qualify anyone to practice as a professional statistician. Does one qualify as a medical practitioner by simply taking few elective medical courses in the university? Similarly, the ability to use statistical software is not a licence to assume the role of data analyst or scientist.

The low level of statistical literacy among the general public and the non-statistical nature of human intuition permits misleading without explicitly producing faulty conclusions. The definition of misleading statistics is weak on the responsibility of the consumer of statistics. Fischer (1979) listed over 100 fallacies in a dozen categories including those of generalization and those of causation. A few of the fallacies are explicitly or

DOI: 10.1201/9781003261148-14

potentially statistical including sampling, statistical nonsense, statistical probability, false extrapolation, false interpolation and insidious generalization. All of the technical/mathematical problems of applied probability would fit in the single-listed fallacy of statistical probability. Many of the fallacies could be coupled to statistical analysis, allowing the possibility of a false conclusion flowing from a blameless statistical analysis, allowing the possibility of a false conclusion flowing from a blameless statistical analysis (Fischer, 1979). One usable definition of "Misuse of Statistics" is using numbers in such a manner that either by intent or through ignorance or carelessness, the conclusions are unjustified or incorrect (Spirer et al., 1998). It is a generalization of misuse of statistics!

A popular slogan expresses statistics as facts from "reliable" figures (data). However, when facts are falsified, it is injurious to the act of statistics and downplays the purpose such figures represent. Darwin (1871) asserts that "false facts are highly injurious to the progress of science, for they often long endure; but false views, if supported by some evidence, do little harm, as everyone takes a salutary pleasure in proving their falseness; and when this is done, one path towards error is closed and the road to truth is often at the same time opened." Statistics abuses typically arise from two major sources: data and methods of analysis.

Data is a complex word that represents the product of a statistical procedure which includes design, collection, editing, processing and dissemination. Hence, just like in the medical profession, statistics has many branches and specifications by experts. The samplers (survey experts, experimental design, etc.) are the custodians of data collection and designs of survey tools and methodologies, etc. While mathematical statisticians develop the methodological algorithms and validate model assumptions, applied statisticians (econometricians, biostatisticians, "environmetricians", data scientist, etc.) put the processes into use to solve real-life problems in their various areas of application (economics, medical, environment, data pattern, etc.). Then computational statisticians develop and inculcate the act of statistical modelling in software for all to enjoy. Hence, all are dealing with the same patient: Data!

There are no statistics without data and there are no useful data without sound statistical procedure. Hence, data handling and processing by an untrained "expert" is synonymous with surgery operation in the hand of a "surgeon" who is in fact a fraudulent or ignorant pretender to medical skill. Similarly, any statistical methodology (prescription) adopted in any study without consulting a well-versed (practising) statistician is synonymous with a patient self-medicating. Strasak et al. (2007) and Indrayan (2007) assert that statistical abuses in medical research can come to play in the process of experimental planning, the conduct of the experiment, data analysis, drawing the logical conclusions and presentation/reporting. Essentially, the contemporary users of data lack the foundational understanding of data and its anatomy – hence the misuse!

According to Gardenier and Resnik (2002), the definition of misuse of statistics comes to play when:

- Statisticians are not in complete agreement on ideal methods/methodological prescription;
- Statistical methods are based on assumptions which are seldom fully met;
- The provisional conclusions have errors and error rates – commonly 5% of the provisional level of significance testing is wrong;
- The data gathering method or design is questionable; and

- The method(s) adopted for analysis is erroneous or in contrast with the data class, e.g. computing the arithmetic mean of gender, location or "unscorable" Likert scale responses rather than proportional mean/average.

Therefore, many misuses of statistics occur because the source is a domain expert, not a statistician. Thus, the domain expert may incorrectly use a method or interpret a result wrongly. Alternatively, misuses occurred when the source is a statistician and not a domain expert (Spirer et al. 1998). In this case, the statistician and the domain expert fail to harness the strength of collaboration. Another cause is that the subject being studied is not well defined. That is, there is a lack of domain experts in the team of researchers.

Lastly, data quality plays the most important role in statistics, thus poor data quality will lead to abuse of statistics regardless of how sound the methodology employed is. Thus, it's like putting the cart before the horse. The data usually leads to the methodology, and the converse is improper. When methodology leads to the data, Andrew Lang (1844–1912) regards this as politicians using statistics in the same way that a drunkard uses lampposts, for support rather than illumination.

The goal of this discourse is to educate our readers; it is to help them appreciate and promote quality statistical outputs but not to point out or name any article or journal for such abuses or misconceptions. Thus, we shall simply address the issue of data and its anatomy. The rest of the work is sectioned as follows; Section 12.2 addresses the type of misuse of data handling. In Section 12.3, we address the anatomy of data in which we take a keen interest in explaining the interconnectivity in the anatomy of data. In Section 12.4, having highlighted the data anatomy, we breakdown the functions associated with some specifications of the statistics profession. Section 12.5 summarizes the work and provides a conclution.

12.2 Abuses/Misuse: Data Handling

There are many ways in which data or statistics could be misused. We highlight some common types of misuse and briefly cite examples where relevant.

12.2.1 Data Misclassification and Ignorance

Misclassification often arises from an erroneous assumption, lack of data understanding (see Section 12.3), or entry and coding of data, among many others. Unfortunately, the garbage in, garbage out (GIGO) nature of computers allows any form of coding and getting the desired output is usually handy even when such a request is invalid or inappropriate. For instance, gender (male=1 and female=0) is nominal data; however, if a researcher ignorantly asks the computer to generate the arithmetic mean of the gender, a result is certain and other "five-numb" (median, and quartiles, even standard deviation [SD], etc.) statistics can be generated as well. Another common example is the Likert scale response; the majority of work we have seen assumes that all Likert scale responses can be scored and obtain the arithmetic mean as well as other statistics. On the contrary, not all Likert scale responses can be scored, hence obtaining arithmetic means or others are erroneous and a misuse. This is because not all Likert scale responses are intended to have a right or wrong response/opinion, or have "equal spacing" between the options.

A good example is an opinion poll about the performance of a sitting President of a country; strongly agree or strongly disagree is neither right nor wrong, hence scoring is inappropriate. The most appropriate will be a proportional average that gives the ratio of agreement to disagreement (binary reporting) or agreement, neutral and disagreement in three levels/classes.

Therefore, when a Likert scale question is "scorable", it is ordinal (and metric after scored) and when scoring is inappropriate, it is a nominal scale. Essentially, there is a need for a good understanding of the data collection, study design and purpose to know the class that a variable should be or class for any research. That Likert scale questions are often time "scorable" does not mean all Likert scale questions should be scored and arithmetic means and SD is possible.

Another example is classification bias. Recently, there has been concern over the globalization of artificial intelligence due to algorithm bias that has been observed. For instance, there is the example of a computer vision algorithm classifying black people as criminals or mismatched for white people. This is pure data bias embedded within the training set for the algorithm. These are just a few examples out of numerous misclassification issues in literature. We must understand that every abuse of statistics is a detriment to our society and the world. Hence, there is a need to educate the producers and users of statistics.

12.2.2 Data Dredging

Dredging is a process typically common in data mining. It involves examining large compilations of data to find a correlation (by all means) without any pre-defined choice of hypothesis specified. This obviously might lead to spurious results since the likelihood of finding correlation among larger compilations (datasets with many variables) of data is quite high. (Meanwhile, statistical significance does not imply practical significance, Moore & McCabe (2003))

Although the process is not entirely invalid, the misuse plays out when the hypothesis arrived at is stated as a fact (without validation on independent data) rather than a suggestion. Hence, to validate or invalidate the process, the hypothesis arrived at must be tested on independent data. The problem is that not everyone understands this and hence abuse and misuse.

12.2.3 Data Manipulation

According to Coase (1960), if you torture the data long enough, it will confess. This famous quote over time has been erroneously interpreted as, if you manipulate data long enough, it will give you what you desire. This implies a practice of including selective reporting or simply making up false data. Hence, data fudging/manipulation has been institutionalized to the point that many journal articles have been subjected to retraction (due to publication bias) as a result of this menace.

Traditionally, data are reviewed and sometimes changes are made in order to ensure the appropriateness of the data and repair of real problems (such as entry error, coding error, outliers, and missingness, among many others) before setting out for analysis. Exploratory data analysis (EDA) is an important ingredient of statistical analysis. However, data manipulation these days implies data fudging and is becoming the order of the day among data scientists and researchers alike, leading to a lack of reproducibility and openness. Neylon (2009), and scientists in general, question the validity of study results that cannot be reproduced by other investigators. However, some scientists refuse to publish their data and methods. Hence, unhealthy data manipulation practices are serious issues in the most honest of statistical analyses and, therefore, an unhealthy practice that poses the most grievous danger to science.

12.2.4 Non-reproducible Statistical Analysis

Every research involves data analysis and reporting. Whenever there is a problem with the design of the study, the problem could be managed (often time) but when the statistical process is abused; every other output from the data may be unreliable. Oftentimes, problems in studies are detected at the analysis stage when the data provided cannot provide an answer to the research objectives. Hence, to overcome this, the majority of researchers will force (torture) the data to conform (confess) with their research objectives (to an offence it never committed) by changing data classes, structure and, even at times, falsifying observations. This is very similar to data fudging and leads to a lack of reproducibility and openness. A recent study finds that more than 70% of researchers have tried and failed to reproduce another scientist's experiments, and more than 50% have failed to reproduce their experiments (Baker, 2016 and Feilden, 2017).

Of course, not all research that fails the test of reproducibility or repeatability is due to data fudging; sometimes it is due to a methodological fault, inefficiency or secrecy. Lack of records or procedure reporting also makes it difficult to reproduce after a long time. However, whether it resulted from methodological inefficiency, or data fudging, the tenets of statistics have been violated.

More so, besides the aforementioned abuses due to data mishandling, there are (several ways) in which statistics are being abused daily by researchers. The bottom line is, any statistics without following the proper statistical process is purely a self-medicated treatment.

12.3 Data Anatomy

In human anatomy, ten organs come into mind; the brain, oesophagus and larynx, lungs, heart, liver, spleen/kidneys, stomach, large intestine and small intestine. The **brain** is arguably the most important organ in the human body because it controls and coordinates actions and reactions. It allows us to think and feel and enables us to have memories and feelings regarding all the things that make us human.

The **oesophagus** functions as a transport point of food and fluid, after being swallowed, from the mouth to the stomach. The **larynx**, commonly called the voice box, houses the vocal folds and manipulates pitch and volume, which is essential for phonation. The **lungs** are the oxygen reservoir. The air we breathe contains oxygen and other gases. Once in the lungs, oxygen is moved into the bloodstream and carried through the body. At each cell in the body, oxygen is exchanged for a waste gas called carbon dioxide. The **heart** is important because it pumps blood around the body, delivering oxygen and nutrients to the cells and removing waste products.

The **liver** has many important metabolic functions. It converts the nutrients in our diets into substances that the body can use, stores these substances, and supplies cells with them when needed. The **kidneys** are powerful chemical factories that perform the function of removing waste products from the body. The **spleen** performs "quality control"; it removes old, malformed, or damaged red blood cells and makes the red blood cells pass through a maze of narrow passages.

The **stomach** houses and stores all that comes from the mouth; it secretes acid and enzymes that digest food. The stomach muscles contract periodically, churning food to enhance digestion. The pyloric sphincter is a muscular valve that opens to allow food to pass from the stomach to the small intestine. The **small intestine** is the "workhorse" of digestion; while food is there, nutrients are absorbed through the walls and into your bloodstream. The **large intestine** has three primary functions: absorbing water and electrolytes, producing and absorbing vitamins, and forming and propelling faeces toward the rectum for elimination.

We illustrate the relationship of human to data anatomy in Figure 12.1.

In data anatomy, statisticians are in the centre of the life cycle. Data is life but without statisticians, there is no life in data. Hence, the lungs (the oxygen reservoir that gives us life) represent the **statisticians**. The brain is the **study design**. It describes controls and coordinates actions (inputs) and reactions (expected outputs). It is the operating manual/tool of the data surgeons. It allows us to design a plan, investigate and have a clear focus of the study and purpose regarding all the things that surround the goal of the project. Without study design, no quality data can be achieved. Meanwhile, during the study design, it is traditional to consider **data ethics** (spleen – data privacy, inclusion, quality control and compliance).

The heart represents **data collection**. It results in the data itself, brings out the work of statisticians and ensures the process of the data collection follows all statistical standards. After collection, **data storage** (the stomach) enables further action such as **data preparation** (the liver). Data preparation helps us to convert the data into usable variables needed for the study. In the course of data preparation, we conduct **data cleaning** (waste removal – kidney) to remove errors due to entry, coding etc. Now, the data are ready (in the small intestine) for **validation and analysis** and eventually kept (in the large intestine) for **reusability**. The larynx and oesophagus represent the **output of the research**. The outcome of any research is a product of the study design; out of the abundance of the mind, the mouth speaketh. The flow of data anatomy is described in Figure 12.2.

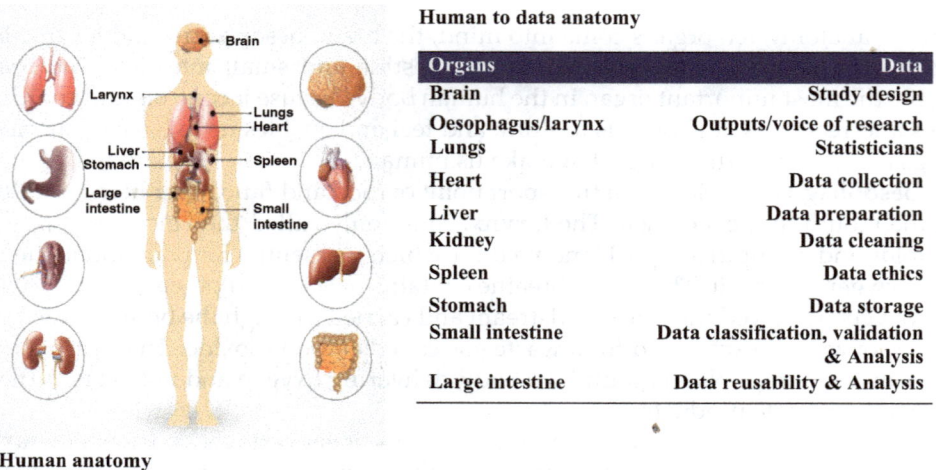

Human to data anatomy

Organs	Data
Brain	Study design
Oesophagus/larynx	Outputs/voice of research
Lungs	Statisticians
Heart	Data collection
Liver	Data preparation
Kidney	Data cleaning
Spleen	Data ethics
Stomach	Data storage
Small intestine	Data classification, validation & Analysis
Large intestine	Data reusability & Analysis

Human anatomy

FIGURE 12.1
Human to data anatomy.

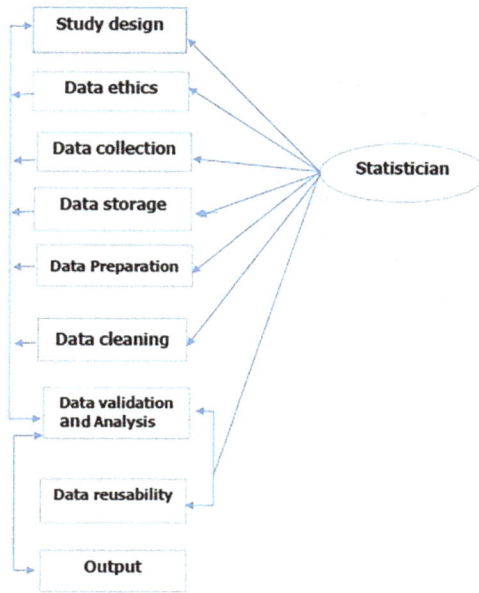

FIGURE 12.2
Flow of data anatomy.

12.4 Data Surgeon and Scientist

Many statisticians have argued that data science is not a new field, but rather another name for statistics (Silver, 2020). Others argue that data science is distinct from statistics because it focuses on problems and techniques unique to digital data (Dhar, 2013). Some have described statistics as a nonessential part of data science (Granville, 2014). Donoho (2015) describes data science as an applied field growing out of traditional statistics. Arguably, the role of data surgeon (statistician) is quite different from and more complicated than the role of a data scientist. One can be a data scientist without a solid background in the ethics of statistics. A data surgeon understands the anatomy of data as a whole while a data scientist typically specialized in some part of the data anatomy. The interest here is not in placing one over the other, but to clearly distinguish the differences. The burning question is: who is a data surgeon? And who is a data scientist?

A data surgeon (statistician) is responsible for the preoperative diagnosis of the patient: data, developing the study design, and providing data collection guidelines. They also provide postoperative advice on data handling, care, processing and treatment (methodological prescription). The data surgeon is also looked upon as the leader of the data gathering team. Essentially, the job description for data surgeons varies depending on what area of specialization they are practising in. Many data surgeons find themselves in a multifaceted career that allows them to put their skills to good use in a combination of workplace settings that promotes systematic collaboration, and career upliftment. Although the workplace settings may vary, one factor remains the same: the statistics profession is one huge privilege and responsibility of leadership, as the ambassador and voice through which data communicates in the society. Consequently, an illegitimate assumption that a person is a statistician without relevant skills and practice pollutes the data anatomy and puts the society in danger.

Just like in human surgery, any mistake with any part of human anatomy could cause death; hence, any abuse within the data anatomy will result in the violation of statistics and invalidate the output of the research. Therefore, a data surgeon (statistician) is the custodian of the ethics of data anatomy.

The first step in becoming a data scientist is to earn a bachelor's degree, typically in a quantitative field (MDS, 2020). Bootcamp coding experience can also be used as an alternate prequalification to supplement a bachelor's degree, and the majority earn a master's degree or a PhD in a quantitative/scientific field. Once qualifications are met, data scientists may later choose to specialize in a sub-field of data science or other associated specializations which include machine learning scientist, data analyst, data consultant, data architect and application architect among many others.

In contrast to statisticians (data surgeons), data scientists are analytical experts who utilize their skills in both technology and social science to find trends and manage data. They use industry knowledge, contextual understanding, and scepticism of existing assumptions to uncover solutions to business challenges. The work of a data scientist typically involves making sense of messy, unstructured data, from sources such as smart devices, social media feeds, and emails that don't neatly fit into a database. They are big data wranglers, gathering and analysing large sets of structured and unstructured data. A data scientist's role combines the subject of statistics, computer science, and mathematics. They analyse, process and model data then interpret the results to create executable plans for companies and other organizations. Hence, data scientists rely on the flow of work of the data surgeons to make good out of their profession. No mistake, data science is an applied branch of statistics.

12.5 Way Forward and Conclusion

Since data is the yardstick of decision science, it is important to safeguard all data processing and gathering, ensuring ethical standards and inclusion. Also, we must ensure that all preoperative diagnosis and postoperative treatment of data is censored by qualified data surgeons. Our world is going through a contemporary data explosion and data abuses are becoming an accepted part of data practices. If this continues, ill will become the order of the pre-operation of data anatomy and algorithms biases, and data misuse and ethical violations will attract little or no sanctions, hence leaving our society in inevitable danger of misinformation, and science will suffer greatly.

We have highlighted some important sources of statistical abuse. This work would look incomplete without proffering some solution or way forward to the problem we have identified. From our submission, some statistical abuses are inadvertent or innocent omissions. Some are due to lack of openness, secrecy and inadequate procedural or record keeping. Some are burned out of desperation to achieve the needed result at the expense of the users or readers. While all others are measurable (mistakes do happen and can be corrected), the last may require a strict law to punish the offenders since it is intentional.

Our search shows that data falsification is a peculiar reason for (the retraction of) some highly regarded articles published by a well-respected scientist. Harvey (2020) cites an example of a notorious case (a Dutch psychologist Diederik Stapel) who admitted to faking data and making up entire experiments published in a substantial number of articles (see Harvey (2020) for more examples). Thus, research fraud through data fabrications is evident as a clear violation of the fundamentals of the peer-review process and research fraud and must be given the utmost focus in peer-review.

Steen (2011a) finds that research fraud has indeed increased in recent years. This apparent increase in the incidence of research fraud (in medicine) is leading to increased harm (to patients) and society (Steen 2011b). Similarly, Bhattacharjee (2013) suggests that scientific fraud has a range of manifestations; some would go unchallenged – or on a continuum of dishonest behaviours that extend from the cherry-picking of data to fit a chosen hypothesis – to outright fabrication.

Consequently, to reduce statistical abuse in science, researchers; must do away with secrecy or its intent in research development, data collection and design. Essentially, every research must be subject to a rigorous peer-review process and open science, with a mandated expression of data collection, processing and manipulation (how the questionnaire is transcribed or coded) methods in the supplementary material. All authors must provide access to research data (raw, manipulated and used), explain the experiment procedure (where applicable) and provide codes leading to figures and diagrams. Where research adopts closed-source software, authors should provide a step-by-step process used. Such will help the reviewers to follow the process and reproduced the work done. Unwillingness to release data (and other pre-research processes) for the peer-review process should lead to automatic rejection without exception.

Finally, all journals should be committed to the integrity of peer-review, and there should be stigmatization for retraction such that journals would at all cost strive to avoid the future event of articles in their journal retracted on the ground of research fraud.

For statistics (as data and profession) to survive the disease of misuse, fraud and abuses, there is an urgent need for statistical literacy; hence statisticians and world statistical bodies must rise to the occasion. Just as we cannot talk about medicine without humans, we cannot talk about statistics without data. Every abuse of statistics is an abuse of humanity. We look forward to a world of anatomical revolution in data processing, gathering, and statistical reporting – a world where statisticians will be honoured and take the rightful position as custodians of data and its anatomy.

References

Afonja, B., Olubusoye, O., Ossai, E. and Ariola, J. (2014). *Introductory Statistics; A Learner's Motivated Approach*. Ibadan, Nigeria: Evans Publisher.

Baker, M. (2016). 1500 scientists lift the lid on reproducibility, available at https://www.nature.com/news/1-500-scientists-lift-the-lid-on-reproducibility-1.19970, accessed 12 July 2020.

Bhattacharjee, Y. (2013). The mind of a con man, *New York Times*, 26 April.

Coase, R. (1960). The problem of social cost. *Journal of Law and Economics*. 3(1): 1–44. doi:10.1086/466560.

Darwin, C. (1871). *The Descent of Man, and Selection in Relation to Sex*. London: John Murray, pp. 200–201, Vol. 1, 1st edition, available at http://darwin-online.org.uk/content/frameset?viewtype=text&itemID=F937.1&pageseq=213, accessed 30 July, 2020.

Donoho, D. (18 September 2015). 50 years of data science, available at http://courses.csail.mit.edu/18.337/2015/docs/50YearsDataScience.pdf, accessed 12 July 2020.

Dhar, V. (2013). Data science and prediction. *Communications of the ACM*. 56(12): 64–73.

Feilden, T. (2017). Most scientist' can't replicate studies by their peers, available at https://www.bbc.com/news/science-environment-39054778, accessed 12 July 2020.

Fischer, D. (1979). *Historians' Fallacies: Toward a Logic of Historical Thought*. New York: Harper & Row. pp. 337–338.

Gardenier, J. and Resnik, D. (2002). The misuse of statistics: concepts, tools, and a research agenda. *Accountability in Research: Policies and Quality Assurance*. 9(2): 65–74.

Granville, V. (2014). Blog, view. Data science without statistics is possible, even desirable, available at https://www.datasciencecentral.com/profiles/blogs/data-science-without-statistics-is-possible-even-desirable, accessed 12 July 2020.

Harvey, L. (2020). Research fraud: a long-term problem exacerbated by the clamour for research grant. *Quality in Higher Education*. 26(3): 243–261.

Indrayan, A. (2007). Statistical fallacies in orthopaedic research. *Indian Journal of Orthopaedics*. 41(1): 37–46.

Lang, A. (31 March 1844–20 July 1912). https://en.wikipedia.org/wiki/Andrew_Lang.

MDS – master's in data science (2020). What is a data scientist? available at https://www.mastersindatascience.org/careers/data-scientist/, accessed 12 July 2020.

Moore, D. S. and McCabe, G. P. (2003). *Introduction to the Practice of Statistics*. USA: W.H. Freeman and Company, 4th edition.

Neylon, C. (2009). Scientists lead the push for open data sharing, research information. *Europa Science*. 41: 22–23.

Silver, N. (2020). What I need from statisticians - Statistics views. www.statisticsviews.com.

Spirer, H., Spirer, L. and Jaffe, A. J. (1998). *Misused Statistics* (revised and expanded 2nd ed.). New York: M. Dekker.

Steen, R. (2011a). Retractions in the scientific literature: is the incidence of research fraud increasing? *Journal of Medical Ethics*. 37(4): 249–253.

Steen, R. (2011b). Retractions in the medical literature: how many patients are put at risk by flawed research? *Journal of Medical Ethics*. 37(11): 688–692.

Strasak, A. M., Zaman, Q., Pfeiffer, K. P., Göbel, G. and Ulmer, H. (2007). Statistical errors in the medical research-a review of common pitfalls. *Swiss Medical Weekly*. 137(3–4): 44–49.

13

The ABC of Successful Statistical Collaborations: Adapting the ASCCR Frame in Developing Countries

O. Olawale Awe
University of Campinas

Eric A. Vance
University of Colorado Boulder

CONTENTS

13.1 Introduction

Collaborative techniques have become important components of solving today's real and complex data problems. The advent of the big data era has made it increasingly necessary to initiate collaborations between statisticians/data scientists and researchers in various fields because collaboration allows two or more experts from similar or different domains to work together with the common goal of achieving what the individuals could not have achieved separately. Many authors have written comprehensively on this burning subject of statistical collaboration and how it can be done successfully. Some outstanding works on statistical consulting and collaboration include the works of Awe (2012), Zahn et al. (2013), Vance (2015), Derr (2000) and Vance and Smith (2019). Awe (2012) described the poor state of statistical consulting and collaboration in developing countries and called for the teaching and expansion of statistical consulting and collaboration techniques across institutions in Africa while advocating for the establishment of statistical consulting centers in all African Universities. Zahn et al. (2013) introduced the POWER (Prepare, Open, Work, End, Reflect) principle for statistical consulting into the literature. This principle helps to ensure effective and successful collaborative and consulting practice. Statistical collaboration has been described as the collaboration between a statistician or data scientist and a domain expert (DE). Vance (2015) described how developments over the past 25 years in computing, funding, personnel, purpose and training have affected academic statistical consulting centers and discussed how these developments and trends point to a range of

DOI: 10.1201/9781003261148-15

potential futures for academic statistical consulting centers. Derr (2000) described the intricate principles of effective communication in statistical consulting. Vance and Smith (2019) define statistical collaboration as "working cooperatively with domain experts to create solutions to research, business, and policy challenges and achieve research, business, and policy goals." They introduced the concept of ASCCR (Attitude, Strategy, Communication, Content and Results) as a useful framework for statistical collaborations. A collaborative statistician is a statistician who has been adequately trained on the technical and non-technical aspects of statistics and how to communicate statistical concepts clearly to non-statisticians. According to Kimball (1957), statistical collaborators are prone to commit errors of the third kind in statistical consulting and collaboration if they are not properly trained.

Indeed, the demand for statistical collaboration has increased tremendously in the 21st century. This is because larger amounts of data have been generated and collected in various areas of knowledge in recent years. These data usually come from sensors, GPS signals, satellite images and social media, among many others. The development of new statistical and computational strategies to analyze these data appropriately in an acceptable time frame becomes essential.

Most widely used data science techniques require data visualization, modeling, big data analysis and machine-learning methods. Widely used statistical learning methods includes linear and non-linear regression, discriminant analysis, cross-validation, bootstrap, clustering (k-means and hierarchical), model selection and regularization methods, splines and generalized additive models, tree-based methods, random forests, singular value decomposition, artificial neural networks and deep learning algorithms which are useful technical methods for the statistician and data scientist to be familiar with when collaborating with non-statisticians. Therefore, it is important for the DEs to seek to collaborate appropriately with data scientists and statisticians to ensure good and optimal results. Statistics has been defined as an inherently collaborative discipline (Ben-Zvi, 2007; Vance, 2015). This is because the elements of statistical methods can best be applied in conjunction with subject matter experts, hereafter referred to as DEs. These methods are not optimally useful if they are not applied for answering research questions of DEs.

Collaboratorating DEs may be researchers from academic, industrial or governmental institutions that reach out for collaboration with a statistician or data scientist.

However, finding the right collaborator or forming a good research partnership may not be easy without prior knowledge of modern trends and standard dos and don'ts. Vance and Smith (2019) wrote a seminal and influential article on the essential elements of successful statistical collaborations which they describe as the ASCCR framework. They also described several fruitful examples based on collaboration by different career sectors. For example, an academic statistician may collaborate with colleagues within the university, through formal statistical support structures created by the institution, or through more ad hoc requests and arrangements. There is no doubt that integration of statistical collaboration laboratories in teaching statistics, which the LISA 2020 Program (Awe, 2012; Awe and Vance, 2014) advocates, provides productive teaching and learning in order to develop and increase students' creative and intellectual resources in today's information society. However, a lot more needs to be expanded on the subject of successful statistical collaborations, especially in the context of developing countries, in an uncomplicated way.

Standard statistical collaboration ethics and techniques would not work well in developing countries for various reasons highlighted in Awe et al. (2015).

To have a useful impact on society, most applied statisticians and data scientists collaborate with various DEs to help understand and refine their research questions, analyze and interpret their data, and present results and recommendations for decision making

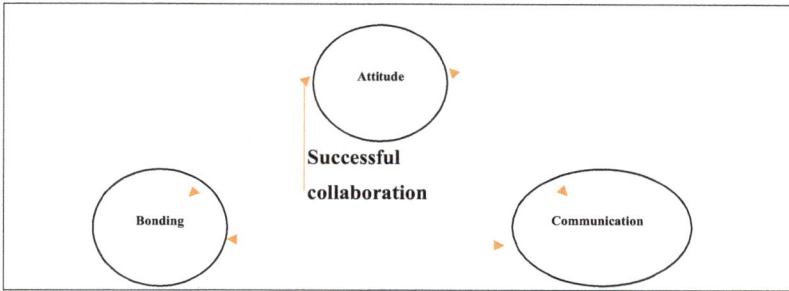

FIGURE 13.1
The interplay of successful collaboration.

and implementation. Hence, for statisticians and data scientists to maximize the impact of their work on society, they must improve the quality of their collaborations by learning and implementing essential collaboration skills. While learning collaboration skills is clearly essential for students of statistics and data science, it is also essential for academics and professional statisticians (Love et al., 2017; Vance and Smith, 2019).

The goal of this chapter is therefore to introduce the ABC of successful statistical collaboration in order to help the statistics and data science communities, especially in developing countries, to understand the principles of collaboration more clearly. Attitude, Bonding (Relationship) and Communication are pivotal and essential parts of effective and successful collaboration. This realization is based on our practical experience as statistical collaborators with vast experience in developing countries for over a decade. Figure 13.1 depicts the model of successful collaboration which we hereafter term as ABC of successful collaborations. This model can be useful in statistical collaboration and other types of research collaborations. Good attitude improves bonding between the two collaborating parties and vice-versa. When you are well bonded with someone in a good relationship, you tend to dispose good attitude toward the person. Also, good attitude promotes good communication and good communication promotes bonding. These three tripod stands are the three elements that surround successful statistical (research) collaborations and can make it seem as simple as "ABC".

The rest of this chapter is organized as follows: The next three sections deal with the ABC of statistical collaboration. It highlights three qualities that stand out from the old stories that have been told about successful collaborations by successful collaborative statisticians. Section 13.2 is based on Attitude, Section 13.3 is based on Bonding, Section 13.4 is based on Communication, while Section 13.5 concludes the chapter.

13.2 Attitude

Attitude is an organized way of thinking or feeling about someone or something, typically one that is reflected in a person's behavior or character (Awe et al., 2015). Successful research collaborations require good attitudes from all concerned parties. It starts from asking good questions (Vance and Smith, 2019) and attending meetings regularly and punctually. Coming late to a collaborative meeting (a usual practice in most developing countries) does not depict a good attitude. It can put off the collaborator. If you need to cancel or cannot make it to a meeting, you must let the other party know about 24 hours before the time of the meeting. It also entails having positive actions, always doing your part of the work and being active and alert during the collaboration. Also, always respond to emails and avoid procrastination. Most or all of these attitudinal elements listed here are lacking in developing countries and must be corrected.

According to Vance and Smith (2019) and Awe et al. (2015), some common attitudes often displayed by statistical collaborators include

- "Primarily focusing on solving the statistics/data problems of the DE"
- "Not listening carefully enough to the DE thereby committing error of the third kind"
- "Interrupting the DE rudely"
- "Pushing forth big statistics methods without thorough explanation to the DE"
- "Not attending meetings regularly and promptly"
- "Not treating the DE with respect"

However, all statistical collaborators have been advised to work ethically and abide by the American Statistical Association's (ASA) Ethical Guidelines for Statistical Practice (COPE, 2018). The Me-You-We attitude can be adopted. There are attitudes about "Me" (What you think about yourself and your role), "You" (Attitudes statisticians have about DEs which can promote or detract from collaborations) and "We" (Attitudes about collaboration itself or how two people should work together or how a statistician and DE should work together). The following are, therefore, recommended as highlights of a good attitude during collaborations:

Attend meetings regularly and promptly; never come late to a meeting, especially the first one

Turn every meeting into an opportunity to make significant contributions to the DE's work

Talk politely, thank your collaborator often for every success.

Interrupt not—it is rude to interrupt the DE. Interrupt politely if you must.

Treat everyone with respect. During a collaborative meeting, start with greeting—Introduce everyone and ensure that everyone in the room has been introduced with smiles, good eye contact and handshakes (Vance and Smith, 2019).

Understand adequately and seek to be adequately understood (Covey, 1989).

Dress corporately when meeting with your collaborator. First impression lasts long.

Explain statistics to solve problems, not to confuse the DE.

13.3 Bonding

In a successful collaboration, there must be a good bonding relationship between the DE and statistician/data scientist. Everyone wants to collaborate with someone with whom they have a good relationship (Goleman, 2006). So, collaborators must develop a smooth relationship from the beginning of the project till the end. Each member of the team must bond together, feel valued and utilized (Boroto and Zahn, 1989). In the ABC model, bonding is usually fed and enabled to grow by having a good attitude and communication. When two researchers are able to bond smoothly and intellectually, it can trigger the production of multiple successful projects by them in a rather seamless fashion. The two-terminal goals of collaboration are (1) to make deep contributions and (2) to cultivate a strong relationship (Vance and Smith, 2019). Relationships are the key to successful collaboration. In a collaborative relationship between the statistician and the DE, there is a longer relationship. Bonding with your collaborator feels easy when it flows out of shared interests, preferences, or experiences. To build a strong relationship and bonding, both parties must act respectfully, express regard and be trustworthy. Vance and Smith (2019) opines that

- Relationships are built during the collaborative process over time.
- Statisticians must respect the skills and values the DE brings to the collaboration.
- Learn and use the language of the DE's discipline.
- Building strong relationships require time, patience and trust.
- The statistician must act trustworthy to gain trust.

The following are some further recommended elements that would help in improving bonding in the collaborative relationship.

Be friendly and welcoming.

Offend not—try as much as possible not to offend your collaborators. Apologize where necessary.

Never hesitate to discontinue a relationship you are not inherently comfortable with.

Determine to make the collaborative relationship work.

Inspire each other.

Note each other's weaknesses and try to complement them.

Go the extra mile. Both parties must be willing to go the extra mile to make the collaboration work.

13.4 Communication

Effective communication is the exchange of information between two parties. It must be done in clear terms for it to be effective. Communication is highly important in every collaboration process. Both the DE and the statistician must be able to communicate clearly on the project. **Communication should be honest in order to develop trust and a good relationship between the statistician and the DE. Teams that fail to communicate effectively will waste time and energy doing things that aren't necessary**. When people talk, listen completely. Most people never listen before talking in some cultures. Communicate with the other person in a manner which you would want to be communicated to yourself. The single biggest problem in communication is the illusion that it has taken place. You must be sure your partner understands what you are trying to communicate. You must be able to communicate the available time you have for the project [this is what Doug Zahn referred to as "Time Conversation" in Zahn (2019)].

The DE must be able to communicate the content and objective of the project clearly. The statistician must seek first to understand and then be understood (Covey, 1989). After the DE has clearly explained the data and the objectives of the project, the statistician or data scientist must be sure to understand and then be able to explain his/her statistical methods while handling back the project to the DE. The two parties must communicate in a way that they understand each other. This is the only way that collaboration can be smooth and successful. Also, both parties must be ready to communicate how far they want to go (wanted conversation). Communicate clearly if you need to cancel a meeting. Communicate methods, results and recommendations well and clearly to each other. Courtesy is key. Always communicate with your collaborators with courtesy during the collaboration. Have good relationship. Be comfortable with each other. Continuity is important but communicate when you are willing to discontinue the project.

Be clear on your objectives and direction. Be up-to-date in researching the topic under consideration and be ready to communicate it clearly as a statistician (Cabrera and McDougall, 2002).

Be specific and definite about your roles, timeline and goals. All these must be communicated between the collaborators. Before a collaboration meeting, transform a few commonly asked questions into great questions and write them down on a sheet of notes. Ask these great questions in the meeting; listen, paraphrase, and summarize the responses; and reflect on how well these questions and your summary elicited useful information for the project and helped strengthen the relationship with the DE. Sometimes you can use diagrams to captivate your thoughts (Vance and Smith, 2019).

Communication is an important skill a collaborative statistician must possess. Collaborative statisticians at LISA "stat labs" in developing countries are being trained to follow Doug Zahn's POWER structure for organizing and facilitating collaboration meetings (Awe et al., 2015; Vance and Smith, 2019; Zahn et al., 2013). Two features of this structure that stand out are the "wanted conversation" and the "time conversation" during the communication phase of statistical collaboration (Boen and Zahn, 1982). The "wanted conversation" establishes what the participants want to get out of the meeting and the "time conversation" establishes how long the meeting will last. We have discovered over the years (Awe and Vance, 2018) that when working with American researchers, these techniques and the ones in Vance and Smith (2019) add helpful structure to the collaborative meetings, but they are a bit more difficult to implement in developing countries perhaps due to several cultural beliefs and patterns (Awe et al., 2015).

During our experience collaborating with researchers (DE) in developing countries, we find that they do not respect the time conversation. A DE is ready to spend the whole day with the statistician and take use up the entire time until his/her problem is solved. Any attempt to try to dismiss the DE after the expiration of the initially allotted time may be termed as being rude. By doing this, there would be no sufficient time to attend to other important projects and there is less productivity. The same thing goes for the able and wanted conversations. "Able conversation" entails telling each other how much you are able and willing to cover or offer before beginning the collaboration. Despite this prior conversation, some DEs would want to stretch the statistician further, thereby breaking the initial agreement on the "able conversation". From experience, these kinds of breaches to initial agreements are common in developing countries. It is not easy to adjourn meetings especially when the goal has not been achieved in a day.

The following rules would help in structuring communication rightly during collaborations.

Communicate your common vision and goals for the project clearly at the beginning.

Open the project with courtesy and courteous greetings (Derr, 2000).

Make everyone feel important during a conversation or meeting.

Maintain a clear and easy mode of communication especially when explaining technical terms.

Understand each other clearly during every conversation.

Never argue with your collaborator.

Interpret with diagrams where necessary.

Create and communicate an atmosphere of freedom with your words.

Ask good and useful questions.

Talk to each other with dignity.

Improve your accent where necessary, especially when communicating with someone from an English-speaking country.

Openly compliment your collaborator.

Never hesitate to communicate gray areas politely.

13.5 Concluding Remarks

This chapter has summarized the ABC of successful statistical collaborations as portrayed by the ASCCR framework of Vance and Smith (2019) with special adaptation to researchers in developing countries. We believe that the contents of this chapter would be useful for the 21st-century statisticians and data scientists. Collaboration skills are useful professional skills that can improve the effectiveness of statisticians and data scientists and thereby increase their impact in the burgeoning world of multidisciplinary research collaborations. While this framework is general and could be applied by students, researchers and professionals who collaborate in any field, we focused specifically on applying this framework to collaborative statisticians and data scientists. It is believed that both the ABC and the ASCCR Frames would help organize and stimulate research and teaching in interdisciplinary research collaborations in both developed and developing countries.

Acknowledgment

Many thanks to two anonymous reviewers.

References

Awe, O. O. (2012). Fostering the practice and teaching of statistical consulting among young statisticians in Africa. *Journal of Education and Practice*, 3(3), 54–59.

Awe, O. O. & Vance, E. A. (2014). Statistics education, collaborative research, and LISA 2020: A view from Nigeria, In K. Makar, B. de Sousa, & R. Gould (Eds.), *Proceedings of the Ninth International Conference on Teaching Statistics (ICOTS9, July, 2014)*, Flagstaff, Arizona, USA. Voorburg, the Netherlands: International Statistical Institute.

Awe, O. O, & Vance, E. A. (2018). Improving the teaching and learning of statistics in Nigerian universities through the LISA 2020 program. *Proceedings of the Tenth International Conference on Teaching Statistics (ICOTS10, July, 2018)*, Kyoto, Japan.

Awe, O. O, Crandell, I. & Vance, E. A. (2015). Building statistics capacity in Nigeria through the LISA 2020 program. *Proceedings of the 60th International Statistical Institute's World Statistics Congress*, Rio de Janeiro, Brazil, 2015.

Ben-Zvi, D. (2007). Using wiki to promote collaborative learning in statistics education. *Technology Innovations in Statistics Education*, 1, 1–18.

Boen, J. R., & Zahn, D. A. (1982). *The Human Side of Statistical Consulting*, Belmont, CA: Lifetime Learning Publications.

Boroto, D. R., & Zahn, D. A. (1989). On becoming valued and utilized. *The American Statistician*, 43, 71–72.

Cabrera, J., & McDougall, A. (2002). *Statistical Consulting*, New York: Springer.

COPE (2018). Ethical Guidelines for Statistical Practice, Technical Report, Committee on Professional Ethics of the American Statistical Association, available at https://www.amstat.org/ASA/Your-Career/EthicalGuidelines-for-Statistical-Practice.aspx.

Covey, S. R. (1989). *The Seven Habits of Highly Effective People: Restoring the Character Ethic*, New York: Simon and Schuster.

Derr, J. (2000). *Statistical Consulting: A Guide to Effective Communication*, Pacific Grove, CA: Duxbury Press.

Goleman, D. (2006). *Social Intelligence: The New Science of Human Relationships*, New York: Random House Publishing Group.

Kimball, A. W. (1957). Errors of the third kind in statistical consulting. *Journal of the American Statistical Association*, 52(278), 133–142.

Love, K., Vance, E. A., Harrell, F. E., Johnson, D. E., Kutner, M. H., Snee, R. D., & Zahn, D. (2017). Developing a career in the practice of statistics: The mentor's perspective. *The American Statistician*, 71, 38–46.

Vance, E. A. (2015). Recent developments and their implications for the future of academic statistical consulting centers. *The American Statistician*, 69(2), 127–137.

Vance, E. A., & Smith, H. S. (2019). The ASCCR frame for learning essential collaboration skills. *Journal of Statistics Education*, 27(3), 265–274.

Zahn, D.A. (2019). *Stumbling Blocks to Stepping Stones: A Guide to Successful Meetings and Working Relationships*, Indianapolis, IN: iUniverse.

Zahn, D., Smith, H., Stallings, J., Stinnett, S., & Vance, E. (2013). Understanding and improving the client-consultant interaction. *Joint Statistical Meetings*, Montreal, Canada.

Part 3

Statistics Education and Women's Empowerment

14

Statistics and Women's Empowerment: Challenges and Opportunities

Pawan Kumar Sharma and Vishal Mahajan
Sher-e-Kashmir University of Agricultural Sciences & Technology of Jammu

Lubna Naz
University of Karachi

CONTENTS

DOI: 10.1201/9781003261148-17

14.1 Introduction

Women make up 48.1% of the total population of India, having a literacy rate of 62% (2011 Census). The status of women in society has seen a continuous change from equal status with men in ancient times through the low points of the medieval period, to the promotion of equal rights in the modern era. In India, women have established their skills in almost all particulars of nation-building. In the political sphere, women have filled the high offices of President, Prime Minister, Speaker of the Lok Sabha, and Leader of the Opposition, and in local government, they have served as Panch/Sarpanches. The lack of statistics remains the most important limitation on extending benefits to women in India. In the absence of proper statistics, it has not been possible to extend the benefits of schemes effectively to women in agriculture, limiting the empowerment of rural women due to their passive participation in the development process. The role of statistics in enhancing the effectiveness of government schemes is discussed in this chapter as a way to recommend suitable policy measures to strengthen farm women in India. The chapter is organized under the following sections.

14.2 Types of Women's Empowerment

Women participate in all spheres of life and attention should be given to all sectors affecting women to advance their empowerment. The family setup and its positioning also affect the economic decision-making of women, such as whether the family is nuclear or joint; whether the family resides in a rural or urban area, all such issues also affect the extent of economic empowerment of women in a family. Further, the economic empowerment depends upon resources, both at individual as well as community level. Such resources include infrastructure for education, training, and skills, access to land, machinery, and credit, and social networking. Above all, leading social norms in a society determine the effectiveness of all these factors in creating economic empowerment of women. Women's empowerment can be further classified at three levels: the household, community, and wider (state, national) arenas. Some aspects of empowerment are discussed below.

14.2.1 Economic Empowerment

The economic empowerment of women entails their capacity to grow and succeed economically. It involves controlling and sharing economic resources, making a choice under multiple economic alternatives, and the power to retain economic gains in an activity. Therefore, economic empowerment reflects the financial independence of women in implementing ideas and generating outcomes. Many wrong economic decisions of men would not be considered to be due to inferior economic thinking, but a wrong economic decision by a woman can very easily expose her to gender bias. Women are allowed to manage the economic affairs of their family, but economic decisions outside the boundaries of their homes are still not under their control.

14.2.2 Socio-Cultural

Socio-cultural empowerment includes freedom in decision-making related to issues such as free uninterrupted movement, access to education and health, and availability of a social space to discuss gender issues such as dowry, drug addiction, and a shift in patriarchal norms.

14.2.3 Interpersonal

Interpersonal empowerment reflects the freedom of women to make decisions on their own behalf related to birth control, marriage, and sexual relations and reproduction.

14.2.4 Legal

Women are protected against criminal acts in every country through a written legal framework. The effectiveness of such legal settings, however, is dependent upon the knowledge of women to exercise the rights given by the rule of law.

14.2.5 Political

Women's political empowerment involves their representation in local, regional, and national government bodies. The reservation of positions for women in the political arena is always recommended to enhance their political empowerment. India reserves 33% of the seats in local municipal bodies for women. Sri Lanka, Bangladesh, and Nepal also follow a reservation policy through legislation for political empowerment of women.

14.2.6 Psychological

The sense of inclusion, self-esteem, and entitlement determine the psychological empowerment of women. Whether a woman has the confidence to live single in society reflects the collective psychology of that society.

14.3 Statistical Measurement of Women's Empowerment

All societies aim at achieving women's empowerment in different spheres as discussed above. Statistics plays a greater role in achieving this by affording greater insight into a given situation and allowing suitable steps to be taken to achieve the same. The United Nations developed two indices for the systematic measurement of gender-related issues, namely, the Gender Development Index (GDI) (1995) and Gender Inequality Index (GII) (2010).

TABLE 14.1

Measurement of the Gender Development Index

Dimensions	Indicators	Dimension Index	Index	
Long and healthy life	Life expectancy Adolescent birth rate	Life expectancy index	• Human Development Index (Female)	Gender Development index=HDI (Female) / HDI (Male)
Knowledge	Expected years of schooling Mean years of schooling	Education index	• Human Development Index (Male)	
Standard of living	GNP per capita (PPP$)	GNI index		

14.3.1 Gender Development Index

The GDI is an extension of the Human Development Index (HDI) to capture gender gaps by accounting for disparities between men and women. Three basic dimensions, health, knowledge, and living standards, are used to construct the index. It is calculated as the ratio of the HDIs for females to those of males (Table 14.1).

Norway, with a GDI value of 0.990, topped the list of 189 countries on the 2019 HDI report. India, although rising one slot to 129th with a GDI of 0.829, remains in the category of countries with medium human development (UNDP, 2019a).

14.3.2 Gender Inequality Index

The GII, introduced in the Human Development Report 2010 of the United Nations Development Program (UNDP), constitutes a useful indicator for measuring gender disparity as a composite measure of inequality between women and men. It includes three dimensions: reproductive health, empowerment, and the labor market. The maternal mortality ratio (MMR) and adolescent birth rate are the health indicators, whereas the share of seats in parliament and the population with secondary education determine empowerment (Table 14.2). Labor force participation measures women's contribution to the labor market.

Switzerland with a GII value of 0.037 ranked first among 162 countries, whereas India (GII value of 0.501) ranked 122nd in 2018 (UNDP, 2019b).

14.4 Statistics for Improving the Ranking of Countries

The universal indices for evaluating gender development and inequality are important indicators of a country's position in the world. The vital statistics for females provide data for comprehensive guidelines to improve the status of women through the introduction and implementation of appropriate policies and programs. Statistics help in deciding the direction of policy wherein steps are taken according to the needs of the nation.

The association of vital statistics and female literacy is well established: Higher literacy rates are associated with lower birth rates and infant mortality rates (Saurabh et al., 2013; Imai et al., 2014; Shetty & Shetty, 2014; Rao & Pingali, 2018).

TABLE 14.2

Measures of the Gender Inequality Index

Dimensions	Indicators	Dimension Index	Index	
Health	Maternal mortality ratio Adolescent birth rate	1. Female reproductive health index	1, 2, and 4 generate Female gender index	Gender Inequality index is a composite measure of the female and male gender indices
Empowerment	Female and male population with at least secondary education	2. Female empowerment index		
		3. Male empowerment index		
	Female and male shares of parliamentary seats		3 and 4 generate Male	
Labor market	Female and male labor force participation rates	4. Female labor market index		
		5. Male labor market index	gender index	

TABLE 14.3

Focus on Women Orientations in Five-Year Plans

Plan	Focus
6th Five-Year Plan	Shift from welfare to developmental issues
7th Five-Year Plan	Raising economic and social status of women
8th Five-Year Plan	Increased emphasis on economic activities
9th Five-Year Plan	From development to empowerment
10th Five-Year Plan	From women alone to gender mainstreaming
11th Five-Year Plan	Proposed to move toward a holistic approach

14.5 Evolution of Policy for Women in India

In India gender mainstreaming started in the Sixth Five-Year Plan when "opportunities for independent employment and income" for women were recognized as a necessary condition for raising their social status. Again, statistics help in deciding the direction of policy for women empowerment in India. The chronology of planning women empowerment in the five-year plans in India is presented in Table 14.3.

14.6 Women in Agriculture

Women are an integral part of farm households, as the majority of farm operations are now handled by them, including plowing, irrigation, sowing, weeding, harvesting, and marketing (Ghosh & Ghosh, 2014). Rao (2006) estimated that 70% of farm work in India is performed by women. In addition, women performed 90% of hoeing and weeding and 80% of the work on food storage and transport. Their economic contributions toward their own household's farm operations are largely underestimated, and statistics can play a greater role in ensuring that they receive their due share in policy planning (Agarwal, 1986). Handling of farm operations by women favors a better nutritional status of their

TABLE 14.4

Organizations Responsible for Women's Empowerment

Tasks for Gender Mainstreaming in Agriculture	Main Ministries	Role of Ministry of Agriculture
Women's empowerment (human capital formation, exposure, leadership, autonomy, self-esteem, and food security)	MoA, MoRD, Social welfare, HRD, Health	Gender focused strategy for agricultural growth (main contributor along with other ministries)
Capacity Building in Agriculture (dissemination of information and technology)	MoA	Various extension and training programs (almost the sole contributor)
Access to Agricultural Inputs (including land, water, and credit besides agri-inputs).	MoRD, MoA, MoEF	Access to agricultural inputs, formation of SHGs, marketing facilities (partial contributor with MoRD and MoEF having a major control over property rights regimes).

TABLE 14.5

Profile of Rajkarni

Age	52 years	
Education	10th	
Farm Land	2.2 acres	
Marital Status	Married	
Social status	Naib Tehsildar	
Address	Village: Said Sohal	
District	Kathua	

families and children (Dillon et al., 2015). The government has set up units with the specific purpose of achieving gender budgeting and maintenance of statistics to forward the benefits to farm women (Table 14.4).

14.7 Success Story of a Farm Woman Entrepreneur

Said Sohal village is a rainfed village selected by the Farm Science Centre of the State Agricultural University at Kathua District. Some of the major problems faced by women entrepreneurs include a patriarchal society, an absence of entrepreneurial aptitude, a lack of marketing and financial provisions, family conflicts, shortages of raw materials, social barriers, lack of information, lack of access to technology, and lack of training. However, Rajkarni, a woman farmer entrepreneur of Negotiated Indirect Cost Rate Agreement, has overcome all the problems to become a successful entrepreneur of Kathua District. Her profile is presented in Table 14.5.

Concerted efforts have been made by the Farm Science Centre to strengthen farming women by ensuring their participation in training programs, partnerships in On-Farm Trials, frontline demonstrations of new technologies, and cgroup activities. This has resulted in building Rajkarni's confidence, who converted herself from a woman farmer to an entrepreneur. Some of the aspects of her entrepreneurship development are discussed below.

14.7.1 Rajkarni as Farm Woman Producing Field Crops

Rakarni is actively engaged in growing field crops during both the kharif and rabi seasons. She managed all the farm operations herself and performed all the scientific operations suggested by the KVK. The areas under different crops that she manages are presented in Table 14.6.

14.7.2 Rajkarni as a Farm Woman Producing Horticultural Crops

Rajkarni also maintains an orchard with citrus, mango, and guava plants. Moreover, vegetable cultivation is an essential component of nutritional kitchen gardening. Details of the horticultural crops maintained in her farm area are presented in Table 14.7.

14.7.3 Rajkarni as a Farm Woman Processing Farm Produce

Processing is the most important aspect of rainfed farming. Rajkarni is actively engaged in processing farm produce, including vegetables and fruits. She has also mobilized neighboring farm women to adopt value addition collaboratively at the village level (Table 14.8).

TABLE 14.6

Cropping Pattern Followed by Rajkarni

Crops	Area (acres)	Production (q)	Net Income
Kharif			
Maize	1.60	26	15000/-
Sorghum	0.10	1.2	12000/-
Rabi			
Wheat	1.60	15	13200/-
Oats	0.10	1.3	12500/-
Total			**52,700/-**

TABLE 14.7

Detail of Horticultural Crops Grown by Rajkarni

Crops	Area (acres)	Production (q)	Net Income
Kharif			
Summer vegetables	0.25	10	20200/-
Rabi			
Winter vegetables	0.25	12	21000/-
Fruit crops(Citrus)	0.20	14	12000/-
Total			**53,200/-**

TABLE 14.8

Details of Processing and Associated Economics

Items Processed	Annual Quantity Produced (q)	Net Income
Pickles		
Vegetables, Lemon, mango	2.0	12000/-
Candy		
Aonla	0.35	15000/-
Total		**27,000/-**

TABLE 14.9

Detail of Group Dynamics

Particulars	Members	Group Activities
Group Dynamics		

Self Help Group	16 women	Social:
		  Problems identification and solutions through consensus
		Economic:
		  Management of finances, micro credit, processing and marketing

14.7.4 Rajkarni as a Farm Woman and Group Leader

Rajkarni has great leadership qualities that have made her successful in bringing all the rural women together to carry out joint efforts in farming, including the adoption of the latest scientific technologies (Table 14.9).

14.7.5 Rajkarni as a Farm Woman Entrepreneur

Besides performing farm activities, Rajkarni also manages a shop in her own village where she undertakes direct sales of milk and milk products produced by her group's members (Table 14.10).

14.7.6 Rajkarni as a Farm Woman and Environmentalist

Being a progressive farm woman, Rajkarni maintains environmental quality by applying sustainable practices (Table 14.11).

TABLE 14.10

Details of Items Sold in Village Shop

Particulars	Activities Conducted
Business	
Owner of sweet shop	• Processing of milk and sale of milk products • Management of funds and accounts of shop • Generated employment for a full time labor

TABLE 14.11

Details of Sustainable Practices Adopted

Particulars	Activities
Conservation Agriculture	
Recycling and reuse of farm outputs as farm inputs	• Vermicompost unit
	• Use of stored farm pond water for vegetable production

TABLE 14.12

Honors Received by Rajkarni

Level	Recognized by
District	KVK Kathua
State	SKUAST-Jammu
National	DD Kisan

14.7.7 Rajkarni as a Recognized Farm Woman Entrepreneur

Rajkarni's efforts and hard work have been well recognized at the state and national levels. She has received several awards as a token of her contributions to boosting agricultural production and cooperation in the Negotiated Indirect Cost Rate Agreement (NICRA) village. A documentary on her efforts has also been covered by DD Kisan. Details of her honors are presented in Table 14.12.

One can imagine the role of statistics and data in highlighting the success of Rajkarni. Much such work in agriculture and the rural sectors has been conducted by women that needs to be documented and authenticated with the relevant statistics. The absence of such data has resulted in underestimating women's contribution, thus restricting the economic empowerment of women.

14.8 Use of Statistics for Implementing Policies for Women

The Ministry of Rural Development is implementing various poverty alleviation and rural development programs. These programs have special components for women. Major schemes having a women's component implemented by the Ministry include the Mahatma Gandhi National Rural Employment Guarantee Act (MGNREGA), Swarnjayanti Gram Swarozgar Yojana (SGSY), now restructured as National Rural Livelihood Mission (Aajeevika), and the Indira Awaas Yojana (IAY). The implementation of these programs is monitored specifically with reference to the coverage of women. While the benefits of these programs flowing to the women can be measured in quantitative terms, for other programs such as Pradhan Mantri Gram Sadak Yojana (PMGSY), it is not always possible to collect segregated data reflecting the direct benefits flowing to the rural women. However, this program does have a significant impact on the living conditions of rural women by providing connectivity through the rural roads, which may enhance the opportunities for girls to gain access to education. Similarly, with better rural roads women may have easier access to health facilities and the local market, which may not only increase their productivity but also their awareness, which goes a long way in changing the traditional social structure and in improving the status of rural women. Similarly, the MGNREGA program guidelines stipulate certain provisions for the creation of specific facilities at the worksites for working women, facilitating their participation in the program. Even under the programs where direct benefits are earmarked for the women, the actual flow of benefits may be much higher, as some of the indirect benefits flowing to the women through the effective implementation of programs are not reflected in the physical progress reports collected from the program implementing agencies. These benefits can only be captured through micro-level or area-specific impact studies with various statistical and research tools.

14.9 National Policy for the Empowerment of Women

In order to improve its overall ranking on the gender-based index, India adopted the National Policy for the Empowerment of Women in 2016 to create an environment enabling women to realize their full potential by providing them equal access in all spheres. The priority areas of the policy are health, education, the economy, agriculture, industry, labor and employment, the service sector, governance and decision-making, and gender budgeting. The policy also targets women in agriculture with the following framework:

- Visibility, entitlement to agricultural services, social protection coverage, etc.
- Training in agriculture and allied subjects to create Krishi Sakhis
- Identifying women as champions of genetic diversity conservation
- Right to immovable property
- Encouraging joint or sole registration of land
- Collective farming through provision of post-harvest, processing, and marketing facilities
- Capturing work and participation in On-Farm Research
- Special package for wives of farmers who committed suicide.

14.9.1 Umbrella Integrated Child Development Services (ICDS)

The Umbrella Integrated Child Development Services, including Anganwadi Services Scheme, Pradhan Mantri Matru Vandana Yojana, National Creche Scheme, Scheme for Adolescent Girls, Child Protection Scheme, and POSHAN Abhiyaan, have been introduced to improve the statistics on the nutritional status of children and women. Community-level planning is recommended, as the impact of the scheme could not be significantly observed in children's nutritional status (Sachdev & Dasgupta, 2001; Dixit et al., 2018).

14.9.2 Mahatma Gandhi National Rural Employment Guarantee Act (MGNREGA)

The MGNREGA guarantees 100 days of employment in a financial year to any rural household whose adult members are willing to do unskilled manual work. It is provided in the Act that when providing employment, priority shall be given to women such that at least one-third of the beneficiaries shall be women who have registered and requested for work under the Act. During the year 2012–2013 (up to December 27, 2012), total employment of 1.3476 billion person-days were reported to have been generated. The employment generated for women was reported as 71.88 crore person-days or 53.34% of total employment generated under this program. To increase participation rates of women workers in MGNREGA, the Ministry has suggested that individual bank/post office accounts must compulsorily be opened in the name of all women MGNREGA workers and their wages directly credited to their own account for the number of days worked by them.

14.9.3 National Rural Livelihood Mission (NRLM)

SGSY has been restructured as the National Rural Livelihood Mission (Aajeevika). A Self Help Group (SHG) of 10–20 women in general (5–20 in difficult areas) is the primary building block of the NRLM institutional design. NRLM should promote SHGs with exclusive women membership. The SHGs and the federation of these SHGs at the village and higher levels shall serve the purpose of providing women members space for self-help, mutual cooperation, and collective action for social and economic development. NRLM is working with groups having exclusive women membership because it recognizes that women are marginalized in the economy, the polity, and society. Thus, building and sustaining the institutions of poor women at various levels would give them social, economic, and political empowerment and thereby bring significant qualitative improvements to their lives. NRLM will especially focus on women-headed households, single women, women victims of trafficking, women with disabilities, and other such vulnerable categories. It is envisioned that by creating such institutions of poor women, NRLM shall facilitate women to assert their rights for inclusion in the economy, for accessing resources, for addressing powerlessness and exclusion, for enabling participation, and most significantly for realizing equity.

An important component of NRLM is the Mahila Kisan Sashaktikaran Pariyojana (MKSP), which aims at supporting women farmers. Primarily, MKSP aims to recognize women farmers, a hitherto unrecognized category, even though most farming activities are almost exclusively handled by women. MKSP also, inter alia, seeks to reduce drudgery for women farmers. During the year 2012–2013 (as of January 22, 2013), out of 569,912 swarojgaris assisted, 477,944 swarojgaris (83.86%) were women.

14.9.4 The Indira Awaas Yojana (IAY)

IAY aims at assisting with the construction of houses to the people below the poverty line in rural areas. Under the scheme, priority is extended to widows and unmarried women. It is stipulated that IAY houses are to be allotted in the name of women members of the household or jointly in the names of the husband and wife. The total number of Dwelling Units sanctioned during the period of 2012–2013 (as of November 22, 2012) was 2,215,637 out of which 1,329,550 (60%) houses were in the name of women and 561,962 (25.36%) were jointly in the name of husband and wife.

14.9.5 e-Trading of Products by Women

Women are mainly confined to production activity, with restricted involvement in marketing functions due to the social norms prevailing in India. Linking women to the market can lead to their empowerment in agricultural domains (Gupta et al., 2017). The northeast is an exception where one can find women as the main driver of marketing functions, both in agriculture and nonagricultural goods. Keeping in view the importance of connecting women to the market for their economic empowerment, the Ministry of Women and Child Development developed Mahila E-haat, a bilingual portal that launched on March 7, 2016. It is a unique direct online marketing platform leveraging technology to support women entrepreneurs/SHGs/NGOs. The products made by women can be showcased to unlimited customers, establishing direct contact between the vendors and buyers.

14.9.6 National Mission on Agricultural Extension and Technology (NMAET)

Keeping in view the contribution of women in agriculture, special provisions in the National Mission on Agricultural Extension and Technology (NMAET) have been reserved for women, including in the Sub-Mission on Agricultural Extension, Sub Mission on Seed and Planting Material, Sub Mission on Agricultural Mechanization, and Sub Mission on Plant Protection and Plant Quarantine. Group activities for women are encouraged to achieve food security through the provision of subsidies for planting material, machinery, and other resources.

14.10 Constraints in Maintaining Statistics on Women in India

Statistics have a great role to play in every sphere of life. In order to attain gender equality, statistics must lead the way to highlight shortcomings in the present system. It is estimated that 50% of the work done by women in India is unpaid. They make up 40% of agricultural labor but have access to 9% of the land. Let us take another perspective where the contribution of owned labor or work done by women in their household would not be included in the national income. Women in farm households perform almost every activity, yet their contribution remains underestimated. In this scenario, it would be difficult to allocate resources in accordance with their contribution. Moreover, women in India still face restrictions on working outside their home or social domain. If the female half of the population could not contribute to economic activity or their work remains outside the domain of economic assessment, improving the GDI or GII will be a very cumbersome task. The increase in female literacy rate along with social changes in terms of women-oriented jobs is a positive sign, and real-time statistics in this regard can help highlight women's share and thus promote a more equitable distribution of resources.

14.11 Conclusion

India has witnessed shifts in women's status in society from equal status with men in ancient times through the low points of the medieval period to the present scenario of promoting their equal rights. In order to make women count in economic activity, we need to count every aspect of work done by women in society. The availability of real-time data on women for measures to support them, such as skill-oriented and vocational training, is imperative. They must be involved in market-oriented activities and financial literacy programs to develop their decision-making abilities. Their access to Information and Communication Technologies (ICT) can help reveal the extent of their involvement in social, political, and economic issues. Gender-disaggregated data on land ownership and inheritance need to be maintained to achieve their economic empowerment. The documentation of successful women entrepreneurs is also an integral part of maintaining statistics. Success stories provide impetus and confidence to other members of society in achieving their goals. The examples of successful women entrepreneurs can serve as building blocks for achieving the large-scale empowerment of women and raising the country's gender-based indices.

References

Agarwal, B. 1986. Women, poverty and agricultural growth in India. *The Journal of Peasant Studies*, 13(4): 165–220.

Dillon, A., McGee, K., Oseni, G. 2015. Agricultural production, dietary diversity and climate variability. *The Journal of Development Studies*, 51(8): 976–995.

Dixit, P., Gupta, A., Dwivedi, L.K., Coomar, D. 2018. Impact evaluation of integrated child development services in rural India: Propensity score matching analysis. *SAGE Open*, 1–7. doi: 10.1177/2158244018785713.

Ghosh, M.M., Ghosh, A. 2014. Analysis of women participation in Indian agriculture. *IOSR Journal of Humanities and Social Science (IOSR-JHSS)*, 19(5): 01–06.

Gupta, S., Pingali, P.L., Andersen, P.P. 2017. Women's empowerment in Indian agriculture: Does market orientation of farming systems matter? *Food Security*, 9: 1447–1463.

Imai, K.S., Annim, S.K., Kulkarni, V.S., Gaiha, R. 2014. Women's empowerment and prevalence of stunted and underweight children in rural India. *World Development*, 62: 88–105.

Rao, E.K. 2006. Role of women in agriculture: A micro level study. *Journal of Global Economy*, 2(2): 107–118.

Rao, T., Pingali, P. 2018. The role of agriculture in women's nutrition: Empirical evidence from India. *PLoS One*, 13(8): e0201115. doi: 10.1371/journal.pone.0201115.

Sachdev, A., Dasgupta, J. 2001. Integrated Child Development Services (ICDS) scheme. *Medical Journal of Armed Forces India*, 57(2): 139–143. doi: 10.1016/S0377-1237(01)80135-0.

Saurabh, S., Sarkar, S., Pandey, D.K. 2013. Female literacy rate is a better predictor of birth rate and infant mortality rate in India. *Journal of Family Medicine and Primary Care*, 2(4): 349–353.

Shetty, A., Shetty, S. 2014. The impact of female literacy on infant mortality rate in Indian States. *Current Paediatric Research*, 18(1): 49–56.

United Nations Development Program (UNDP). 2019a. Human Development Report 2019. http://hdr.undp.org/sites/default/files/hdro_statistical_data_table4.pdf.

United Nations Development Program (UNDP). 2019b. Human Development Report 2019. http://hdr.undp.org/sites/default/files/hdro_statistical_data_table5.pdf.

15

Women in STEM, Progress and Prospects: The Case of a Ghanaian University

Atinuke Adebanji and Salami Suleiman

KNUST-Laboratory for Interdisciplinary Statistical Analysis (KNUST-LISA)

CONTENTS

15.1 Introduction

The role of women in society has evolved from passivity to active involvement in contemporary times and can no longer be overlooked. It has been agreed that the development and growth of a nation would not be possible unless women are brought into the mainstream of national development. This is because women make an important contribution to economic development and further make significant improvements to life [1]. In most regions of the world, women make up half the national workforce, may earn more college and graduate degrees than men, especially in the social-related fields of study, and by some estimates represent the largest single economic force in the world. Yet a gender gap persists in Science, Technology, Engineering, and Mathematics (STEM) more than in other professions.

The 2012 US economic projections indicated a need for approximately 1 million more STEM professionals than the United States could produce at the current rate over a decade for the country to retain its historical preeminence in science and technology. To achieve this feat, a projected 34% increase in undergraduate STEM degrees is required [2]. Much of this shortfall in STEM training could be filled by well-trained and highly skilled female scientists. If so many STEM jobs are required to sustain a developed economy, it would be logical to postulate that the STEM workforce needed to pull the Ghanaian economy out of the present lower-middle-income status in which it finds itself will run into extremely high numbers.

In addition to the statistics in the United States, the Census Bureau of Statistics shows an appreciable increase of women in STEM jobs over the past four decades. However, it is

DOI: 10.1201/9781003261148-18

reported that this increase has not been consistently maintained. There was an increase from 7% to 23% in the number of women in STEM jobs in the United States, but the last two decades saw only a mere 3% growth [2]. Questions have, therefore, been raised regarding this staggering progress, and importantly on the paucity of women in science careers. Even though the issue amounts to possible differences at the high extremes of ability distributions, the available evidence points to the fact that social and cultural forces may contribute to this. It is, therefore, critical to overcome these factors and erase the assumption that women are not wanted or incompetent. One of the ways to overcome these inherent contributors to the paucity of women in science is through empowerment.

Girls' and women's empowerment through science and technology has the potential to enable them to realize their potential, shape their life in accordance with their aspirations, and strengthen the advancement of science and eventually wealth accumulation at both the individual and national levels.

The global framework for collective action to reduce poverty and improve the lives of poor people was codified in the Millennium Development Goals (MDGs), which later evolved into the Sustainable Development Goals (SDGs). The MDGs specified poverty eradication as their first goal. Poverty has been said to have a woman's face, and it is universally agreed that empowering women is the best way to fight poverty. Interestingly, the SDGs also have empowered women for development as the 5th of its 17 goals. This implies that achieving the SDGs seems to be an elusive destination if innovative approaches toward empowering and inspiring women are not conceptualized and pursued aggressively, not only for the eradication of poverty but more importantly for the protection of Planet Earth.

A negative correlation has been shown between a country's developmental index and student attitudes to science [10]. In Ghana, science and technology play a critical role in the economic development of the country, and both men and women are needed to be actively involved in science and technology advancement. Female scientists all over the world, however, are known to face more difficult conditions in pursuit of their profession than their male counterparts and report lower levels of productivity in terms of publications in academics [1].

The advent of the COVID-19 pandemic with its attendant mitigation and suppression measures, which have resulted in unpleasant outcomes of a financial and economic downturn for millions of households, has been a source of additional pressure on women and girls. It has been shown that COVID-19 has disproportionately affected women, further widened gender inequalities, and impacted the professional output of women more than men because of home and childcare responsibilities ([12], [13] and [14]).

The pandemic has brought to the fore the pressing need for highly skilled manpower in information technology as the world quickly adapts to technology-driven modus operandi for education, trade, finance, and social networking. There are also more opportunities to work from home, which affords more women the opportunity of less rigid domestic arrangements.

Science, Technology, and Innovation (STI) plays a strategic role in Africa's development, and this was corroborated by the heads of government of African states by signing the 10-year African science agenda, dubbed the Science, Technology and Innovation Strategy for Africa (STISA-2024), in 2014. It is hoped that STISA-2024 would help streamline the STI activities of African states and move the continent from the present state of low science and technological prowess to states with highly skilled and knowledge-based societies [7].

Post-independence Ghana had a leader (Dr. Kwame Nkrumah) who believed that national development was only feasible if science and technology were given the pride of place in national strategic planning. Some institutions were established as a result, one of them being the College of Science and Technology in 1951 (now Kwame Nkrumah University of

Science and Technology). Despite the initial post-independence technological and scientific capacity-building efforts, Ghana has not metamorphosed into a country with STI as the main driver of all socio-economic activities. Ghana presently does not place a sufficient premium on STI as evidenced in the proportion of the national budget allocated to STI fluctuating between 0.3% and 0.35% of the Gross Domestic Product (GDP). This is below the minimum benchmark of 1% set by Organisation of African Unity (OAU) in 1980 [3].

A definitive national policy on STI that spells out the vision, goals, objectives, and investment priorities in research and development is still lacking in Ghana. Previous policy documents such as the 2000 and 2004 papers did not reach the implementation stage. The current national STI policy of May 2017 identifies three areas of concern for building Ghana's STI capacity, along with 13 constraints on STI applications. The document, however, fails to identify the low representation of women in STEM careers as a possible encumbrance of national technological development. Section 4.7 on *Promoting the Participation of Women in Science and Technology*, however, acknowledges the low representation of women in science and technology careers and proposes that special incentive and motivation packages for female science students be introduced by the Ministry of Environment, Science, Technology and Innovation (MESTI) to improve female representation. It is unclear, however, what exactly these measures would be [3].

The population of Ghana is 52% female, but there are fewer women involved in the learning, teaching, and practice of STEM-related fields than their male counterparts. This is primarily because women have traditionally not been encouraged to embrace STEM courses at all levels of educational learning.

15.2 The Problem

The problem of female participation in STEM careers has been studied by several researchers. A growing literature reveals gender stereotyping against women in STEM programs as students and their career as possible explanations. Such stereotypes are known to be pervasive [5]. Again, there are relatively few successful female scientists (thankfully now increasing), technologists, engineers, and mathematicians to serve as role models and mentors for young and upcoming females in such programs. Yet others have also debated the lack of resources as a key contributing factor to the paucity of women in STEM. A failure of more women to pursue STEM careers and to maintain their presence in high positions requires a deliberate and concerted effort on all frontiers. To gain a better understanding of this phenomenon, an analysis of the total enrollment in postgraduate STEM programs at KNUST was performed.

15.3 Results

The total female enrollment from 2000 to 2017 into STEM postgraduate programs is presented in Figure 15.1. The plot depicts a noisy pattern with a gentle upward climb from 2012 to 2017. The data will be disaggregated by gender and compared across Engineering, Life Sciences, and Physical Sciences.

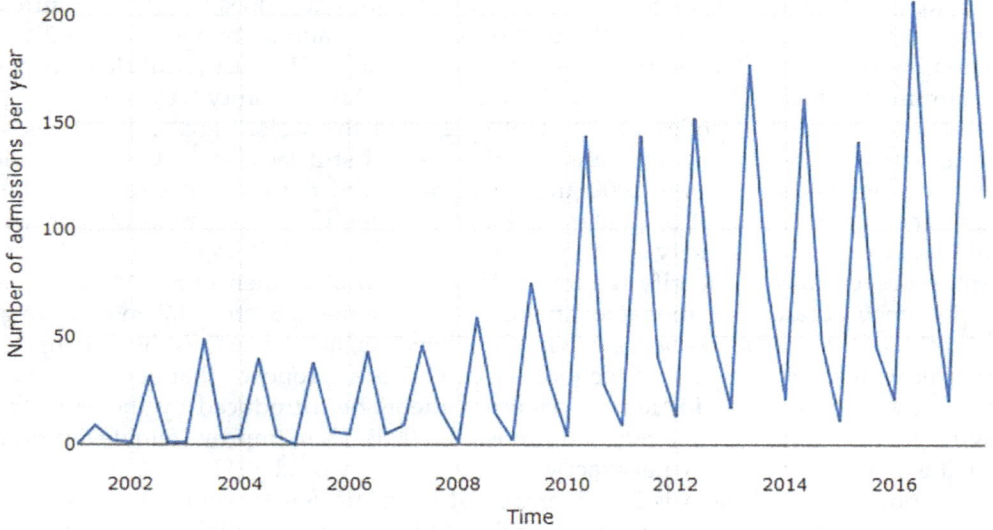

FIGURE 15.1
Total female enrollment in STEM postgraduate programs.

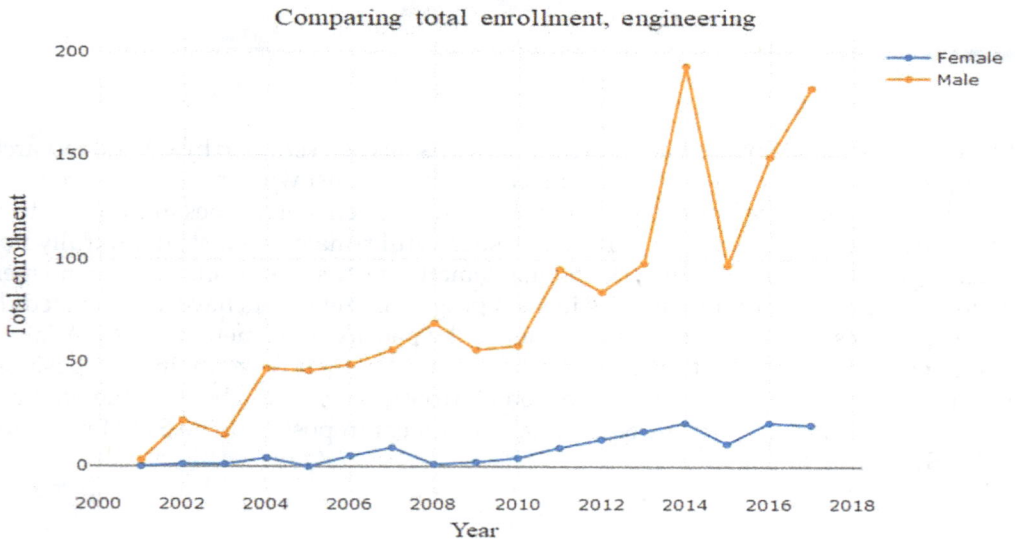

FIGURE 15.2
Postgraduate enrollment in Engineering programs.

The trend of time plots of total postgraduate enrollment in Engineering and Physical Science programs from 2000 to 2017 (disaggregated by gender) is presented in Figures 15.2 and 15.3, respectively. The male enrollment in engineering shows a very rapid climb, while the female enrollment continues to hover close to the horizontal axis. The female

Comparing Total Enrollment, Physical Sciences

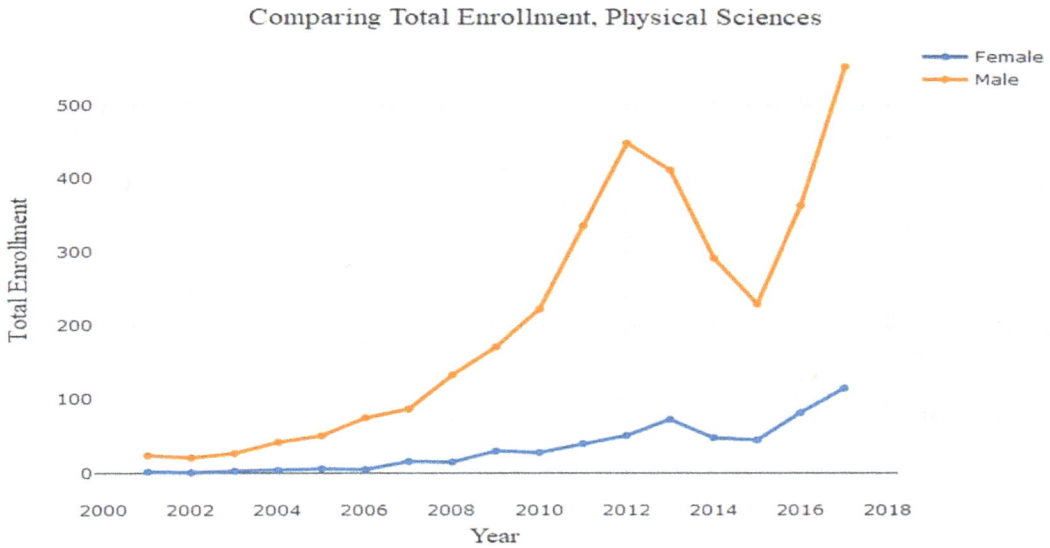

FIGURE 15.3
Postgraduate enrollment in Physical Science programs.

Comparing total enrollment, life sciences

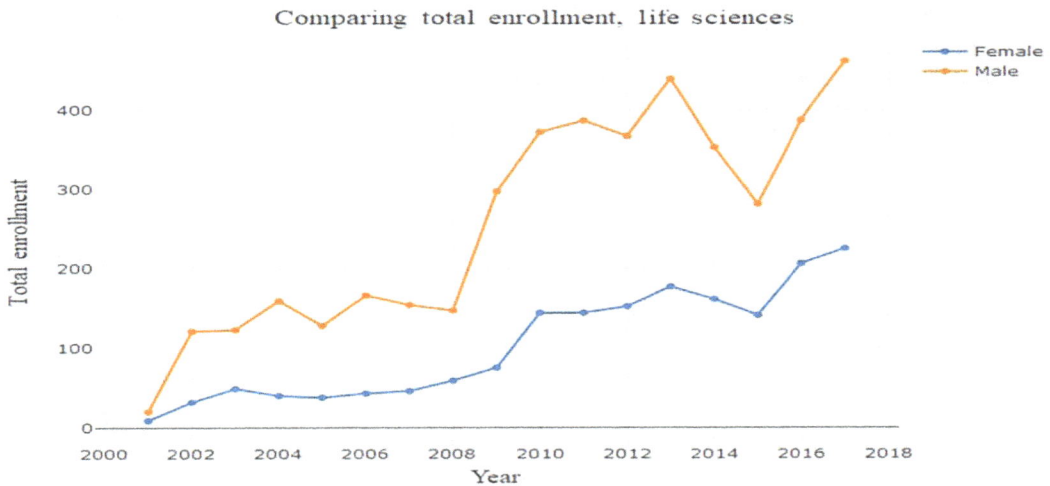

FIGURE 15.4
Postgraduate enrollments in Life Science programs.

postgraduate enrollment in Physical sciences programs follows the same pattern observed in the plot for Engineering but with greater female participation. The enrollment in Life Sciences programs for the same period shows greater participation of females, but still not at par with males. Health Sciences is the driver of female participation in the life sciences. The plot showed a climb in 2010 and another climb in 2016 (Figure 15.4).

15.4 Linear Trend Model for Female Enrollment

A trend-line model is fitted to the female data to determine the suitability of describing the enrollment data by a single regression line, yielding a significant negative intercept ($p < 1.37e{-}05$) and an equally significant time component ($p < 1.29e{-}05$), both significant at the .001 level of significance. The coefficient of determination shows that 32% of female enrollment is explained by time. The result of the linear trend model is presented in Table 15.1, and the predicted regression line is shown in Figure 15.5.

The time-on-time trend plot for total female enrollment in Figure 15.6 shows a steady progression in total enrollment that, however, is not observed across disciplines. Life sciences are the highest gainer of all the STEM categories.

No seasonal pattern was observed in the data when decomposed into the respective components. A trend is observed as shown in Figure 15.7.

TABLE 15.1

Linear Trend Model Estimates

Coefficients	Estimate	Std. Error	t-value	Pr(> |t|)
(Intercept)	−13631.547	2820.204	−4.834	1.37e–05 ***
time(tm)	6.808	1.404	4.851	1.29e–05 ***

Signif. codes: 0 '***' 0.001 '**' 0.01 '*' 0.05 '.' 0.1 ' ' 1
Residual standard error: 49.18 on 49 degrees of freedom
Multiple R²: 0.3244, Adjusted R²: 0.3106
F-statistic: 23.53 on 1 and *df=*49, *p*-value: 1.289e–05

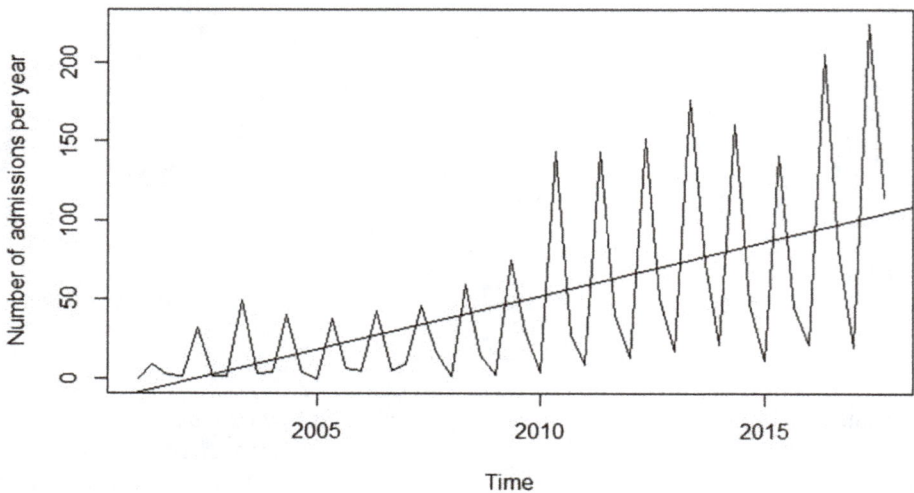

FIGURE 15.5
Regression line for total female enrollment.

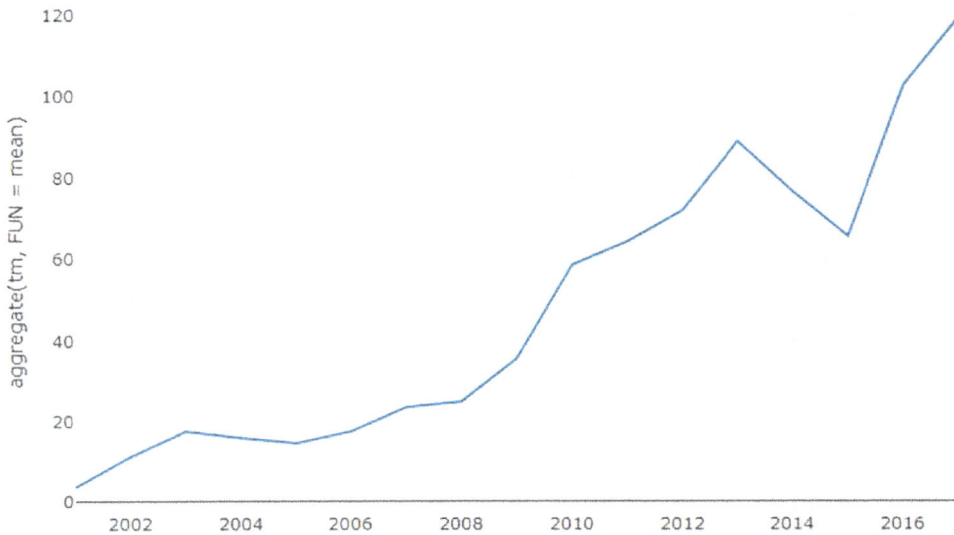

FIGURE 15.6
Total female enrollments in year-on-year trend plot.

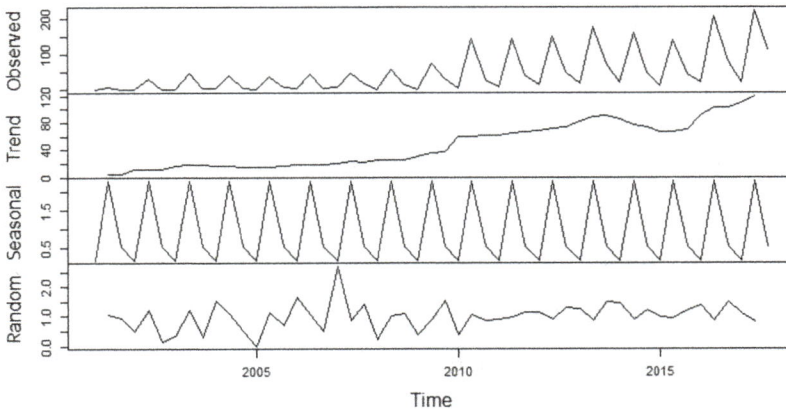

FIGURE 15.7
Decomposition of time series plot.

15.5 Profile Analysis

The profile analysis of the gender-disaggregated enrollment data for engineering, life sciences, and physical sciences showed a significant difference in the profiles of the trend plots (Table 15.2). Tests for parallelism, equal levels, and flatness were all rejected at the

TABLE 15.2

Profile Analysis of Program Enrollment by Gender

Data Summary	Female	Male
ENG	8.18	85.77
LIFE	102.41	359.00
PHY	33.18	238.06

Hypothesis Tests

Test for parallel profiles

Multivariate Test Statistic		Approx.F	num.df	den.df	-value	
Wilks	0.6061	10.0729	2		31	0.0004
Pillai	0.3938	10.0729	2		31	0.0004
Hotelling-Lawley	0.6498	10.0729	2		31	0.0004
Roy	0.6498	10.0729	2		31	0.0004

Test for equal levels

	df	Sum Sq	Mean Sq	F-value	$Pr(> F)$
group	1	274441	274441	23.04	3.56e–05 ***
Residuals	32	381250	11914		

Signif. codes: 0 '***' 0.001 '**' 0.01 '*' 0.05 '.' 0.1 '·' 1.
Test for flat profiles: p-value < 7.543644e–121.

0.001 significance level. Results further show an average female-to-male ratio of 1:10, 1:3.5, and 1:7 for engineering and the life and physical sciences, respectively.

15.6 Conclusion

In this study, female enrollment in postgraduate STEM programs in one Ghanaian university was analyzed for three categories, Engineering, Life Sciences, and Physical Sciences, from 2000 to 2017 using linear trend analysis, and the results show a good fit with a significant slope. A further comparison of the profiles of females to males showed a highly significant difference in the profiles, as also reported in [2], [4], and [6]. The revised STI policy for Ghana was developed to achieve a national outcome where **STI** capability underpins the sustainable production and processing of natural resource endowments, driven by a strengthened knowledge base to participate actively in the production of higher technology goods and services for local consumption and export [3]. This vision, though laudable, will pose a daunting task if the female half of the population continues to be underrepresented in high-skill STEM professions and is not provided the requisite opportunities and support to make a meaningful contribution to national development.

References

Campion P., Shrum W. (2004); Gender and science in development: Women scientists in Ghana, Kenya, and India. *Science, Technology, & Human Values*, 29(4):459–485.

David B., Tiffany J., Langdon D., McKittrick G., Khan B., Doms M. (2011); Women in STEM: A Gender Gap to Innovation. *US Department of Commerce, Economics and Statistics Administration* (04-11).

Draft National STI Policy for Ghana (2017); Ministry of Environment, Science, Technology and Innovation (MESTI); http://mesti.gov.gh/wp-content/uploads/2017/07/Draft-National-STI-Policy-Document-10-July-2017.pdf (accessed on 02/02/2018).

Global Health (2020); Gender based inequalities A=amplified by COVID-19 pandemic. https://globalhealth.ie/gender-based-inequalities-amplified-by-covid-19-pandemic/.

Measuring Gender Equality in Science and Engineering: The SAGA Science, Technology and Innovation Gender Objectives List (STI GOL); UNESCO, 2016.

Miller D.I., Eagly A.H., Linn M.C. (2015); Women's representation in science predicts national gender-science stereotypes: Evidence from 66 nations. *Journal of Educational Psychology*, 107(3):631.

National Accreditation Board (as of 4th May 2017) Science, Technology and Innovation Strategy for Africa 2024, African Union, 2014; https://au.int/sites/default/files/newsevents/working documents/33178-wd-stisa-english__final.pdf (accessed on 01/02/2018).

Plan International (n.d.); How will COVID-19 affect girls and young women? https://plan-international.org/emergencies/covid-19-faqs-girls-women.

Sadler P.M., Sonnert G., Hazari Z., Tai R. (2012); Stability and volatility of STEM career interest in high school: A gender study. *Science Education*, 96(3):411–427.

Smyth F.L., Nosek B.A. (2015); On the gender–science stereotypes held by scientists: Explicit accord with gender-ratios, implicit accord with scientific identity. *Frontiers in Psychology*, 6:Article 415.

Tytler R., Osborne J. (2012); Student attitudes and aspirations toward science. In: *Second International Handbook of Science Education*, edited by Fraser, B.J, Tobin, K.G and McRobbie, C.J. Dordrecht, Netherlands: Springer, pp. 597–625.

United Nations: The World's Women 2015: Trends and Statistics. New York: United Nations, Department of Economic and Social Affairs, Statistics Division 2015, Sales No. E.15.XVII.8.

World Economic Forum: Why are women more than men suffering during the COVID-19 pandemic? https://www.weforum.org/agenda/2020/05/what-the-covid-19-pandemic-tells-us-about-gender-equality/.

16

Cross-Sectional Data Analysis of Female Labor Force Participation Using Factor Analysis: A Case Study of Bihar State, India

Aviral Pandey

A N Sinha Institute of Social Studies

CONTENTS

16.1 Introduction

One of the greatest puzzles in Indian labor market trends in recent years is that while the economy has recorded a high growth rate of more than 6%–7% for the past 15 years, the labor participation rate of women has decreased. In particular, in Bihar, one of India's underdeveloped states, the decrease in the female labor force participation rate is significant in the rural areas, while the state's economy has recorded remarkable growth, with growth rates sometimes better than that of India as a whole over the same period. While the female labor force participation rate in India **decreased** from 14.8% to 13.4% in urban areas and rural areas from 24.9% to 18.1% in 2004 and 2011, respectively, Bihar recorded a drop from 6.8% to 5.4% in urban areas and a significant drop from 13.8% to 5.8% in rural areas in the same time period (Figure 16.1).

While women tend to engage in activities that are not captured or recorded in the labor statistics such as domestic work, women's participation in the labor market would positively affect the national and state economy, and their earnings could also be a means of promoting women's empowerment; overall, women's labor force participation may be an important factor rectifying inequality between men and women. As the economy has recorded particularly strong growth, the central question is why the labor force participation rate of women has declined in Bihar. While giving priority to a women's agenda, the present government is also emphasizing female empowerment via affirmative action (such as reservations in jobs and politics for women) and other indirect policy measures (such as liquor prohibition and campaigns against dowry and child marriage in Bihar).

DOI: 10.1201/9781003261148-19

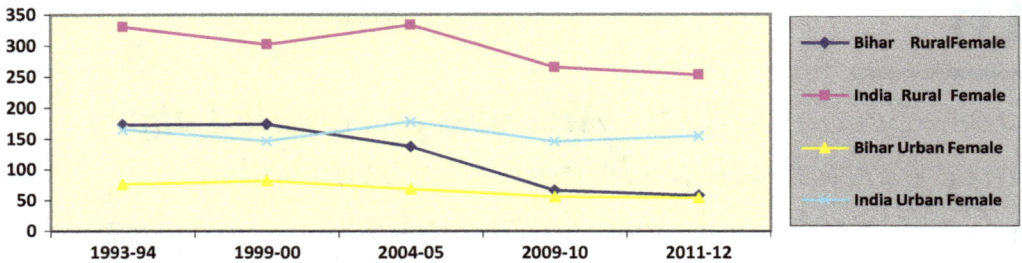

FIGURE 16.1

Female labor force participation in India and Bihar. (Author's compilation using reports of various employment and unemployment surveys of NSSO.)

The government is also focusing on self-help-group-based development models for the empowerment of rural women in the state. However, the impact of the recent interventions in terms of female participation in labor market still has not been investigated properly. There is a need for evidence-based research that can assess the impacts of affirmative action on the aspirations of women and their choice of work in the labor market. In addition, understanding the level of quality of women's employment is also important for future policy measures.

16.2 The Current Status of the Economy of Bihar

Bihar had the slowest growth of all the states of India during the nineties. The growth rate of the economy in Bihar was only 2.7%, 3.4% points lower than the national average for India from 1991 to 2001. However, after 2004 Bihar saw a significant improvement in income growth. After Nitish Kumar (of the JDU-led government) became chief minister of the state, matters began to change. The economy of Bihar continues on a path of high growth that is assisting in the development transformation of the state (see Table 16.1). The real Gross State Domestic Product (GSDP) growth rate remains impressive at 10.5% in

TABLE 16.1

Gross State Domestic Product (GSDP) of Bihar at Constant (2011–2012) Prices (2011–2012 to 2018–2019, Rs. Crore)

Sl. No.	Item	2011– 2012	2012– 2013	2013– 2014	2014– 2015	2015– 2016	2016– 17	2017– 2018 (P)	2018– 2019 (Q)
1	Primary	62,265	68,256	60,902	59,926	62,524	68,764	73,150	73,613
2	Secondary	45,341	39,339	50,282	56,247	56,325	64,384	67,427	71,666
3	Tertiary	134,092	144,015	149,478	153,245	167,240	178,960	202,397	229,307
4	Total GSVA	241,698	251,609	260,662	269,418	286,090	312,108	342,974	374,585
5	Taxes on Products	17,169	19,588	22,638	26,793	30,900	30,532	34,732	42,722
6	Subsidies on products	11,724	14,347	13,650	16,729	20,501	19,636	20,937	22,956
7	GSDP	247,144	256,851	269,650	279,482	296,488	323,004	356,768	394,350
8	Per Capita GSDP (Rs)	23,525	24,068	24,874	25,379	26,504	28,424	30,906	33,629

Source: Author's compilation of data from the Economic Survey of Bihar.

2018–2019 with improved growth performance in the service sector, including trade, repair, hotels and restaurants, road and air, and industry, including manufacturing, energy, gas, water supply and other utility services, and construction in 2018–2019. There has been a strong rebound in the services sector, with continuous double-digit growth for 2 years since 2017–2018. The growth of the economy in Bihar was associated with the lowest inflation (1.25%) in India from October 2018 to October 2019. There are reasons to believe that the considerable decline in the incidence of poverty witnessed in the past decade is holding its ground and sustaining its momentum in the wake of continuing efforts on the part of the state government to address distributional concerns in society through affirmative action and government interventions to promote job and self-help organizations. Indeed, Bihar's economy is poised to take off and consolidate its recent gains under strong leadership, which is to some extent informed by evidence-based policymaking and the implementation of public programs. Still, there remain challenges in terms of governance, leakages in the implementation of public programs, and low women's work participation. Here, it is important to identify the factors that affect women's work participation in the districts of Bihar.

16.3 Literature Review

The female labor force rate (FLPR) has historically been lower in northern states such as Bihar (Dasgupta and Verick 2016). It seems that the persistence of discriminatory feudal structures (which do not allow the markets to function independently) caused growth to disproportionately benefit the dominant castes and gender in the state. Thus, under the influence of feudal and patriarchal structures, women in the state lack a sense of ownership in the family and society. Simultaneously, the institutional capacity for governance has also been very limited during the colonial and postcolonial periods in Bihar. Under such circumstances, the pace of social change was also slower in the colonial and immediate postcolonial periods in Bihar (Witsoe 2013). Witsoe (2013) also finds that due to caste empowerment policies (the Jai Prakash Movement and constitution of the Mandal Commission) "democratic participation" has increased in Bihar (although its benefits have not been distributed evenly between men and women), but it radically threatened the patronage of the state by systematically weakening its institutions and disrupting its development projects. Bihar saw democratically endorsed "non-governance" for almost six decades. Conditions have thus been more difficult for women in Bihar.

Recent literature, such as Lahoti and Swaminathan (2013), find that states like Bihar, Orissa, and West Bengal have experienced the largest decline in FLPR and the largest decline in the share of female workers in agriculture in India since 1983. It has been said that the FLPR declined because the market has not taken care of gender and the present form of development has not been employment-intensive in nature. Thus, it seems that in absence of employment opportunities and employability of female workers (due to low skill and literacy), women have left the labor market in Bihar (Ghosh and Mukhopadhyay 1984, Lahoti and Swaminathan 2013). Improvements in income and increases in female participation in education have also been reported as factors responsible for the decline in female labor participation in Bihar, especially in rural areas. Besides this, Bihar has also seen an increase in the share of women workers in nonagriculture sectors (still the lowest of all the states).

Studies have also reported the role of male migration in the female labor force participation in Bihar. Based on data collected from 36 villages of Bihar, Rodgers (2012) discusses how factors such as male migration affected the FLPR in Bihar in the preceding 30 years (1980–2010). The massive migration of male labor has tightened local labor markets and contributed to the higher participation of women in the labor force, especially in rural areas (agriculture). However, employment opportunities for women outside agriculture are extremely limited in rural Bihar, though there have been a few openings in the education and health sectors for better-educated women. Rodgers also finds that caste appears to be a stronger determinant of female labor force participation than class, while class is a stronger determinant of work status than caste in Bihar. Most of the explanations of low female labor force participation in India and Bihar focus on supply-side factors like cultural norms and household characteristics that frown on women working outside the home. Less attention has been given to demand-side explanations, that is, the availability of suitable jobs (Chand, 2014). Studies also suggest that more than half of the decline in female LFPR can be explained by a deficit of suitable jobs at the local level (Chatterjee et al. 2015, Rodgers 2012). Kannan and Raveendran (2012) and Chand (2014) also suggest that poor agricultural performance and low diversification of jobs in rural areas are responsible for this. However, one cannot ignore that the employment scenario for women has changed significantly in Bihar in recent years, and a study in this respect is required.

It is also widely accepted that situations have significantly changed since the emergence of Nitish Kumar's coalition government and his agenda of "Sushashan" (good governance) in Bihar. Nitish Kumar started focusing on the most deprived, or the OBCs and SCs, and on women in Bihar. He also introduced certain schemes that achieved success beyond expectations. One was to provide a bicycle for every schoolgirl in ninth grade. More than 20,000 Self-Help Groups (SHGs) with two hundred thousand women from deprived sections were created in the state during Nitish's first term. Employment for women was also generated in different ways. Of the two hundred thousand primary teachers appointed in the Nitish's first 5 years, 50% were women (Sinha 2011). Through the announcement of reservations for women in Panchayats, a 35% reservation for women in police jobs and creation of a special women's battalion of the Bihar Military Police (Singh 2015), and an anti-dowry campaign, the present government has tried to send a clear-cut message. However, in the last 7 years since 2010, the data show that crimes against women have increased in Bihar. This may have a significant bearing on the labor market equilibrium and gender dimensions of employment and unemployment in Bihar. Also, since 2005, special emphasis has been placed on the agriculture sector in Bihar, which has led to and encouraged technology-based production. This is clearly evident in the increase in the number of tractors and harvesters in the last 10 years in the state. On the other hand, Mahatma Gandhi National Rural Employment Guarantee Act (MNREGA) has also emerged as an attractive option for the labor force, especially for women in the rural areas of Bihar, which has also affected the rural labor market, as the wages have changed in both sectors (agriculture and nonagriculture). More female students are also attending formal education, which has, in turn, affected their aspirations and their participation in the labor market. Most studies have concentrated on household and social characteristics to explain the low/declining FLPR, while ignoring the role of the technological changes in agriculture production systems and changing aspirations of women in Bihar.

These changes have been noted that despite the great importance of the women's agenda in the priorities of the present government in Bihar, as shown in its emphasis on female empowerment through such measures as reservations in employment and politics for women, the prohibition of liquor, and campaigns against dowry and child marriage in

Bihar. As around 89% of the total population of Bihar lives in the rural areas, it is necessary to identify the impacts of changes in policies and government support mechanism on the female population of rural Bihar. However, the impact of the recent changes in terms of female participation in labor market still has not been investigated properly. There is a need for evidence-based research that can also reveal how the efforts of the present government are affecting the aspirations of rural women and their choice of work in the labor market. Besides this, understanding the quality of women's employment is also important. Here, it is also imperative to suggest policies to increase female participation in the labor market in Bihar.

16.4 The Employment Scenario in Bihar

The creation of livelihood opportunities and the increase of decent employment for the growing labor force are essential ingredients for inclusive development. Bihar has made notable improvements in the male workforce participation (WPR) (more than 60% in both rural and urban areas), though it remains lower than the national average of 72% in rural areas and 69% in urban areas in 2018. The unemployment rate (UR) in Bihar, 6.8% in rural and 9% in urban areas in 2017–2018, was higher than the national level.

Although the share of agriculture in the state's GSDP has declined, it continues to absorb around 50% of the total workforce. This leads to disguised employment in the form of lower incomes and productivity of labor in rural Bihar. Similarly, a rising share of marginal workers in the total workforce, 38% in 2011, is another area of concern. The state government is trying to help rural as well as urban youth to find suitable and decent employment in nonfarm sectors within or outside the state. The role of the Jeevika scheme has been notable in meeting the reskilling needs of displaced and disadvantaged groups. The government is also trying to provide ample alternative employment opportunities by promoting agro-based and food processing industries in the state.

Increases in employment opportunities in the organized sector are fundamental to ensure quality, security, and good governance in the labor market. In Bihar, the share of self-employed workers among the total male workforce is around 56%; only 12% of male workers are regular wage/salary holders, and around 32% of male workers perform casual labor. Contrariwise, 33% of female workers are regular wage/salary holders in the state (higher than the national rate).

The employment situation is not uniform across the districts of Bihar. Normally, census data are used to analyze district-wide employment in India and the states. As data for the present situation are not available, we have used data from the 2011 Census to analyze the situation of employment in Bihar by district. Table 16.2 shows the existing level of differences across districts in terms of worker population ratios. In 18 districts of Bihar, less than 20% of the total population is employed. This shows a high dependency of the population on the limited number of employed people in the state. In all 38 districts of Bihar, less than 30% of the population is working.

The low WPR is also reflected in the high UR in the state. A gender-wise comparison of the work population ratio can be seen in Tables 16.2 and 16.3. Table 16.3 shows the distribution of districts in the gap in male and female work population ratios, showing that the gap is highest in the Kishanganj, Sitamarhi, and Vaishali Districts. Patna (the capital of Bihar), with the highest per capita income, is also experiencing a high gap in the male and

TABLE 16.2

Worker/Population Ratio (Main Workers), 2011

Name	WPR (Total)	WPR (Male)	WPR (Female)	WPR (Male) – WPR (Female)
Nalanda	25.75	36.47	14.13	22.34
Jamui	25.30	33.65	16.24	17.42
Araria	24.77	36.71	11.81	24.89
Madhubani	23.94	35.19	11.79	23.40
Purnia	23.78	35.35	11.22	24.13
Patna	23.49	36.43	9.05	27.37
Sheikhpura	23.26	33.33	12.43	20.90
Arwal	23.01	34.87	10.23	24.64
Gaya	22.93	32.40	12.83	19.58
Nawada	22.91	32.84	12.34	20.51
Kishanganj	22.74	38.58	6.06	32.52
Jehanabad	22.37	34.30	9.41	24.89
Purba Champaran	22.21	34.44	8.66	25.79
Madhepura	22.16	31.98	11.37	20.62
Sitamarhi	22.01	35.79	6.68	29.12
Katihar	21.74	34.72	7.61	27.10
Samastipur	21.29	34.60	6.69	27.91
Supaul	21.20	30.84	10.84	20.00
Sheohar	21.00	33.51	6.99	26.52
Vaishali	20.49	33.77	5.64	28.13
Pashchim Champaran	20.27	30.43	9.09	21.35
Lakhisarai	20.17	31.00	8.15	22.85
Begusarai	19.92	31.32	7.18	24.14
Buxar	19.25	31.37	6.11	25.26
Saharsa	19.12	29.78	7.35	22.43
Muzaffarpur	18.96	30.00	6.69	23.30
Aurangabad	18.91	28.99	8.02	20.97
Bhojpur	18.77	30.39	5.96	24.43
Darbhanga	18.57	30.06	5.97	24.09
Banka	18.39	27.68	8.14	19.54
Khagaria	18.37	28.53	6.91	21.62
Rohtas	18.33	30.16	5.43	24.74
Bhagalpur	17.58	27.95	5.80	22.15
Kaimur (Bhabua)	17.50	27.77	6.34	21.43
Munger	17.06	27.29	5.39	21.90
Gopalganj	14.17	23.69	4.85	18.84
Siwan	13.88	23.98	3.65	20.32
Saran	13.80	23.55	3.58	19.97

Source: Author's compilation from the 2011 Census.

female WPRs in Bihar. The correlation between male and female WPRs (0.55, significant at the 0.05 level) shows that women's and men's work participation levels are trending in similar directions in the districts of Bihar. The higher variation in WPR (female) than WPR (male) may be a reason that there is a high gender gap in WPR in the districts of Bihar and why the gap between WPR (male) and WPR (female) is positively correlated with the total

TABLE 16.3

Districts Classified by Gap in Male and Female Work Force Participation Rates

Category	Districts
High (32 to 28)	Kishanganj, Sitamarhi, Vaishali
Medium (less than 28 to 23)	Samastipur, Patna, Katihar, Sheohar, Purba Champaran, Buxar, Araria, Jehanabad, Rohtas, Arwal, Bhojpur, Begusarai, Purnia, Darbhanga, Madhubani, Muzaffarpur
Low (less than 23)	Lakhisarai, Saharsa, Nalanda, Bhagalpur, Munger, Khagaria, Kaimur (Bhabua), Pashchim Champaran, Aurangabad, Sheikhpura, Madhepura, Nawada, Siwan, Supaul, Saran, Gaya, Banka, Gopalganj, Jamui

WPR (0.63, significant at the 0.05 level), but the correlation between WPR (male) and WPR (female) shows that districts in the state where WPR (male) is high are also experiencing high WPR (female). Thus, we may conclude that to reduce the gender gap in work participation rates, female employment should be increased in districts where the WPR (female) is on the lower side.

16.5 Female Labor Force Participation in Bihar: A District-Level Analysis

This analysis was conducted at the district level to identify the main correlates of WPR. An important change taking place in Bihar is improvement in access to roads and urbanization, but how road development and urbanization affect livelihood is an important issue. Due to increasing government expenditures on road construction, access to quality roads has improved in Bihar. Similarly, with the absence of opportunity in rural areas, people are shifting to the urban areas in Bihar. After road construction, a new kind of agglomeration can be seen in different districts of Bihar and the number of rural towns has been increasing in Bihar. Such changes impact WPR in the state.

Further, in a male-dominated society (shown in the NFHS data by a higher preference for sons and higher mortality of female infants than male infants in Bihar) where women are not treated as the principal earning holders, it is essential to identify the determinants of female labor force participation. Most of the variables included in this analysis are for the year 2011 and are taken from the population census, the Economic Survey of Bihar, and the NSS (National Sample Survey).

Most studies have used regression analysis to explain the variance in cross-sectional data related to WPR across locations (country/state/district/province). Factor analysis (Kremelberg 2011) was used in this chapter to clarify the dynamics of variables that interact with each other and explain the variance in dependent variables (WPR in the present case). The variables considered in the factor analysis include the following: HHS: household size; CHWOM: the proportion of children to women; WFPR: main WPR rate; LIT: literacy rate; SC: percentage of scheduled caste population; OTW: percentage of workers engaged in nonhousehold manufacturing and services; CUL: percentage of the workforce engaged as cultivators; AGLB: percentage of the workforce engaged as agricultural laborers; HHW: percentage of workers in household industries; FMR: female–male ratio in the population; and MDPI: multidimensional poverty index. In addition to the rural-specific variables, we considered URBN, the percentage of the population in the urban areas in

TABLE 16.4

Results of Factor Analysis (District)

Variable/Factor	Factor 1	Factor 2	Factor 3
WPR	−0.21	0.02	0.92
WPR MALE	−0.19	0.05	0.97
WPR FEMALE	−0.16	−0.06	0.61
HHS	0.80	0.04	−0.10
CHWOM	−0.80	−0.26	0.22
SC	0.49	−0.14	0.19
CUL	0.26	−0.47	−0.21
AGLB	−0.60	−0.41	0.48
HHW	0.21	0.08	−0.01
OTW	0.44	0.70	−0.39
LIT	0.89	0.23	−0.21
FMR	0.16	−0.19	−0.13
URBAN	0.27	0.90	0.09
MDPI	−0.82	−0.45	0.29
PCI	0.20	0.86	0.09
Eigenvalue	6.18	3.10	2.36
Explained Variation	0.44	0.22	0.17

Note: Number of observations: 38 (number of districts in Bihar).

the district. Table 16.4 shows the findings of factor analysis. Literacy has the strongest association with the underlying latent variables in Factor 1, with a factor loading of 0.89. The multidimensional poverty situation, household size, children per women population, percentage of agriculture workers in total workers, and percentage of scheduled caste households of total households are also associated with Factor 1. The findings of the factor analysis do not support the view that urbanization would raise the work participation rate (Factor 1 from Table 16.4).

R indicates rural areas, HHSZ: household size; CHILD-WOM: proportion of children to women; WFPR: main WPR rate; LIT: literacy rate; SC: percentage of scheduled caste population; OTHERACT: percentage of workers engaged in nonhousehold manufacturing and services; CUL: percentage of the workforce engaged as cultivators; AGLAB: percentage of workforce engaged as agricultural laborers; MFGHH: percentage of workers in household manufacturing; F/M: female–male ratio in the population, BPL: percentage of households below the poverty line; URBN: percentage of the population in the urban areas; AVMPCE: average monthly per capita consumption expenditure; INEQ: inequality as the difference between the minimum and maximum values of the consumption expenditure. The variables are for 2011, 2011–2012, or 2010–2011.

It was expected that the male work participation rate would be positively related to urbanization, but that is not the case in Bihar. Bihar is also different from other states in having the lowest urbanization of all the states of India. Thus, the negative relationship between urbanization and work participation rates is a serious phenomenon. The analysis also shows an emerging phenomenon in Bihar where the factor loading of the male work participation rate is higher than the female work participation on Factor 1. However, Factors 2 and 3 show a negative association with the level of urbanization. The male and female participation rates indicate a negative association with both Factors 1 and 2, which reflects the substitutability between male and female workers, such that female workers appear to be employed only when male workers are not available. Conversely, the presence of male workers tends to reduce the absorption of female workers.

Multidimensional poverty and worker participation (total, male, female) in the labor market are positively associated, while growth reduces both poverty and female work participation. Again, cultivation raises women's work participation while other nonfarm activities reduce it. This suggests that in agriculture-dependent households there is a need for both women and men to contribute labor, while rural nonfarm activities are not productive enough to attract large numbers of workers or are not geared to absorbing workers on a large scale. A rise in the rural female–male ratio does not reduce women's work participation rate, which is in fact in accordance with accepted views. On the other hand, a higher child–woman ratio increases work participation, implying that women and men from households with more children are forced to join the labor market in order to meet the minimum consumption requirements. These contrasting findings emerging from Factors 1 and 3 can be explained on the ground that what Factor 1 reveals is a much stronger and more evident phenomenon, while the findings for Factor 3 reveal that certain new features according to the theory are emerging simultaneously though not so evidently.

16.6 Conclusion and Ways Forward

Declining women's work participation in one of the fastest-growing states of India, Bihar, is an issue meriting serious discussion. Since 2004, the state of Bihar has experienced a double-digit growth. Several factors, including changes in government, have been identified as explanatory factors of this trend in Bihar. Unfortunately, "the fruits of higher growth" are not inclusive in nature and women are not receiving their due share of the growth of Bihar. A decline in the gender gap in employment is one of the important indicators of inclusive growth. Unfortunately, Bihar is experiencing a high decline in female work participation. The situation of women's employment is also not uniform across the districts of Bihar. In most of the districts, less than 10% of the total female population is working in any activities. Here, an analysis has been performed using cross-sectional data to determine the roles of different factors in explaining the variance in WPR in Bihar based on one of the important technique known as "factor analysis." The analysis shows that female workers seem to be employed only when male workers are not available. Women are not able to get good jobs in Bihar. Thus, the female WPR is a "Push Factor" phenomenon in Bihar.

References

Chand, Ramesh (2014). From Slowdown to Fast Track: Indian Agriculture since 1995. National Centre for Agricultural Economics and Policy Research working paper 1/2014, New Delhi.

Chatterjee, Urmila, Murgai, Rinku, Rama, Martin (2015). Job Opportunities along the Rural-Urban Gradation and Female Labor Force Participation in India. Policy Research working paper no. WPS 7412. World Bank Group.

Dasgupta, Sukti, Verick, Sher Singh (2016). *Transformation of Women at Work in Asia: An Unfinished Development Agenda*. Geneva/New Delhi: ILO/SAGE India.

Ghosh, Bahnisikha and Mukhopadhyay, Sudhin K (1984). Displacement of the Female in the Indian Labor Force. *Economic and Political Weekly*, XIX(47): 1998–2002.

Kannan, K. P., Raveendran, G. (2012). Counting and Profiling the Missing Labor Force. *Economic & Political Weekly*, 47(6): 43–59.

Kremelberg, David (2011). *Factor Analysis, Practical Statistics: A Quick and Easy Guide to IBM® SPSS® Statistics, STATA, and Other Statistical Software*. Sage Publication.

Lahoti Rahul, Swaminathan Hema (2013). Economic Growth and Female Labor Force Participation in India. Working paper no. 414. IIM Bangalore.

Rodgers, Janine (2012). Labor Force Participation in Rural Bihar: A Thirty-Year Perspective Based on Village Surveys. IHD Working paper NO. WP 04/2012.

Singh, Santosh (2015). *Ruled or Misruled: The Story and Destiny of Bihar*. Bloomsbury Publishing India Ltd.

Sinha, Arun (2011). *Nitish Kumar and Rise of Bihar*. Penguin Group.

Witsoe, Jeffrey (2013). Democracy against Development: Lower-Caste Politics and Political Modernity in Postcolonial India. University of Chicago Press.

17

Determining Variation in Women's Labor Force Participation in a Fragile Region of Sub-Saharan Africa Using Cochran's Q Statistic

Albert Ayorinde Abegunde, Jacob Wale Mobolaji, Lasun Mykail Olayiwola, and Akanni Akinyemi
Obafemi Awolowo University

Adeyemi Adewale Alade
University of Lagos

Ore Fika
Erasmus University

Ifeoma Betty Ezike
Central Bank of Nigeria

CONTENTS

DOI: 10.1201/9781003261148-20

17.1 Introduction

Labor force participation (LFP) among women in fragile communities of Sub-Saharan Africa (SSA) is associated with many challenges. These include victimization, criticism, cultural and religious biases, and unabated post-conflict violence against women (International Labor Office, 2014). Others include gender inequality, unfavorable policies, domestic violence, and overburdening of women that are working by extended relatives (United Nations Development Fund for Women, 2000). Added to these is the perception that LFP among women is a taboo or less important to the socioeconomic development of communities. Such perception is influenced by an African tradition that sees married women as full-time home keepers (Aromolaran, 2004). All these challenges might have contributed to why the gender gap among the economically active females per 100 males was 77 in the wake of the millennium in the region and dropped to about 53 in 2008 and a little below 50 in 2014 but rose fairly above 50 in 2019 (International Labor Office, 2014; World Economic Forum, 2020).

A series of gender-related development strategies have been adopted to boost women's participation in the labor force. Some of these strategies include pro-women government policies and gender advocacies by researchers, nongovernmental organizations (NGOs), and foreign institutions. Along this line, Verick (2014) observed that there were policies toward increased women's involvement in LFP to raise the income level of residents in the region. These policies have yielded low results because of the generally low number of people that are engaged in gainful employment in SSA.

Several socioeconomic changes in the region have also positively aided LFP among women. Notable among these are civilization and an upward trend in urbanization. The former enhances female liberalization while the latter accommodates modern ideas about feminism. However, civilization in SSA is accompanied by a quest for independence. Thus, as communities seek freedom and claim their rights over joint resources, they engage in intercommunal conflicts which later plunge them into economic recession. Implicit to this is that conflicts are common among developing countries that were once colonized by western nations. This may explain why about 82% of the 80 countries that had recently passed through violent conflicts in the world are located in the global south with Africa taking the lead (Abegunde, 2011). The pathetic aspect of the conflict is that women and children are more vulnerable during and after its occurrence. Thus, many SSA countries experience conflicts and wallow in poverty with rising unemployment among their women.

The failure of past strategies and the lack of statistical information to persuade policy-makers that women could participate actively in labor have worsened the condition in developing countries. This is worrisome as many policy decisions taken are not based on data information. This is evident in past research works on women in fragile areas. Some authors have claimed that most women do take to informal economic activities with very poor returns (Handley et al., 2009).

Some scholars opined that as conflicts in Africa negatively impact the women in the labor force, it has also propelled some of them to participate in it (Mujahid et al., 2013). Some also contended that the indifference of the SSA governments to post-conflict reconstruction does not aid women's productivity in fragile communities while others believed that since women and children are vulnerable during the conflict, its negative effects on them are far-fetching, affecting the vulnerable to contribute meaningfully to economic peace-building (Abegunde, 2011). Analyses of the results of the works of these previous authors were carried out qualitatively. Methodologically, none of these past studies has adopted Cochran's Q test to drive their arguments to the conclusion (Cochran, 1950).

The current study on LFP among SSA women in post-conflict communities of Southwestern Nigeria is an attempt to establish a position on this, using a statistical tool of relevance. This is necessary because assertions on women's participation or disempowerment have not been well determined in the literature of the study area using data collected before, during, and after conflict incidence. Such three levels of data are better determined using Cochran's Q. Test. Such research is due to violent conflict in the area like the Nigerian environment. According to the Internal Displacement Monitoring Centre report of 2016, out of the 33.3 million people critically affected by violent conflict in the world, most of them were in the Middle East and SSA, and Nigeria was rated number 5 on the top list of such countries (Albuja et al., 2014). In the parabola, the country realized an 89% increase in the estimated size of her economy between the year 1990 and 2010 and claimed to have had the largest economy in Africa with an estimated nominal GDP of USD 510 billion which was well above that of South Africa's USD 352 billion in the year 2014 (Central Bank of Nigeria, 2014). As United Nations Economic Community for Africa (1972) had long asserted that an average woman in Nigeria spent an estimated 2,600 hours while an average man spent about 1,800 hours in informal jobs that employed over 70% of the country's labor force, the concern of this study is to statistically probe women's participation in the labor force in the country, positioning her post-conflict communities as the case study. Such a level of women's productivity has not been recently confirmed in places affected by intercommunity violent conflicts in the study area, which is the thrust of the current study.

There are different forms of violent conflict across towns in SSA. The least explored has been intercommunity conflict, which is most common than other forms in Southwestern Nigeria (Abegunde, 2011). The role of women in LFP has not been included in the country's policies on peace-building and post-conflict economic reconstruction. Considering gender experiences during and after conflicts in SSA where women have a low level of education and widowers' households' burdens become heavy (Abegunde, 2011), despite their sudden change in their marital, trade/professional, and income statuses due to displacement; women participation in the labor force could assume unimaginable dimension. The present study is interested in using Cochran's Q statistic to establish relationships between the levels of women's participation in the labor force in the study area. This is to provide information on policy formulation on women's economic role in post-conflict peace-building in the study area.

17.1.1 The Cochran's Q Statistic

Cochran's Q statistic is used to analyze success rate data. It tests the hypothesis that several related dichotomous variables have the same mean. The statistic focuses on the dependent variables which are measured on the same individual or matched individuals in three or more samples of measurements. Cochran's Q test requires dichotomous nominal data.

Cochran's Q statistic was developed by Cochran (1950) to measure the variations in the outcome of a repeated categorical variable with binary response. The tool analyses the success rate of the outcome data, testing the hypothesis that the outcome variables have the same success rate for the matched sets over time. The statistic focuses on the outcome variables measured on the same individual or matched individuals in two or more measurements. Cochran's Q test is based on the assumption that the observations are independent of one another and randomly selected from a large population, and the sample size is large enough.

The major limitation of Cochran's Q statistic is that it only determines the occurrence of a change; it does not evaluate the extent of the change. To evaluate the extent of the change, a multiple McNemar's test, though cannot measure interaction effect will be required. Also, Cochran's Q statistics do poorly detect the actual heterogeneity among studies as significant in meta-analysis. It does not allow a control group because it only works on dependent observations.

17.1.2 Conflict and Women Participation in the Labor Force

Information on women in conflict areas are found in the literature. Some of these past studies related women in conflict areas to sexuality and gender, marital status, natural disaster and households, deaths by gender, work-family, displacement, professions, and girls in the households (Slegh et al., 2014), among others. Each of these can be said to have direct or indirect linkage with the labor market during or after conflicts. Population displacement during the conflict is linked with LFP because it resulted in the loss of primary and change in secondary occupations and skills, especially among women. The effects of conflicts through deaths of breadwinners can also result in socioeconomic vulnerability, affecting women.

Conflicts can also skew or disorganize gender positions in the labor market, just as ethnic discrimination affects occupational performance and placement of the afflicted minority. In times past in Australia, married women participating in the labor force were open to disincentives, especially on maternity leave and this was enshrined in law and custom until the late 1960s (Kelley, 2008). In like manner, government and companies' jobs paid women lower salaries and wages than their male counterparts during this era, and sometimes, young women's jobs were terminated upon marriage until these practices were legally abolished. This practice is still in vogue in some developing nations, though with a new look that cannot be easily eradicated by law due to cultural and religious interwoven.

Occupationally, some professions in Africa are mostly practiced by women, yet with lucrative returns. This is because African culture places women in subordinate positions in all spheres, including occupation and participation in the labor force (Aremu, 2010). Such trades include clay-pot making, tying and dyeing, weaving, and hairdressing (Aremu, 2010). In places where women are scared to work in the open due to conflict, these professions suffer in the labor market. In the same vein, where there are not enough men to occupy some 'masculine' professions, the market is skeptical of absorbing able female counterparts, causing a fall in per capita income among the populations. Even when women occupy professions that are accepted by the norms of society, literature has shown that work-family conflict does take place when such jobs demand more commitment, affecting her family work activity. Such interference may result in household violence, work stress, or gender discrimination that could negatively affect the employee's behavior within the family domain. This can automatically affect gender participation in the labor force and sometimes bidirectional in their effects. Some of these household conflicts can be violent, having long-term effects.

Violent conflict generally limits women's freedom of movement to seek for economic subsistence. Due to insecurity, women in certain regions of the world have developed weak or no interest in certain jobs that are located in sensitive places that are clouded with fear of safety. In Darfur, violence against internally displaced women occurred when they moved out beyond certain perimeters to seek livelihood (Natsios, 2012). In some places, rape has been employed as part of the weapons of war. The extended effects of conflict on women, resulting in a lack of employment opportunities also lead to commercial and exploitative sex and other shadowed jobs for income generation to meet basic needs. In some cases, some men disown their wives after a conflict, discriminating against them for being abused and forcing them to face economic crises in the post-conflict era. Thus, violent conflict is linked with households, resulting in a sexual conflict that affects women's participation in the labor force, negatively influencing their contributions to economic development.

On the other hand, women in post-conflict areas of developing nations tackle occupational challenges by involving in small-scale businesses and subsistence farming as constructive approaches to LFP in the post-conflict era. As far back as the end of the Second World War, Australia's women participation in the labor force rose suddenly as many of them were engaged in clerical and blue-collar jobs, resulting in a drastic reduction in domestic service as over 90% of their young women have been actively participating in the labor force. This scenario is common to other developed nations, although on average, available information on women's participation in the labor market between 1980 and 2008 across the world only showed a minor increase from 50.2% to 51.7% (International Labor Office, 2014).

As gender participation in the labor force increases, many women in displaced households face tough times with low earnings, social discrimination, and poor welfare condition due to their late entry into the labor market, low pay, compared to what men in their cadres earn, and limited time and ability to generate income due to household commitments. In some other instances, women in the labor force in conflict environment change trade often, are forced to continual migration due to protracted conflicts, sometimes hastily and are rarely able to sell their assets. Household violence does occur after the crisis in post-conflict communities, with women bearing most of the social and economic effects and being forced to engage in different kinds of jobs to cope with living.

17.1.3 Women in Post-conflict Development

There are several studies on women's positions in post-conflict development. Some of them related women, development, and conflict with transition economies, labor market, informality and poverty, LFP at households, local, regional and international levels, while others examined women's participation in the labor force and conflicts with victims' violence and millennium development goals, sex peace and security, reconstruction, development, and gender health (True, 2013; Vecchio et al., 2014).

Some of the past works have recognized the centrality of women to post-conflict development by empowering them economically through employment in blue-collar jobs and provision of necessary incentives for informal business activities. An example was the experiences of women in the post-conflict economy of Uganda, after the violent crises that lasted for about 20 years and destroyed the social and economic fabrics of the region (Sarensen, 1998). Before the war, women were confined to the domestic economy, making LFP skew toward men. Apart from women's low number in the formal sector, International Alert (2012) reported that the region had the tradition of engaging women in subsistence farming in rural areas or confining them to petty agro-trading to meet only household needs, while men faced cash cropping. During the conflict, many women became widows and some men that were alive were disabled, depressed, or lost contact with household members. Many men lost their investments and some abandoned cash cropping for menial works, fearing that the peace agreement might collapse. These provided platforms for women to rise to the challenge, unchallenged. The outcome of women's intervention in the economy was encouraging that the government backed them up through incentives, gender-oriented policies, and absorption into politics and governance. As displaced people returned, they met many women in large-scale agriculture and trade, controlling the economy. Some of those women who seized the opportunity and gained monetary power and joined the politics waded influence and became relevant in the country's economic reconstruction programs.

17.2 Theoretical Framework

Gender equality through women empowerment has been directly linked with sustainable economic development and of more relevance to this is the feminist development theories that were postulated to position women in the developmental limelight, criticized gender-related discriminatory actions, traditions, policies, and practices, and justified the position of women in sustainable community economic development through LFP (Verick, 2014).

Women in development feminist model advocated for women inclusion in all developmental programs, dealt with a primitive way of handling women, removed subordination, elaborated more on modernization theory of economic development and believed in women liberalization to increase their participation in the labor force and promote economic development (McDonald et al., 2000). This model, though first postulated after the Second World War and applied in developed nations, has found relevance in developing countries because gender inequality has been observed to be most common among the poor communities of the world. This does not mean that its relevance across the globe has been weakened. For instance, it was employed in a survey across the world to understand the role of women in development in the year 2004. According to a study, remittance from women migrants in developed nations was isolated as the key source of income for households in developing countries as the women annually remit hundreds of billion dollars to the global south (van Naerssen et al., 2015).

The need for women's impact in a society polarized by men gave birth to women and developed feminist model. This theory went beyond mere women's inclusion and attacked a male-dominated world by seeking for 'women-only' kingdom that will produce radical feminism in a gender polarized world. The third model went further to reveal exploitations against women, irrespective of their inclusion or radical involvement in LFP. According to the gender and development feminist model, working women are always saddled with more responsibilities than their male counterparts because they are saddled with household chores due to their cultural, biological, and societal positions. This results in double works with less pay and produces 'gender slavery'.

Along this line, though not really in support of the gender and development model, Oyeyemi (2007) contended that feminist theories are western European perspectives that are alien to African culture and did not represent the realities of African women who need not work to sustain a livelihood. It was also argued that such theories contradict African women's comfort because they are not excluded from public works and services, despite their domestic responsibilities. These authors opined that the side effect of feminist models is that they place unnecessary burdens on African women who ought to live without the stress of life while depending on their agile male counterparts as breadwinners. While Oyeyemi (2007) was trying to protect African women by condemning feminist theories, the author did not consider adverse situations like violent conflicts where men could be killed and African women would not be able to enjoy the full support of their household breadwinners. Thus, this current study adopted women in the development feminist model, identified with the women in fragile communities of Southwestern Nigeria that have lost properties and relatives to conflicts and appraised their contributions to the labor force, irrespective of their challenges. Awakening to this concern resulted in this study.

17.2.1 The Study Area

Southwestern Nigeria is predominantly and traditionally dominated by a large group of people called the 'Yorubas'. Many arguments have been placed to explain the origin of the Yoruba tribe. Some authors argued that Ife is the homeland of all the Yorubas while others argued that the Yorubas migrated from Egypt due to certain conflicts as common to many empires in the past (Olatunji, 1996). Whichever the argument, the Yorubas in modern times have grouped themselves by their major towns into Oyo, Ijebu, Ondo, Ife, Ijesa, Egba, Ekiti, Akoko, Awori, Egun, Owe, Yagba, and Owo. These correspond approximately with their communal groupings, and the dialects differ, though closely related (Ayo, 2002). Each of these groupings was an empire on its own, warring against another until the advent of the colonial rule that promulgated law against violent conflict and illegal carrying of arms.

Before the colonial administration, wars were not only among empires but also between neighboring towns who either wished to claim supremacy over the other and subject the inferior to paying tributes and servitude or desired to expand territory by taking over the land and property of discrete settlements. Thus, as noted by Ayo (2002), quests for socio-economic needs by the people to meet necessities of life resulted in bullying, exploitation, and conflict among communities. Although these were brought under control during colonial rule, depressed economy, failure of the state to meet the needs of the people, and intermittent political upheavals have erupted the neglected feuds and fueled discords between neighboring communities in the country. Thus, before the wake of the millennium, inter-communal conflicts in Southwestern Nigeria were common among some discrete communities that are closely located to each other. These include Emure and Ise (Ekiti State), Ife and Modakeke (Osun State), Iju and Itaogbolu; Ilaje, and Arogbo (Ondo State). Others are Irawo-Owode and Irawo-Ile; Oke-Iho and Iseyin (Oyo State). Recent conflicts in these towns lasted for about a decade and concurrently occurred and spanned between 1995 and 2005, while the cold war continues thereafter. During the conflicts, many lives were lost, property destroyed and many households were displaced with women bearing most of the effects as some of them lost their family members and belongings. This affected their economies until recently when both men and women in the region worked on their

FIGURE 17.1
Map of Southwestern Nigeria showing conflict zones. (Author's FieldWork, 2013.)

own, without government interventions to rebuild their communities. The involvement of women in LFP to cope with this challenge prompted this study.

17.3 Data for the Study

This study centered on intercommunal conflicts among communities and focused on four pairs of such settlements, each selected from affected states in Southwestern Nigeria as reflected in Figure 17.1. Thus, communities considered in this work were Emure and Ise (Ekiti State), Ife and Modakeke (Osun State), Iju, and Itaogbolu (Ondo State) and Irawo-Ile and Irawo-ode (Oyo State). These eight settlements were found of recurrent intercommunity violent conflicts in Nigeria (Internal Displacement Monitoring Centre, 2016) and were purposively selected to form the sample frame.

17.3.1 Sampling Frame and Sample Size

A multistage sampling technique was adopted to select the respondents for this study. Since households were the targets of the data source, information on political wards in

TABLE 17.1

Sample Size in Communities that Had Experienced Intercommunal Conflicts in Southwestern Nigeria

State	Name of Settlements	Nature of Settlement	Number of Wards	Number of Wards Selected	Number of Buildings in the Selected Wards	Number of Women Sampled	Number of Questionnaire's Collected
Ekiti	Emure-Ekiti	Rural	11	6	619	31	31
	Ise-Ekiti	Rural	12	6	672	34	34
Oyo	Irawo-Ile	Rural	2	1	362	19	19
	Irawo-Owode	Rural	1	1	300	15	15
Ondo	Iju	Rural	5	3	483	25	24
	Itaogbolu	Rural	6	3	341	18	17
Osun	Ife	Urban	19	10	1,340	67	22
	Modakeke	Rural	4	2	768	38	19
Total			**60**	**32**	**4,885**	**247**	**181**

each of the eight settlements selected for this study were as reflected in Independent National Electoral Commission (INEC) enumerated area document and a past study on demographic and health survey of Nigeria {National Population Commission (NPC) [Nigeria] and ICF International (NPC and ICFI), 2014}.

The information revealed that there were 60 political wards in the selected settlements. These officially include Iju (5), Itaogbolu (6), Emure-Ekiti (11), Ise-Ekiti (12), Irawo-Ile (2), Irawo- Owode (1), Ife (19), and Modakeke (4). Using a multistage sampling technique, the study selected approximately 50% of the wards in each of the sampled communities. Information from the INEC on the total number of residential buildings in the randomly selected 32 wards was updated through a reconnaissance survey. Table 17.1 shows that there were 4,885 houses within the sample frame.

The study further sampled 5% of the houses across the board. It should be noted that a residential building in Nigeria can host multiple households but the current study targeted any available oldest female in each of the selected buildings for questionnaire administration through a systematic sampling method. The choice of 5% sample size was influenced by two reasons. First, the work of Spiegel et al. (2000) contended that a 3% sample size is accepted for empirical studies that are to be conducted in a semi-homogenous environment. Concerning Spiegel, Schiller, and Srinivasan's (2000) opinion, literature has confirmed that all communities in Southwestern Nigeria speak a common Yoruba Language; they have similar cultures and are influenced by the same socioeconomic lifestyle (Olatunji, 1996). Second, there had been some studies that were conducted in fragile communities that were of fairly low sample size due to poor responses from the aggrieved and conflict-affected residents (Osioro, 2014). This is what Osioro (2014) referred to as 'Number under fire' when he worked on the 'Challenges of gathering quantitative data in highly violent settings'. The study further made it clear many respondents in conflict areas are under grieve, suspecting researchers, and unwilling to dialogue. Thus, researchers are cautioned to be careful in discountenancing questionnaire that was not fully filled by these aggrieved respondents because there may not be a better alternative. This warning was heeded in the current study.

A strategy put in place to alleviate the challenge of data collection for this study was by contacting the community leaders in every neighborhood/ward where the samples were drawn. The neighborhood community leaders (NCLs) were targeted to fill the ethical forms for two reasons. First, literature revealed that communities in Southwestern Nigeria are governed at the grassroots by NCLs who are traditionally called chiefs (Abegunde, 2011). These chiefs monitor their communities' sociocultural affairs and represent the interests of the people within and outside their immediate neighborhoods. Second, interviewing residents in fragile communities can pose threats to researchers. Attempts to contact relevant local indigenous authorities have been identified as one of the means to avoid such danger during the survey (Krogstad, 2012). The NCLs are the nearest authorities to the people. They have the local power to advise their residents on the positions they should take on issues that relate to their immediate environment.

17.3.2 Data Collection Technique

Before the conduct of the survey, two types of ethical clearance forms were prepared. The first set was designed for the NCLs of each of the selected wards in the study area, while the second set was for the women that were to be interviewed in the sampled households. The researchers for the current study made inquiries about the names and contact addresses of the NCLs and contacted each of them in their respective houses to intimate them with the research intent and to seek their permission to interview the women in their localities. Where the researchers could not locate an NCL, a phone call was made to seek his consent. In the course of making the calls to the NCLs that were not physically available, requests were further made for physical contacts with their secretaries or deputies to stand in for them to convey the intention of the study to their residents.

This delegated authority from the NCLs further empowered their representatives to sign the required ethical forms. All the available and contacted NCLs gave their consent.

Reconnaissance revealed that all the contacted NCLs could speak the English Language. This assisted the researchers in the current study to express the purpose of the study to these NCLs. They all granted ethical permission. Following the researchers' requests, each of the NCLs provided a local community security officer who assisted to interpret the information contained in the questionnaire to illiterate respondents where necessary in his neighborhood. Each of these local officers also followed the researchers to the selected houses in his neighborhood. They expressed the concerns of the researchers on LFP among women about the past intercommunal conflicts. Despite these, some of the women declined to sign the consent forms prepared for the respondents before they could be interviewed. Many of them claimed that the remembrance of the past conflicts could have traumatic effects on them. The opinions of such women were respected and were not interviewed.

This study gave priority to interviewing old married women because they were expected to be more engaged in LFP than their young counterparts. Traditionally, young married and single women are not permitted to engage in outdoor works in SSA (Aromolaran, 2004). Where an old married woman was not available due to displacement, divorcement, or death during or after the conflicts, the oldest young adult married or a single woman in the house was taken in lieu. This is because literature has shown that both married and single women do engage in LFP in a conflict environment (Aromolaran, 2004). Although the study targeted 247 (5% of the study population) women as the sample size, Table 17.1 reflects that only 181 (73.3% of the study sample size) of them were either available for questionnaire administration or agreed to be interviewed. Information in Table 17.1 reveals that such low response was recorded in urban communities than in their rural counterparts. Worthy to note here is that not all the questions in the questionnaire were answered by every woman that was interviewed. However, all necessary data that were needed to achieve the goal of the current study were collected during the survey.

17.3.3 Data Needs

The sociodemographic data required from respondents for the study include their age, religion, income level, marital and education statuses, and community grouping of the place residence. Community grouping in this study was either considered to be rural or urban. This was based on how settlements in the study area were classified in the Nigerian Population Census of 2006 and reflected in the demographic and health information of the study area in 2014 (NPC and ICFI, 2014). Considering the age of respondents, women in the current study were considered as either young adults (below 30 years), mid old adults (30–39 years), fairly old adults (40–49 years), or old adults (above 49 years). The salary scale that was set by the Nigerian government for her workers at the time of this study was used to determine respondents' income levels. Thus, the minimum salary scale for the study is ≠18,000 (Nigerian Naira) (United Nations Common Systems, 1997). The ≠18,000 is equivalent to $50 (USA Dollars). Three educational groups were set for the study. The least applied to women that had below secondary (primary) education while the highest was for those that had tertiary education and the mid group was for women with secondary education. Women that were living with their husbands during the study were considered married while those living alone, with or without children were treated as being single. Women that divorced were treated separately.

Other pieces of information that were collected for the current study focused on the variation in, rate, and level of LFP among the women; types of jobs that they were engaged in among others. Data on these were collected based on the women's experiences before, during, and after the intercommunal conflicts in the study area. Information from the reconnaissance survey revealed that there were four types of jobs that were common among women. These are civil service, farming, trading, and artisan. Those who were considered as engaging in menial jobs were respondents who pan-handled, fetched domestic water from streams and fire-wood to make living or those that swept floors of other residents' houses or wiped dirt from their cars' screens without being instructed but were sometimes remunerated for their services on a compassionate ground. This study also recognized women that were engaged in any job or vacation for a living. They were classified as unemployed in the study. The current study also recognized the women that were employed in formal private jobs and retirees. Most of the women in the study were Christians; thus, respondents' religions were categorized as Christianity and others.

17.3.4 Analysis of Data

Data collected were analyzed using R statistical package and presented in tables. Chi-Squared test was conducted to examine the relationship between women's LFP and their sociodemographic characteristics. Cochran's Q test was employed to investigate the variations in the proportion of women in LFP before, during, and after conflict. In order to account for interactions among the explanatory variables and draw inferences for the study at multivariate level, multivariable analysis was conducted using binary logistic regression. In conducting the analysis, the current study determined the women's poverty status by grouping their income into two levels—those that were considered poor and not poor in the study area. A woman that earned above three US Dollars per day in her LFP was considered not poor and vice-versa (International Labor Office, 2014).

In conducting this analysis using Cochran's Q statistic, it was hypothesized that several related dichotomous variables have the same mean. The statistic focuses on the dependent variables which are measured on the same individual or matched individuals in three or more samples of measurements. Cochran's Q test requires a dichotomous nominal data. It is used under the assumption that the observations are independent of one another and randomly selected from a large population, and the sample size is large enough.

Given the outcome variable(s) with k treatment in which each k treatment is independently applied to s subjects and each outcome measured as a success (1) or as a failure (0), the test statistic is given as

$$T = \frac{k(k-1)\sum_{j=1}^{k}\left(X_j - \frac{N}{k}\right)^2}{\sum_{i=1}^{s} X_i(k - X_i)}$$

Where k is the number of measurements, X_j is the column total for the jth measurement, s is the number of subjects, X_i is the row total for the ith subject, and N is the total.

The major limitation of Cochran's Q statistic is that it only determines the occurrence of a change; it does not evaluate the extent of the change. To evaluate the extent of the change, a multiple McNemar's test, though cannot measure interaction effect will be required. Also, Cochran's Q statistics do poorly detect the actual heterogeneity among studies as significant in meta-analysis. It does not allow a control group because it only works on dependent observations.

In order to use the R statistical package to conduct Cochran's Q test, having read the data into the R memory and named "mydata", the R package executing Cochran's Q test was activated using the command below:

Library (RVAideMemoire)

All relevant explanatory variables including age, education, marital status, and place of residence were all converted to factor variables in R:

mydata$age <- factor(mydata$age)
mydata$education <- factor(mydata$education)
mydata$marriage <- factor(mydata$marriage)
mydata$residence <- factor(mydata$residence)

The Cochran's Q test was then conducted using the R command below:

Cochran.qtest (lfp~conflict | respondents, data=mydata, alpha=0.05)

To compute the multiple comparison tests, McNemar test was used as follows:

Library (rcompanion)
Multiple<- pairwiseMcnemar (lfp~conflict | respondents, data=mydata,
      test="permutation", Method="fdr")

The binary logistic regression analysis was conducted using R programming command as follows:

The package in which the logit command is executed was first activated using the following command:

Library (aod)

In order to ensure that R recognizes all categorical variables appropriately, the variables age, education, marital status, and place of residence were converted to a categorical variable as follows:

```
mydata$age <- factor(mydata$age)
mydata$education <- factor(mydata$education)
mydata$marriage <- factor(mydata$marriage)
mydata$residence <- factor(mydata$residence)
```

The logistic regression results were obtained for women's LFP before and after conflict separately using the R command below:

Before conflict:

```
mylogit    <-glm(lfpbefore ~ age+education+marriage+residence,    data=mydata,
    family="binomial")
summary (mylogit)
```

After conflict:

```
mylogit    <-glm(lfpafter ~ age+education+marriage+residence,    data=mydata,
    family="binomial")
summary (mylogit)
```

In order to generate the odds ratio and 95% confidence interval for the regression estimates of each of the explanatory variables, the R command below was used:

```
exp (cbind (OR=coef (mylogit), confint (mylogit)))
```

17.4 Findings

17.4.1 Sociodemographic Characteristics of the Women in Post-conflict Communities of Southwestern Nigeria

Table 17.2 reflects the sociodemographic characteristics of the women in this study. According to the table, most of the women were within the labor force age 30–39 (24.3%), 40–49 (32.0%), and 50–59 years (20.4%). Only a few of them were between 15–29 and 60–64 years old (11.6% each). A larger percentage of them had secondary (34.3%) and tertiary education (39.8%); more than half (69.6%) of them were married while a little below

TABLE 17.2

Sociodemographic Characteristics of the Women that Participated in the Labor
Force during Conflict in Southwestern Nigeria

Sociodemographic Characteristics	Frequency (N=181)	Percentage
Age Group (years)		
<30	21	11.6
30–39	44	24.3
40–49	58	32.0
50–59	37	20.4
≥60	21	11.6
Educational Background		
Illiteracy	25	13.8
Primary	22	12.2
Secondary	62	34.3
Tertiary	72	39.8
Marital Status		
Single	54	29.8
Married	126	69.6
Divorced	1	0.6
Income		
<N18,000	146	80.7
N18000–N59,999	34	18.8
N60000–N99999	1	0.6
Residence		
Rural	140	77.3
Urban	41	22.7
Religion		
Christianity	145	80.1
Islam	35	19.3
Traditional religion	1	0.6

one-third (29.8%) were single. More than three-quarter of the women were Christians
(80.1%), from rural areas (77.3%) and were poor, earning below N18,000 (95 dollars) (80.7%)
per month while about one-fifth of them were Muslims (19.3%) and hailed from urban
communities (22.7%).

17.4.2 Sociodemographic Characteristics of Respondents about LFP in Post-conflict Communities of Southwestern Nigeria

Table 17.3 shows the result of the chi-square analysis of the factors determining women's
LFP before the conflict. As shown in the table, women's participation in the labor force
significantly varied with age ($\chi^2=11.940$; $p<0.05$). The percentage of LFP was higher among
older women age 40–49 years (75.9%) and 30–39 years (75.0%) than younger women below
30 years old (38.1%). The level of LFP of women in the study area decreased with increasing level of education, peaked among women with primary or no formal education (97.8%)
and lowest among women with tertiary education (50.0%). Hence, there was a significant
relationship between LFP and women's level of education ($\chi^2=28.632$; $p<0.05$). Similarly,

TABLE 17.3

Sociodemographic Characteristics of Women about LFP in Post-conflict Communities of Southwestern Nigeria

Sociodemographic Characteristics	Unemployed N (%)	Employed N (%)	χ^2	p
Age Group (years)				
<30	13 (61.9)	8 (38.1)	11.940	0.008
30–39	11 (25.0)	33 (75.0)		
40–49	14 (24.1)	44 (75.9)		
≥50	32 (54.2)	26 (45.8)		
Educational Background				
Primary or none	1 (2.2)	46 (97.8)	28.632	<0.001
Secondary	15 (24.2)	47 (75.8)		
Tertiary	36 (50.0)	36 (50.0)		
Marital Status				
Single	29 (53.7)	25 (46.2)	19.887	<0.001
Married	24 (26.1)	102 (80.9)		
Residence				
Rural	35 (25.0)	105 (75.0)	4.075	0.044
Urban	17 (41.5)	24 (58.5)		
Religion				
Christianity	45 (31.0)	100 (69.0)	1.892	0.169
Others	7 (19.4)	29 (80.6)		

Likelihood ratio used due to less than 5 expected counts in some of the cells.

marital status varied significantly with the LFP of women in the study area ($\chi^2=19.887$; $p<0.05$). The LFP rate was higher among employed married women (80.9%) than among unmarried ones. Participation was higher among rural women (75.0%) than urban women (58.5%). Hence, there was a significant relationship between women's place of residence and LFP ($\chi^2=4.075$; $p<0.05$). On the contrary, religion had no significant relationship with women's LFP in post-conflict areas of Southwestern Nigeria. This contradicts the view of Guiso et al. (2003) that religion influences participation in the labor force.

17.4.3 Women LFP in Different Economic Activities by Their Sociodemographic Characteristics before the Conflicts in Southwestern Nigeria

Table 17.4 reveals the economic activities the women in the study area were involved in by their sociodemographic characteristics before the conflict. As shown in the table, a larger proportion of those in rural communities was involved in civil service (33.6%) and trading (33.6%), whereas urban women were involved more in trading (36.6%), other informal jobs (29.3%) and civil service (24.4%) than under any of the identified economic activities.

The majority of the younger women were unemployed (61.9%), followed by a handful of those who were involved in trading (28.6%) and minority in civil service and other jobs (4.8%). However, none of the younger women who were below 30 years of age was a trader or an artisan. On the contrary, the mid-aged adults (30–39 years) were more in trading (36.4%) and civil service (29.5%) than in other economic activities. On the contrary, older

TABLE 17.4

Women LFP in Different Economic Activities by Their Sociodemographic Characteristics before the Conflicts in Southwestern Nigeria

Labor Force Participation by Sociodemographic Characteristics	Civil Service	Trading	Farming	Artisan	Unemployed	Others	Total	χ^2
Residence								
Rural	33.6	33.6	10.0	8.6	9.3	5.0	140	18.27**
Urban	24.4	36.6	0.0	2.4	7.3	29.3	41	(LR)
Age								
Young adult (<30)	4.8	28.6	0.0	0.0	61.9	4.8	21	55.09***
Mid adult (30–39)	29.5	36.4	4.5	11.4	15.9	2.3	44	(LR)
Older adult (40–49)	39.7	37.9	6.9	6.9	3.4	5.2	58	
≥50	35.6	31.1	15.6	2.2	6.7	8.8	58	
Education								
Primary or None	17.0	51.0	17.0	12.8	2.2	0.0	47	46.44***
Secondary	21.0	35.5	4.8	6.5	24.2	8.0	62	(LR)
Tertiary	50.0	22.2	4.2	4.2	12.5	7.0	72	
Marital Status								
Single[a]	24.1	18.4	5.6	1.9	44.4	5.6	54	60.95***
Married	31.5	34.3	7.7	7.2	13.8	5.5	126	(LR)

Note: Only the number and percentage that responded to each of the variables among the respondents are reported.

LR – Likelihood ratio used due to small cell sizes.

[*] $p < 0.05$; ** $p < 0.01$; *** $p < 0.001$.

[a] The unmarried, separated, divorced, and widowed were merged.

adults (40–49 years) were the most women who engaged in civil service (39.7%) followed by those involved in trading (37.9%). Similarly, women who were 50 years or more were more civil servants (35.6%), traders (31.1%), and a few in farming activities (15.6%)

More than half of the respondents who had primary or no formal education were into trading (51.0%) with a few of them in civil service (17.0%)) and farming (17.0%). Only 2.2% of them were unemployed. Of the women who had secondary education, more than one-third of them were into trading (35.5%) with a few of them in civil service (21.0%). Unemployment was highest among them (24.2%) compared to women of another educational status. Women with tertiary education were the most in civil service (50.0%) compared to women of another educational status. This is because qualitative education increases women's productivity and makes them employable above the less privileged. Although, the general problem of unemployment and lack of desirable jobs for graduates still pose a challenge to LFP among women in SSA (especially in conflict areas) (International Alert, 2012) as only half of those with tertiary education (50.0%) were engaged in civil service which is the main source of white-collar job in Africa. Thus, a larger proportion of the single women in fragile communities of Southwestern Nigeria were unemployed (44.4%) followed by those who worked as civil servants (24.1%) and traders (18.4%). The married were more involved in trading (34.3%) and civil service (31.5%) than in other economic activities, except for a few (13.8%) of them that were unemployed. Findings also revealed that 22.2% of the women in the study area were traders, a few were farmers and artisans (4.2% each), 7.0% were involved in other economic activities, while the rest were unemployed (12.4%).

TABLE 17.5

Testing for Variation in LFP before, during, and after Conflicts in Southwestern Nigeria

	Before Conflicts	During Conflicts	After Conflicts		
	N (%)	*N* (%)	*N* (%)	Cochran's Q	*p*
Unemployed	50 (29.8)	56 (33.3)	41 (24.4)	7.938	0.019
Employed	118 (70.2)	112 (66.7)	127 (75.6)		

Note: Only the number and percentage that responded among the respondents are reported.

TABLE 17.6

Multiple Comparison Test of the Variations in Women's Labor Force Participation over the Three Conflict Periods

	Percentage Difference in Women's Labor Force Participation Rate	
	(%)	McNemar's χ^2 (*p*-value)
Before conflicts vs During conflict	−4.98	0.146
Before conflicts vs After conflict	+7.69	0.248
During conflicts vs after conflict	+13.34	0.015

17.4.4 Variation in LFP before, during, and after Conflicts in Southwestern Nigeria

On a general note, women's LFP rate was 70.2% before the conflicts in the study area. As reflected in Table 17.5, the participation rate during the conflicts decreased to 66.7%, while it rose again to 75.6% after the conflicts. The level of women's participation in labor force in various situations was subjected to Cochran's Q statistical test which confirmed that there was a significant variation in the level of women's participation before, during, and after the conflict (Cochran's Q =7.938; $p<0.05$). This confirmed the findings of UNICEF (2005) that conflict-affected female participation in the labor force in the study area.

The result of the multiple comparisons in Table 17.6 indicated that the variations in women's LFP rate at the three conflict periods were significant only between conflict and post-conflict periods. As shown in the result, there was over 13% increase in women's LFP after the conflict compared to the conflict period. The participation rates were similar, though with a negligible difference, between the preconflict and conflict period (about 5% decline) and between the preconflict and post-conflict period (about 8% increase).

Further analysis of variations in the level of women's LFP was carried out by their various sociodemographic characteristics. As reflected in Table 17.7, there was a significant variation in rural women's level of LFP before, during, and after conflicts in Southwestern Nigeria (Cochran's Q=3.852. $p<0.05$). However, the participation rate significantly varied among urban women (Cochran's Q=10.000; $p<0.05$). Specifically, the participation rate significantly varied only among mid-adult women (Cochran's Q=10.714. $p<0.05$), whereas there was no significant variation in the level of participation of women of younger and older adults before, during, and after the conflict. Meanwhile, the level of participation during the conflict reduced across age groups, except among women age 50 and above.

There was no significant variation in the level of LFP before, during, and after conflict among the women who did not have more than secondary education, except among the ones with tertiary education whose LFP level significantly varied before, during, and

TABLE 17.7

Testing for Variation in LFP before, during, and after the Conflicts by Sociodemographic Characteristics of Women in Southwestern Nigeria

Labor Force Participation by Sociodemographic Characteristics	Before Conflicts	During Conflicts	After Conflicts	Cochran's Q	p
Residence					
Rural	95 (74.2)	89 (69.5)	99 (77.3)	3.852	0.146
Urban	23 (57.5)	23 (57.5)	28 (70.0)	10.000	0.007
Age					
Young adult (<30)	8 (38.1)	7 (33.3)	9 (42.9)	3.000	0.223
Mid adult (30–39)	33 (75.0)	28 (63.6)	38 (86.4)	10.714	0.005
Older adult (40–49)	44 (75.9)	43 (74.1)	45 (77.6)	0.857	0.651
Very old ≥50	33 (73.3)	34 (75.6)	35 (77.8)	0.857	0.651
Education					
Not more than Primary	46 (97.9)	48 (93.6)	39 (93.6)	2.000	0.368
Secondary	47 (75.8)	36 (66.1)	37 (71.0)	3.600	0.165
Tertiary	36 (50.0)	39 (52.8)	60 (66.7)	19.077	<0.001
Marital Status					
Single	26 (48.1)	23 (42.6))	35 (64.8)	11.700	0.003
Married	103 (81.1)	100 (78.7)	101 (79.5)	1.167	0.558

Note: Only the number and percentage employed among the respondents are reported.

after the conflict (Cochran's Q=19.077; $p<0.05$). Also, the level of LFP before, during, and after conflict varied significantly among unmarried women (Cochran's Q=11.700; $p<0.05$), whereas there was no significant variation in participation rate among the married ones.

17.4.5 The Difference in Poverty Status of Respondents by LFP Status before, during, and after Conflicts in Southwestern Nigeria

A Cochran's Q test was carried out to establish the variations between poverty status and those of LFP among women before, during, and after the conflicts in the study area. As reflected in Table 17.8, there were high significant variations in the statuses of poor women who were either employed (Cochran's Q=30.296, $p<0.05$) or lacked jobs (Cochran's Q=18.091. $p<0.05$) in the labor force before, during, and after the conflicts in Southwestern Nigeria (Cochran's Q=3.852. $p<0.05$), with the former higher than the latter. This clearly reveals that even if the employment rate possibly improved after conflicts, the kinds of economic activities engaged in by the women who claimed to be employed before, during, and after the conflicts might not have raised their living standard well above their counterparts who were unemployed during these periods respectively. This suggests that there was no significant difference between unemployed women in the study area who ought to participate in the labor force but did not and their counterparts who claimed to be employed but were poorly remunerated and hence were not living in a special economic class above others due to low pay. This result agreed with the works of Ndubisi (2013) who observed that women in Nigeria have long hours of work (mostly domestic) with little or no pay. This is a challenge to SSA economy in the present economic regression era, especially in fragile areas like post-conflict communities of Southwestern Nigeria.

TABLE 17.8

Difference in Poverty Status of Respondents by LFP Status before, during, and after Conflicts in Southwestern Nigeria

	Poverty Status	Before Conflicts	During Conflicts	After Conflicts	Cochran's Q
Employed	Poor	89 (78.1)	97 (85.1)	74 (64.9)	30.296***
	Not poor	25 (21.9)	17 (14.9)	40 (35.1)	
Unemployed	Poor	57 (85.1)	59 (88.1)	44 (65.7)	18.091***
	Not poor	10 (14.9)	8 (11.9)	23 (34.3)	

* $p < 0.05$; ** $p < 0.01$; *** $p < 0.001$.

TABLE 17.9

Occupational Status of Women Whose Spouses Were Lost to Conflicts in Southwestern Nigeria

	Occupation after Conflicts	
Occupation before Conflicts	After Conflicts ($n=133$)	After Conflicts ($n=133$)
Civil servant	42 (32.3)	54 (40.6)
Trading	45 (33.8)	45 (33.8)
Farming	10 (7.5)	9 (6.8)
Artisan	8 (6.0)	9 (6.8)
Retired	1 (0.8)	2 (1.5)
Unemployed	21 (15.0)	11 (8.3)
Others	6 (4.5)	3 (2.3)

Note: Only 133 respondents indicated that they lost their spouses to conflict in the study area.

17.4.6 Occupational Status of Women Whose Spouses Were Lost to Conflicts in Southwestern Nigeria

In most conflict environments, individuals' participation in economic activities is always paralyzed by violence and its effects, and women are not exempted. Table 17.9 portrays the occupational status of women whose spouses were lost to conflicts in the study area. According to the table, 32.3% of the women were civil servants before the conflicts but rose to 40.6% after the conflicts. Those who were unemployed before the conflicts (15.0%) also reduced to 8.3% after conflicts. This indicates that women who could afford staying unemployed while their husbands were alive were now forced to look for jobs to make living after losing husbands to conflicts. Hence the reduction in unemployment, increase in the proportion of civil servants and artisans (from 6.0% to 6.8%) in the study area clearly indicated that LFP among women in the study area improved after conflicts. Results in Tables 17.8 and 17.9 indicated that the improvement in LFP might not have had expected positive effect on the living standard of the women in the study area.

17.4.7 Occupational Change of Economically Active Women as a Result of Conflicts in Southwestern Nigeria

Findings on occupational change of women due to conflict in the study area were based on 93% of the sampled 181 respondents that attended to the questions on the subject during

TABLE 17.10

Occupational Change of Economically Active Women as a Result of Conflicts in Southwestern Nigeria

Occupation before Conflict	Occupations	During Conflicts	After Conflicts
Civil servants (*n*=36)	Civil servants	32 (88.9)	33 (91.7)
	Trading	1 (2.8)	1 (2.8)
	Retired	1 (2.8)	1 (2.8)
	Unemployed	2 (5.6)	1 (2.8)
	Civil servants	4 (7.7)	3 (5.8)
	Trading	45 (86.5)	45 (86.5)
Trading (*n*=52)	Farming	1 (1.9)	1 (1.9)
	Unemployed	2 (3.8)	–
	Private formal job	-	1 (1.9)
	Civil servants	5 (18.5)	6 (22.2)
	Trading	2 (7.4)	3 (11.1)
Farming (*n*=27)	Farming	12 (44.4)	11 (40.7)
	Artisan	5 (18.5)	5 (18.5)
	Retired	2 (7.4)	1 (3.7)
	Unemployed	1 (3.7)	1 (3.7)
Artisan (*n*=5)	Artisan	5 (100.0)	5 (100.0)
	Civil servants	–	1 (3.7)
Private formal job (*n*=27)	Trading	–	1 (3.7)
	Unemployed	2 (7.4)	–
	Private formal job	25 (92.6)	25 (92.6)
	Civil servants	–	5 (23.8)
Unemployed (*n*=21)	Trading	–	1 (4.8)
	Unemployed	19 (90.5)	11 (52.4)
	Private formal job	2 (9.5)	4 (19.0)

Note: Only respondents who revealed their occupational change during and after conflict are reported.

the interview. According to Table 17.10, of all the women who were civil servants before the conflict, 11.1% of them could no longer continue as civil servants but as traders (2.3%), unemployed (5.6%), and retirees (2.8%). However, the proportion in civil service rose to 91.7% while unemployment reduced to 2.8% after the conflict. The proportion of traders reduced to 86.5% during and after the conflicts. During the conflicts, about 3.8% became unemployed but were all fixed up in other jobs after the conflicts. The effect of the conflicts appears to be heavier on farmers as reflected in the table. Of all who were farmers before the conflicts, those engaged in farming during (44.4%) were fairly lower than those who practiced it after (40.7%) the conflict respectively. While 3.7% were unemployed during and after the conflicts, 18.5% and 22.2% joined civil service while others went into other occupations. According to the table, there was no occupational change among the artisans. Of those who were involved in private formal jobs before the conflicts, 92.6% continued with the job during conflicts while the rest (7.4%) were jobless. However, the unemployed went into civil service and trading after conflicts. Only a few (9.5%) of the unemployed women got private formal jobs during conflicts. However, unemployment reduced from 90.5% during to 52.4% after conflicts while 23.8% went into civil service, trading (4.8%), and private formal jobs (19.0%). It became vivid from the table that women whose income could not sustain their households or whose means of livelihood got affected by the conflicts

TABLE 17.11

Effect of Conflict on Income Level of Women Whose Spouses Was lost to Conflicts in Southwestern Nigeria

Income Categories	Before Conflicts	After Conflicts
<N18,000	110 (82.7)	91 (68.4)
N18000–N49,000	19 (14.3)	35 (26.3)
N50000–N99000	4 (3.0)	6 (4.5)
≥N100,000	0 (0.0)	1 (0.8)
Total	133	133

Note: One US Dollar was equivalent of about 149 Nigeria Naira in 2015.
   : Note that 48 respondents did not respond to earning any monthly income.

must have been those who changed their job statuses. Worthy to note here is that some women who were unemployed before stood up to the challenge and sought for jobs after the conflicts.

17.4.8 Effect of Conflicts on the Income Level of Women Whose Spouses Were Lost to the Conflict in Southwestern Nigeria

Considering the monthly income of the women in Table 17.11, the level of poverty was very high among them in the fragile communities despite that over 60% of them were engaged in the labor force. As reflected in the table, women's earnings were better after than it was before the conflicts. In the latter era, the percentage of the highest-paid women was 3.0, earning between N50000–N99000 ($217.50–236.99), while over four-fifths (82.7%) of them earned below N18000 ($79.00) accordingly. In the contrary, after conflicts, a woman (0.8%) earned ≥N100,000 (about $435.0), 4.5% earned between N50000–N99000 ($217.50–236.99) and about one-quarter (26.3%) earned between N18000–N49000 ($79.00–217.49), while those earning below N18000 ($79.00) were about two-thirds (68.4%) of the respondents. The marginal difference among those with high pay is an indication that change in the earnings might not have had significant improvement in the women's standard of living in the study area, as revealed in Table 17.8.

17.4.9 Determinants of Women's LFP during and after Conflicts in Southwestern Nigeria

Table 17.12 shows the multivariate analysis of the predictors of LFP of women in conflict environments using logistic regression analysis. During conflicts, as shown in the table, the odds of LFP increased by the ages of women in the study area with 30–39 years (OR=3.3; $p<0.05$), 40–49 years (OR=5.2; $p<0.05$) and 50 years and above (OR=5.1; $p<0.05$) accordingly, relative to younger women who were below 30 years old (reference category). Hence, older women were more likely to participate in the labor force than younger women during conflicts in fragile communities of Southwestern Nigeria. Women who had secondary education (OR=0.1; $p<0.05$) and tertiary education (OR=0.001; $p<0.05$) were significantly less likely to be employed during conflicts, relative to women with primary or no formal education (reference category). Similarly, urban women were less likely to be involved in economically productive activities (OR=0.9; $p>0.05$) during conflicts, compared to their rural counterparts. During conflicts also, the odds of LFP were higher (OR=6.5; $p<0.05$) among married women, relative to the unmarried ones (reference category).

TABLE 17.12

Multivariate Analysis of Determinants of Women's LFP during and after Conflicts Using Binary Logistic Regression

	During Conflicts		After Conflicts	
	B	Adjusted OR (95% C.I)	B	Adjusted OR (95% C.I)
Age Group (years)				
<30 RC	–	–	–	–
30–39	1.21	3.3 (0.8–14.0)	3.40	29.8*** (4.5–197.0)
40–49	1.64	5.2* (1.1–24.6)	2.57	13.0** (2.1–79.8)
≥50	1.62	5.1* (1.0–25.2)	2.57	13.1** (2.1–83.0)
Educational Background				
Primary or none RC	–	–	–	–
Secondary	−1.98	0.1* (0.0–0.8)	−1.50	0.2 (0.0–2.3)
Tertiary	−3.95	0.0*** (0.0–0.1)	−4.03	0.0** (0.0–0.2)
Marital Status				
Single RC	–	–	–	–
Married	1.87	6.5** (2.0–21.1)	1.16	3.2 (0.8–12.2)
Residence				
Rural RC	–	–	–	–
Urban	−0.14	0.9 (0.3–2.3)	−0.27	0.8 (0.3–2.1)

OR, Odds Ratio; C.I., Confidence Interval; RC, Reference Category (RC=1).
*.$p<0.05$; ** $p<0.01$; *** $p<0.001$.

Similarly, after the conflicts, the odds of LFP were much higher among older women of age 30–39 years (OR=29.8; $p<0.05$), 40–49 years (OR=13.0; $p<0.05$) and 50 years and above (OR=13.1; $p<0.05$), relative to younger women who were below 30 years old (reference category). Hence, older women who were still agile were more likely to participate in the labor force than younger women during conflicts. Women who had secondary education (OR=0.2; $p>0.05$) and tertiary education (OR=0.001; $p<0.05$) were much less likely to be employed during conflicts, relative to those with primary or no formal education (reference category). Also, urban women were less likely to be involved in economically productive activities (OR=0.8; $p>0.05$) during conflicts, compared to their rural counterparts. The odds of LFP after the conflicts were higher (OR=3.2; $p>0.05$) among married women, relative to the unmarried ones (reference category).

17.5 Discussion and Conclusion

A survey on LFP was carried out among women ($n=153$) in conflict-affected areas of Southwestern Nigeria, focusing on four pairs of communities that had experienced intercommunal clashes for about a decade before this study. Three pairs of these communities were rurally based while only a pair was urban. Findings revealed variations in the percentages of women unemployed before (29.8), during (33.3%), and after (24.4%) the conflicts. This is an indication that though conflicts are expected to have a negative impact on women participation in LFP in SSA because its development is always influenced by levels of predominant wars (Blattman, 2010), there could still be gender coping mechanisms to improve employment rate, even better than the periods before it struck. Thus, the numbers of women that were economically engaged after conflicts in this study were more than those that had jobs before it. This is despite the fact that more than two-thirds ($n=133$) of them had lost their breadwinners to conflicts before this study was conducted and they had to change their jobs to cope with economic situations in their fragile communities. These findings agreed with the work of Cai (2010) that conflict in Africa targets breadwinners of households but differed in gender's capability to cope because it did not deter the affected women in the study area to rise to the challenge of contributing to their local economy.

A cursory look at the unemployed women in the study area showed that most of them were in their young ages of the labor force. This could have been due to engagement in domestic works. As noted by (Aromolaran, 2004), SSA n young women who are married are not expected to engage in any outdoor work that can contribute to the economy, except domestic services. On the contrary, married older women are fairly free to fetch for their children. The latter opinion could have accounted for why there were older women who were into civil service, trading, and farming than their younger and mid-age counterparts in the study area. This further agreed with the work of Assaad (2009) who found that there were older women in the labor force of Egypt than younger ones. This experience was also claimed to be common to developed nations like the United States, not limited to Africa only (United States Bureau of Labor Statistics, 1994). However, this differed from the findings of Jessen (2012) in some experimented villages in India where young women in the rural areas chose to enter into the labor market early to be economically buoyant before marriage, not minding the delay effect of this on their marital commitment. Thus, the opinion that young women are not engaged in LFP in developing nations cannot be generalized.

This study observed that the rate of LFP was higher among women in rural areas (69.5%) than those in urban communities (57.5%). Several factors could have been responsible for this. Occupationally, a reasonable proportion of women who participated in the labor force during conflicts in the rural areas of this study were farmers and artisans. During conflicts, there seem to be an easy entry (and exit) into these occupations, unlike jobs that require specialized services which may delay employment into labor market because the rate of securing a formal job is very low in developing nations (Golub and Hayat, 2014). Also, land for farming is always readily available in rural areas of Africa, unlike in urban environment where many social and economic land-uses compete for spaces and heighten land values. This must have contributed to why participation was higher among rural women (74.2%) than urban women (57.5%) in the study area. Thus, farming (0.0%) and artisan (2.4%) were not well represented in urban areas of this study as civil service (24.4%) and trading (36.6%) were. With the general low level of education among women in African communities, especially in rural areas, many of them are expected to take into menial jobs, trades, or vocations that require little or no investment and skill, resulting in poor economic returns (Kuepie et al., 2009). Such LFP cannot improve the living standard of women that were involved even if the number of those that were engaged in it increases. Thus, the high significant variations in the statuses of poor women who were either employed (Cochran's $Q=30.296$, $p<0.05$) or lacked jobs (Cochran's $Q=18.091$. $p<0.05$) in the labor force before, during, and after the conflicts is an indication that LFP among women in Southwestern Nigeria did not create differences in standard of living between those that were engaged in the labor force and the unemployed.

The value of education among women in conflict communities also influenced the level of variation in their LFP before, during, and after the conflict among those who did not have more than secondary education, except among the ones with tertiary education whose LFP rate significantly varied before, during, and after the conflict (Cochran's $Q=19.077$; $p<0.05$). This is because qualitative education offered specialized training with certificates that can be tendered anywhere to secure another job with reasonable pay, even during displacement. This may not be so with those who are less privileged educationally. This study also agreed with the opinion of Tuwora and Sossoub (2008) that the level of LFP of women decreased with increasing level of education; peaked among women with primary or no formal education (97.6%) and lowest among women with tertiary education (50.0%). This made the relationship between LFP and women's level of education to be significant ($\chi^2=28.632$; $p<0.05$). It is evident in this study that the few women who were educated must have found solace in civil service after the conflicts since the number of women in such job were lower than what it was before and during the conflict in the study area.

In this study, LFP rate before, during, and after the conflicts also varied significantly among unmarried women (Cochran's $Q=11.700$; $p<0.05$) because they were not as tied to family responsibilities as their married counterparts. Culturally, married women in Africa are open to unending family responsibilities with husbands, children, grandchildren, and in-laws. This factor may not allow married women in Africa the flexibility desired in LFP. Despite this limitation, the study showed that some women in conflict areas of Southwestern Nigeria attempted to cope with the challenge through changing jobs, especially in civil service and trading. The increments observed in the number of women in some of these jobs and reduction in the number that was not employed after the conflict indicated that there was an upward change among females in the labor market in the fragile communities. This was why the number of those that earned low income before the conflicts reduced after it, and those with better pay increased, although slightly. This means that though the conflicts led to loss of breadwinners in many of the studied households, most women stepped up after the conflicts from low economic state the conflicts met them. This is because the percentage of those that earned low monthly income (below N18000 or $79.00) after conflicts fairly reduced (68.4%) compared to the period before their occurrence (82.7%); although without significant change in the higher earnings and residents' standard of living. This agrees with the view of Collier (2004) that civil crises cannot be as terrible as world war and that the former will always bring with it some economic elements that would promote *growth* in low-income communities: to the level they had not been, even before conflicts. This assertion is a justification for this study.

The study noted that women's contribution to LFP in post-conflict communities of SSA could have yielded minimal economic benefit but gender intervention in the fragile communities positively impacted economic condition, at least better than the period before conflicts struck. Leaning on the test result of Cochran's Q statistic, the study concluded that women's contribution to LFP in the fragile communities improved women's level of employment, especially among the low-income earners, better than the period before conflicts struck. Since most of the women in the study area earned low wages, gender participation in the fragile local communities in Southwestern Nigeria could be said to have positively impacted many who could have been more afflicted by the effect of crises after conflicts. Further studies can delve into men's participation in labor force to also confirm the role of men in the economic reconstruction of the study area.

References

Abayomi, A.A. (2014) Sociological implications of domestic violence on children's development in Nigeria. *Journal of African Studies and Development* 6(1): 8–13.

Abegunde, A.A. (2011) Educational behaviour of residents living in inter-communal conflict zones of Southwestern Nigeria. *International Journal of Economics and Sustainable Development* 2(4): 103–114.

Albuja, S., Arnaud, E., Caterina, M., Charron, G., Foster, F., Glatz, A., Hege, S., Howard, C., Klos, J., Kok, F., Kritskiy, V., Pagot, A., Ruaudel, H., Rushing, E.J., Turner, W., Walicki, N. and Wissing, M. (2014) *Global Overview 2014 People internally displaced by conflict and violence*. Internal Displacement Monitoring Centre Norwegian Refugee Council Chemin de Balexert 7–9 CH-1219 Châtelaine (Geneva). www.internal-displacement.org.

Aremu, J.O. (2010) Conflicts in Africa: Meaning, causes, impact and solution. *An International Multi-Disciplinary Journal, Ethiopia* 4(4), Serial No. 17: 549–560.

Assaad, R. (2009) *The Egyptian Labor Market Revisited*. American University in Cairo Press, pp. 15–16.

Ayo, S.B. (2002) *Public Administration and the Conduct of Community Affairs Among the Yoruba in Nigeria*. Institute for Contemporary Studies, Oakland, CA.

Blattman, C. (2010) Post-conflict recovery in Africa. In Aryeetey, E., Devarajan, S., Kanbur, R. and Kasekende, L. (Eds.), *The Micro Level Entry for the Oxford Companion to the Economics of Africa*. Oxford University Press, Oxford.

Cai, L. (2010) The relationship between health and labor force participation: Evidence from a panel data simultaneous equation model. *Labor Economics* 17: 77–90.

Cochran, W.G. (1950) The comparison of percentages in matched samples. *Biometrika* 37(3/4): 256–266. doi: 10.2307/2332378.

Collier, P. (2004) *Development and Conflict*. Centre for the Study of African Economies, Department of Economics, Oxford University.

Golub, S. and Hayat, F. (2014) Employment, unemployment, and underemployment in Africa. WIDER Working Paper. World Institute for Development Economics Research. United Nations University.

Guiso, L., Sapienza, P. and Zingales, L. (2003) People's opium? Religion and economic attitudes. *Journal of Monetary Economics* 50: 225–282.

Handley, G., Higgins, K., Sharma, B., Bird, K. and Cammack, D. (2009) *Poverty and poverty reduction in sub-Saharan Africa: An overview of the issues*. Overseas Development Institute 111 Westminster Bridge Road London SE1 7JD.

Internal Displacement Monitoring Centre (2016) Global Report on Internal Displacement, Grid 2016. Norwegian Refugee Council. http://www.internal-displacement.org/.

International Alert (2012) *Fighting for Speech. Women's Participation in Burundi's Democratic Transition*. International Alert and EASSI, London, p. 12.

International Labor Office (2014) *Risk of a Jobless Recovery?* International Labor Office (ILO), Geneva.

Jessen, R. (2012) Do labor market opportunities affect young women's work and family decisions? Experimental evidence from India. *The Quarterly Journal of Economics*. doi: 10.1093/qje/qjs002.

Kelley, J. (2008) Trends in women's labor force participation in Australia: 1984–2002. *Social Science Research* 37: 287–310.

Kuepie, M., Nordman, C.J. and Roubaud, F. (2009) Education and earnings in urban West Africa'. *Journal of Comparative Economics* 37: 491–515.

Krogstad, E.G. (2012) Security, development, and force: Revisiting police reform in Sierra Leone. *African Affairs* 111(443): 261–280. doi: 10.1093/afraf/ads004.

McDonald, M., Connelly, M.P., Murray, L.T. and Parpart, J.L. (2000) Feminism and development: theoretical perspectives. In Barriteau, E., Connelly, M.P. and Parpart, J.L. (Eds.), *Theoretical Perspectives on Gender and Development*. International Development Research Centre (IDRC), Ottawa, pp. 51–160.

National Population Commission (NPC) [Nigeria] and ICF International. 2014. *Nigeria Demographic and Health Survey 2013*. NPC and ICF International, Abuja, Nigeria, and Rockville, MD, USA.

Natsios, A.S. (2012) *Sudan, South Sudan, and Darfur: What Everyone Needs to Know*. Oxford University Press, New York.

Ndubisi, N.O. (2013) Role of gender in conflict handling in the context of outsourcing service marketing. *Journal of Psychology and Marketing* 30(1): 26–35.

Olatunji, O.O. (1996) *The Yoruba: History, Culture and Language* (J.F. Odunjo Memorial Lectures). University Press, Ibadan.

Oyeyemi, H. (2007) *The Opposite House*. Bloomsbury, London.

Sarensen, B. (1998) Women and post-conflict reconstruction: Issues and sources. WSP Occasional Paper No. 3. United Nations Research Institute for Social Development Programme for Strategic and International Security Studies.

Slegh, H., Barker, G. and Levtov, R. (2014) Gender Relations, Sexual and Gender-Based Violence and the Effects of Conflict on Women and Men in North Kivu, Eastern Democratic Republic of the Congo. Results from the International Men and Gender Equality Survey, (IMAGES). Promundo-US and Sonke Gender Justice, Washington, DC, and Capetown, South Africa.

True, J. (2013) *Women, Peace and Security in Post-Conflict and Peace-Building Contexts*. Norwegian Peace Building Resource Centre, Norway.

Tuwora, T. and Sossoub, M-A. (2008) Gender discrimination and education in West Africa: Strategies for maintaining girls in school. *International Journal of Inclusive Education*, Taylor & Francis, 12(4): 363–379. doi: 10.1080/13603110601183115.

UNICEF (2005) *The Impact on Women and Girls in West and Central Africa and the UNICEF Response of Conflict*. The United Nations Children's Fund (UNICEF), New York.

United States Bureau of Labor Statistics (1994). *The American Work Force: 1992–2005*. United States Bureau of Labor Statistics

van Naerssen, T., Smith, L., Davids, T. and Marchand, M.H. (Eds.) (2015) *Women, Gender, Remittances and Development in the Global South*. Ashgate, Farnham, UK. https://wol.iza.org/uploads/articles/220/pdfs/feminization-of-migration-and-trends-in- remittances.pdf (Accessed on November 5, 2021).

Vecchio, N., Mihala, G., Sheridan, J., Hilton, M.F., Whiteford, H. and Scuffham, P.A. (2014) A link between labor participation, mental health and class of medication for mental well- being. *Economic Analysis and Policy* 44: 376–385.

Verick, S. (2014) *Female Labor Force Participation in Developing Countries*. International Labor Organization, India, and IZA, Germany.

World Economic Forum (2020) Global gender gap report 2020. World Economic Forum Office, CH-1223 Cologny/Geneva Switzerland. http://reports.weforum.org/global-gender-gap-report-2020/dataexplorer.

Part 4

Statistical Literacy and Methods across Disciplines

18

Understanding Uncertainty in Real-Life Scenarios through the Concepts of Probability and Bayesian Statistics

Richa Vatsa and Sandeep K. Maurya

Central University of South Bihar

CONTENTS

18.1 Introduction

In real life, we come across two types of circumstances, those that present repetitive results. These conditions are explained by the laws of physics and pure sciences, and are called deterministic. The other circumstance is random, in that we may expect it to result in different possible consequences when replicated; thus, it adheres to the uncertainty of outcomes. Uncertainty is a law of nature. We experience it at almost every moment of our life. Many decisions, simple to complex, are taken to make our life comfortable and secure in various circumferences. These decisions become challenging when taken in the face

DOI: 10.1201/9781003261148-22

of uncertainty. Therefore, in such scenarios, probability statements are needed to choose optimal decisions. In statistics, uncertainty can be measured with the concept of probability theory in general.

Historically, from the dawn of civilization, humans have been interested in understanding uncertainty and probability through games of chance and gambling. However, the advent of probability as a mathematical discipline is relatively recent. Ancient Egyptians, about 3500 B.C., were using the astragalin, a four-sided die-shaped bone found in the heels of some animals, to play a game now called hounds and jackals. The ordinary six-sided die was created about 1600 B.C. and since then has been used in all kinds of games. The ordinary deck of playing cards, probably the most popular tool in games and gambling, is much more recent than dice. However, it is not known where and when dice originated. Real progress started in France in 1654 when Blaise Pascal (1623–1662) and Pierre de Fermat (1601–1665) exchanged several letters in which they discussed general methods for the calculation of probabilities. In the seventeenth century, beginning with James Bernoulli (1654–1705) and Abraham de Moivre (1667–1754), and after that Pierre-Simon Laplace (1749–1827), Siméon Denis Poisson (1781–1840), and Karl Friedrich Gauss (1777–1855), the theory of probability and its applications grew rapidly in many different directions.

Let us consider a few examples to understand probability: (1) If a student takes a mathematics test, what is the chance of her passing the test? (2) If a fair coin is tossed twice, what is the chance of getting heads in both trials? (3) A lot consists of 10 useful articles, 4 with minor defects, and 2 with major defects. One article is chosen at random. What is the chance that it has no defects? Answers to all these problems of uncertainty can be found with the concept of probability, which shall be explained later in this chapter.

One can move from uncertainty to certainty with increasing information regarding a particular phenomenon. For example, the concept of conditional probability allows us to reduce uncertainty with increased knowledge. Suppose some prior knowledge, personal judgments, or experts' opinions are available about the related events that could affect our decision in life. In that case, inferences concerning uncertainty combined with prior information may be drawn through the theory of Bayesian statistics.

Bayesian statistics provides uncertainty measurements of the model assumptions around the facts gathered regarding a particular phenomenon. Suppose there is no reason to accept that there is a 50% chance that a student will pass a test. With Bayesian statistics, we may find probabilistic statements around the chances of her passing a test given the tutor's prior judgment of her ability to pass the test and the results of other tests she took.

In our daily life, we often update our belief or judgment in light of available facts and act accordingly. For example, suppose that a person is down with fever and headache. A medical practitioner speculates that it is viral fever. This expert's prior judgment is revised based on the appropriate medical diagnoses, and medication is prescribed. In other words, the test results are examined with the field expert's prescription. Thus, prior knowledge is required in some life scenarios to make better decisions.

Further, the Bayesian approach is suitable for real-life scenarios that cannot be replicated as laboratory experiments. Should I carry an umbrella during a trip in case it rains in the coming days? What is the chance that a particular volcano will erupt soon? These are problems of prediction quantified based on current observations. However, such real-life events may not be replicated as laboratory experiments. Therefore, they may not be supported by the non-Bayesian definitions of probabilities. Nevertheless, the uncertainty around these events may be quantified based on subjective or experts' opinions.

The concepts of probability and Bayesian statistics have found applications in many applied sciences and social sciences to understand uncertainty in real-life problems.

Therefore, this chapter aims to acquaint readers with a basic yet effective way of learning and dealing with these statistical concepts.

The chapter is organized as follows: Section 18.2 deals with basic definitions and terminology probability theory. Section 18.3 describes the concept of probability along with its properties of conditional and independent events. The technique of Bayes' theorem is discussed in Section 18.4. The application of Bayes' theorem for random variables and parameters is detailed in Section 18.5. The concept of choosing a prior density in Bayesian statistics is described in Section 18.6. Bayesian estimation procedures are explained in Section 18.7. Bayesian predictive analysis can be found in Section 18.8. An application of Bayesian statistical concepts with a real-life example is presented in Section 18.9. The conclusion of the whole chapter is summarized in Section 18.10.

18.2 Basic Definitions and Terminologies

We may define a life experiment as either deterministic or nondeterministic. A deterministic experiment results in a unique outcome when replicated under certain conditions. For example, the laws of gravitation describe the phenomena of a falling body quite precisely under certain conditions. Nevertheless, most real-life situations with random outcomes, such as the life of an electric bulb or the face of a fair coin when tossed, are called nondeterministic, random, or probabilistic experiments.

An experiment repeated under the same environmental conditions may be said to be a random experiment. All the possible outcomes of a random experiment are known in advance; however, any particular performance of the experiment is unknown. The set of all possible outcomes of an experiment is called the sample space of the experiment, denoted by S. The sample space can be either countable or uncountable, according to the number of elements in the space.

For example, (1) if an experiment consists of flipping a fair coin, then the sample space is $S = \{H, T\}$, where H denotes head and T stands for tail of the coin; (2) if an experiment involves a fair coin tossed twice, the corresponding sample space is $S = \{HH, HT, TH, TT\}$, and (3) if an experiment consists of throwing a fair die, the related sample space is $S = \{1, 2, 3, 4, 5, 6\}$.

An event, a vital terminology in probability theory, is any collection or sample of outcomes of a random experiment. It is a subset of the sample space. It may be either simple, if it contains only one outcome, or compound if it contains more than one outcome. The performance of a random experiment is called a trial.

For the above examples, let us define some events. Suppose E_1 and E_2 are the simple events referring to getting a head and getting a tail in Experiment (1), respectively; E_3 and E_4 stand for getting two heads and two tails, individually, in Experiment (2); E_5 denotes getting one head and one tail in Experiment (2). Further, event E_6 relates to getting a number among 1–6 in Experiment (3); events E_7, E_8, and E_9 refer to getting an odd number, an even number, and a prime number in Experiment (3), respectively. On the other hand, tossing a coin once or two times, and throwing a die are the trials of the random experiments.

All the possible outcomes in a trial together are called an exhaustive event. Events are called mutually exclusive if one of them happening restricts the occurrence of the others at the same time. The outcomes of a trial are said to be equally likely if the occurrence or nonoccurrence of any of them is not preferential to others. The number of outcomes that entail the occurrence of an experiment is called favorable events. The events are said to be independent if the occurrence or nonoccurrence of any of them does not affect the performance of the other events.

In the above random experiments, the exhaustive events are: getting a head and getting a tail in Experiment (1); events *HH, HT, TH,* and *TT* in Experiment (2); and 1, 2, ..., 6 in Experiment (3). The events E_1 and E_2, E_3, and E_4, and events E_7 and E_8 are mutually exclusive in their respective experiments. Event E_6 is also mutually exclusive, as only one number appears on the uppermost face of a die at a time.

In Experiment (1), getting a head and getting a tail are equally likely outcomes because the coin is fair. Similarly, in Experiment (3), getting a particular number is also an equally likely outcome as the die is fair. All the above-defined events $E_1, E_2, E_3, E_4, E_5, E_6, E_7, E_8$, and E_9 are favorable events. In Experiment (1), getting a head and getting a tail when tossing a fair coin two times are independent. Similarly, if we repeat Experiment (3) two times, then getting a particular number on the first throw is independent of throwing it the second time, as the die is fair.

An event can be treated as a set of simple events. Thus, all the operations and laws of set theory can be applied to an event.

18.3 Probability

The probability of an event is a quantification of its chance of happening on a scale of 0–1. With known probabilities of events of a random experiment (or of the many possible choices in a decision-making scenario), it becomes easier to choose the most probable action or choice to minimize loss or maximize gain in a life full of uncertainties.

There are different ways of calculating probabilities. The very basic one is called classical probability. This approach requires the outcomes of a random experiment to be equally likely. It may be defined thus: If there are n mutually exclusive, exhaustive, and equally likely cases out of which m are favorable to an event, then the probability of that event happening is defined by the fraction $\frac{m}{n}$. In other words, if a random experiment with mutually exclusive and equally likely outcomes is repeated several times, then classical probability is the ratio of the number of equally likely outcomes to the number of trials of the random experiment.

Let us consider the example of a student taking a mathematics test, with event P: passing the test, and F: failing the test. In this case, the sample space is $S=\{F, P\}$, with $n=2$. The elements of S represent the possible test results. The favorable event is P, i.e., $m=1$. From the above definition, the probability of passing the test is $\frac{1}{2}$. Further, let us impose another restriction. That is, the test is conducted twice, and the student is declared to have passed if she passes both tests. Then the sample space is $S=\{FP, PF, FF, PP\}$, i.e., $n=4$. The elements of S represent the consecutive results of the test when conducted twice. The favorable event is PP, i.e., $m=1$. Thus, the probability of passing the test both times is $\frac{1}{4}$.

Similarly, the probability of getting a head in a trial of tossing a fair coin is $\frac{1}{2}$. However, if the coin is tossed twice, the sample space is $S=\{HH, HT, TH, TT\}$, with $n=4$; the favorable event is HH, i.e., $m=1$. Thus, the probability of getting heads in both trials is $\frac{1}{4}$.

Example 18.1

Let us consider another example. Suppose some complex components are assembled in a plant that uses two different assembly lines, L_1 and L_2. Line L_1 uses older equipment than L_2, so it is somewhat slower and less reliable. Suppose on a given day that line L_1 has assembled 8 components, of which 2 have been identified as defective and 6 as nondefective. On the other hand, L_2 has produced 1 defective and 9 nondefective components. Then find the probability of selecting (1) a component from line L_1, (2) a component from line L_2, (3) a defective component, (4) a nondefective component, (5) a defective component from line L_1, and (6) a nondefective component from line L_2.

Let us define B for defective and B' for nondefective events and summarize all the information given in Table 18.1. From the table, we can calculate the required probabilities as

$$P(\text{a component selected from } L_1) = P(L_1) = \frac{n(L_1)}{N} = \frac{8}{18} = 0.44,$$

$$P(\text{a component selected from } L_2) = P(L_2) = \frac{n(L_2)}{N} = \frac{10}{18} = 0.56,$$

$$P(\text{a defective component selected}) = P(B) = \frac{n(B)}{N} = \frac{3}{18} = 0.17,$$

$$P(\text{a non-defective component selected}) = P(B') = \frac{n(B')}{N} = \frac{15}{18} = 0.83,$$

$$P(\text{a defective item selected from line } L_1) = P(L_1 \cap B) = \frac{n(L_1 \cap B)}{N} = \frac{2}{18} = 0.11,$$

$$P(\text{a non-defective component selected from } L_2) = P(L_2 \cap B') = \frac{n(L_2 \cap B)}{N} = \frac{1}{18} = 0.06.$$

Note: The term $n(\cdot)$ denotes the number of favorable events and N stands for the total number of exhaustive events (the number of components lined up in the plant).

According to Mood et al. (1974), the classical definition of probability has its limitations. Outcomes of a random experiment/scenario are required to be equally likely, which may not be possible in practice. Further, if the total number of trials or the number of outcomes of a random experiment is infinite, the classical definition fails. To deal with these limitations, we may approach the relative frequency approach of probability. It is defined as $\frac{m}{\lim_{n \to \infty} n}$. This definition no longer requires the outcomes to be equally likely.

TABLE 18.1

Table for Events in Example 18.1

		Condition		
		B'	Total	
Line	L_1	2	6	8
	L_2	1	9	10
Total		3	15	18

18.3.1 Counting Rule

The rule of counting is a basic need in calculating probability. It would be quite time-consuming and cumbersome to manually count the favorable number of events and the sample sizes. There are some easy yet effective methods of counting objects mathematically.

Product rule: Let a small community consist of 8 women, each of whom has 2 children. Suppose the women and their children are sitting together in a hall. If one woman and one of her children are to be chosen as mother and child of the year, there are $8 * 2 = 16$ possible choices. Let us understand how this is found with the counting rule.

Let a set or an event consists of some pairs of objects. If the first element or object of the pair can be selected in m ways, and for each of these m ways, the second element of the pair can be chosen in n ways, then the total number of pairs will be $m * n$.

There is an alternative explanation of this rule with two stages of operations. If the first stage can be performed in any one of m ways, and for each such way there are n ways to perform the second stage, then there are $m * n$ ways of carrying out the two stages in sequence.

The above principle can be generalized. Let there be k experiments to be performed, such that the kth experiment may result in any of m_k possible outcomes. If for each of these m_k possible outcomes, there are m_{k+1} possible outcomes of the $(k+1)$th experiment, then, there is a total of $m_1, m_2, ..., m_k$ possible outcomes of k experiments. For example, let a box contain 5 black, 3 green, 7 red, and 6 white balls. A sample of 4 balls is chosen in such a way that it contains one of each color. Then, $5 * 3 * 7 * 6 = 630$ sets of samples of 4 balls are possible in the experiment.

Permutations and combinations: Let us consider a group of nn distinct individuals or objects (here the term "distinct" means that there is some characteristic that differentiates any particular individual or object from any other). A natural query may arise: How many ways are there to select a subset of size r from the group?

Let there are three types of invitation letters: a, b, and c available, and we have to find an ordered arrangement of posting these letters to the respective addresses. There are 6 ordered arrangements possible: *abc, acb, bca, bac, cab,* and *cba.* These types of arrangements are known as permutations. With n objects, $n! = n * (n-1) * (n-2) * ... * 3 * 2 * 1$ permutations are possible. In above example there are $3! = 3 * 2 * 1 = 6$ possible permutations.

Further, the number of permutations of r individuals/objects formed from the total of n individuals/objects in a group is defined as $^nP_r = \dfrac{n!}{(n-r)!}$.

An unordered subset is called a combination. Let there be n objects, out of which different groups of r objects are to be chosen in no particular order. The whole way of selection, defined by the combination, is $\begin{pmatrix} n \\ r \end{pmatrix} = {}^nC_r = \dfrac{n!}{(n-r)! * r!}$ and is read as "n choose r."

Let a box contain 10 balls and draw a sample of 4 balls from the box. There are $\begin{pmatrix} 10 \\ 4 \end{pmatrix} = {}^{10}C_4 = \dfrac{10!}{4! * 6!} = 210$ possible samples of 4 balls to be drawn.

Let us consider an example where an article is chosen randomly with no defect from a lot consisting of 10 useful articles, 4 with minor defects, and 2 with major defects. In this case, the total number of exhaustive events is $n = \begin{pmatrix} 16 \\ 1 \end{pmatrix} = 16$, and the number of favor-able events is $m = \begin{pmatrix} 10 \\ 1 \end{pmatrix} = 10$. Thus the required probability is $\dfrac{10}{16} = \dfrac{5}{8}$.

The permutation concept is applied to the selection of ordered arrangements without replacement. On the other hand, for unordered selections without replacement, the combination rule is used. For ordered selections with replacement, an object is replaced to the group after it is chosen. That is, the selection of an object is allowed more than once. Thus, r objects out of a group of n are selected $\underbrace{n * n * \ldots * n}_{r \text{ times}} = n^r$ ways.

18.3.2 Axiomatic Definition of Probability

The computation of probability for a real-life scenario may be cumbersome with the classical and relative frequency approaches. Therefore, it is necessary to build probability models that are based on certain assumptions regarding the events. This approach of defining probability with axioms is called the axiomatic approach of probability. It may not provide the actual probability of an event; however, it achieves the goal by closely approximating it based on the following axioms.

Axiom 18.1: For any event E, $0 \le P(E) \le 1$,

Axiom 18.2: For sample space S, $P(S) = 1$.

Axiom 18.3: For any sequence of mutually exclusive events; $E1$, $E2$, ..., (i.e., for any two Ei and Ej, $\left(\text{for any two } E_i \text{ and } E_j, \ i \ne j, \ E_i \cap E_j = \varphi \right)$, $P\left(U_{i=1}^{\infty} E_i \right) = \sum_{i=1}^{\infty} P(E_i)$.

The term (E) is called the probability of the event E. Axiom 18.1 implies that the probability of any events always lies between the limits 0 and 1. Axiom 18.2 means that the probability of the whole sample space is 1. Axiom 18.3 states that whether the collection is finite or infinite, the probability of the union of the disjoint events is equal to the sum of their individual probabilities.

Theorems based on the probability of events: Certain theorems are given below (without proof) based on these axioms of probability. These theorems help define the probabilities of the events using the relationships between the outcomes of a random experiment.

1. The probability of the complementary event, denoted by E^c, E' or $\overline{E}$, is $1 - P(E)$.
2. The probability of the null event is $(\varphi) = P(Sc) = 1 - P(S) = 0$.
3. For any two events A and B, the union of these two can be defined as $(A \cup B) = P(A) + P(B) - P(A \cap B)$. This theorem is called the additive law of probability.
4. For any two events A and B, $P(A^c \cap B) = P(B) - P(A \cap B)$.
5. For any two events A and B such that $A \subset B$, $P(A) \le P(B)$.

In Example 18.1, one can easily check that $P(B') = 1 - P(B)$ satisfies Theorem 18.1. Moreover, $P(S) = P(L_1) + P(L_2) = 1$ fulfills Axiom 18.2 as the events L_1 and L_2 are exhaustive events in S. Based on the above theorems, we can also find, rather than compute manually, $P(L_2) = 1 - P(L_1) = 1 - 0.44 = 0.56$.

Further, the probability that a component is either from line 1 or defective, with Axiom 18.3 and Theorem 18.3, $P(L_1 \cup B) = P(L_1) + P(B) - P(L_1 \cap B) = 0.44 + 0.17 - 0.11 = 0.5$.

Further with Theorem 18.4, $P(L_2 \cap B) = P(\overline{L}_1 \cap B) + P(B) - P(L_1 \cap B) = 0.17 - 0.11 = 0.06$.

18.3.3 Subjective Probability

The subjective approach to probability helps measure uncertainty in real-life scenarios that may not be repeated as random experiments, such as an event of rain or natural calamity. A real-life scenario may not happen or be repeated under similar conditions. In this case, the classical, relative frequency, and axiomatic approaches to the probability concept do not work. For example, natural calamity events and no natural calamities are neither equally likely nor can either happen under similar conditions every time. There may be different known and unknown reasons for their occurrence that may not function uniformly. The frequency approach to counting the happenings and the probabilities of such events do not pertain here. The subjective approach of probability is based on one's belief, judgment, or past knowledge about the happening of an event. For example, the chances of rain on a particular future date may be quantified based on the experts' opinion about it. One may use some conditional information regarding the event, such as weather information like cloudy, sunny, windy, and humid.

The subjective approach replaces the counting rule or frequency approach with subjective belief or knowledge. However, it follows the axioms and theorems of probabilities on subjective probabilities.

18.3.4 Conditional Probability

The probability of an event may depend on the extra information available. We may say that the occurrence of an event is reliant on the occurrence of other events. Conditioning our knowledge on such information revises the probability of the event, and it is termed as the conditional probability. Symbolically, the probability $P(E)$ of any event E is called the unconditional probability of that event. If we have any additional information, defined by event A, on the happening of event E, then, one may be interested in examining how the information "an event A has occurred before E" affects the probability assigned to E. Event E is called the conditional event.

The probability of E given A is denoted by $P(E|A)$, and is called the conditional probability of happening of E given A. It is computed as $P(E|A) = \dfrac{P(E \cap A)}{P(A)}; P(A) > 0.$

Example 18.2

For Example 18.1, we may be interested in finding the probability that a component is selected from line L_1 given that it is defective (B). Taking the reference from Table 18.1 and Example 18.1, we can compute

$$P(\text{a component selected from line } L_1 \text{ given that it is defective})$$

$$= P(L_1|B) = \frac{P(L_1 \cap B)}{P(B)} = \frac{\left(\dfrac{2}{18}\right)}{\left(\dfrac{3}{18}\right)} = \frac{2}{3} = 0.667.$$

$$P(\text{a component selected from line } L_2 \text{ given that it is non-defective})$$

$$= P(L_1|B'), \text{ and, } P(L_1|B') = \frac{P(L_1 \cap B')}{P(B')} = (9/18)/(15/18) = 0.6.$$

We can compare Examples 18.1 and 18.2 and see that the probabilities of L_1 and L_2 increased with extra information provided by B and B', respectively.

Multiplicative rule of probability: The above-mentioned definition of conditional probability yields the multiplicative rule of probability for two events, given as $P(E \cap A) = P(E|A)P(A); P(A) > 0$. Alternatively, it can also be written as $P(E \cap A) = P(A|E)P(E); P(E) > 0$.

The term $P(E|A)$ represents the conditional probability of E when A has already happened, and similarly, $P(A|E)$ stands for the conditional probability of A when E has already occurred.

From the above example, one can verify that $P(L_1 \cap B) = P(L_1|B)P(B) = \left(\frac{2}{3}\right) * \left(\frac{3}{18}\right) = \frac{2}{18}$.

18.3.5 Probability of Independent Events

We learned that the conditional probability $P(E|A)$ revises the probability of an event E when the subsequent information is supplied by event A. It is different from the unconditional probability $P(A)$ of event A. However, it may also be possible that the subsequent information supplied by the other event A is not relevant to E, i.e., A is independent of E. In such cases, the probability of a conditional event is the same as the probability of an unconditional event.

Symbolically, two events E and A are said to be independent if $P(E|A) = P(E)$; otherwise, they are called dependent events. This definition may present the multiplication rule for independent events as $P(E \cap A) = P(E|A)P(A) = P(E)P(A)$. Also, it is clear from Example 18.1 that $P(L_1) * P(B) = \frac{8}{18} * \frac{3}{18} \neq \frac{2}{18} = P(L_1 \cap B)$. Hence, these events are dependent. To understand the concept of independent events more clearly, let us consider another example.

> **Example 18.3**
>
> Suppose that 30% of branded washing machines require service while under warranty; on the other hand, only 10% of its dryers need such service. If someone purchases both a washer and a dryer of the same brand, what is the probability that both machines will need warranty service?
>
> Let A denote the event that the washer needs service while under warranty, and B is defined analogously for the dryer. Then $P(A) = 0.3$ and $P(B) = 0.1$.
>
> Assuming that the two machines will function independently of one another, the desired probability is $P(A \cap B) = P(A) * P(B) = 0.3 * 0.1 = 0.03$.
>
> It is straightforward to show that A and B are independent if and only if A^c and B are independent, or A and B^c are independent, or A^c and B^c are independent.

18.3.6 Some Basic Results on Probability

Let A, B, and C be arbitrary events. The probability of events defining some relations of A, B and C, are presented as below:

1. Only A occurs $\Rightarrow P(A \cap B^c \cap C^c)$.
2. Both A and B occur, but C does not occur $\Rightarrow P(A \cap B \cap C^c)$.
3. All three events occur $\Rightarrow P(A \cap B \cap C)$.

4. At least one occurs $\Rightarrow P(A\cup B\cup C)=1-P\left(A^{c}\cup B\cup C^{c}\right)$

5. Exactly one occurs $\Rightarrow P\left(\left(A\cap B^{c}\cap C^{c}\right)\cup\left(A^{c}\cap B\cap C^{c}\right)\cup\left(A^{c}\cap B^{c}\cap C\right)\right)$

6. Exactly two occur $\Rightarrow P\left(\left(A\cap B\cap C^{c}\right)\cup\left(A\cap B^{c}\cap C\right)\cup\left(A^{c}\cap B\cap C\right)\right)$

7. At least two occur $\Rightarrow P\left(\left(A\cap B\cap C^{c}\right)\cup\left(A\cap B^{c}\cap C\right)\cup\left(A^{c}\cap B\cap C\right)\cup\left(A\cap B\cap C\right)\right)$

8. None occur $\Rightarrow P(A\cup B\cup C)^{c}=P\left(A^{c}\cap B^{c}\cap C^{c}\right)$

Let us consider another example to understand the above basic results.

Example 18.4

The odds that three independent critics will favorably review a book on statistics are 3 to 2, 4 to 3, and 2 to 3. What is the probability that, of the three reviews: (i) All will be favorable? (ii) The majority of the reviews will be favorable? (iii) Exactly one review will be favorable? (iv) Exactly two reviews will be favorable? (iv) At least one of the reviews will be favorable?

Suppose that A, B, and C denote the events that the book is favorably reviewed by the first, the second, and the third critic, respectively. We can compute the required probabilities with the given odds ratios in the following manner.

Let there be $(m+n)$ mutually exclusive and exhaustive cases, out of which m cases are favorable to the happening of an event E and n are favorable to nonhappening of the event E. Then, the odds in favor of is defined as $m:n$ or m to n, and the probability of happening of E is defined as the fraction $\dfrac{m}{(m+n)}$.

With the given information, the probabilities of the defined events are found as

$$P(A)=\frac{3}{5};\ P(B)=\frac{4}{7};\ \text{and}\ P(C)=\frac{2}{5},$$

$$\Rightarrow P\left(A^{c}\right)=\frac{2}{5};\ P\left(B^{c}\right)=\frac{3}{7},\ \text{and}\ P\left(C^{c}\right)=\frac{3}{5},$$ according to Theorem 18.1 on probabilities.

i. The probability that all critics will be favorable is

$$P(A\cap B\cap C)=P(A)*P(B)*P(C)=\frac{3}{5}*\frac{4}{7}*\frac{2}{5}=\frac{24}{175}$$

(since all the events A, B and C are independent of each other).

ii. The probability that the event that the majority, i.e., at least two reviews are favorable, is $P\left(\left(A\cap B\cap C^{c}\right)\cup\left(A\cap B^{c}\cap C\right)\cup\left(A^{c}\cap B\cap C\right)\cup\left(A\cap B\cap C\right)\right)$. Since all the events A, B, and C are mutually exclusive, by Axiom 18.3, the union of the mutually exclusive events can be written as the sum of their probabilities. Therefore, the above probability is equal to

$$P\left(A\cap B\cap C^{c}\right)\cup\left(A\cap B^{c}\cap C\right)\cup\left(A^{c}\cap B\cap C\right)\cup\left(A\cap B\cap C\right)$$

$$=P(A)*P(B)*P\left(C^{c}\right)+P(A)*P\left(B^{c}\right)*P(C)+P\left(A^{c}\right)*P(B)*P(C)+P(A)*P(B)*P(C)$$

$$=\frac{3}{5}*\frac{4}{7}*\frac{3}{5}+\frac{3}{5}*\frac{3}{7}*\frac{2}{5}+\frac{2}{5}*\frac{4}{7}*\frac{2}{5}+\frac{3}{5}*\frac{4}{7}*\frac{2}{5}=\frac{94}{175}$$

iii. Similarly, the probability that exactly one review will be favorable can be obtained as

$$P\big(\big(A\cap B^c\cap C^c\big)\cup\big(A^c\cap B\cap C^c\big)\cup\big(A^c\cap B^c\cap C\big)\big)$$

$$= P\big(A\cap B^c\cap C^c\big)+\big(A^c\cap B\cap C^c\big)+\big(A^c\cap B^c\cap C\big)$$

$$= P(A)*P\big(B^c\big)*P\big(C^c\big)+P\big(A^c\big)*P(B)*P\big(C^c\big)$$

$$+P\big(A^c\big)*P\big(B^c\big)*P(C)=\frac{3}{5}*\frac{3}{7}*\frac{3}{5}+\frac{2}{5}*\frac{4}{7}*\frac{3}{5}+\frac{2}{5}*\frac{3}{7}*\frac{2}{5}=\frac{63}{175}$$

iv. In the same way, the probability that exactly two reviews will be favorable can be obtained as:

$$P\big(\big(A\cap B\cap C^c\big)\cup\big(A\cap B^c\cap C\big)\cup\big(A^c\cap B\cap C\big)\big)$$

$$= P\big(A\cap B\cap C^c\big)+P\big(A\cap B^c\cap C\big)+P\big(A^c\cap B^c\cap C\big)$$

$$= P(A)*P(B)*P\big(C^c\big)+P(A)*P\big(B^c\big)*P(C)+P\big(A^c\big)*P(B)*P(C)$$

$$=\frac{3}{5}*\frac{4}{7}*\frac{3}{5}+\frac{3}{5}*\frac{3}{7}*\frac{2}{5}+\frac{2}{5}*\frac{4}{7}*\frac{2}{5}=\frac{70}{175}$$

v. The probability that at least of one the reviews will be favorable is

$$P(A\cup B\cup C)=1-P\big(A^c\cap B^c\cap C^c\big)=1-P\big(A^c\big)*P\big(B^c\big)*P\big(C^c\big)$$

$$=1-\frac{2}{5}*\frac{3}{7}*\frac{3}{5}=\frac{157}{175}$$

Thus, with the above-explained axioms, theorems, and concepts of probabilities, we can avoid using the technique of counting the number of favorable cases or making a subjective judgment about every event, making it easier to compute probabilities of simple or compound events of real lives.

Examples 18.1 and 18.3 are taken from Devore (2012). For more details about the probability concept, see Ross (2014), Meyer (1965), Mood et al. (1974), Miller and Miller (2014), and Devore (2012).

18.4 Bayes' Theorem

Bayes' theorem (or rule, or law) is a mathematical formula that is widely used in almost all applied fields of study for finding reverse or inverse conditional probabilities. If two events have a cause and effect relationship, it measures the uncertainty in cause given the effect. Let us consider the following example to understand Bayes' theorem.

Example 18.5

Let the chances that a person will have a specific disease be 0.001. The test designed to diagnose the disease has a true-positive rate of 95% and a false-negative rate of 1%. If a person undergoes the test with positive results, what are the chances of having the disease? Symbolically, if A is the incidence of disease and event B is the test being positive, then $P(A) = 0.001$, $P(B|A) = 0.95$, and $P(B|A^c) = 0.01$. What then is $P(A|B)$?

By Bayes' rule,

$$P(A|B) = \frac{P(A \cap B)}{P(B)} = \frac{P(B|A)P(A)}{P(B)}$$

Event A may be thought of as a cause and event B as an effect.

Using the tree diagram in Figure 18.1, we can compute $P(B) = P(A \cap B) + P(A^c \cap B)$ $= 0.00095 + 0.00999 = 0.01094$.

(Note that tree diagrams in probability are displays of the multiplicative rule of probabilities. On the diagram, the values written on lines are the probabilities of the events sequenced by the lines. The columns on the RHS of the diagram stand for the joint probabilities of the events displayed, and their calculated results with the multiplicative rule of probabilities.)

Therefore, the inverse (conditional) probability is found to be $P(A|B) = \dfrac{0.00095}{0.01094} = 0.087$.

Thus, with additional information through the diagnosis test, the probability of occurrence of the disease increases.

An R-code for the computation of inverse probability based on Bayes' theorem for the above example is as follows:

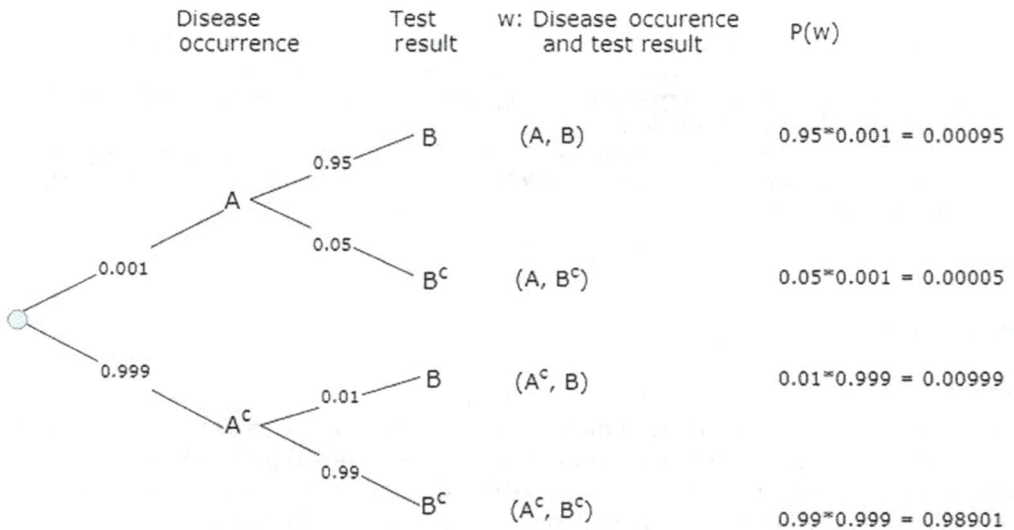

FIGURE 18.1
Tree diagram for calculating the probability of occurrence of disease and the result of the diagnostic test.

```
################################################################
inverse.prob = function (a, b, c) {
likelihood.prior = b*a;
m.likelihood = b*a + c*(1-a)
return(likelihood.prior /.likelihood)}
inverse.prob(0.001, 0.95, 0.01)
# result
[1] 0.08683729
################################################################
```

In this example, the probability of event A was given based on the experts' knowledge or past research. Thus, Bayes' theorem is a method to update an expert's knowledge about an unknown in light of the observations.

Example 18.6

Let us consider an example of playing cards. Suppose there are two stacks of 9 cards each. The first stack has 3 aces, 4 kings, and 2 queens; the second contains 2 aces, 3 kings, and 4 queens. A card is randomly chosen from the first stack and added to the second. If a randomly selected card from the second stack is a queen, what is the probability that the first card was also a queen?

Let A_i, K_i, and Q_i be the events of drawing an ace, a king, and a queen from the ith stack. We are interested in finding $P(Q_1|Q_2)$. Bayes' formula to calculate this inverse probability for the above example is as follows:

$$P(Q_1|Q_2) = \frac{P(Q_1 \cap Q_2)}{P(Q_2)} = \frac{P(Q_2|Q_1)P(Q_1)}{P(Q_2)}, \text{ and}$$

$$P(Q_2) = P(Q_2 \cap Q_1) + P(Q_2 \cap A_1) + P(Q_2 \cap K_1).$$

We can use the calculations in the tree diagram displayed in Figure 18.2 to find this inverse probability. The marginal probability $P(Q_2)$ can be found by adding the respective joint probabilities of Q with the cause events A_1, K_1, and Q_1; it is calculated as equal to $\frac{1}{9} + \frac{2}{15} + \frac{8}{45} = \frac{19}{45}$. The joint probability $P(Q_2 \cap Q_1) = \frac{1}{9}$, as given in the diagram.

Therefore, the inverse probability $P(Q_1|Q_2) = \left(\frac{1}{9}\right) / \left(\frac{19}{45}\right) = \frac{5}{19}$ is based on Bayes' rule.

Thus, we can see that the information about the card drawn from the stack increases the probability of a queen having been drawn from the first stack, if only slightly.

The same result for more than two cause events can be obtained using the following R-code to find inverse probability based on the Bayes' theorem.

```
###################### R-environment #########################
inverse.prob = function (r1, r2, r3, a, b){
likelihood.prior = r1*a;
m.likelihood = r1*a + r2*b + r3*(1-a-b)
return(likelihood.prior/m.likelihood)}
inverse.prob(0.5, 0.4, 0.4, 0.222, 0.333)
# r e s u l t
[1] 0.2629086
################################################################
```

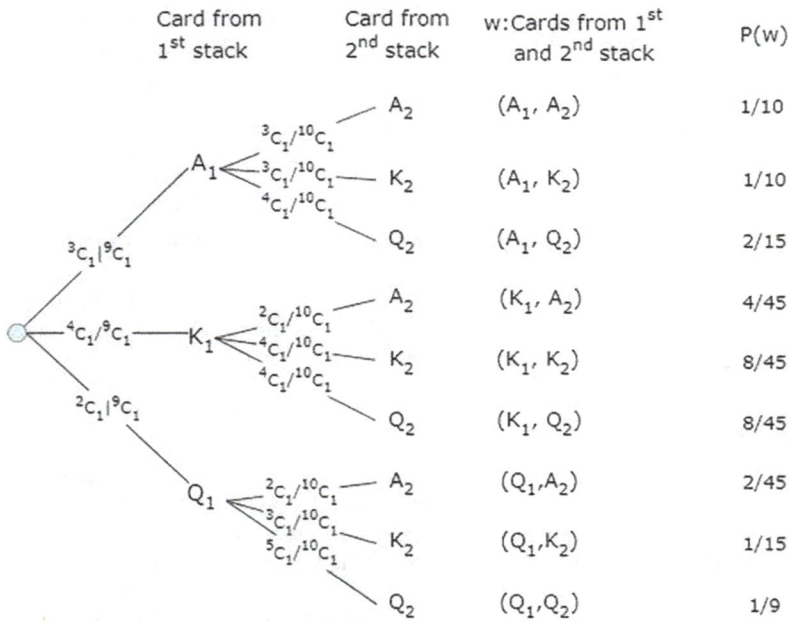

Card from 1st stack	Card from 2nd stack	w: Cards from 1st and 2nd stack	P(w)
	A_2	(A_1, A_2)	1/10
A_1	K_2	(A_1, K_2)	1/10
	Q_2	(A_1, Q_2)	2/15
	A_2	(K_1, A_2)	4/45
K_1	K_2	(K_1, K_2)	8/45
	Q_2	(K_1, Q_2)	8/45
	A_2	(Q_1, A_2)	2/45
Q_1	K_2	(Q_1, K_2)	1/15
	Q_2	(Q_1, Q_2)	1/9

Branch probabilities (first stack): $^3C_1|^9C_1$ (to A_1), $^4C_1|^9C_1$ (to K_1), $^2C_1|^9C_1$ (to Q_1).
Branch probabilities (second stack) from A_1: $^3C_1/^{10}C_1$, $^3C_1/^{10}C_1$, $^4C_1/^{10}C_1$; from K_1: $^2C_1/^{10}C_1$, $^4C_1/^{10}C_1$, $^4C_1/^{10}C_1$; from Q_1: $^2C_1/^{10}C_1$, $^3C_1/^{10}C_1$, $^5C_1/^{10}C_1$.

FIGURE 18.2
Tree diagram for calculating probabilities of cards drawn at first and the second stacks.

Example 18.6 considers the classical definitions of probability, whereas Example 18.5 applies the subjective probability concept to quantify marginal (prior) probabilities. Thus, Bayes' theorem finds applications through both the classical and the Bayesian approaches to statistics. While the classical approach calculates the inverse/reverse probability of causes given the effects based on the facts, the Bayesian notion allows for an update of prior beliefs or knowledge about causes in the light of facts or observation of effects. We can now present a generalized definition of Bayes' theorem as follows:

Definition: Let $C_1, C_2, \ldots, C_K$ mutually exclusive and exhaustive events with nonzero probabilities. Suppose any event E, with $P(E) > 0$, occurs after any of the C_i's have already happened.

Then the inverse/reverse probability can be computed as $\forall i = 1: K, P(C_i|E) = \dfrac{P(E|C_i)P(C_i)}{P(E)},$

and $P(E) = \sum_{i=1}^{K} P(E \cap C_i).$

Historical note on Bayes' theorem: The discovery of Bayes' theorem is attributed to Thomas P. Bayes. His work on "Bayes' Theorem" was posthumously published by his apprentice Richard Price in 1764. He introduced the rule based on the definition and relationship between marginal, conditional, and joint probabilities as $(A|B) = \dfrac{P(A \cap B)}{P(B)}$, where event A is followed by event B. He applied the expression to find a probabilistic statement of an unknown cause event lying in an interval given an observation about an effect event. He used the subjective probability approach to define the marginal (prior) probability of an unknown cause, considering equal probability of all its possible values. However, his work on the theorem was restrictedly applied to an estimation problem related to the billiard table. More historical details of Thomas Bayes' work on Bayes' theorem can be found in Stigler (1982).

Unaware of Thomas Bayes' published work, Pierre-Simon Laplace published his first work on inverse probability in 1774 and found the generalized definition of the Bayes' rule. Like Bayes, Laplace also considered equal probabilities to define marginal (prior) probabilities of cause events. However, his formula for Bayes' theorem may also be applied beyond the equally likely concept of probabilities. A detailed description of Laplace's work can be found in Stigler (1986). Both Bayes and Laplace considered estimation problems of a cause given observations on effects. Nevertheless, Laplace went beyond computing posterior probabilities of causes to define predictive posterior probabilities of new (future) observations on effects, accounting for the cause's posterior uncertainty.

For a detailed discussion of both definitions of Bayes' theorem, see Dale (1982) and Fienberg (2006).

18.5 Bayes' Theorem for Random Variables and the Parameter Concept

A random variable (r.v.) represents (or rather numerically assigns values to) similar specific problems in studies, experiments, or surveys. For example, a r.v. X is assumed to represent performance on an exam with the exhaustive events pass and fail, assigning them the values 1 and 0, respectively. The same random variable may be used to refer to problems under study with only two (dichotomous) outcomes. The distribution of r.v. X modeled around the parameter θ assigns the values of X probabilities as some function of θ. The inverse probability, in the case of random variables, can then be defined in the following manner.

If a set of observations around X has been obtained, it may be of some interest to find the most probable values of θ that gave rise to the observations. A conditional probability of θ, given the realizations around X, defines its inverse probability for the case of random variables. Let a r.v. X follow the distribution $f(x|\theta), \theta \in \Theta$. Assume that a set of observations $\underline{x} = \{x_1, x_2, \ldots, x_n\}$ has been found on X. Then, considering θ a random quantity, the (inverse) conditional probability density of θ given $\underline{x}$ may be computed by Bayes' theorem as,

$$P(\theta|\underline{x}) = \frac{P(\underline{x}|\theta)P(\theta)}{P(\underline{x})}$$

$P(\theta)$ is the density of θ assumed before observing data $\underline{x}$ on X; it is called a prior density of θ over Θ. Similarly, the conditional density $P(\theta|\underline{x})$ is called the posterior density of θ given $X = \underline{x}$. Thus, the measurement of uncertainty in θ, $P(\theta)$, is updated in light of data $\underline{x}$ to give the posterior uncertainty measurement $P(\theta|\underline{x})$.

The notation $P(\underline{x})$, also denoted $m(\underline{x})$, is called the marginal likelihood of $X = \underline{x}$ after integrating or summing out the uncertainty of θ in $P(\theta, \underline{x}) = P(\theta)P(\underline{x}|\theta)$. It is defined symbolically a,

$$P(\underline{x}) = \begin{cases} \int_{\Theta \in \Theta} P(\underline{x}|\theta)P(\theta)d\theta, & \text{if } \theta \text{ is continuous.} \\ \sum_{\theta \in \Theta} P(\underline{x}|\theta)P(\theta), & \text{if } \theta \text{ is discrete.} \end{cases}$$

Further, it is a good medium to check whether prior model assumptions are appropriate; hence, it is also referred to as model evidence. The term $P(x|\theta)$ may be read as the joint density of $X = x$ given θ. It is also treated as the likelihood of θ given $X = x$, and is termed $L(\theta|x)$. With this definition, the theorem may be rewritten as $P(\theta|x) = \dfrac{L(\theta|x)P(\theta)}{P(x)}$.

Furthermore, $P(x) = \begin{cases} \displaystyle\int_{\Theta} L(\theta|x)P(\theta)d\theta, & \text{if } \theta \text{ is continuous.} \\ \displaystyle\sum_{\Theta} L(\theta|x)P(\theta), & \text{if } \theta \text{ is discrete.} \end{cases}$

As $P(x)$ is not a function of θ, it may be treated as constant w.r.t. θ. Thus, we may also write Posterior $\propto$ Likelihood $\times$ Prior.

The likelihood term should not be confused with a density of parameters. It only refers to the likelihood of each value of θ in Θ generating a set of data observed on X. If θ is multidimensional, the marginal posterior density of a parameter component θi given $X = x$, is found by summing or integrating the joint posterior density of θ given $X = x$ w.r.t other components of θ except θi. Symbolically, if the term $/i$ stands for all components of θ vector except i, the marginal posterior density of θi is found as:

$$P(\theta_i|x) = \begin{cases} \displaystyle\int_{\theta/i} P(\theta|x)d\theta_{/i}, & \text{if } \theta_{/i} \text{ is continuous.} \\ \displaystyle\sum_{\theta/i} P(\theta|x), & \text{if } \theta_{/i} \text{ is discrete.} \end{cases}$$

18.6 Why Is θ Random?

Unlike in the frequentist approach to statistics, treating θ as random is the key concept of Bayesian statistics. The Bayesian paradigm is based on the subjective notion of probability. There may be different values of an unknown (parameter) under different hypotheses, thus accounting for its uncertainty. This uncertainty may be quantified based on experts' judgment, personal belief, or past studies about the unknown. Assigning a prior density to unknown parameters has always been a matter of debate among Bayesian statisticians that thus gives rise to a branch of study under Bayesian statistics called prior elicitation.

Let us consider some examples to calculate posterior densities under different prior assumptions.

Example 18.7

Suppose a coin is spun 250 times on its edge, resulting in the head facing up 140 times and the tail 110 times. The nature of the coin is unknown. Based on this information, we have to compute the posterior probability of getting a head given the result of the experiment.

There can be three steps to find the posterior probabilities.

1. Set a prior density.
2. Define the likelihood.

3. Compute the posterior density.

Step 1: Suppose there is no prior information on the type of coin spun. In this case, we may consider all the values of the chance of getting head, say θ in (0, 1), equally probable. Therefore, we may assign the standard uniform density as the prior density of θ. That is, $\theta \sim (0, 1)$, a priori.

Step 2: If the total number of trials is 250, $^{250}C_{140}$ combinations of 140 heads and 110 tails as the sequences of outcomes are possible. Let X denote the number of heads, then $P(X = 140|\theta) = {}^{250}C_{140}\theta^{140}(1-\theta)^{110}$, $0 < \theta < 1$. The term $(1-\theta)$ is the probability of getting a tail. The likelihood of θ, $L(\theta|X = 140)$, is then $L(\theta|X = 140) = {}^{250}C_{140}\theta^{140}(1-\theta)^{110}$, $0 < \theta < 1$.

Step 3: By Bayes' law, the posterior density of θ is computed as,

$$(\theta|X = 140) = \frac{L(\theta|X = 140)P(\theta)}{\int_{\theta} L(\theta|X = 140)P(\theta)d\theta} = \frac{\theta^{140}(1-\theta)^{110}}{\int_{\theta}\theta^{140}(1-\theta)^{110}d\theta} = \text{Beta}(\theta; 141, 111).$$

An equivalent R-code to compute the above posterior density is as follows:

```
###############################################################
step = 0.01
theta = seq (0,1, by=step) ptheta1 = dunif (theta, 0, 1)
n= 250
r = 140
likeli = theta^r * (1-theta)^(n-r) joint1 = likeli *ptheta1
ml = sum(joint1 *step) post1 = joint1/ml
###############################################################
```

Suppose before conducting the experiment that we learned that the coin is biased toward heads. This means that $\theta > 0.5$ values are more probable a priori. Given this piece of information, let us set our prior density of θ as $P(\theta) = \text{Beta}(\theta; a=4, b=2.5)$. As a result, the posterior density $P(\theta|X = 140)$ comes out as $\text{Beta}(\theta; 144, 112.5)$. The R-code to compute this posterior density is as follows:

```
###############################################################
step = 0.01
theta = seq (0, 1, by=step)
ptheta2 = dbeta(theta, 4, 2.5)
n= 250
r = 140
likeli = theta ^ r * (1-theta) ^ (n-r) joint = likely * ptheta2
ml = sum(joint *step) post2 = joint /ml
###############################################################
```

Alternatively, suppose we were told that the coin is lopsided toward tails. An appropriate prior density to quantify this information may be $P(\theta) = \text{Beta}(\theta; a=2.5, b=4)$. The posterior density corresponding to this prior density is obtained as $\text{Beta}(\theta; a=142.5, b=114)$. In the R-code, we just have to change the prior density type to get the posterior density in the third case. We can compare these results using the posterior probability plots shown on the left side of Figure 18.3. We can see from the figure that the posterior densities almost overlap with each other. That is, the effect of different prior densities on the resultant posterior is near negligible. This happens in the case of strong likelihoods and with large data sets.

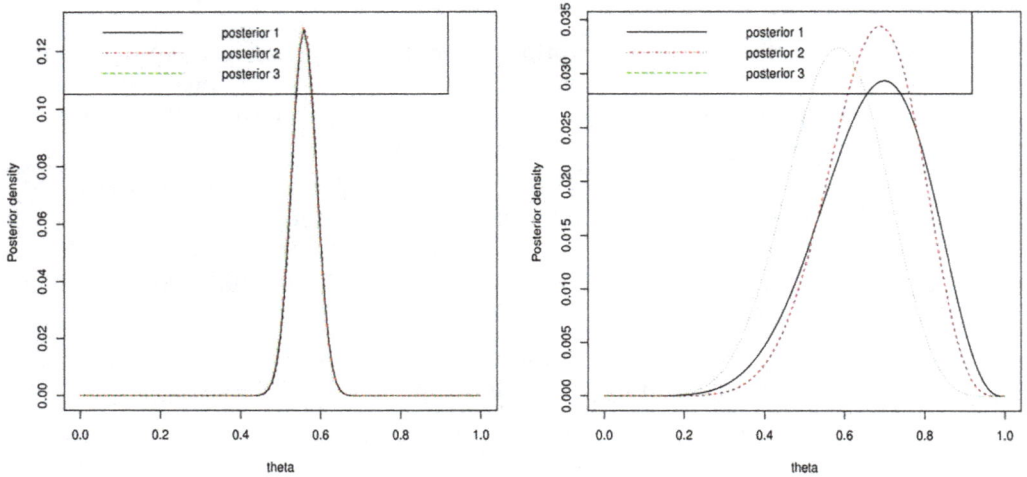

FIGURE 18.3
Comparison of posterior densities for different choices of priors with the number of trials and the number of heads 250 and 140, left; 10 and 7, right.

Let us consider a data set of size 10 only. If the experiment was conducted 10 times and resulted in 7 heads and 3 tails, were the choices of different prior densities effective on the posterior densities? The answer lies on the right side of Figure 18.3. We can see that the three posterior densities are separate from the prior density choices. Therefore, in the case of weak data, it is essential to take sufficient care when choosing prior densities. The next section describes the approaches to choose an appropriate prior based on the information available about an unknown parameter.

The R-syntax to produce the graph is as follows:

```
##################################################################
matplot (theta, cbind(post1, post2, post3), type= " l ", ylab =
"Posterior density ") legend ("topright ", legend = c ("posterior 1",
"posterior 2", "posterior 3"),
      col = 1:3, lty = 1:3)
##################################################################
```

18.7 Choosing a Prior

The process of choosing an appropriate prior by quantifying knowledge, no-knowledge, or little knowledge about the parameter θ before observing data is the first step in calculating posterior uncertainty. In the case of weak or insufficient data, a prior density plays an important role in posterior inference about θ. On the other hand, with sufficient and strong datasets, prior density has only a substantial effect on the inference a posteriori. Therefore, one should be careful while choosing a prior density for a meaningful inference about θ.

There are mainly two broad categories of approaches to determining a prior density, the subjective and objective approaches. The subjective approach is a way to quantify experts' judgment, one's prior belief, or knowledge about θ. On the other hand, objective prior densities reflect no or little prior information about θ.

18.8 Subjective Approach

Quantifying a prior density around θ based on experts' knowledge or judgment is called the prior elicitation. The subjective approach of choosing a prior provides ways to turn the experts' opinion into probabilistic statements. According to Berger (1985), there are four methods of the subjective approach based on the type of subjective knowledge available about $\theta \in \Theta$, namely, the relative likelihood approach, histogram approach, matching of a given functional form, and CDF determination.

The *relative likelihood* approach is applied when one value of θ is taken as less or more likely than the other values. If this type of information is available, a curve joining the comparative likelihoods of the values of θ may be sketched and matched with a suitable density curve. The *histogram approach* is useful when θ is continuous in Θ. The parametric space Θ may be binned into intervals. Each interval may then be assigned a probability based on the subjective information available. However, it may be challenging to determine probabilities for tail areas, at least for unbounded Θ. Further, it may be hard to decide a suitable number of intervals of θ if Θ is large. For the determination of a prior density for the two approaches, nonparametric methods, e.g., kernel density estimation, may be applied.

In the *matching of a given functional form* approach, a functional form of the prior density is assumed and matched with a suitable density. To ultimately determine the density, one needs to specify the hyper-parameters (parameters of the prior density). If it is reasonable to guess or find the mean and variance of θ a priori, one can find hyper-parameters relating to these two moments. However, the variance of a parameter may not always be easy to know or guess. In such cases instead of specifying moments, one can look for information on different fractiles of the density. Based on the values of these fractiles, relevant hyper-parameters may then be decided.

The *CDF determination method* is based on sketching the different fractiles $(Z(\alpha), \alpha)$ and then matching a density function of a suitable form. In practice, the CDF determination method has mostly been applied under the name "bisection method." The method is applied as follows: Extract experts' information about the second and third quantiles and match them with a chosen density to determine its hyper-parameters. If the other quantile values also match the corresponding properties of the density, declare the density the working prior density. Otherwise, drop the first choice of the density and proceed with a similar density match of the quantiles. For details about the bisection method, please see Dey and Liuy (2007) and Oakley and O'Hagan (2007).

18.9 Objective Approach

Objective priors are also named vague priors, noninformative priors, and weak priors to suit the case of no or little prior information about θ. The simplest prior in this category is

Laplace's prior. This prior assumes no preference among the values of θ; that is, the probability of a specified value of θ is a constant. Symbolically, $P(\theta) \propto \text{constant} > 0$, $\theta \in \Theta$. If θ is finite, $P(\theta)$ can be taken as a uniform density, assigning equal weight to each value of $\theta \in \Theta$. On the other hand, if Θ is unbounded, the prior is improper; that is, it integrates or sums to infinity. However, if the data are strong, the resultant posterior would be a proper density (integrating or summing to 1, thus satisfying the necessary criterion of a density). Laplace's prior is easy to use and it reflects no prior information about θ. However, it does not follow the invariance property. The no prior information property does not translate into re-parameterizations of θ.

The Jeffreys prior overcomes the lack of invariance in Laplace's prior. It is defined as

$P_J(\theta) \propto |I(\theta)|^{1/2}$, $\theta \in \Theta$. The term $I(\theta) = -E\left(\dfrac{\partial^2}{\partial \theta^2} \log f(X|\theta)\right)$ is the Fisher information of θ.

If θ is multidimensional, $I(\theta)$ is a matrix, and the determinant of $I(\theta)$ is considered in the definition.

However, if the parameter is scalar, the absolute of (θ) is applied in the definition of the Jeffreys prior density. There are three forms of the Jeffreys prior depending on Θ.

$$P_j(\theta) \propto \begin{cases} \text{constant}, & \text{if } \theta \in (\infty, -\infty) \\ \dfrac{1}{\theta}, & \text{if } \theta \in (0, -\infty) \\ \dfrac{1}{\theta(1-\theta)}, & \text{if } \theta \in (0, 1) \end{cases}$$

The Jeffreys prior borrows the functional relationship of θ with X and information about how the data model is sensitive to the changes in θ through the use of Fisher information. Most importantly, it ensures the invariance of the no-information criterion under re-parametrization. However, the Jeffreys prior becomes complicated as the dimension of θ increases.

The reference priors are found by maximizing a (Kullback–Leibler) divergence between posterior and prior densities, ensuring the maximum effect of the data on the posterior. These priors are invariant and are suitable for the parameters partitioned into a "parameter of interest" and "nuisance" categories. More details on the reference prior can be found in Berger and Bernardo (1991).

A uniform or Laplace's prior over entire Θ may constitute an improper prior and may result in a proper posterior density only in the case of strong data. However, with locally uniform priors, cases of improper posterior density can be avoided. This prior density assumes equal probabilities over a finite region of Θ where the likelihood of θ is strong; otherwise, it places a zero weight outside the region. It ensures the properties of a density are proper and thus always result in a proper posterior density.

18.10 Conjugate Prior

A class of priors for a class of likelihoods is called conjugate if a resulting posterior density (for a prior density and a likelihood of their respective classes) also belongs to the same class of the corresponding prior. Conjugate priors are the most useful as they provide ease

TABLE 18.2

TABLE 18.2

List of Conjugate Priors for Different Likelihoods under the Exponential Family of Densities with Corresponding Posterior Densities

Likelihood	Conjugate Prior Density	Posterior Density
Binomial	Beta	Beta
Negative binomial	Beta	Beta
Poisson	Gamma	Gamma
Normal, with unknown mean	Normal	Normal
Normal, with unknown variance	Inverse gamma	Inverse gamma
Normal, with unknown mean and variance	Normal-inverse gamma	Normal-inverse gamma

of computation of posterior densities, at least for lower-dimensional problems. Bayesian computation problems that involve the computation of posterior densities and their summaries are the key initial points in Bayesian inference. The computation problems in Bayesian inference generally involve complexities due to the lack of a standard form of prior densities and/or multidimensionality of θ. However, the complexities may be eased with a standard chosen form of prior densities, at least for lower-dimensional parameters.

Conjugate priors exist for the exponential family of densities. The conjugate priors with resulting posterior densities for different likelihoods of the exponential family of densities are summarized in Table 18.2.

Conjugate priors may belong to either informative or noninformative priors. If substantial knowledge about θ is available, it may be incorporated in conjugate priors through the hyper-parameters. However, in the case of no or little information about θ, the hyper-parameters must be chosen to reflect this situation. For example, one may choose large prior variances to impose "no prior information" about θ.

For more details on choosing a prior, Berger (1985), Lee (2012), Gelman et al. (2006), and Gosh et al. (2006) may be considered.

18.11 Bayesian Interval Estimation

It is often convenient to deal with the most probable values of the unknowns rather than just placing their uncertainty measurements (probabilities). For example, given a sample of observations of the IQs of certain age groups of students, there is a 95% chance that the average IQ may lie in the interval 100–120. This makes more sense than having the probabilities of different values of the average IQ only. An interval estimate provides a summarized estimate of the uncertainty in an unknown. **Interval estimation** aims to find the shortest possible interval within which an unknown may lie with a specified probability.

In the Bayesian approach, the interval estimation of θ is very straightforward. It only requires finding subsets of θ that have a specified posterior probability and are termed credible intervals. A $100(1-\alpha)\%$ credible interval $(a, b) \in \Theta$ is defined such that,

$$\int_a^b P(\theta|\underline{x})d\theta \geq (1-\propto), \quad \text{if } \theta \text{ is continuous, or}$$

$$\sum_{\theta \in (a,b)} P(\theta|\underline{x}) \geq (1-\propto), \quad \text{if } \theta \text{ is discrete.}$$

There can be many possible choices of 100 $(1-\alpha)$% credible intervals $(a, b) \in \Theta$ satisfying the above property. The aim is to choose the shortest possible interval with the most probable values of θ as its members. Such shortest possible credible intervals (or regions) containing the most probable values of θ are called highest posterior density or HPD intervals (or regions). Thus, a 100 $(1-\alpha)$% HPD interval $(a, b) \in$ can be defined for the following two conditions:

$$P(a|\underline{x}) = P(b|\underline{x}), \text{ and } P(\theta|\underline{x}) \geq P(\theta'|\underline{x}), \quad \text{for any } \theta' \notin (a,b)$$

(Given observations $\underline{x}$, the terms a and b are some functions of $\underline{x}$, i.e., $a \equiv a(\underline{x})$, and $b \equiv b(\underline{x})$.)

If $P(\theta|\underline{x})$ is symmetric and unimodal, HPD intervals are equivalent to equal tail credible intervals. In such a case, the 100$(1-\alpha)$% HPD interval (a, b) is defined for

$$P(\theta < a|\underline{x}) = P(\theta < b|\underline{x}) = \frac{\alpha}{2}.$$

18.12 Predictive Posterior Density

Quantifying uncertainty in future events is often desirable for government administration, public sectors, private sectors, industrialists, and even individuals for maximum gain or minimum loss in decision-making. For example, predictions of the unemployment rate would help the government open a number of new jobs. Knowing the chances of drought in the coming agricultural months would help farmers prepare for necessary actions to handle crop yield loss.

In Bayesian inference, prediction of a future observation is also subject to uncertainty measurements. The posterior density of a future or a new observation X_{new} given the current observations $\underline{x}$ is called a predictive posterior density and is computed as $P(X_{new}\underline{x}) = \int_{\Theta} P(X_{new}|\theta,\underline{x})P(\theta|\underline{x})d\theta = \int_{\Theta} P(X_{new}|\theta)P(\theta|\underline{x})d\theta$. The independently and identically distributed assumption of observations given θ allows us to set $P(X_{new}|\theta,\underline{x}) = P(X_{new}|\theta)$ in the definition of $P(X_{new}|\underline{x})$.

The posterior density $P(\theta|\underline{x})$ must be computed beforehand. Uncertainty in future observations may be further summarized with credible intervals or HPD intervals defined in the same manner as for unknown parameters.

The prediction of future events in the Bayesian inference is found by summing over the uncertainty of unknown parameters. However, in the frequentist approach, the prediction of a future event is found by plugging in the most suitable point estimate (the m.l.e. or unbiased estimate) of unknown parameters. Thus, it underestimates the uncertainty in prediction.

18.13 Application of Bayesian Statistics to Real-Life Data

For a possible application of Bayesian concepts, a data subset of the IQs of 40 women aged 20–25 years is borrowed from an R dataset IQ Guessing at https://vincentarelbundock. github.io/Rdatasets/datasets.html. The study of women's IQ may find applications in assessing academic growth, evaluating job performances, understanding health issues, and creating or finding suitable jobs for women of certain IQ groups. Therefore, we are interested in presenting Bayesian inferences on (1) the average IQ of the population of women and (2) the unknown IQ of a new woman from the population.

Let x denote the iid IQ observations of women, following a Gaussian distribution with unknown population average θ and precision τ. We consider the Jeffreys prior on θ and τ, $P(\theta, \tau) = P(\theta)P(\tau) \propto \dfrac{1}{\tau}$, $\theta \in (-\infty, \infty)$, $\tau \in (0, \infty)$. The joint posterior density $P(\theta, \tau | x)$ calculated with Bayes' law and is recognized as a normal-Gamma density.

The marginal posterior density of θ, $P(\theta | x)$, is found by integrating $P(\theta, \tau | x)$ w.r.t τ, yielding the $\sqrt{n}(\theta - x)/S_{n-1} \sim t_\vartheta$ distribution. It implies that the posterior distribution of θ given $\bar{x}$ is approximately normal with mean $\bar{x}$ and variance S_{n-1}^2. The terms $\bar{x}$, S_{n-1}^2, n, ϑ, and t_ϑ refer to the sample mean of data x, square root of sample variance, sample size, degrees of freedom (df) equal to $n-1$, and t-distribution with ϑ df, respectively.

The 95% HPD interval of θ is equal to the *equal tail credible interval* with the symmetric t_ϑ distribution, and is given as $\left[\overline{x} + S_{n-1} \times \dfrac{t_{\vartheta, \frac{\alpha}{2}}}{\sqrt{n}} \right]$. The subscript $\alpha = 1 - 0.95 = 0.05$, and the term $t_{\vartheta, \frac{\alpha}{2}}$ is the $\frac{\alpha}{2}$th quantile of the distribution.

We apply a numerical integration method to find the posterior density and the HPD interval for general cases when analytical solutions may not be tractable. The method provides a good approximation to the true results for low-dimensional problems. We approximated the 95% HPD interval of θ as [121.9, 128.06] with the numerical integration method. On the other hand, the corresponding result using the formula obtained as [120.1161, 129.8339].

We present below the R-syntax for computing and plotting the marginal posterior probabilities of θ and finding its 95% HPD intervals.

```
###Compute Posterior and predictive posterior densities with Numerical
Integration####
####set up prior###################################
theta = seq (70, 190, 0.01) # set possible range of theta tau =
seq(0.0001, 1, 0.01) # set possible range of tau
# 2D lattice of theta and tau
par1 = matrix(theta,length(theta),length(tau)) par2 = matrix(tau,
length(theta),length(tau)) log_prior = -log(par2) # log prior density

####data ####################################
data<-read.csv("IQGuessing.csv") # Import IQGuessing
x = data$TrueIQ # store True IQ sample observations in x n = length(x) #
no. of observations
xbar = mean(x); xvar = var(x) # sample mean and sample variance

## compute likelihood############## log_like = -0.5*n*(log(2*pi)
- log(par2))
for(i in 1:n){log_like = log_like -0.5*par2*(x[i]-par1)*(x[i]-par1)}

#### Compute posterior ##################### log_like_prior = log_
prior + log_like # log of joint density like_prior = exp(log_like_prior)
# joint density
norm_const = sum(rowSums(like_prior)) # normalizing constant of posterior
joint_posterior = like_prior/norm_const # joint posterior density
```

```
post_theta = rowSums(joint_posterior)

#### compute HPD interval ##################### credMass = 0.95 # size
of credible region
sortedPost = sort(post_theta, decreasing=T) # sort the posterior
probabilities # find the index of minimum of the sorted posterior
probabilities
# cumulatively adding to 0.95
HDIheightIdx = min(which(cumsum(sortedPost)>=credMass)) HDIheight =
sortedPost[HDIheightIdx] # posterior probabilities at index # indices of
highestposterior density adding to 0.95
indices = which(post_theta >= HDIheight)
# lowest theta value at the lowest indices of highest posterior density
HPD_Lni = min(theta [indices])
# largest theta value at the largest indices of highest posterior density
HPD_Uni = max(theta[indices])

####HPD interval using formula from the standard form of posterior
density #### alpha = 0.05; nu = n-1
# Lower and upper quantile of t-density with nu d.f. qL = qt(alpha/2,nu);
qU = qt(1-alpha/2,nu)
# HPD lower and upper bounds
HPD_Lf = xbar + sqrt(xvar)*qL/sqrt(n); HPD_Uf = xbar + sqrt(xvar)*qU/sqrt(n)

####print HPD results####
show(data.frame(HPD_Lni, HPD_Uni)); show(data.frame(HPD_Lf,HPD_Uf)) ####
plotting of posterior density of theta#########
plot(theta, post_theta, ylab = "posterior density", type = "l") #########
####################End##################################
```

The predictive posterior distribution of a new unknown observation X_{new} is computed as follows:

$$P(X_{new}|\underline{x}) = \int_{\theta,\tau} P(X_{new}|\theta,\tau)P(\theta,\tau|\underline{x})d\theta d\tau.$$

It may be difficult to compute the above predictive posterior distribution analytically. However, we could compute the joint predictive posterior distribution of X_{new} and θ given $\underline{x}$, $P(X_{new},\theta|\tau,\underline{x}) \propto \tau^{\frac{\vartheta}{2}} * \exp\left(-\frac{\vartheta}{2}*S_{n-1}^2\right) * \exp\left(-n*\frac{\tau}{2}*(\theta-\bar{x})^2\right) * \exp\left(-\frac{\tau}{2}*(X_{new}-\theta)^2\right).$

Integrating $P(X_{new},\theta|\tau,\underline{x})$ over the uncertainty in τ, $P(X_{new},\theta|\underline{x}) \propto \frac{\vartheta}{2} * (\vartheta*S_{n-1}^2)^{\frac{\vartheta}{2}} * \left[\vartheta*S_{n-1}^2 + n*(\theta-\bar{x})^2 + (X_{new}-\theta)^2\right]^{-\left(\frac{\vartheta}{2}+1\right)}.$

Further, we may compute the marginal predictive posterior distribution of X_{new} given $\underline{x}$, $P(X_{new}|\underline{x})$, by evaluating the integral over $P(X_{new},\theta|\underline{x})$ w.r.t θ using a numerical integration method. To do so, we first consider the values of X_{new} on grid points, then we approximate the integration w.r.t θ in the definition of $P(X_{new},\theta|\underline{x})$ with summation over the considered grid values of θ.

The 95% HPD interval of X_{new} obtained with the numerical integration method is [94.32, 155.63]. The plots of the numerically obtained marginal posterior density of θ given x and of the predictive posterior density of X_{new} given $\underline{x}$ are shown in Figure 18.4.

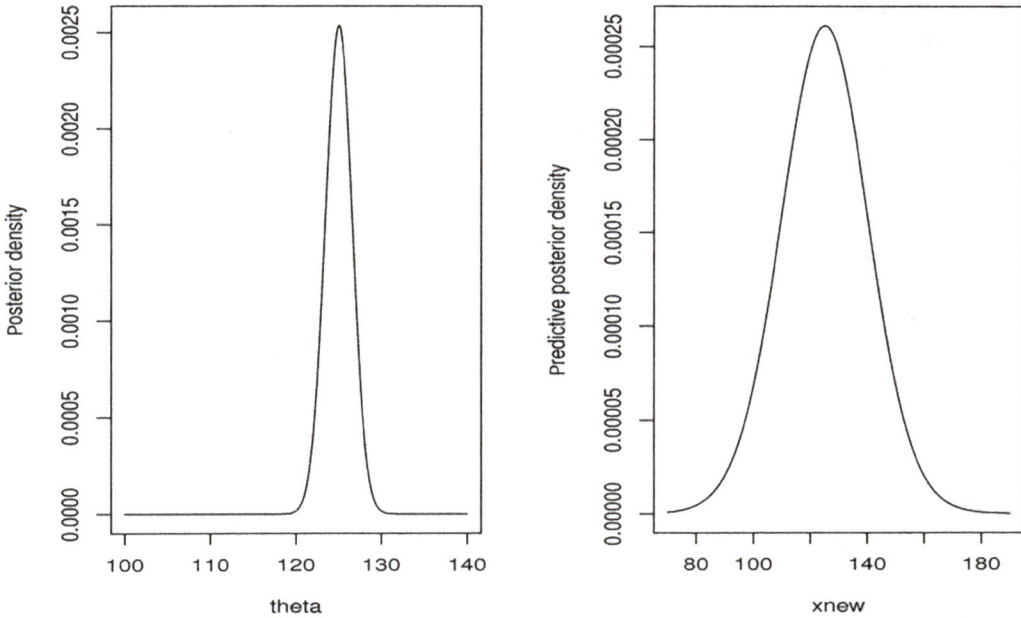

FIGURE 18.4
(Left) Posterior density of θ given x, and (right) predictive posterior density of X_{new} given $\underline{x}$ obtained with a numerical integration method.

```
########### R-code for prediction #################################
xnew = seq(70,190, 0.01) ### assume range of xnew on grid
unknown1 = matrix(xnew,length(xnew),length(theta))
unknown2 = matrix(theta,length(xnew),length(theta))

####calculate joint predictive posterior density of xnew and theta #####
A = nu*xvar; B = nu*0.5; C = n*(unknown2-xbar)*(unknown2-xbar)
D = (unknown1-unknown2)*(unknown1-unknown2)
log_joint_pred = -0.5*log(pi)+log(B)+ B*log(A)-(B+1)*log(A+C+D)
joint_pred = exp(log_joint_pred)
joint_pred = joint_pred/sum(rowSums(joint_pred))
pred_xnew = rowSums(joint_pred)

##### compute HPD interval############################
credMass = 0.95
sortedPred = sort(pred_xnew, decreasing=T)
HDIheightIdx = min(which(cumsum(sortedPred)>=credMass)) HDIheight =
sortedPred[HDIheightIdx]
HDImass = sum(pred_xnew[pred_xnew >= HDIheight])
indices = which(pred_xnew >= HDIheight)
HPD_xLni = min(xnew[indices]); HPD_xUni = max(xnew[indices])

#print HPD intervals
show(data.frame(HPD_xLni, HPD_xUni))
#### plot the predictive posterior density####################
plot(xnew,pred_xnew, ylab= "predictive posterior density", type = "l")
##################################################################
```

Given the IQs of 40 women of age-group 20–25, which range from 101 to 154, and an assumed noninformative prior density over the average IQ of women, its posterior density is symmetric with a peak around 125. The average IQ of women, approximately, lies within 120 to 129 with a 95% chance.

The IQ of a new woman based on the above information has a nearly symmetric density with a peak at around 125. There is a 95% chance that the new women's IQ, approximately, lies within 94–156.

Bayesian computation (evaluation of posterior densities, posterior estimates, posterior predictive distribution) to the problems with multidimensional unknown parameters may be analytically intractable. Simulation-based methods, e.g., MCMC techniques, and functional approximations such as the Gaussian approximation, Laplace approximation, INLA, and variational Bayes approximations, are widely used in solving Bayesian computation problems. However, descriptions of these techniques are beyond the scope of this article.

18.14 Conclusion

Students usually find it difficult to deal with statistical concepts concerning uncertainty. Many textbook sources on these notions either provide heavy theory or are mostly applied in nature. In this chapter, we have presented a combination of simple theoretical and applied approaches to understanding uncertainty through probability and basic Bayesian concepts with suitable examples. We used relevant plots, diagrams, and tables to explain the quantification of uncertainty in real-life scenarios. We also provided R-codes to find Bayesian probabilities that may be difficult to compute manually. We hope that the chapter will provide a basic yet effective platform for readers to understand and apply probability and Bayesian statistical concepts to deal with uncertainty in real-life scenarios. We referenced several useful books for the explanation of mathematical definitions of the probability and Bayesian statistics concepts, including Ross (2014), Isaac (2013), Meyer (1965), Miller and Miller (2014), Hogg et al. (2005), Mood et al. (1974), Gelman et al. (2006), Lee (2012), Berger (1985), and Gosh et al. (2006).

References

Berger, J. O. (1985). *Statistical Decision Theory and Bayesian Analysis* (2nd ed.). Springer Series in Statistics. New York: Springer.

Berger, J. O. and Bernardo, J. (1991). On the development of reference priors. *Bayesian Statistics* 4, 35–60.

Dale, A. (1982). Bayes or Laplace? An examination of the origin and early applications of Bayes' theorem. *Archive for History of Exact Sciences* 27 (1), 27–47.

Devore, J. L. (2012). *Probability and Statistics for Engineering and the Sciences*. Boston, MA: Cengage Learning.

Dey, D. K. and Liuy, J. (2007). A quantitative study of quantile based direct prior elicitation from expert opinion. *Bayesian Analysis* 2 (1), 137–166.

Fienberg, S. E. (2006). When did Bayesian inference become Bayesian? *Bayesian Analysis* 1 (1), 1–40.

Gelman, A., Carlin, J. B., Stern, H. S. and Rubin, D. B. (2006). *Bayesian Data Analysis*. Boca Raton, FL: Chapman and Hall/CRC.

Gosh, J., Delampady, M. and Samanta, T. (2006). *An Introduction to Bayesian Analysis: Theory and Methods* (1st ed.). New York: Springer-Verlag.

Hogg, R., McKean, J. and Craig, A. (2005). *Introduction to Mathematical Statistics*. Upper Saddle River, NJ: Pearson Education.

Isaac, R. (2013). *The Pleasures of Probability*. New York: Springer Science & Business Media.

Lee, P. M. (2012). *Bayesian Statistics: An Introduction* (4th ed.). Hoboken, NJ: Wiley Publishing.

Meyer, P. L. (1965). *Introductory Probability and Statistical Applications*. New Delhi: Oxford and IBH Publishing.

Miller, I. and Miller, M. (2014). *John E. Freund's Mathematical Statistics with Applications* (8th ed.). Hoboken, NJ: Pearson.

Mood, A. M., Graybill, F. A. and Boes, D. C. (1974). *Introduction to the Theory of Statistics* (3rd ed.). McGraw-Hill Series in Probability and Statistics. New York: McGraw-Hill Book Company.

Oakley, J. E. and O'Hagan, A. (2007). Uncertainty in prior elicitations: a nonparametric approach. *Biometrika* 94 (2), 427–441.

Ross, S. (2014). *A First Course in Probability*. Upper Saddle River, NJ: Pearson.

Stigler, S. M. (1982). Thomas Bayes' Bayesian inference. *Journal of RSS* 145 (2), 250–258.

Stigler, S. M. (1986). Laplace's 1774 memoir on inverse probability. *Statistical Science* 1 (3), 359–363.

19

Generalized Dual to Exponential Ratio Type Estimator for the Finite Population Mean in the Presence of Nonresponse

Sara Zahid, Asifa Kamal, and Mahnaz Makhdum

Lahore College for Women University

CONTENTS

19.1 Introduction

Data collection through surveys is a formidable task in developing countries. Almost two-third portion of enumerators reported research articles based on survey data in developing countries that had reported an average response rate of 69% (Lupu & Michelitch, 2018). Data fabrication is also observed in developing countries; one reason might be the nonresponse rate (Lupu & Michelitch, 2018). The collection of data is more complex in developing countries due to illiteracy and prevalence of a variety of languages due to different ethnicities in these countries, increasing the nonresponse rate (Lee & Perez, 2014). The translation of a questionnaire from 2 to 10 languages was observed in developing countries (Lupu & Michelitch, 2018). In spite of the adoption of stringent translation measures, it is hard to produce similar questionnaires across languages, which affects the response rate (Davidov & De Beuckelaer, 2010, Heath et al., 2009).

The problem of nonresponse occurs in two different forms, item nonresponse, and unit nonresponse. Unit nonresponse means complete failure in obtaining data from a sampling unit, and item nonresponse arises if a respondent does not provide information on a certain item or items of questionnaire. Item nonresponse can be handled at the analysis stage by imputing missing information, but unit nonresponse should be addressed at the data collection stage. Due to increased nonresponse, revisits are common to handle the issue of nonresponse. Lupu and Michelitch (2018) reported that 67% of enumerators revisited the household at least once. Some had been contacted again as many as five times.

DOI: 10.1201/9781003261148-23

Hansen and Hurwitz (1946) were the first to discuss the complexity of nonresponse using the idea of revisiting the sampling unit, which can be successfully used in developing countries to control the unit nonresponse rate at data collection stage. According to Hansen and Hurwitz's (1946) strategy, a portion of nonrespondents is readdressed by exerting extra efforts to collect information. An estimator was developed that comprised the information on respondents contacted in the first wave of data collection and information provided by nonrespondents re-contacted in the second trial of data collection. Hansen and Hurwitz (1946) constructed an estimator when a response is missing for the variable under study. Sampling experts further exploited that idea and used auxiliary information to improve the precision of nonresponse estimators. Many researchers have expanded the idea and proposed different estimators when nonresponse is expected for both study variable and auxiliary variables or for either, like Cochran (1977), Rao (1986), Khare and Srivastava (1993, 1995, 1997), Singh and Kumar (2008), Sing et al. (2009), Kumar and Bhougal (2011), Singh et al. (2009), Chanu and Singh (2015), Pal and Singh (2016), and Zubair et al. (2018).

Current research aims to develop a new generalized estimator of the population mean in simple random sampling without replacement (SRSWOR) in the presence of nonresponse. Two auxiliary variables are also incorporated in the proposed estimator to improve its precision.

Hansen and Hurwitz (1946) recommended a double sampling plan for the estimation of the population mean in the presence of nonresponse; a two-step process was suggested to handle non response.

i. A simple random sample of size n is chosen from population N and information is collected through a mailed questionnaire.

ii. A subsample of size $r = n_2 / k$ where $(k < 1)$ from nonrespondents n_2 is selected and approached to collect information via face-to-face interview.

In this strategy of data collection, the population is assumed to consist of a response stratum of size N_1 and nonresponse stratum of size N_2. Assume that n_1 respond and $n_2 = n - n_1$ are those who do not respond. Let

$$\bar{Y} = \sum_{i=1}^{N} \frac{y_i}{N}, \ \bar{Y}_1 = \sum_{i=1}^{N} \frac{y_i}{N_1}, \ \bar{Y}_2 = \sum_{i=1}^{N} \frac{y_i}{N_2},$$

$$S_y^2 = \sum_{i=1}^{N} \frac{(y_i - \bar{y})^2}{N-1}, \ S_{y_1}^2 = \sum_{i=1}^{N_1} \frac{(y_i - \bar{y})^2}{N_1 - 1}, \ S_{y_2}^2 = \sum_{i=1}^{N_2} \frac{(y_i - \bar{y})^2}{N_2 - 1}$$

The unbiased estimator proposed by Hansen and Hurwitz (1946) in case of nonresponse on a study variable is

$$\bar{y}^* = (n_1 / n)\bar{y}_1 + (n_2 / n)\bar{y}_{2r}$$

The variance of Hansen and Hurwitz (1946) is given as follows:

$$\mathrm{Var}(\bar{y}^*) = \left(\frac{1-f}{n}\right)S_y^2 + \frac{W_2(k-1)}{n}S_{y_2}^2 \tag{19.1}$$

The following notations have been used regarding the bias and mean square error (MSE) of the proposed estimators:

$$e_o^* = \frac{\bar{y}^* - \bar{Y}}{\bar{Y}}; \quad e_1 = \frac{\bar{x} - \bar{X}}{\bar{X}}; \quad e_2 = \frac{\bar{z} - \bar{Z}}{\bar{Z}}; \quad e_1^* = \frac{\bar{x}^* - \bar{X}}{\bar{X}}; \quad e_2^* = \frac{\bar{z}^* - \bar{Z}}{\bar{Z}}$$

$$E\left(e_o^*\right) = E(e_1) = E(e_2) = E\left(e_1^*\right) = E\left(e_2^*\right) = 0$$

$$E\left(e_o^*\right) = \theta C_y^2 + \lambda C_{y2}^2, E\left(e_1^*\right) = \theta C_x^2 + \lambda C_{x2}^2, E\left(e_2^*\right) = \theta C_z^2 + \lambda C_{z2}^2$$

$$E\left(e_1^2\right) = \theta C_x^2, E\left(e_2^2\right) = \theta C_z^2$$

$$E\left(e_o^* e_1\right) = \theta \rho_{yx} C_y C_x, \quad E\left(e_o^* e_2\right) = \theta \rho_{yz} C_y C_z, E(e_1 e_2) = \theta \rho_{xz} C_x C_z$$

$$E\left(e_1^* e_2\right) = E\left(e_1 e_2^*\right) = \theta \rho_{xz} C_x C_z, \quad E\left(e_o^* e_1^*\right) = \theta \rho_{yx} C_y C_x + \lambda \rho_{y2x2} C_{y2} C_{x2}$$

$$E\left(e_o^* e_2^*\right) = \theta \rho_{yz} C_y C_z + \lambda \rho_{y2z2} C_{y2} C_{z2}$$

$$E\left(e_1^* e_2^*\right) = \theta \rho_{xz} C_x C_z + \lambda \rho_{x2z2} C_{x2} C_{z2}$$

where $\theta = \dfrac{1-f}{n}, \quad \lambda = \dfrac{w_2(k-1)}{n} \quad g = \dfrac{n}{N-n}$

The notation used for the dual-transformation is given below:

$$\bar{x}_d = (1+g)\bar{X} - g\bar{x}, \quad \bar{x}_d^* = (1+g)\bar{X} - g\bar{x}^*$$

$$\bar{z}_d = (1+g)Z - g\bar{z}, \quad \bar{z}_d^* = (1+g)Z - g\bar{z}^*$$

i. Cochran (1977) proposed the following ratio and product estimator for the mean of population $\bar{Y}$:

$$t_{1(1)} = \bar{y}^*\left(\frac{\bar{X}}{\bar{x}^*}\right), \quad t_{1(2)} = \bar{y}^*\left(\frac{\bar{x}^*}{\bar{X}}\right) \tag{19.2}$$

The MSE of the Cochran (1977) estimators are

$$\text{MSE}\left(t_{1(1)}\right) = \bar{Y}^2\left[\theta\left(C_y^2 + C_X^2 - 2\rho_{yx}C_y C_x\right) + \lambda\left(C_{y2}^2 + C_{X2}^2 - 2\rho_{y2X2}C_{y2}C_{X2}\right)\right]$$

$$\text{MSE}\left(t_{1(2)}\right) = \bar{Y}^2\left[\theta\left(C_y^2 + C_X^2 - 2\rho_{yx}C_y C_x\right) + \lambda\left(C_{y2}^2 + C_{X2}^2 - 2\rho_{y2X2}C_{y2}C_{X2}\right)\right]$$

ii. Srivenkataramana (1980) developed a dual-to-ratio and dual-to-product type estimator in the presence of nonresponse under different situations given by

$$t_{2(1)} = \bar{y}^* \left(\frac{\bar{x}_d}{X} \right), \quad t_{2(2)} = \bar{y}^* \left(\frac{\bar{x}_d}{\bar{x}^*} \right)$$

$$t_{2(3)} = \bar{y}^* \left(\frac{X}{\bar{x}_d} \right), \quad t_{2(4)} = \bar{y}^* \left(\frac{\bar{x}^*}{\bar{x}_d} \right)$$

(19.3)

The MSEs of the Srivenkataramana (1980) estimators are

$$\text{MSE}\left(t_{2(1)}\right) = \bar{Y}^2 \left[\theta \left(C_y^2 + g^2 C_x^2 - 2g\rho_{yx} C_y C_x \right) + \lambda C_{y2}^2 \right]$$

$$\text{MSE}\left(t_{2(2)}\right) = \bar{Y}^2 \left[\theta \left(C_y^2 + g^2 C_x^2 - 2g\rho_{yx} C_y C_x \right) + \lambda \left(C_{y2}^2 + g^2 C_{x2}^2 - 2g\rho_{y2x2} C_{y2} C_{x2} \right) \right]$$

$$\text{MSE}\left(t_{2(3)}\right) = \bar{Y}^2 \left[\theta \left(C_y^2 + g^2 C_x^2 - 2g\rho_{yx} C_y C_x \right) + \lambda C_{y2}^2 \right]$$

$$\text{MSE}\left(t_{2(4)}\right) = \bar{Y}^2 \left[\theta \left(C_y^2 + g^2 C_x^2 - 2g\rho_{yx} C_y C_x \right) + \lambda \left(C_{y2}^2 + g^2 C_{x2}^2 - 2g\rho_{y2x2} C_{y2} C_{x2} \right) \right]$$

iii. Rao (1986) proposed ratio and product estimators for the estimation of the population mean under nonresponse using a single auxiliary variable.

$$t_{3(1)} = \bar{y}^* \left(\frac{\bar{X}}{\bar{x}} \right), \quad t_{3(2)} = \bar{y}^* \left(\frac{\bar{x}}{\bar{X}} \right)$$

(19.4)

The MSEs of the Rao (1986) estimators are

$$\text{MSE}\left(t_{3(1)}\right) = \bar{Y}^2 \left[\theta \left(C_y^2 + C_x^2 - 2\rho_{yx} C_y C_x \right) + \lambda C_{y2}^2 \right]$$

$$\text{MSE}\left(t_{3(2)}\right) = \bar{Y}^2 \left[\theta \left(C_y^2 + C_x^2 - 2\rho_{yx} C_y C_x \right) + \lambda C_{y2}^2 \right]$$

iv. Singh et al. (2009) proposed different exponential ratio product estimators for the estimation of population mean under nonresponse using a single auxiliary variable.

$$t_{4(1)} = \bar{y}^* \exp \left[\frac{\bar{X} - \bar{x}}{\bar{X} + \bar{x}} \right], \quad t_{4(2)} = \bar{y}^* \exp \left[\frac{\bar{X} - \bar{x}^*}{\bar{X} + \bar{x}^*} \right]$$

$$t_{4(3)} = \bar{y}^* \exp \left[\frac{\bar{x} - \bar{X}}{\bar{x} + \bar{X}} \right], \quad t_{4(4)} = \bar{y}^* \exp \left[\frac{\bar{x}^* - \bar{X}}{\bar{x}^* + \bar{X}} \right]$$

(19.5)

The MSEs of the Singh et al. (2009) estimators are

$$\text{MSE}\left(t_{4(1)}\right) = \bar{Y}^2 \left[\theta \left(C_y^2 + \frac{C_x^2}{4} - \rho_{yx} C_y C_x \right) + \lambda C_{y2}^2 \right]$$

$$\text{MSE}\left(t_{4(2)}\right) = \bar{Y}^2\left[\theta\left(C_y^2 + \frac{C_x^2}{4} - \rho_{yx}C_yC_x\right) + \lambda\left(C_{y2}^2 + \frac{C_{x2}^2}{4} - \rho_{y2x2}C_{y2}C_{x2}\right)\right]$$

$$\text{MSE}\left(t_{4(3)}\right) = \bar{Y}^2\left[\theta\left(C_y^2 + \frac{C_x^2}{4} + \rho_{yx}C_yC_x\right) + \lambda C_{y2}^2\right]$$

$$\text{MSE}\left(t_{4(4)}\right) = \bar{Y}^2\left[\theta\left(C_y^2 + \frac{C_x^2}{4} - \rho_{yx}C_yC_x\right) + \lambda\left(C_{y2}^2 + \frac{C_{x2}^2}{4} + \rho_{y2x2}C_{y2}C_{x2}\right)\right]$$

v. The Singh et al. (2009) estimators after applying the dual transformation are

$$t_{5(1)} = \bar{y}^* \exp\left[\frac{\bar{x}_d - \bar{X}}{\bar{x}_d + \bar{X}}\right], \quad t_{5(2)} = \bar{y}^* \exp\left[\frac{\bar{X} - \bar{x}_d}{\bar{X} + \bar{x}_d}\right] \tag{19.6}$$

The following are the expressions of the respective MSEs.

$$\text{MSE}\left(t_{5(1)}\right) = \bar{Y}^2\left[\theta\left(C_y^2 + g^2\frac{C_x^2}{4} - g\rho_{yx}C_yC_x\right) + \lambda C_{y2}^2\right]$$

$$\text{MSE}\left(t_{5(2)}\right) = \bar{Y}^2\left[\theta\left(C_y^2 + g^2\frac{C_x^2}{4} - g\rho_{yx}C_yC_x\right) + \lambda C_{y2}^2\right]$$

vi. Chanu and Singh (2015) proposed an exponential dual to ratio estimator of finite population mean in the presence of nonresponse. Estimators along with their MSE are given as

$$t_{6(1)} = \bar{y}^*\left[\alpha_1 \exp\left(\frac{\bar{X} - \bar{x}}{\bar{X} + \bar{x}}\right) + (1 - \alpha_1)\exp\left(\frac{\bar{x}_d - \bar{X}}{\bar{x}_d - \bar{X}}\right)\right] \tag{19.7}$$

$$\text{MSE}\left(t_{6(1)(\text{opt})}\right) = \bar{Y}^2\left[\theta\left(C_y^2 - k_{yx}^2C_x^2\right) + \lambda C_{y2}^2\right]$$

where

$$\alpha_1 = \frac{g}{g-1} - \frac{2k_{yx}}{g-1} \quad = \rho_{yx}\frac{c_y}{c_x}$$

$$t_{6(2)} = \bar{y}^*\left[\alpha_2 \exp\left(\frac{\bar{X} - \bar{x}^*}{\bar{X} + \bar{x}^*}\right) + (1 - \alpha_2)\exp\left(\frac{\bar{x}_d - \bar{X}}{\bar{x}_d + \bar{X}}\right)\right] \tag{19.8}$$

$$\text{MSE}\left(t_{6(2)(\text{opt})}\right) = \bar{y}^2 \left[\theta C_y^2 + \lambda C_{y2}^2 - \frac{\left(\theta \rho_{yx} C_y C_x + \lambda \rho_{y2x2} C_{y2} C_{x2}\right)^2}{\theta C_x^2 \bar{x}_d + \lambda C_{x2}^2} \right]$$

where

$$\alpha_2 = \frac{g}{g-1} - \frac{2}{(g-1)} \left[\frac{\theta \rho_{yx} C_y C_x + \lambda \rho_{y2x2} C_{y2} C x2}{\theta C_x^2 + \lambda C_{x2}^2} \right]$$

19.2 The Proposed Estimator

Following Yasmeen et al. (2015), a generalized dual to exponential ratio type estimator under non response is proposed. The bias and MSE expression of the proposed estimator has been derived.

$$t_s = \delta \bar{y}^* \left(\frac{\bar{X}}{\bar{x}^*} \right)^\beta \exp\left[\alpha \frac{\bar{Z} - \bar{z}^*}{\bar{Z} + \bar{z}^*} \right] + (1-\delta)\bar{y}^* \left(\frac{\bar{x}^*}{\bar{X}} \right)^\beta \exp\left[\alpha \frac{\bar{z}_d^* - \bar{Z}}{\bar{z}_d^* + \bar{Z}} \right] \tag{19.9}$$

where α and β are generalized constants and ω and δ are optimization constants. Using the above notations in the estimator, we obtain:

$$t_s = \delta \bar{Y}\left(1+e_o^*\right) \left(\frac{\bar{X}}{\bar{X}\left(1+e_1^*\right)} \right)^\beta \exp\left[\alpha \frac{\bar{Z} - \bar{Z}\left(1+e_2^*\right)}{\bar{Z} + \bar{Z}\left(1+e_2^*\right)} \right]$$

$$+ (1-\delta)\bar{Y}\left(1+e_o^*\right) \left(\frac{(1+g)\bar{X} - g\bar{x}^*}{\bar{X}} \right)^\beta \exp\left[\alpha \frac{(1+g)\bar{Z} - g\bar{z}^* - \bar{Z}}{(1+g)\bar{Z} - g\bar{z}^* + \bar{Z}} \right]$$

$$t_s = \delta \bar{Y}\left(1+e_o^*\right)\left(1+e_1^*\right)^{-\beta} \exp\left[\frac{-\alpha e_2^*}{2}\left(1+\frac{e_2^*}{2}\right)^{-1} \right]$$

$$+ (1-\delta)\bar{Y}\left(1+e_o^*\right)\left(1-ge_1^*\right)^\beta \exp\left[\frac{-\alpha ge_2^*}{2}\left(1-\frac{ge_2^*}{2}\right)^{-1} \right]$$

$$t_s = \delta \bar{Y}\left(1+e_o^*\right)\left(1 - \beta e_1^* + \frac{\beta^2 e_1^{*2}}{2} + \frac{\beta e_1^{*2}}{2} \right)\exp\left[\frac{-\alpha e_2^*}{2} + \frac{\alpha e_2^{*2}}{4} \right]$$

$$+ (1-\delta)\bar{Y}\left(1+e_o^*\right)\left(1 - g\beta e_1^* + \frac{\beta^2 g^2 e_1^{*2}}{2} + \frac{\beta g^2 e_1^{*2}}{2} \right)\exp\left[\frac{-\alpha ge_2^*}{2} - \frac{\alpha g^2 e_2^{*2}}{4} \right]$$

Neglecting the terms of power two or greater, it is possible to re-write the estimator as

$$t_s = \delta \bar{Y}\left(1+e_o^*\right)\left(1 - \beta e_1^* + \frac{\beta^2 e_1^{*2}}{2} + \frac{\beta e_1^{*2}}{2}\right)$$

$$\left[1 + \left(-\frac{\alpha e_2^*}{2} + \frac{\alpha e_2^{*2}}{4}\right) + \frac{\left(-\frac{\alpha e_2^*}{2} + \frac{\alpha e_2^{*2}}{4}\right)}{2!}\right]$$

$$+ (1-\delta)\bar{Y}\left(1+e_o^*\right)\left(1 - g\beta e_1^* + \frac{\beta^2 g^2 e_1^{*2}}{2} + \frac{\beta g^2 e_1^{*2}}{2}\right)$$

$$\left[1 + \left(-\frac{\alpha g e_2^*}{2} - \frac{\alpha g^2 e_2^{*2}}{4}\right) + \frac{\left(-\frac{\alpha g e_2^*}{2} - \frac{\alpha g^2 e_2^{*2}}{4}\right)}{2!}\right]$$

Simplifying the above equation yields

$$t_s - \bar{y} = \bar{Y}\left[e_o^* + (\delta g - \delta - g)\left\{\frac{\alpha}{2}e_2^* + \beta e_1^* + \frac{\alpha}{2}e_o^* e_2^* + \beta e_o^* e_1^*\right\}\right.$$

$$+ (\delta g^2 + \delta - g)\left\{\frac{\alpha}{4}e_2^{*2} + \frac{\beta}{2}e_1^{*2}\right\} + (\delta + g^2 - \delta g^2)\left\{\frac{\alpha^2}{8}e_2^{*2} + \frac{\beta^2}{2}e_1^{*2} + \frac{\alpha\beta}{2}e_1^* e_2^*\right\}\right]$$

(19.10)

Applying the expectation on both sides of equation (19.10), we get

$$E\left(t_2 - \bar{Y}\right) = \bar{Y}\left[(\delta g - \delta - g)\left\{\frac{\alpha}{2}\left(\theta_{\rho yz}C_y C_z + \lambda_{\rho y2 z2}C_{y2}C_{z2}\right) + \beta\left(\theta_{\rho yx}C_y C_x + \lambda_{\rho y2 x2}C_{y2}C_{x2}\right)\right\}\right.$$

$$+ (\delta g^2 + \delta - g)\left\{\frac{\alpha}{4}\left(\theta C_z^2 + \lambda C_{z2}^2\right) + \frac{\beta}{2}\left(\theta C_x^2 + \lambda C_{x2}^2\right)\right\}$$

$$+ (\delta + g^2 - \delta g^2)\left\{\frac{\alpha^2}{8}\left(\theta C_z^2 + \lambda C_{z2}^2\right) + \frac{\beta^2}{2}\left(\theta C_x^2 + \lambda C_{x2}^2\right)\right.$$

$$+ \frac{\alpha\beta}{2}\left(\theta_{\rho xz}C_x C_z + \lambda_{\rho x2 z2}C_{x2}C_{z2}\right)\right\}\right]$$

The bias of t_s is derived as

$$\text{Bias}(t_s) = \bar{Y}\left[\theta\left[\left(\delta g^2 + \delta - g\left(\frac{\alpha}{4}C_z^2 + \frac{\beta}{2}C_z^2\right) + \left(\delta - g^2 - \delta g^2\right)\left(\frac{\alpha^2}{8}C_z^2 + \frac{\beta^2}{2}C_x^2 + \frac{\alpha\beta}{2}\rho_{xz}C_xC_z\right)\right)\right.\right.$$

$$+ \left(\delta g + \delta - g\right)\left(\frac{\alpha}{2}\rho_{yz}C_yC_z + \beta\rho_{yx}C_yC_x\right)$$

$$+ \lambda\left[\left(\delta + g - \delta g^2\right)\left(\frac{\alpha^2}{8}C_{z_2}^2 + \frac{\beta^2}{2}C_{x_2}^2 + \frac{\alpha\beta}{2}\rho_{x_2z_2}C_{x_2}C_{z_2}\right)\right.$$

$$+ \left[\left(\delta + g - \delta g^2\right)\left(\frac{\alpha^2}{8}C_{z_2}^2 + \frac{\beta^2}{2}C_{x_2}^2 + \frac{\alpha\beta}{2}\rho_{x_2z_2}C_{x_2}C_{z_2}\right)\right.$$

$$\left.\left.\left.+ \left(\delta g + \delta - g\right)\left(\frac{\alpha}{2}\rho_{y_2z_2}C_{y_2}C_{z_2} + \beta\rho_{y_2x_2}C_{y_2}C_{x_2}\right)\right]\right]\right] \tag{19.11}$$

After taking the square of both sides of equation (19.10) and applying expectation we get

$$E\left(\left(t_s - \bar{Y}\right)\right)^2 = \bar{Y}^2\left[E\left(e_0^{*2}\right) + \left(\delta g - \delta - g\right)^2\beta^2 E\left(e_1^{*2}\right) + \left(\delta g - \delta - g\right)^2\frac{\alpha^2}{4}E\left(e_2^{*2}\right)\right.$$

$$\left.+ 2\beta\left(\delta g - \delta - g\right)E\left(e_0^*e_1^*\right) + \alpha\left(\delta g - \delta - g\right)E\left(e_0^*e_2^*\right) + \alpha\beta\left(\delta g - \delta - g\right)^2 E\left(e_1^*e_2^*\right)\right]$$

Or

$$\text{MSE}(t_s) = \bar{Y}^2\left[\theta\left(C_y^2 + \left(\delta g - \delta - g\right)^2\left(\beta^2 C_x^2 + \frac{\alpha^2}{4}C_z^2 + \alpha\beta_{\rho_{xz}}C_xC_z\right)\right.\right.$$

$$+ \left(\delta g - \delta - g\right)\left(2\beta\rho_{yx}C_yC_x + \alpha\rho_{yz}C_yC_z\right)\right)$$

$$+ \lambda\left(C_{y_2}^2 + \left(\delta g - \delta - g\right)^2\left(\beta^2 C_{x_2}^2 + \frac{\alpha^2}{4}C_{z_2}^2 + \alpha\beta\rho_{x_2z_2}C_{x_2}C_{z_2}\right)\right.$$

$$\left.\left.+ \left(\delta g - \delta - g\right)\left(2\beta\rho_{y_2x_2}C_{y_2}C_{x_2} + \alpha\rho_{y_2z_2}C_{y_2}C_{z_2}\right)\right] \tag{19.12}$$

Let $\left(\delta g - \delta - g\right)$ be equal to τ; equation (19.12) is then written as follows

$$\text{MSE}(t_s) = \bar{Y}^2\left[\theta\left\{C_y^2 + \tau^2\left(\beta^2 C_x^2 + \frac{\alpha^2}{4}C_z^2 + \alpha\beta\rho_{xz}C_xC_z\right) + \tau\left(2\beta\rho_{yx}C_yC_z + \alpha\rho_{yz}C_yC_z\right)\right\}\right.$$

$$\left.+ \lambda\left\{C_{y_2}^2 + \tau^2\left(\beta^2 C_{x_2}^2 + \frac{\alpha^2}{4}C_{z_2}^2 + \alpha\beta\rho_{x_2z_2}C_{x_2}C_{z_2}\right) + \tau\left(2\beta\rho_{y_2x_2}C_{y_2}C_{x_2} + \alpha\rho_{y_2z_2}C_{y_2}C_{z_2}\right)\right\}\right]$$

$$\tag{19.13}$$

Differentiating (19.13) with respect to τ, the optimum value of τ is obtained as

$$\tau = \frac{-\theta\left(2\beta\rho_{yx}C_yC_x + \alpha\rho_{yz}C_yC_z\right) - \lambda\left(2\beta\rho_{y2x2}C_{y2}C_{x2} + \alpha\rho_{y2z2}C_{y2}C_{z2}\right)}{2\left[\theta\left(\beta^2C_x^2 + \dfrac{\alpha^2}{4}C_z^2 + \alpha\beta\rho_{xz}C_xC_z\right) + \lambda\left(\beta^2C_{x2}^2 + \dfrac{\alpha^2}{4}C_{z2}^2 + \alpha\beta\rho_{x2z2}C_{x2}C_{z2}\right)\right]} = \tau_0 \quad (19.14)$$

When the optimal value (19.14) is entered in equation (19.13), the minimum *MSE* of tt_{tv} denoted by

$$\mathrm{MSE}_{\min}\left(t_s\right) = \bar{Y}^2\theta\left\{C_y^2 + \tau_o^2\left(\beta^2C_x^2 + \frac{\alpha^2}{4}C_z^2 + \alpha\beta\rho_{xz}C_xC_z\right) + \tau_o\left(2\beta\rho_{yx}C_yC_x + \alpha\rho_{yz}C_yC_z\right)\right\}$$

$$+ \bar{Y}^2\lambda\left\{C_{y2}^2 + \tau_o^2\left(\beta^2C_{x2}^2 + \frac{\alpha^2}{4}C_{z2}^2 + \alpha\beta\rho_{x2z2}C_{x2}C_{z2}\right) + \tau_o\left(2\beta\rho_{y2x2}C_{y2}C_{x2} + \alpha\rho_{y2z2}C_{y2}C_{z2}\right)\right\}$$

$$(19.15)$$

19.3 Some Special Cases of the Generalized Estimator

In this section, special cases of the proposed generalized estimator are discussed. The special cases for estimator t_s are obtained in Table 19.1 by assuming various values of constants.

TABLE 19.1

Some Special Cases of t_s

δ	α	β	Resulting Estimators	Names
0	α	β	$t_{s(1)} = \bar{y}^*\left[\dfrac{\bar{x}_d^*}{\bar{X}}\right]^\beta \exp\left[\alpha\dfrac{\bar{z}_d^* - \bar{Z}}{\bar{z}_d^* + \bar{Z}}\right]$	Proposed estimator 1
1	α	β	$t_{s(2)} = \bar{y}^*\left[\dfrac{\bar{X}}{\bar{x}^*}\right]^\beta \exp\left[\alpha\dfrac{\bar{Z} - \bar{z}^*}{\bar{Z} + \bar{z}^*}\right]$	Proposed estimator 2
1	1	1	$t_{s(3)} = \bar{y}^*\left[\dfrac{\bar{X}}{\bar{x}^*}\right] \exp\left[\dfrac{\bar{Z} - \bar{z}^*}{\bar{Z} + \bar{z}}\right]$	Proposed estimator 3
δ_o	1	1	$t_{s(4)} = \delta\bar{y}^*\left[\dfrac{\bar{X}}{\bar{x}^*}\right] \exp\left[\dfrac{\bar{Z} - \bar{z}^*}{\bar{Z} + \bar{z}^*}\right] + (1-\delta)\bar{y}^*\left[\dfrac{\bar{x}_d^*}{\bar{X}}\right] \exp\left[\dfrac{\bar{z}_d^* - \bar{Z}}{\bar{z}_d^* + \bar{Z}}\right]$	Proposed estimator 4
1	-1	-1	$t_{s(5)} = \bar{y}^*\left[\dfrac{\bar{x}^*}{\bar{X}}\right] \exp\left[\dfrac{\bar{z}^* - \bar{Z}}{\bar{z}^* + \bar{Z}}\right]$	Proposed estimator 5
0	1	1	$t_{s(6)} = \bar{y}^*\left[\dfrac{\bar{x}_d^*}{\bar{X}}\right] \exp\left[\dfrac{\bar{z}_d^* - \bar{Z}}{\bar{z}_d^* + \bar{Z}}\right]$	Proposed estimator 6
δ_o	-1	-1	$t_{s(7)} = \delta\bar{y}^*\left[\dfrac{\bar{x}^*}{\bar{X}}\right] \exp\left[\dfrac{\bar{z}^* - \bar{Z}}{\bar{z}^* + \bar{Z}}\right] + (1-\delta)\bar{y}^*\left[\dfrac{\bar{X}}{\bar{x}_d^*}\right] \exp\left[\dfrac{\bar{Z} - \bar{z}_d^*}{\bar{Z} + \bar{z}_d^*}\right]$	Proposed estimator 7

(Continued)

TABLE 19.1 (*Continued*)

Some Special Cases of t_s

δ	α	β	Resulting Estimators	Names
0	0	1	$t_{s(8)} = \bar{y}^* \left[\dfrac{\overline{x_d^*}}{\overline{X}} \right]$	Proposed estimator 8
0	0	−1	$t_{s(9)} = \bar{y}^* \left(\dfrac{\overline{X}}{\overline{x_d^*}} \right)$	Proposed estimator 9
1	0	−1	$t_{1(1)} = \bar{y}^* \left(\dfrac{\overline{X}}{\overline{x}^*} \right)$	Cochran (1977)
1	0	−1	$t_{1(2)} = \bar{y}^* \bar{y}^* \left(\dfrac{\overline{x}^*}{\overline{X}} \right)$	Cochran (1977)

19.4 Efficiency Comparison

In this section, the MSE of the proposed estimator t_s is compared with the MSEs of certain existing estimators, and conditions are determined under which the proposed estimator performs better than an existing similar type of estimators developed under the category of nonresponse estimators.

$$\text{Var}\left(\bar{y}^*\right) = \bar{Y}^2 \left(\theta C_y^2 + \lambda C_{y2}^2 \right) \tag{19.16}$$

19.5 Proposed Estimator

$$\text{MSE}(t_s) = \bar{Y}^2 \left[\theta H + \lambda I \right] \tag{19.17}$$

where

$$H = C_y^2 + \tau^2 \left(\beta^2 C_x^2 + \frac{\alpha^2}{4} C_z^2 + \alpha\beta\rho_{xz}C_xC_z \right) + \tau\left(2\beta\rho_{yx}C_yC_x + \alpha\rho_{yz}C_yC_z \right)$$

$$I = C_{y2}^2 + \tau^2 \left(\beta^2 C_{x2}^2 + \frac{\alpha^2}{4} C_{z2}^2 + \alpha\beta\rho_{x2z2}C_{x2}C_{z2} \right) + \tau\left(2\beta\rho_{y2x2}C_{y2}C_{x2} + \alpha\rho_{y2z2}C_{y2}C_{z2} \right)$$

Now we compare:

i.
$$MSE(t_s) - MSE\left(\bar{y}^*\right) \leq 0 \qquad \text{if}$$
$$\theta\left(H - C_y^2\right) + \lambda\left(I - C_{y2}^2\right) \leq 0$$

ii.
$$MSE(t_s) - MSE(t_{1(1)}) \leq 0 \qquad \text{if}$$
$$\theta\left(H - C_y^2 - C_x^2 + 2\rho_{yx}C_yC_x\right) + \lambda\left(I - C_{y2}^2 - C_{x2}^2 + 2\rho_{y2x2}C_{y2}C_{x2}\right) \leq 0$$

iii.
$$MSE(t_s) - MSE(t_{2(2)}) \leq 0 \qquad \text{if}$$
$$\theta\left(H - C_y^2 - gC_x^2 + 2g\rho_{yx}C_yC_x\right) + \lambda\left(I - C_{y2}^2 - g^2C_{x2}^2 + 2g\rho_{y2x2}C_{y2}C_{x2}\right) \leq 0$$

iv.
$$MSE(t_s) - MSE(t_{3(1)}) \leq 0 \qquad \text{if}$$
$$\theta\left(H - C_y^2 - C_x^2 + 2\rho_{yx}C_yC_x\right) + \lambda\left(I - C_{y2}^2\right) \leq 0$$

v.
$$MSE(t_s) - MSE(t_{4(2)}) \leq 0 \qquad \text{if}$$
$$\theta\left(H - C_y^2 - \frac{C_x^2}{4} + \rho_{yx}C_yC_x\right) + \lambda\left(I - C_{y2}^2 - \frac{C_{x2}^2}{4} + \rho_{y2x2}C_{y2}C_{x2}\right) \leq 0$$

vi.
$$MSE(t_s) - MSE(t_{5(1)}) \leq 0 \qquad \text{if}$$
$$\theta\left(H - C_y^2 - g^2\frac{C_x^2}{4} + g\rho_{yx}C_yC_x\right) + \lambda\left(I - C_{y2}^2\right) \leq 0$$

vii.
$$MSE(t_s) - MSE(t_{6(1)}) \leq 0 \qquad \text{if}$$
$$\theta\left(H - C_y^2\right) + \lambda\left(I - C_{y2}^2\right) + \frac{\left(\theta\rho_{yx}C_yC_x + \lambda\rho_{y2x2}C_{y2}C_{x2}\right)^2}{\theta C_x^2 + \lambda C_{x2}^2} \leq 0$$

19.6 Empirical Study

The performance of the proposed estimators $t_{s(4)}$ and $t_{s(7)}$ using the optimum value, δ_o, is investigated in this section through a theoretical and simulation study. Existing similar types of estimators in the literature, i.e., Cochran (1977), Srivenkataramana (1980), Sing et al. (2009), and Chanu and Singh (2015), are used to assess the performance of the new suggested estimators.

The percentage relative efficiency (PRE) of the proposed ratio type estimators $t_{s(4)}$ and other similar types of estimators is computed with Hansen and Hurwitz's (1946) estimator and reported in Table 19.3. Similarly, the PRE of the proposed product type estimator $t_{s(7)}$ is documented in Table 19.4. A description of the variables for Populations I and II is as follows:

Population I:
Source: Sarndal et al. (1992)
Y = Revenues from 1985 municipal taxation (in millions of kronor)
X = Total number of seats in the municipal council
Z = Real estate values according to 1984 assessment (in millions of kronor)

Population II:
Source: Damodar (2004, p. 433)
Y = Average miles per gallon
X = Top speed, miles per hour
Z = Cubic feet of cab space

19.7 Simulation Study

The simulation study was carried out by generating 5000 simulations from a multivariate normal distribution using Populations I and II. R is used to generate the simulations (Table 19.2).

19.8 Results and Conclusion

Nonresponse is the major data collection challenge when conducting surveys in many developing countries. A methodology has been developed to address the issue of

TABLE 19.2

Parameter Values for Populations I and II

Parameters	[a]Population I		[b]Population II	
	10%	20%	10%	20%
N	284	284	81	81
N	113	113	13	13
$\bar{Y}$	245.08803	245.08803	33.87901	33.87901
$\bar{X}$	47.53521	47.53521	112.45679	112.45679
$\bar{Z}$	3077.52465	3077.52465	98.76543	98.76543
C_y	2.4331361	2.4331361	0.2999365	0.2999365
C_{y_2}	1.3632090	1.1024274	0.1776737	0.1638439
C_x	0.2324984	0.2324984	0.1255599	0.1255599
C_{x_2}	0.2465933	0.2559517	0.1539601	0.1347936
C_z	1.5419948	1.5419948	0.2258027	0.2258027
C_{z_2}	1.2133013	0.9347162	0.3603855	0.3355736
ρYX	0.5805598	0.5805598	−0.6884256	−0.6884256
$\rho Y_2 X_2$	0.9260839	0.8643259	−0.5950331	−0.4256172
ρYZ	0.9358796	0.9358796	−0.3658085	−0.3658085
$\rho Y_2 Z_2$	0.8189937	0.8410244	−0.4828859	−0.2063441
ρXZ	0.6770815	0.6770815	−0.04264551	−0.04264551
$\rho X_2 Z_2$	0.7350704	0.7656953	−0.75896633	−0.65094496

[a] When relationship is positive between study and auxiliary variables.
[b] When relationship is negative between study and auxiliary variables.

nonresponse, and an empirical study has been conducted to check its efficiency. It is evident from Tables 19.3 and 19.4 that the proposed estimators outperform existing estimators. It can be also be observed that the PRE of both proposed estimators decreases as k increases. Contrariwise, the PRE also declined as the percentage of nonresponses increased. The simulation results endorse the results of theoretical MSEs (Tables 19.5 and 19.6).

TABLE 19.3

Percent Relative Efficiencies (PREs) for Population I

Estimators	Nonresponse 10%			Nonresponse 20%		
—	$k = 2$	$k = 3$	$k = 4$	$k = 2$	$k = 3$	$k = 4$
$\bar{y}^*$	100	100	100	100	100	100
$t_{1(1)}$	112.5815	113.7335	114.8023	113.3144	115.1142	116.7587
$t_{2(1)}$	107.0544	106.6991	106.3778	106.9410	106.4975	106.1072
$t_{2(2)}$	108.2436	108.9729	109.6450	108.7350	109.8919	110.9386
$t_{3(1)}$	110.7144	110.1572	109.6550	110.5363	109.8419	109.2333
$t_{4(1)}$	105.3249	105.0608	104.8216	105.2407	104.9108	104.6199
$t_{4(2)}$	106.2081	106.7481	107.2442	106.5811	107.4431	108.2193
$t_{5(1)}$	103.2048	103.0489	102.9075	103.1551	102.9602	102.7880
$t_{6(1)}$	147.1343	143.9315	141.1362	146.0992	142.1665	138.8521
$t_{6(2)}$	156.6405	162.3766	168.0289	153.7614	157.8324	162.2653
$r_{s(4)}$	532.8669	511.2482	496.3061	514.6963	486.2154	468.8684

PRE = var $(\bar{y}^*)/$MSE$(\cdot)*100.$

TABLE 19.4

Percent Relative Efficiencies (PREs) for Population II

Estimators	Nonresponse 10%			Nonresponse 20%		
—	$k = 2$	$k = 3$	$k = 4$	$k = 2$	$k = 3$	$k = 4$
$\bar{y}^*$	100	100	100	100	100	100
$t_{1(2)}$	165.6423	164.4243	163.3126	160.2744	154.8286	150.3196
$t_{2(3)}$	111.0644	110.5924	110.1590	110.7293	109.9948	109.3544
$t_{2(4)}$	111.9107	112.2172	112.5024	111.6247	111.6635	111.6978
$t_{3(2)}$	162.6126	158.7777	155.3855	159.8751	154.1311	149.3928
$t_{4(3)}$	130.6462	129.1198	127.7382	129.5604	127.2197	125.2225
$t_{4(4)}$	132.9309	133.4796	133.9916	131.6082	130.9728	130.4170
$t_{5(2)}$	104.9028	104.7048	104.5221	104.7623	104.4527	104.1809
$t_{s(7)}$	218.9375	198.5144	183.6082	203.7293	179.3828	164.1510

PRE = var $(\bar{y}^*)/$MSE$(\cdot)*100.$

TABLE 19.5

Percent Relative Efficiencies (PRE's) for Population I

Estimators	Nonresponse 10%			Nonresponse 20%		
$-$	$k = 2$	$k = 3$	$k = 4$	$k = 2$	$k = 3$	$k = 4$
$\bar{y}^*$	100	100	100	100	100	100
$t_{1(1)}$	111.1850	112.9143	113.7347	111.3569	114.8217	110.4864
$t_{2(1)}$	106.6550	105.5745	106.1257	105.4445	107.6963	103.1752
$t_{2(2)}$	107.4053	107.1948	108.9673	107.4234	109.7946	106.8966
$t_{3(1)}$	110.0289	108.4350	109.2575	108.2433	111.6272	104.7373
$t_{4(1)}$	105.0409	104.2158	104.6347	104.1165	105.8218	102.4174
$t_{4(2)}$	105.6027	105.4335	106.7419	105.5854	107.3998	105.2034
$t_{5(1)}$	102.9961	102.5683	102.7935	102.5159	103.4306	101.4739
$t_{6(1)}$	135.6761	139.5006	138.3064	140.4151	140.0680	118.0636
$t_{6(2)}$	139.3942	145.1163	164.2044	165.0003	144.4856	149.3066
$t_{s(4)}$	530.6415	515.9086	499.5570	513.0776	487.9018	467.9661

PRE $= \mathrm{var}\left(\bar{y}^*\right)/\mathrm{MSE}(\cdot) * 100$.

TABLE 19.6

Percent Relative Efficiencies (PRE's) for Population II

Estimators	Nonresponse 10%			Nonresponse 20%		
$-$	$k = 2$	$k = 3$	$k = 4$	$k = 2$	$k = 3$	$k = 4$
$\bar{y}^*$	100	100	100	100	100	100
$t_{1(2)}$	179.9040	170.6789	165.4803	179.5803	155.8567	169.6743
$t_{2(3)}$	139.0544	112.6789	116.3961	133.3008	110.8976	117.8961
$t_{2(4)}$	143.0722	116.9876	135.4069	148.8035	116.8765	110.3544
$t_{3(2)}$	163.6501	159.4567	157.3181	157.4553	154.8976	149.6978
$t_{4(3)}$	132.5429	117.6788	114.1222	128.3667	116.8943	148.3928
$t_{4(4)}$	135.7113	133.8976	129.4488	141.0507	130.2245	125.3335
$t_{5(2)}$	105.8373	104.6705	105.7867	110.0152	103.6754	104.1800
$t_{s(7)}$	291.9805	190.7865	189.6987	280.9209	252.6754	165.1510

PRE $= \mathrm{var}\left(\bar{y}^*\right)/\mathrm{MSE}(\cdot) * 100$.

References

Chanu, W., and Singh, B. (2015). Improved exponential ratio cum exponential dual to ratio estimator of finite population mean in presence of non-response. *Journal of Statistics Applications & Probability*, 4(1):103.

Cochran, W. G. (1977). *Sampling Techniques*. John Wiley and Sons: New York.

Damodar, N. G. (2004). *Basic Econometrics*. The McGraw Hill: New York.

Davidov, E., and De Beuckelaer, A. (2010). How harmful are survey translations? A test with Schwartz's human values instrument. *International Journal of Public Opinion Research*, 22:485–510.

Hansen, M. H., and Hurwitz, W. N. (1946). The problem of non-response in sample surveys. *Journal of the American Statistical Association*, 41(236):517–529.

Heath, A., Martin, J., and Spreckelsen, T. (2009). Cross-national comparability of survey attitude measures. *International Journal of Public Opinion Research*, 21:293–315.

Khare, B., and Srivastava, S. (1993). Estimation of population mean using auxiliary character in presence of non-response. *National Academy Science Letters*, 16:111–111.

Khare, B., and Srivastava, S. (1995). Study of conventional and alternative two phase sampling ratio, product and regression estimators in presence of nonresponse. *Proceedings–National Academy of Sciences India Section A*, 65:195–204.

Khare, B., and Srivastava, S. (1997). Transformed ratio type estimators for the population mean in the presence of nonresponse. *Communications in Statistics–Theory and Methods*, 26(7):1779–1791.

Kumar, S., and Bhougal, S. (2011). Estimation of the population mean in presence of nonresponse. *Communications for Statistical Applications and Methods*, 18(4):537–548.

Lee, T., and Perez, E.O. (2014). The persistent connection between language-of-interview and Latino political opinion. *Political Behavior*, 36:401–425.

Lupu, N., and Michelitch, K. (2018). Advances in survey methods for the developing world. *Annual Review of Political Science*, 21:195–214.

Pal, S., and Singh, H. (2016). Finite population mean estimation through a two-parameter ratio estimator using auxiliary information in presence of non-response. *Journal of Applied Mathematics, Statistics and Informatics*, 12(2):5–39.

Rao, P. (1986). Ratio estimation with sub sampling the non-respondents. *Survey Methodology*, 12(2):217–230.

Singh, H. P., and Kumar, S. (2008). A regression approach to the estimation of the finite population mean in the presence of non-response. *Australian & New Zealand Journal of Statistics*, 50(4):395–408.

Singh, R., Kumar, M., Chaudhary, M. K., and Smarandache, F. (2009). Estimation of mean in presence of non-response using exponential estimator. *Infinite Study*.

Srivenkataramana, T. (1980). A dual to ratio estimator in sample surveys. *Biometrika*, 67(1):199–204.

Yasmeen, U., Noor ul Amin, M., and Hanif, M. (2015). Generalized exponential estimators of finite population mean using transformed auxiliary variables. *International Journal of Applied and Computational Mathematics*, 1(4):589–598.

Zubair, M., Ali, A., Nasir, W., and Rashad, M. (2018). An exponential estimator in presence of non-response. In *16th International Conference on Statistical Sciences*, volume 32, pages 209–222.

20

A Comprehensive Tutorial on Factor Analysis with R: Empirical Insights from an Educational Perspective

O. Olawale Awe
University of Campinas
Global Humanistic University

Philip Olu Jegede
Obafemi Awolowo University

James Cochran
University of Alabama

CONTENTS

20.1 Introduction

Factor analysis (FA) is an important multivariate statistical approach that is commonly used by researchers in a variety of fields. It is a collection of statistical procedures for identifying groups of correlated variables (factors) in a multivariate study. FA attempts

DOI: 10.1201/9781003261148-24

to explain the variability among several interdependent observed (manifest) variables in terms of fewer variables that are termed constructs or latent variables. The main idea behind FA is that several observed variables have similar patterns of responses because they are all associated with a latent variable.

For example, the pattern of responses to the items of a scale relating to the dietary pattern, weight, and nutrition awareness will likely be similar because they are all associated with a latent or unobserved variable that represents *health status*. The number of factors extracted from a data set can be as few as one or as many as the number of observed variables in the data, but will ultimately depend on the extent of the variation of the observed variables the factors explain. Generally speaking, FA often involves eliciting a dataset each from two different samples sequentially or dividing one data set randomly into two mutually exclusive and collectively exhaustive subsamples. In the first stage of an FA, the first sample provides responses that are explored to extract items to detect structure in the relationships between the observed variables. In the second stage, the results from stage one are applied to the second subsample of data to verify the relationship identified in stage one. This will be explained in detail later.

FA was first primarily used by educationists and psychologists (Glass and Hopkins, 1996). However, its practical use, which dates back over 100 years, has expanded to a wide range of other disciplines such as medicine, engineering, sociology, pharmacy, physics, astronomy, and agriculture in the last few decades (Mulaik, 2010). The literature is awash with articles on FA and its applications. The works of researchers such as Williams et al. (2010), Yong and Pearce (2013), Hogarty et al. (2005), and Luo et al. (2019) provide detailed background knowledge on FA. Other work of interest in education, nursing, and medicine includes Hoban et al. (2005), Muhlenberg and Berge (2001), Watson (2004), Bryant et al. (1999), Fisher and King (2010), and Baynton (1992). In education and clinical contexts, for instance, FA can be used in the development, evaluation, and refinement of tests, scales, and measures that can be useful in their own right to researchers (Williams et al., 2010).

However, with the advent of modern statistical software packages that simplify data analysis, many studies have employed various software such as SPSS, Stata, SAS, Minitab, and AMOS. In this tutorial, we make use of R statistical software, which, in our experience, is still new to many researchers in developing countries. This chapter, therefore, attempts to provide novice researchers in developing countries with a simplified approach to exploratory and confirmatory factor analysis (CFA) with R. As understanding of statistics continues to grow in the developing countries, especially in this era of multivariable high frequency and big data analytics, we believe that it is timely to offer an uncomplicated tutorial on this important topic to the burgeoning global statistical readership.

R is a free software environment for statistical computing, data analysis, and graphics. It compiles and runs on a wide variety of UNIX platforms, Windows, and Mac OS. It has been projected to be the preferred software for the future (alongside Python) because of its highly extensive nature. It is the software of choice for statisticians, data scientists, and researchers in developed countries and it is highly acceptable in many top-rated international journals (Luo et al., 2019). Hence, researchers in developing countries need to use R for data analysis. More details on the basics of R and how to download it can be found at https://www.r-project.org/ (a detailed tutorial on R is beyond the scope of this work). R is relatively easy to install, use, and implement because of its highly interactive nature. It is assumed that the readers of this chapter have basic knowledge of R but lack the technical know-how of FA and how to implement it in R (which is the major concern of this paper). There are two major types of FA, namely, exploratory factor analysis (EFA) and CFA; both are covered in this tutorial.

In addition to this non-exhaustive introductory section, the rest of this paper is structured as follows: Section 20.2 covers EFA, Section 20.3 covers CFA, Section 20.4 provides an archetypical illustrative example encountered during collaborative projects in a developing country, and Section 20.5 concludes this tutorial.

20.2 Exploratory Factor Analysis

As the name implies, EFA is used to explore data to find hidden factors underlying a set of observed variables. Usually, these hidden factors cannot be measured directly, but are considered as latent (natural) groupings of observed variables. EFA is particularly useful in the dimension reduction of many large inter-correlated variables as a statistical technique that can be used to identify the latent relational structure among a set of variables in order to narrow them down to a smaller set of variables (factors). It can be used to handle large data (large numeric matrices consisting of more than 200 items with many variables). Any variables uncorrelated with any other variables should be omitted, and no variables that are perfectly correlated with each other should be included in the analysis (Luo et al., 2019). One of each pair of perfectly correlated variables should be removed or the two variables may be combined if appropriate. EFA is considered the method of choice for analyzing and interpreting self-reporting questionnaires in the field of education and other fields (Barendse et al., 2012). It is a precursor to CFA (Gerbing and Hamilton, 1996).

20.2.1 Steps Involved in Conducting EFA

There are five basic steps in conducting EFA:

20.2.1.1 Determine Whether the Data Are Suitable for Factor Analysis

After collecting the data to be used for FA, certain assumptions must be verified and the data must be checked for sample-size adequacy. Although there are several sample-size recommendations in literature, common consensus suggests a sample size of at least 200 observations (Hogarty et al., 2005). As a rule of thumb, the sample size/variables ratio should range between 3:1 and 20:1 (MacCallum et al., 1999). The researcher should first conduct the Kaiser–Meyer–Olkin (KMO) measure of sampling adequacy and Bartlett's test of sphericity to determine the suitability of the respondent data for FA (Williams et al., 2010). The KMO index lies between 0 and 1; a KMO index of 0.5 and above is acceptable for FA (Jegede et al., 2015; Yong and Pearce, 2013). KMO and MSA (individual measures of sampling adequacy for each item) test whether there are a significant number of factors in the dataset. KMO technically tests the ratio of item correlations to partial item correlations (Halpin et al., 2014).

If the partial correlations are similar to the raw correlations, it means the item does not share substantial variance with other items. These tests are especially recommended when the sample size to a variable ratio of the data is less than 5:1. Bartlett's test of sphericity must be significant (with probability value less than 0.05) for the data to be adequately suitable for FA. Bartlett's sphericity test is useful for testing the hypothesis that correlations between variables are greater than would be expected by chance: Technically, it tests if the matrix is an identity matrix. The p-value should indicate with statistical significance that the null hypothesis that all off-diagonal correlations are zero should be rejected. Note that variables (items) with MSA values below 0.5 indicate that the item does not belong to a group and may be removed from the FA. For sample-size adequacy (generally): a sample size of 50 is very poor, 100 is poor, 200 is fair, 300 is good, 500 is very good, and more than 1,000 is considered excellent (see MacCallum et al., 1999; Williams et al., 2010). Also, all variables must be correlated. A correlation matrix should be used to assess the suitability of individual variables in the analysis. If no correlations go beyond a 0.3 coefficient among the variables, then the researcher should reconsider whether FA is the appropriate statistical method to use on the data (Williams et al., 2010). Correlation among the variables does not however imply causality (Awe, 2012).

20.2.1.2 Extract Initial Factors

The next straightforward exercise is to extract the factors, but how the factors are to be extracted must be determined. There are various methods that can be used to extract factors, including Principal Component Analysis (PCA), Principal Axis Factoring (PAF), Maximum Likelihood, Image Factoring, Alpha Factoring, and Generalized Least. The methods of PCA and PAF are the most commonly used in literature (Bollen, 2019; Luo et al., 2019), especially when there are up to 30 variables or items in the study. Once the latent factors have been derived, the axes that represent these factors within the multidimensional space can be rotated about the origin. This is done in an effort to reduce minimize the complexity of the factors (which are linear combinations of the original variables) and make the resulting structure more interpretable.

There are two major classifications of rotations: orthogonal rotations, which maintain orthogonality (perpendicularity or zero correlations) between the extracted factors, and oblique rotations, which allow the extracted factors to have nonzero correlations. If the factors theoretically allow for interdependence, the latter may be used. The former includes the varimax, quartermax, equimax, and orthogonal procrustes (or orthogonal target) rotations; the latter includes promax, oblimin, direct oblimin, promax, and oblique procrustes (or oblique target) rotations.

20.2.1.3 Determine the Exact Number of Factors to Retain

After extracting the initial factors, you must determine the number of factors to retain. There is no established procedure for this step; EFA is heuristic and exploratory in nature, and this often leads the researcher to explore many options. Three primary methods for determining the final number of factors to retain include (a) scree plots, (b) eigenvalues, and (c) parallel analysis. Williams et al. (2010) suggest that these multiple techniques all be used to explore the appropriate factors to be extracted before deciding on the number of factors to be retained, all the more so if the scree plot is messy and difficult to interpret, where alternative extraction methods should be employed. Most researchers make this decision on the basis of the scree plot, which is often confusing. The scree plot connects the eigenvalues (representing variances explained by each factor, so that sometimes the sums of squared factor loadings are used instead) for many possible factors from maximum to minimum. The appropriate number of factors to retain is that of the last value on the x-axis before the sudden downward inflection of the plot. Alternatively, some researchers retain all the factors with eigenvalues in excess of 1. More recently, parallel analysis has been adjudged to be the best method for deciding the number of factors to extract

(Barendse et al., 2012). It is not often used in the literature because it is not yet embedded in most commercial statistical software packages such as SPSS, SAS, and Stata. However, it is available in R software (which we shall demonstrate in Section 20.4 of this work).

20.2.1.4 Rotate (Spread Variability) Factors

This step helps to ensure that no original variable is strongly related to more than one factor to the greatest extent possible (Bollen, 2019). The researcher should also examine items that do not load or are unable to be assigned to a factor, and then make a decision whether such items should be discarded. After the successful factor extraction, Cronbach's Alpha (α) can be calculated to assess whether the original variables that make up each factor consist of a unidirectional additive score (usually Cronbach's α must exceed 0.7 to constitute a reliable scale). If there is cross-loading (a variable load on more than one factor), the factor with the highest loading is generally selected.

20.2.1.5 Interpreting and Naming the Factors

This aspect involves examining which variables are attributed to each factor and then giving the factors descriptive, meaningful names or themes. When we interpret the extracted factors, adequate names (meaning) of factors are necessary and must be done after conducting CFA (which we discuss in the next section). For instance, in a study, if a factor has five variables that all pertain to social status, the researcher would label the factor "Social Factor." Note that a well-defined factor should have at least three high-loading variables (if only one or two high loading(s) exist, the factors might have been overextracted or multicollinearity may exist). Fact or scores can be used to construct scales or as potential predictors in a regression model for some response of interest (Bartholomew, 2007); in the latter case, this strategy is frequently employed in conjunction with orthogonal methods to eliminate multicollinearity from the resulting regression model.

20.3 Confirmatory Factor Analysis

CFA is a multivariate statistical technique that is often conducted after EFA to assess how well the measured variables represent the number of constructs obtained (Kieffer, 1999). CFA specifies how a set of observed variables are related to some underlying latent factor or factors. In EFA data is simply explored for information about the number of factors required to represent the original variables and how all measured variables are related to every latent variable, but in CFA the researcher can specify the number of factors to extract and which measured variable is related to which factor (latent variable). CFA is simply a tool used to confirm or reject the measurement theory (Jegede et al., 2015). In most cases, variables and factors from EFA are fed into a structural equation model, which is then used in CFA (Nora and Cabrera, 1993).

The goal of CFA is to explain the relationships among the observed variables by specifying a latent (unobserved) structure connecting them. Generally, a CFA would be expected to answer a question such as, "Do my survey questions accurately measure some factors?" It is a form of structural equation modeling that assumes a sufficient sample size ($n > 200$), the correct *a priori* model specification, and data that are collected by a random sample.

Statistical software such as AMOS, LISREL, EQS, and SAS are used for CFA. In AMOS, visual paths are manually drawn on the graphic window and analysis is performed (Adelodun et al., 2013). In LISREL, CFA can be performed graphically as well as from the menu (Adelodun and Awe, 2013), while in SAS, CFA can be performed using code in the software. However, we use R software in this work. We will guide the readers through how to run a CFA in R using the *lavaan* package in R, cover how to interpret the output, and discuss how to write up the results. It is assumed that the reader has some basic experience with and knowledge of R.

20.3.1 Steps in Conducting a Confirmatory Factor Analysis

In this section we discuss the steps that are often followed when conducting CFA.

20.3.1.1 Defining the Individual Constructs

We first define the individual constructs of the model. This involves a pretest to evaluate the construct items using EFA and a confirmatory test of the measurement model that is conducted using CFA.

20.3.1.2 Designing the Overall Measurement Model Theory

In CFA, we should consider the concept of unidimensionality regarding between-construct error variance and within-construct error variance. At least four constructs and three items per construct should be present in the research. The measurement model must be specified. Most commonly, the value of the loading estimate should be one per construct.

20.3.1.3 Specifying the Model (Structural Equation Model)

The model must be specified *a priori* from the knowledge gleaned from EFA. This is what would be fed into the system for the software to produce results for CFA before model validity can be determined.

20.3.1.4 Assessing the Measurement Model Validity

Assessing the measurement model validity occurs when the theoretical measurement model is compared with reality to see how well the data fit. For example, the factor loading of an inclusive variable should be greater than or equal to 0.5 (Bollen, 2019). The chi-square statistic and other goodness-of-fit statistics such as RMR, GFI, NFI, CFI, TLI, RMSEA, SIC, BIC, etc., are some key indicators that help in measuring the model validity. All these will be explained further in Section 20.4 because they are important statistics that must be computed while conducting CFA.

20.4 Illustrative Example and Implementation with R

This section contains an illustrative empirical example on how to run EFA and CFA with R.

20.5 Description of Data for Exploratory Factor Analysis

Suppose we develop a 25-item scale to measure the mathematics resilience of engineering undergraduates in an African university. After the item generation, the scale was administered to 271 students in several numerate disciplines (departments) from the said university, who were to rate themselves on a 7-point scale regarding their perceived resilience in solving mathematics problems. EFA was conducted on the obtained responses.

20.6 Data Analysis and Results of EFA

This section highlights a practical step-by-step FA of the data described above with R. Although there are several packages that can be used to perform FA in R, we use the pack> ages "psych" in this tutorial.

To do FA in R, you need to first load the following package:

```
>install.packages("psych")
>library(psych)
```

Then read in your data using the following code:

```
mydata=read.csv("efa.csv", header=T)
```

#Note that it is better to create a working directory on your system from where you can read your data. (Here we suppose the name of the data to be used is "efa.csv").
   Run the following to view your data

```
>head(mydata)
```

Start by performing the KMO measure of sampling adequacy and Bartlett's test of sphericity using the codes below:

```
> KMO(mydata)

Kaiser-Meyer-Olkin factor adequacyCall: KMO(r = mydata)
Overall MSA = 0.95MSA for each item =
Item5 Item6 Item8 Item9 Item13 Item14 Item23 Item26 Item28 Item34
Item350.97 0.97 0.95 0.95 0.94 0.98 0.92 0.96 0.97 0.97 0.94
Item37 Item39 Item40 Item41 Item42 Item43 Item44 Item45 Item46
Item47 Item310.94 0.98 0.97 0.98 0.95 0.98 0.97 0.97 0.96 0.94 0.79
Item33 Item36 Item380.83 0.89 0.93

> cortest.bartlett(mydata) # Bartlett's sphericity test.
```

```
$chisq
[1] 5410.664 (The test is significant)
> res <- fa.parallel(mydata)
>scree (mydata)
```

Bartlett's test of sphericity is significant (with a probability value less than 0.05) for the data to be adequately suitable for FA. Bartlett's sphericity test is useful for testing the hypothesis that correlations between variables are greater than would be expected by chance.

Next, perform a parallel analysis to determine the appropriate number of factors to extract;
    Parallel analysis suggests that the number of factors=4 and the number of components=2

Also, confirm the number of factors via the scree plot by running the following simple code:

```
>scree (mydata)
```

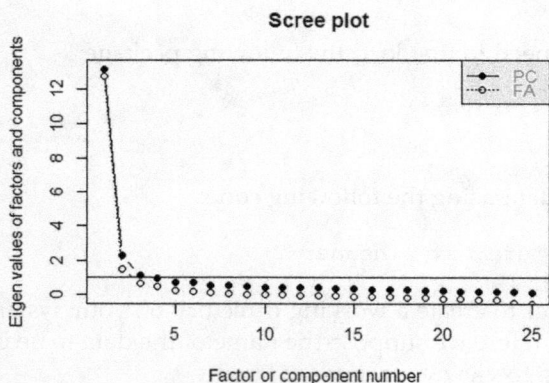

The scree plot suggests we extract 3 clear factors.
    Next, extract and rotate the number of factors using varimax rotation.
```
>
> res <- fa(mydata, fm="minres", nfactors=4, rotate="varimax")
> print(res)
Factor Analysis using method = minres
Call: fa(r = mydata, nfactors = 4, rotate = "varimax", fm = "minres")
```
    Standardized loadings (pattern matrix) based upon correlation matrix

	MR1	MR4	MR3	MR2	h2	u2	com
Item5	0.35	0.16	0.50	0.38	0.54	0.46	3.0
Item6	0.36	0.32	0.51	0.21	0.54	0.46	3.0
Item8	0.35	0.25	0.66	-0.02	0.62	0.38	1.8
Item9	0.32	0.32	0.72	0.15	0.75	0.25	1.9
Item13	0.36	0.47	0.39	0.09	0.51	0.49	3.0

(Continued)

	MR1	MR4	MR3	MR2	h2	u2	com
Item14	0.45	0.48	0.35	0.25	0.61	0.39	3.4
Item23	0.27	0.60	0.36	0.10	0.57	0.43	2.1
Item26	0.43	0.57	0.31	0.10	0.62	0.38	2.5
Item28	0.35	0.49	0.40	0.14	0.54	0.46	3.0
Item34	0.45	0.58	0.27	0.23	0.67	0.33	2.7
Item35	0.56	0.61	0.08	0.15	0.71	0.29	2.1
Item37	0.43	0.60	0.19	0.28	0.66	0.34	2.5
Item39	0.66	0.33	0.27	0.17	0.64	0.36	2.0
Item40	0.72	0.33	0.32	0.15	0.75	0.25	2.0
Item41	0.58	0.29	0.25	0.31	0.58	0.42	2.6
Item42	0.75	0.32	0.21	0.19	0.74	0.26	1.7
Item43	0.64	0.42	0.28	0.12	0.67	0.33	2.2
Item44	0.74	0.29	0.31	0.15	0.75	0.25	1.8
Item45	0.71	0.37	0.27	0.13	0.73	0.27	1.9
Item46	0.77	0.25	0.32	0.19	0.80	0.20	1.7
Item47	0.78	0.23	0.27	0.12	0.75	0.25	1.5
Item31	0.01	0.13	0.01	0.61	0.39	0.61	1.1
Item33	0.11	0.03	0.01	0.80	0.66	0.34	1.0
Item36	0.16	0.12	0.16	0.79	0.68	0.32	1.2
Item38	0.29	0.14	0.21	0.70	0.65	0.35	1.6

	MR1	MR4	MR3	MR2
SS loadings	6.53	3.70	3.02	2.90
Proportion Var	0.26	0.15	0.12	0.12
Cumulative Var	0.26	0.41	0.53	0.65
Proportion Explained	0.40	0.23	0.19	0.18
Cumulative Proportion	0.40	0.63	0.82	1.00

Mean item complexity=2.1
Test of the hypothesis that 4 factors are sufficient.

The degrees of freedom for the null model are 300 and the objective function was 20.74 with chi-square of 5410.66
The degrees of freedom for the model are 206 and the objective function was 1.77

The root mean square of the residuals (RMSR) is 0.03 The df corrected root mean square of the residuals is 0.03

The harmonic number of observations is 271 with the empirical chi-square 119.26 with prob<1. The total number of observations was 271 with Likelihood chi-square=457.49 with prob<3.8e–21

Tucker-Lewis index of factoring reliability=0.928
RMSEA index=0.067 and the 90% confidence intervals are 0.059 0.076
BIC=−696.55
Fit based upon off-diagonal values=1 Measures of factor score adequacy

	MR1	MR4	MR3	MR2
Correlation of (regression) scores with factors	0.92	0.85	0.86	0.92
Multiple R square of scores with factors	0.85	0.73	0.74	0.85
Minimum correlation of possible factor scores	0.71	0.45	0.49	0.70

#Explore further with three factors and cutoff of 0.5 using the following codes:

```
>res <- fa(mydata, fm="minres", nfactors=3, rotate="varimax")
> print(res$loadings, cutoff = 0.5)
```

Loadings:

MR1	MR3	MR2
Item5		
Item6	0.586	
Item8	0.641	
Item9	0.733	
Item13	0.582	
Item14	0.552	
Item23	0.628	
Item26	0.570	
Item28	0.597	
Item34	0.518	0.543
Item35	0.632	
Item37	0.511	
Item39	0.681	
Item40	0.733	
Item41	0.595	
Item42	0.779	
Item43	0.674	
Item44	0.749	
Item45	0.739	
Item46	0.769	
Item47	0.777	
Item31		0.609
Item33		0.807
Item36		0.790
Item38		0.708

	MR1	MR3	MR2
SS loadings	7.166	5.353	2.966
Proportion Var	0.287	0.214	0.119
Cumulative Var	0.287	0.501	0.619

Factor Analysis

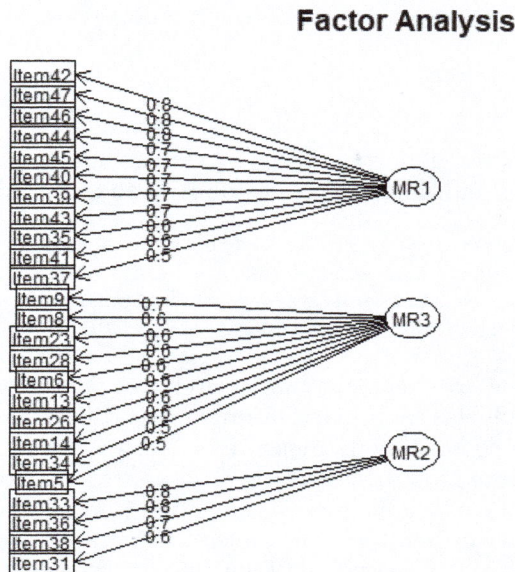

FIGURE 20.1
Factor analysis diagram.

It appears clear that there are three distinct factors in the data. Note that item 5 would be excluded since it has no loadings. Hence, the latent factors are as follows (Figure 20.1):

Factor 1: Item 35, Item 37, Item 39, Item 40, Item 41, Item 42, Item 43, Item 44, Item 45, Item 46, and Item 47.

Factor 2: Item 6, Item 8, Item 9, Item 13, Item 14, Item 23, Item 26, Item 28, and Item 34.

Factor 3: Item 31, Item 33, Item 36, and Item 38.

Naming the factors: The researcher may now name these factors as appropriate. For instance, Factor 1 contains items that pertain to hard work, so we name it "Diligence." Factor 2 contains items that pertain to perseverance or persistence, so we name it "Persistence." Factor 3 contains items that have to do with materials used, so we name it "Access to materials."

20.7 Data and Results of Confirmatory Factor Analysis

After extracting the latent factors through EFA, the next step is to conduct a CFA to validate the scales. There are several software available for conducting CFA models. In this tutorial, we use lavaan, an R package for CFA and structural equation modeling.

Set up

If you do not already have lavaan installed, you will need to do that first:

```
install.packages("lavaan")
library(lavaan)
```

To build a CFA model in lavaan, you will save a string with the model details. Each line is one latent factor, with its indicators following the =~ (read this symbol as "is measured by").

```
cfa.model <- ` diligence =~ Item35 + Item37 + Item39 + Item40 + Item41+
Item42+ Item43+ Item44+ Item45+Item46 + Item47
persistence =~ Item6 + Item8+ Item9+Item13+ Item14+ Item23+ Item26+ Item28
+ Item34 access =~ Item31+Item33+Item36 +Item 38 `
```

In the code above, there are three latent factors referring to the mathematics resilience of engineering undergraduates in an African university: diligence, persistence, and access. The latent factors are never directly measured (which is what it means for them to be latent), but we assume the 25 variables we did observe are indicators of those latent factors.

To estimate the model in lavaan, the easiest method is to use the CFA function. It comes with several defaults for estimating CFA models, including the assumption that you will want to estimate covariances among all of your latent factors (that is, you do not have to write those covariances into the model above). You can proceed to run a basic CFA by using the code

```
fit = cfa(cfa.model, data=mydata)
```

There are numerous options for controlling the way the model is interpreted, estimated, and presented. One popular one is by using the "sem" function. So you can simply replace "cfa" by "sem" in the code above to have:

```
fit = sem(cfa.model, std.lv=TRUE,
  missing="fiml", orthogonal = TRUE, data=mydata)
```

Note also other options like "fiml" and "orthogonal" to cater for missing values and orthogonality in the data respectively.

You can get most of the information you want about your model from one summary command:

```
summary(fit, fit.measures=TRUE, standardized=TRUE)
```

This produces a lot of output, from which you have to select and interpret the most important information.

In this tutorial, we will consider and explain the output piece by piece, and then use parameterEstimates(fit) to pull out parts of the summary() output individually.

We will explain a few of the most useful and widely-used fit indices to measure the suitability of your model in the following results using maximum likelihood estimation method.

```
> cfa.model <- ` diligence =~ Item35 + Item37 + Item39 + Item40 + Item41+
Item42 + Item43 + Item44 + Item45+Item46 + Item47
+ persistence =~ Item6 + Item8 + Item9 +Item13 + Item14 + Item23 + Item26
+ Item28 + Item 34
+ access =~ Item31 + Item33 +Item36 + Item38'
```

```
> fit=cfa(cfa.model,data=mydata)
> summary(fit,fit.measures=T,standardized=T)
```

Results	
lavaan 0.6-6 ended normally after 51 iterations	
Estimator	ML
Optimization method	NLMINB
Number of free parameters	51
Number of observations	271
Model Test User Model	
Test statistic	747.883
Degrees of freedom	249
P-value (chi-square)	0.000
Model Test Baseline Model	
Test statistic	5417.235
Degrees of freedom	276
P-value	0.000

User Model versus Baseline Model:

Comparative Fit Index (CFI)	0.903
Tucker-Lewis Index (TLI)	0.892

Loglikelihood and Information Criteria:

Loglikelihood user model (H0)	–11337.429
Loglikelihood unrestricted model (H1)	–10963.488
Akaike (AIC)	22776.859
Bayesian (BIC)	22960.567
Sample-size adjusted Bayesian (BIC)	22798.861

Root Mean Square Error of Approximation:

RMSEA	0.086
90% confidence interval—lower	0.079
90% confidence interval—upper	0.093
P-value RMSEA <= 0.05	0.000

Standardized Root Mean Square Residual:

SRMR	0.059

Parameter Estimates:

Standard errors	Standard
Information	Expected
Information saturated (h1) model	Structured

Latent Variables:

	Estimate	Std.Err	z-value	P(>|z|)	Std.lv	Std.all
diligence =~						
Item35	1.000			1.496	0.766	
Item37	0.931	0.073	12.772	0.000	1.393	0.727
Item39	1.101	0.076	14.409	0.000	1.647	0.804
Item40	1.196	0.076	15.717	0.000	1.790	0.861
Item41	1.031	0.079	13.094	0.000	1.542	0.742
Item42	1.098	0.071	15.503	0.000	1.642	0.852
Item43	1.073	0.073	14.687	0.000	1.605	0.816
Item44	1.154	0.073	15.723	0.000	1.726	0.862
Item45	1.158	0.075	15.532	0.000	1.732	0.853
Item46	1.176	0.073	16.119	0.000	1.759	0.879
Item47	1.203	0.078	15.399	0.000	1.800	0.848
persistence =~						
Item6	1.000			1.520	0.708	
Item8	0.943	0.087	10.813	0.000	1.434	0.681
Item9	0.975	0.079	12.288	0.000	1.482	0.775
Item13	0.930	0.082	11.338	0.000	1.414	0.714
Item14	1.021	0.083	12.346	0.000	1.552	0.779
Item23	0.841	0.073	11.545	0.000	1.279	0.728
Item26	0.910	0.074	12.268	0.000	1.383	0.774
Item28	0.916	0.079	11.651	0.000	1.393	0.734
Item34	0.992	0.079	12.548	0.000	1.508	0.792
access =~						
Item31	1.000			1.147	0.581	
Item33	1.295	0.143	9.079	0.000	1.485	0.752
Item36	1.463	0.152	9.630	0.000	1.678	0.848
Item38	1.432	0.152	9.436	0.000	1.643	0.807

Covariances:

	Estimate	Std.Err	z-value	P(>|z|)	Std.lv	Std.all
diligence ~~						
persistence	2.043	0.251	8.156	0.000	0.898	0.898
access	0.820	0.151	5.444	0.000	0.478	0.478
Persistence ~~						
access	0.828	0.157	5.277	0.000	0.475	0.475

Variances:

	Estimate	Std.Err	z-value	P(>|z|)	Std.lv	Std.all
.Item35	1.578	0.143	11.041	0.000	1.578	0.414
.Item37	1.732	0.155	11.167	0.000	1.732	0.472
.Item39	1.488	0.137	10.870	0.000	1.488	0.354
.Item40	1.113	0.107	10.421	0.000	1.113	0.258

(Continued)

	Estimate	Std.Err	z-value	P(>|z|)	Std.lv	Std.all
.Item41	1.937	0.174	11.122	0.000	1.937	0.449
.Item42	1.017	0.097	10.517	0.000	1.017	0.274
.Item43	1.292	0.120	10.797	0.000	1.292	0.334
.Item44	1.033	0.099	10.418	0.000	1.033	0.257
.Item45	1.119	0.107	10.504	0.000	1.119	0.272
.Item46	0.913	0.089	10.204	0.000	0.913	0.228
.Item47	1.269	0.120	10.559	0.000	1.269	0.282
.Item6	2.305	0.212	10.855	0.000	2.305	0.499
.Item8	2.378	0.217	10.964	0.000	2.378	0.536
.Item9	1.460	0.140	10.460	0.000	1.460	0.399
.Item13	1.917	0.177	10.825	0.000	1.917	0.490
.Item14	1.563	0.150	10.431	0.000	1.563	0.394
.Item23	1.454	0.135	10.759	0.000	1.454	0.471
.Item26	1.283	0.123	10.470	0.000	1.283	0.401
.Item28	1.658	0.155	10.724	0.000	1.658	0.461
.Item34	1.355	0.131	10.322	0.000	1.355	0.373
.Item31	2.588	0.241	10.715	0.000	2.588	0.663
.Item33	1.697	0.185	9.185	0.000	1.697	0.435
.Item36	1.104	0.163	6.779	0.000	1.104	0.282
.Item38	1.446	0.180	8.015	0.000	1.446	0.349
diligence	2.238	0.302	7.413	0.000	1.000	1.000
persistence	2.311	0.352	6.566	0.000	1.000	1.000
access	1.316	0.268	4.905	0.000	1.000	1.000

```
> inspect(fit, what="std", "r2")
$lambda
```

	dilgnc	prsstn	access
Item35	0.766	0.000	0.000
Item37	0.727	0.000	0.000
Item39	0.804	0.000	0.000
Item40	0.861	0.000	0.000
Item41	0.742	0.000	0.000
Item42	0.852	0.000	0.000
Item43	0.816	0.000	0.000
Item44	0.862	0.000	0.000
Item45	0.853	0.000	0.000
Item46	0.879	0.000	0.000
Item47	0.848	0.000	0.000
Item6	0.000	0.708	0.000
Item8	0.000	0.681	0.000
Item9	0.000	0.775	0.000
Item13	0.000	0.714	0.000
Item14	0.000	0.779	0.000
Item23	0.000	0.728	0.000
Item26	0.000	0.774	0.000

(Continued)

	dilgnc	prsstn	access
Item28	0.000	0.734	0.000
Item34	0.000	0.792	0.000
Item31	0.000	0.000	0.581
Item33	0.000	0.000	0.752
Item36	0.000	0.000	0.848
Item38	0.000	0.000	0.807

$theta

	Item35	Item37	Item39	Item40	Item41	Item42	Item43	Item44	Item45	Item46
Item35	0.414									
Item37	0.000	0.472								
Item39	0.000	0.000	0.354							
Item40	0.000	0.000	0.000	0.258						
Item41	0.000	0.000	0.000	0.000	0.449					
Item42	0.000	0.000	0.000	0.000	0.000	0.274				
Item43	0.000	0.000	0.000	0.000	0.000	0.000	0.334			
Item44	0.000	0.000	0.000	0.000	0.000	0.000	0.000	0.257		
Item45	0.000	0.000	0.000	0.000	0.000	0.000	0.000	0.000	0.272	
Item46	0.000	0.000	0.000	0.000	0.000	0.000	0.000	0.000	0.000	0.228
Item47	0.000	0.000	0.000	0.000	0.000	0.000	0.000	0.000	0.000	0.000
Item6	0.000	0.000	0.000	0.000	0.000	0.000	0.000	0.000	0.000	0.000
Item8	0.000	0.000	0.000	0.000	0.000	0.000	0.000	0.000	0.000	0.000
Item9	0.000	0.000	0.000	0.000	0.000	0.000	0.000	0.000	0.000	0.000
Item13	0.000	0.000	0.000	0.000	0.000	0.000	0.000	0.000	0.000	0.000
Item14	0.000	0.000	0.000	0.000	0.000	0.000	0.000	0.000	0.000	0.000
Item23	0.000	0.000	0.000	0.000	0.000	0.000	0.000	0.000	0.000	0.000
Item26	0.000	0.000	0.000	0.000	0.000	0.000	0.000	0.000	0.000	0.000
Item28	0.000	0.000	0.000	0.000	0.000	0.000	0.000	0.000	0.000	0.000
Item34	0.000	0.000	0.000	0.000	0.000	0.000	0.000	0.000	0.000	0.000
Item31	0.000	0.000	0.000	0.000	0.000	0.000	0.000	0.000	0.000	0.000
Item33	0.000	0.000	0.000	0.000	0.000	0.000	0.000	0.000	0.000	0.000
Item36	0.000	0.000	0.000	0.000	0.000	0.000	0.000	0.000	0.000	0.000
Item38	0.000	0.000	0.000	0.000	0.000	0.000	0.000	0.000	0.000	0.000

	Item47	Item6	Item8	Item9	Item13	Item14	Item23	Item26	Item28	Item34
Item35										
Item37										
Item39										
Item40										
Item41										
Item42										
Item43										
Item44										
Item45										
Item46										
Item47	0.282									

(Continued)

	Item47	Item6	Item8	Item9	Item13	Item14	Item23	Item26	Item28	Item34
Item6	0.000	0.499								
Item8	0.000	0.000	0.536							
Item9	0.000	0.000	0.000	0.399						
Item13	0.000	0.000	0.000	0.000	0.490					
Item14	0.000	0.000	0.000	0.000	0.000	0.394				
Item23	0.000	0.000	0.000	0.000	0.000	0.000	0.471			
Item26	0.000	0.000	0.000	0.000	0.000	0.000	0.000	0.401		
Item28	0.000	0.000	0.000	0.000	0.000	0.000	0.000	0.000	0.461	
Item34	0.000	0.000	0.000	0.000	0.000	0.000	0.000	0.000	0.000	0.373
Item31	0.000	0.000	0.000	0.000	0.000	0.000	0.000	0.000	0.000	0.000
Item33	0.000	0.000	0.000	0.000	0.000	0.000	0.000	0.000	0.000	0.000
Item36	0.000	0.000	0.000	0.000	0.000	0.000	0.000	0.000	0.000	0.000
Item38	0.000	0.000	0.000	0.000	0.000	0.000	0.000	0.000	0.000	0.000

	Item31	Item33	Item36	Item38
Item35				
Item37				
Item39				
Item40				
Item41				
Item42				
Item43				
Item44				
Item45				
Item46				
Item47				
Item6				
Item8				
Item9				
Item13				
Item14				
Item23				
Item26				
Item28				
Item34				
Item31	0.663			
Item33	0.000	0.435		
Item36	0.000	0.000	0.282	
Item38	0.000	0.000	0.000	0.349

`$psi`

	dilgnc	prsstn	access
diligence	1.000		
persistence	0.898	1.000	
access	0.478	0.475	1.000

```
#fitmeasures(fit)
inspect(fit,"r2")
```

Item35	Item37	Item39	Item40	Item41	Item42	Item43	Item44	Item45	Item46	Item47
0.586	0.528	0.646	0.742	0.551	0.726	0.666	0.743	0.728	0.772	0.718

Item6	Item8	Item9	Item13	Item14	Item23	Item26	Item28	Item34	Item31	Item33
0.501	0.464	0.601	0.510	0.606	0.529	0.599	0.539	0.627	0.337	0.565

Item36	Item38
0.718	0.651

CFI (Comparative fit index): Measures whether the model fits the data better than a more restricted baseline model. Higher is better, with a value greater than 0.9. Thus, this model is a good fit since the CFI is above 0.90.

AIC (Akaike's information criterion): Attempts to select models that are the most parsimonious and suitable representations of the observed data. Lower values of AIC are better.

BIC (Schwarz's Bayesian information criterion): This is quite similar to AIC but a little more conservative, also used to select models that are the most parsimonious representations of the observed data. Lower values are better.

TLI (Tucker-Lewis index): This index is similar to CFI, but it penalizes overly complex models (making it more conservative than CFI). Measures whether the model fits the data better than a more restricted baseline model. Values greater than 0.9 indicate a good fit. In this example, we obtained a TLI value of approximately 0.90 so it passes for a good fit.

RMSEA (Root mean square error of approximation): The "error of approximation" refers to residuals. Instead of comparing to a baseline model, it measures how closely the model reproduces data patterns (i.e., the covariances among indicators). Lower values are better. It comes along with a 90% CI in lavaan, as can be seen in the results above (and in other major SEM software). The p- value printed with this result summarizes the test of the hypothesis that RMSEA is less than or equal to 0.05 (a cutoff sometimes used for "close" fit); here, our RMSEA exceeds 0.05 (it is 0.086, with a 90% CI from 0.07 to 0.09), so the p-value indicates lack of significance. Hence we conclude that this model fits well. Factor loadings can be interpreted like a usual regression coefficient and the factored variables can be used as explanatory variables for various response variables in separate regression models. Figure 20.2 represents the path diagram of the model in this study. It indicates the number of variables loaded under each of the three factors identified.

These are essentially the most important fit measures to look for in CFA or SEM. However, you must note that if your model fits well, it does not necessarily mean it is a "good" model, or that it reflects truth or reality. Knowing the theory behind your model is crucial to deciding whether or not a model is reasonable. Sometimes, a researcher may wish to compute modification indices, which tells how model fit, would change if you add new parameters to the model. Since it is a CFA model and not exploratory (you already know what variables you want to include in the model after conducting EFA), if you make the changes these indices suggest, you run a serious risk of over-fitting your data and reducing the

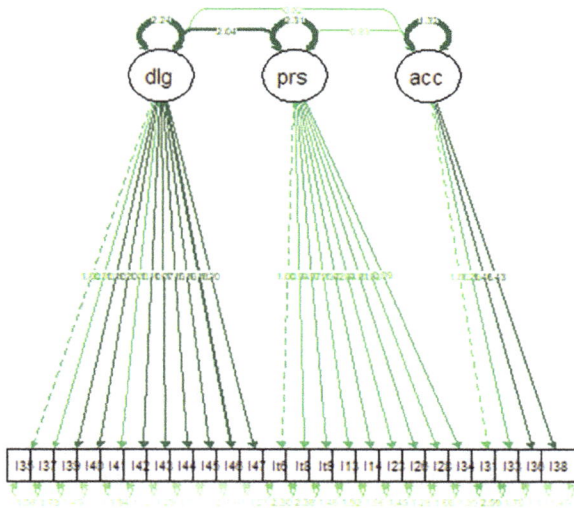

FIGURE 20.2
Path diagram of the model.

generalizability of your results. Instead, we recommend using modification indices mostly as another description in cases where your model is not fitting well as a way to examine the residuals of the model.

20.8 Concluding Remark

In this paper, we presented the basics of FA. We have demonstrated a step-by-step procedure for conducting both exploratory and CFA with practical example R codes for an empirical application in the field of education. We believe the lessons learned from this tutorial will be useful for students, researchers, practitioners, and professionals across several disciplines, and they will help correct some misconceptions about FA that are commonly held by novices.

References

Adelodun, O.A., and Awe, O.O. (2013). Using LISREL program for empirical research. *Transnational Journal of Science and Technology*, 3(8), 1–14.

Adelodun, O.A., Obilade, T.O., and Awe, O.O. (2013). Large type fit indices of mathematics adult learners: A covariance structure model. *Mathematical Theory and Modeling*, 3(12), 80–84.

Awe, O.O. (2012). Correlation or causality: New Evidence from cross-sectional statistical analysis of auto-crash data in Nigeria. *Transnational Journal of Science and Technology* 2(7), 48–63.

Barendse, M.T., Oort, F.J., Werner, C.S., Ligtvoet, R., and Schermelleh-Engel, K. (2012). Measurement bias detection through factor analysis. *Structural Equation Modeling: A Multidisciplinary Journal*, 19(4), 561–579.

Bartholomew, D.J. (2007). Three faces of factor analysis. In *Factor Analysis at 100*, pp. 23–36. Routledge.

Baynton, M. (1992). Dimensions of "control" in distance education: A factor analysis. *American Journal of Distance Education*, 6(2), 17–31.

Bollen, K.A. (2019). When good loadings go bad: Robustness in factor analysis. *Structural Equation Modeling: A Multidisciplinary Journal*, 1–10.

Brown, E.C., Aman, M.G., and Havercamp, S.M. (2002). Factor analysis and norms for parent ratings on the Aberrant Behavior Checklist-Community for young people in special education. *Research in Developmental Disabilities*, 23(1), 45–60.

Bryant, F.B., Yarnold, P.R., and Michelson, E.A. (1999). Statistical methodology: VIII. Using confirmatory factor analysis (CFA) in emergency medicine research. *Academic Emergency Medicine*, 6(1), 54–66.

Fisher, M.J., and King, J. (2010). The self-directed learning readiness scale for nursing education revisited: A confirmatory factor analysis. *Nurse Education Today*, 30(1), 44–48.

Gerbing, D.W., and Hamilton, J.G. (1996). Viability of exploratory factor analysis as a precursor to confirmatory factor analysis. *Structural Equation Modeling: A Multidisciplinary Journal*, 3(1), 62–72.

Glass, G., and Hopkins, K. (1996). *Statistical Methods in Education and Psychology*.

Halpin, P.F., da-Silva, C., & De Boeck, P. (2014). A confirmatory factor analysis approach to test anxiety. *Structural Equation Modeling: A Multidisciplinary Journal*, 21(3), 455–467.

Hoban, J.D., Lawson, S.R., Mazmanian, P.E., Best, A.M., and Seibel, H.R. (2005). The self- directed learning readiness scale: A factor analysis study. *Medical Education*, 39(4), 370–379.

Hogarty, K.Y., Hines, C.V., Kromrey, J.D., Ferron, J.M., and Mumford, K.R. (2005). The quality of factor solutions in exploratory factor analysis: The influence of sample size, communality, and overdetermination. *Educational and Psychological Measurement*, 65(2), 202–226.

Jegede, O.P., Faleye, B.A., and Adeyemo, E.O. (2015). Factor analytic study of Lecturer's Teaching Assessment Scale in Obafemi Awolowo University, Nigeria. *World Journal of Education*, 50(3), 121–130.

Jegede, P.O. (2007). Factors in computer self-efficacy among Nigerian College of Education teachers. *Journal of Psychology in Africa*, 17(1–2), 39–44.

Kieffer, K.M. (1999). An introductory primer on the appropriate use of exploratory and confirmatory factor analysis. *Research in the Schools*, 6(2), 75–92.

Luo, L., Arizmendi, C., and Gates, K.M. (2019). Exploratory factor analysis (EFA) programs in R. *Structural Equation Modeling: A Multidisciplinary Journal*, 26(5), 819–826.

MacCallum, R.C., Widaman, K.F., Zhang, S., and Hong, S. (1999). Sample size in factor analysis. *Psychological Methods*, 4(1), 84.

Mulaik, S.A. (2010). Factor analysis at 100, *Structural Equation Modeling: A Multidisciplinary Journal*, 17(1), 150–164.

Muilenburg, L., and Berge, Z.L. (2001). Barriers to distance education: A factor-analytic study. *American Journal of Distance Education*, 15(2), 7–22.

Nora, A., and Cabrera, A.F. (1993). The construct validity of institutional commitment: A confirmatory factor analysis. *Research in Higher Education*, 34(2), 243–262.

Watson, B. (2004). Making sense of factor analysis: The use of factory analysis for instrument development in health care research. *Nurse Researcher*, 11(3), 91–93.

Williams, B., Onsman, A., and Brown, T. (2010). Exploratory factor analysis: A five-step guide for novices. *Australasian Journal of Paramedicine*, 8(3), 1–13.

Yong, A.G., and Pearce, S. (2013). A beginner's guide to factor analysis: Focusing on exploratory factor analysis. *Tutorials in Quantitative Methods for Psychology*, 9(2), 79–94.

21

Retrieval of Unstructured Datasets and R Implementation of Text Analytics in the Climate Change Domain

Olusesan Michael Awoleye and Albert Ayorinde Abegunde
Obafemi Awolowo University

O. Olawale Awe
University of Campinas

CONTENTS

21.1 Introduction

The disruptive innovation we witness today in communication and information retrieval (IR) is premised on the development of emerging technologies, which, in turn, is powered by web technology and the Internet (Awoleye *et al.*, 2014; Pegoraro, 2014; Vitolo *et al.*, 2015). In the era of globalization, SM has transformed the way we live and has become

DOI: 10.1201/9781003261148-25

indispensable for effective communication among people, peers, families, friends, and organizations globally (Onyijen *et al.*, 2019). The adoption of SM and its effective use has gone beyond personal use for communication, information exchange, and pleasures; SM is now being employed to revolutionize communication and collaboration in the scientific sphere. It was reported that Facebook has over 2 billion users globally, which is about a third of the world population (Oltulu *et al.*, 2018). In the same vein, Twitter was also noted to have over 328 million users generating over 500 million tweets daily (Wasim, 2017).[1] Other platforms in this category are LinkedIn with 610 million monthly users, Pinterest with 250 million monthly visitors, and Instagram with 100 million, among others.[2] The avalanche of data generated by users on various SM platforms such as Facebook, Twitter, YouTube, Instagram, LinkedIn, Pinterest, and Flickr cannot be overemphasized. The volume of data generated on these platforms by users daily has necessitated the need to employ machine learning techniques to filter useful information from such big data. It is in this regard that this work has employed machine learning techniques to harvest and analyze climate change data in Twitter over Nigeria geolocation.

Climate change is important at this time because scenes of flood, storms, wildfire, and other natural disasters fed by global warming are indications of how much weather and climate can affect our lives. In the context of Nigeria, it is therefore important to explore how Nigeria's climate has been changing. This includes, but is not limited to, increased temperature, variable rainfall, rises in sea levels and flooding, drought, desertification, land degradation, and likely extreme weather effects that affect humans and the ecosystem. Thus, the research employed a qualitative study using opinions on Twitter as a source of data for the analysis (Kumar and Bala, 2016; Bose *et al.*, 2019). In this regard, natural language processing (NLP) procedures have been used extensively to explore big data in similar research (Procter *et al.*, 2013; Agerri *et al.*, 2015; Baltas *et al.*, 2016). It also has some related various programming implementations, including R, Python, Java, and Matlab, that are largely object-oriented. The choice of R programming for the implementation of the unstructured dataset used to illustrate text mining in this document is premised on its robustness, versatility, scalability, reproducibility, and level of community support (McMurdie and Holmes, 2013; Vitolo *et al.*, 2015). It is interesting to note that these features limit the existing statistical software packages in handling the required analysis in this context. R also provides a solution to addressing these challenges using the step-by-step approach involved in text mining of small to a large dataset, as explained in this document. R as an open-source software also has support for other software libraries: packages and functions, extending the functionality of the base R language and core packages. This will further help to demystify the procedures with concise steps. In addition, it is important to reiterate the inclusion of documentation and examples, which are usually released alongside these packages, which reduces the herculean task of learning programming languages, especially for people without prior programming knowledge. The Comprehensive R Archive Network now has over 10,000 packages that are published under scrutiny for procedural conformity and interoperability. This document, thus, enunciates some related machine learning concepts as background to provide a better understanding of the R implementation.

---

[1] https://www.dsayce.com/social-media/tweets-day/.
[2] https://www.internetworldstats.com/social.htm.

21.2 Social Media as a Veritable Source of Data

Social media (SM) does not have a universal definition; among many definitions advanced by researchers, Kaplan and Haenlein (2010) refer to it as a virtual social world. In the same vein, Drury (2008) described SM as online resources that people use to *share* 'content': video, photos, images, text, ideas, insight, humor, opinion, gossip, news—the list goes on. These resources include blogs, vlogs, social networks, message boards, podcasts, public bookmarking, and wikis. In this document, therefore, we define the concept as emerging and disruptive technologies that provide a medium for the creation and exchange of information among a community connected by beliefs, interests, ideologies, and careers. Part of the major benefits that the emerging technologies have brought to mankind are their ability to propagate the expression of opinions, thoughts, and feelings (Ahmed, 2017) on products, processes, ideologies, or beliefs. Businesses, both public and private, and other organizations use SM to showcase or to promote their businesses and products as well as to get feedback from users and customers. As millions of users make use of these platforms daily, they generate huge amounts of data. These data are usually not in any unique form or structure that may make it easily accessible using search queries for their retrieval. These largely unstructured datasets tend to contain useful knowledge and information that may be somehow latent. This is where the relevance of machine learning has come to the fore. Apart from the fact that SM has been used widely for communication in our daily lives, it has also been employed as a tool during crises and extreme circumstances (Ahmed, 2017), such as outbreaks of infectious diseases such as bird flu, Lassa fever, Ebola, and coronavirus, and other natural disasters like hurricanes and typhoons (Takahashi *et al.*, 2015). Different SM platforms are designed to achieve different purposes of social networking. For example, Facebook is used to connect to friends and families and to discover what is going on in the world and to share and express one's own opinion on it. Unlike Facebook, which uses both text and pictures, Instagram specifically relies on visuals such as photos and videos. Twitter, in the same vein, is a microblogging technology that allows one to send and receive short messages/posts called tweets that are usually about 140 characters long and can include links, hashtags, special characters, etc. It is important to state that data from Twitter and others are used by academic researchers examining how people use the tools in different circumstances (Ahmed, 2017). For example, the recent wave currently sweeping America and other developed countries regarding the police brutality and inequality in the social and justice system in America was first promoted on SM. This started when passers-by filmed and later uploaded a video of how American police tortured and eventually killed one African-American George Floyd. This brewed rage and eventually led to several days of street protests. A hashtag #BlackLivesMatter was then created on Twitter to assist them to collate opinions and posts relating to fighting their cause for freedom. Hashtags have been used considerably in research, especially on Twitter. For example, Chae (2015) explored the hashtag #Supplychain to gain insights into the concept of supply chain management. In the same vein, the hashtags #YolandaPH and #Haiyan were used in a research carried out by Takahashi *et al.* (2015) regarding a typhoon disaster in the Philippines, while #EbolaOutbreakAlert was used to investigate the outbreak of the Ebola pandemic (Ahmed and Bath, 2015).

21.3 Web Crawling

Web crawling can be described as a program that most search engines use to find what is new on the Internet; it is sometimes referred to as a web crawler or a spider. For example, Google's web crawler is known as GoogleBot.[3] The bot crawls the web and collects documents to build a searchable index for the different search procedures. The program starts at a website and follows every hyperlink on each page. One can say that everything on the web will eventually be found and collected, because the bot hops from one website to another. In a practical sense, a list of websites to be crawled or ignored could be itemized in a text file, e.g., 'robots.txt', which is sometimes called robot exclusion protocol. This is synonymous with an access control list in computer networking, which is a file that specifies the ports, protocols, IP addresses, etc., to be allowed or disallowed within a networking environment (Ferraiolo *et al.*, 1999).

21.4 Web Crawler Architectures

It is not sufficient for a web crawler to have a good strategy; it is also desirable for it to have a highly optimized architecture. The architecture shown in Figure 21.1 is a typical one of a web crawler.

Traditional web crawling employed a set of instructions provided in a given text file such as robot.txt, which are sometimes referred to as a set of policies. At the end of the

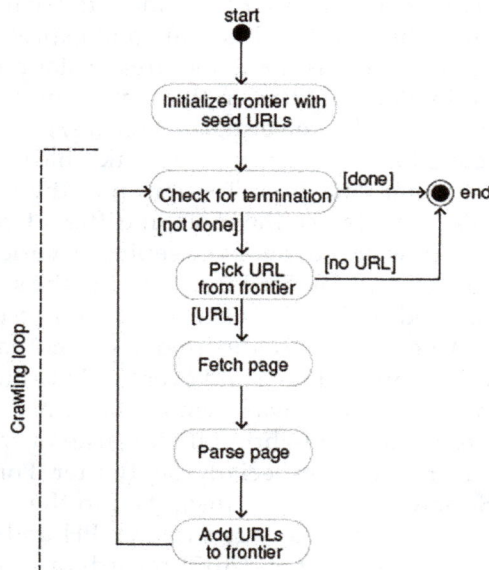

FIGURE 21.1
Crawling procedure. (Pant *et al.*, 2004.)

---

[3] http://www.wpthemesplanet.com/2009/09/how-does-web-crawler-spider-work/ retrieved 12 June, 2010.

file, a marker is provided to show the end of the file; otherwise, the crawler moves to the next line to execute the instruction. This is recursively parsed while reaping the relevant page(s) specified, parsing them, and adding URL to the frontier. This process is recursive and continues until the end of the file is reached. It is worth stating here that Shkapenyuk and Suel (2002) reported the insistence of several scholars of the necessity of crawling important pages first. The authors further mentioned the possibility of crawling pages on a particular topic or of a particular type. Re-crawling (refreshing) pages to optimize the overall 'freshness' of a collection of pages and scheduling of crawling activities over time was also noted to be of utmost importance.

21.5 Vector-Based Model of IR

The main focus in the IR field is to be able to effectively search for information relevant to the user's needs within a given gamut of data (Oren, 2005). Several search methods are available, but one of the most popular paradigms for indexing and searching is the vector-based model of IR. One of the vector-based models is a family of variants of a very widely used scheme referred to as term frequency, inverse document frequency (*tf.idf*) methods (Salton, 1989). These schemes employ a small number of documents, collections, and query features to provide a measure of relevance for each document relative to the user's query. For example, in a recommender system, assume that N is the total number of documents that can be recommended to users and n is the total number of documents containing the index term, f is the raw term frequency, and mf is the maximum raw term frequency.

Figure 21.2 shows a function that integrates these parameters as $(f/mf) \times \log \dfrac{N}{n}$.

Since N is the total number of documents that can be recommended to users, let us also assume that the keyword k_i appears in n_i of them (Adomavicius and Tuzhilin, 2005) and $f_{i,j}$ is the number of times keyword k_i appears in document d_j. Then $TF_{i,j}$, the term frequency (or normalized frequency) of keyword k_i in document d_j, is defined as

$$TF_{i,j} = \frac{f_{i,j}}{\max_z f_{z,j}}$$

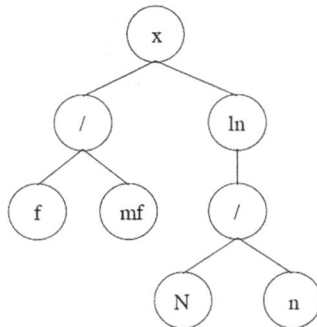

FIGURE 21.2
A parse tree of td.idf indexer. (Oren, 2002.)

where the maximum is computed over the frequencies $f_{z,j}$ of all keywords k_z that appear in the document d_j. However, keywords that appear in many documents are not useful for distinguishing a relevant document and a nonrelevant one. Therefore, the measure of inverse document frequency (IDF_i) is often used in combination with simple term frequency ($TF_{i,j}$). The inverse document frequency for keyword k_i is usually defined as

$$IDF_i - \log \frac{N}{n_i}.$$

Then the *TF-IDF* weight for keyword k_i in document d_j is defined as

$$w_{i,j} = TF_{i,j} \times IDF_i$$

and the content of document d_j is defined as $\text{Content}(d_j) = (w_{1j}, \ldots, w_{kj})$.

21.6 Document Indexing

Document indexing is the process of transforming document text into a representation of text, comprising three steps: tokenization, filtration, and stemming. During tokenization, terms are lowercased and punctuation marks removed. Rules must be in place so that digits, hyphens, and other symbols can be parsed properly. Tokenization is followed by filtration. During filtration, commonly used terms and terms that do not add any semantic meaning (stopwords) are removed. In most IR systems, survival terms are further reduced to common stems or roots, this is known as stemming. Thus, the initial content of length l is reduced to a list of terms (stems and words) of length l' (i.e., $l' < l$). Raw text in a collection is first parsed into generalized words called tokens. Tokenization first lowercases the terms and includes the removal of punctuations, spaces, and other symbols that made the structure found in the article headers (e.g., CSS).

Noun phrases were used as tokens since this has been proven in the literature to be useful by Lang (1995). Thereafter, a vector of token counts for the document is created. This is the size of the total vocabulary without reckoning with tokens not occurring in the documents, which could be zero. This approach is generally known as the *bag-of-words* model and is the basis for this representation. This approach does not capture the order of the tokens in the document; it is assumed that it captures the necessary information needed for our filtering purposes. The next step is referred to as filtration, which is the process of removing the frequently occurring (used) terms sometimes called stopwords: 'the', 'of', 'and', etc.; this is done since either their inclusion or their removal does not impact retrieval efficiency. Stopwords are filtered out before or after the processing of natural language, as they do not seem to add much meaning to a sentence when they are ignored or excluded from it.

21.6.1 Indexing Using LUCENE

Lucene uses a combination of a Vector Space Model (VSM) and a Boolean model to determine the relevance of a document to a given query. In IR, documents are represented as vectors (Polettini, 2004) and the term weighting techniques serve as a determinant of the level of success or failure of the vector space method (model). The main idea behind the VSM is the more times a query term appears in a document relative to the number of times the term appears in all documents in the entire collection, the more relevant the document is to the query. Lucene uses the Boolean model to first narrow down the document that needs to be scored based on the use of Boolean logic in the query specification. Lucene draws its popularity and/or strength from the good results and applicability to nonstructural texts. Two issues are considered when weights are to be assigned to terms especially in the IR domain: (1) the local information from individual documents and (2) the global information from a collection of documents. Salton (1998) is a leading study on this point by presenting a VSM, commonly known as the 'term vector model'. The weighting scheme is presented as follows:

$$(\text{Term Weight})w_i = tf_i * \log\left(\frac{D}{df_i}\right) \tag{21.1}$$

where

- tf_i = term frequency (term counts) or the number of times a term i occurs in a document.
- df_i = document frequency or the number of documents containing term i.
- D = number of documents in the database.

21.7 Local Weights

The equation shows that the weight of a term (w_i) increases with term counts (tf_i). This is observed as being vulnerable to term repetition abuses in the model; Garcia (2006) described it as an adversarial practice known as keyword spamming.

Given a query q,

1. for documents of equal lengths, those with more instances of q are favored during retrieval; and
2. for documents of different lengths, longer documents are favored during retrieval since these tend to contain more instances of q.

21.8 Global Weights

In equation (21.1), the $\log(D/df_i)$ term is known as the inverse document frequency (IDF_i), a measure of the volume of information associated with term i within a set of documents. Inspecting the df_i/D ratio, this is the probability of retrieving from D a document containing term i. The probability is simply inverted in the equation and the log is taken. The result is then pre-multiplied by tf_i.

21.9 Methods of Accessing Datasets Online

This section describes four main techniques for retrieving data into the R environment: (1) using open-source and proprietary software, (2) directly using the application program interface (API) with some R codes, (3) pointing directly to the data via a given web page, and (4) using files from other sources such as comma-separated files (CSV) or text files. Before this, the first thing that must be done is setting the RStudio environment; this includes getting and setting the working directory. This makes working with files much easier when one intends to retrieve or save a given file from/to the local drive. The following R codes could be used to achieve this.

```
# Set current working directory.
setwd("/Analysis/climate") #the parameters in parentheses represents the
file path

# Get and print current working directory.
getwd()
```

21.9.1 Using Open-Source and Proprietary Software

There are many ways in which data could be crawled online and fetched into the R programming environment. One such way is by using Twitter Archiving Google Sheet as designed by Hawksey (2014). This is a Google spreadsheet that only requires the users to specify the keyword, hashtags, or combination of hashtags to be retrieved. Logical operations such as 'OR' or 'AND' are allowed; this searches for either of the keywords/hashtags or combines the given search terms in the query, respectively. There are other open-source software that can be used to fetch data from Tweeter as well, such as Mozdeh and Chorus. Other proprietary software includes NodeXL, RapidMiner, and Discover Text. These require Twitter API (Kim *et al.*, 2020), which necessitates that a user registers a Twitter account and requests the authorization code, which consists of consumer_key and the consumer_secret key. These are computer-generated series of alphanumeric codes.

21.9.2 Retrieving Data from Twitter Using API

```
#This section installs and load the twitter package to
#the RStudio environment
        install.packages("twitteR")
        library(twitteR)

#This sets up the API permission procedures which must
#be generated from a twitter account
consumer_key<- "PxwadCtG0tGdDlAe2Y18yXqaZ"
consumer_secret<- "fQpST1zaYXhMby3CH19LKZdYVO60R47v1P4wevKLnoBGSGQAWe"
access_token<- "45821565-MH8DOws1M3ol6chaKEX9nGOn5Oa7ceSvhYwtWcVzM"
access_secret<- "3RbCpTkMrGdCX7ewMmMhhaI62WAkVrL6qnZZueVqoBJe" setup_
twitter_oauth(consumer_key, consumer_secret, access_token, access_secret)

# The following line of codes requests 2000 tweets which
# contains #ClimateChange hashtag
```

	text	replyToSN	created	truncated	retweetCount	isRetweet	retweeted	longitude	latitude
1	RT @CanadaFP: Minister Champagne spoke to Foreign Minister Casten Nemra of the #M₹	NA	11/06/2020 0:42	FALSE	1	TRUE	FALSE	NA	NA
2	RT @AnikaMolesworth: We must transition quickly & effectively away from	NA	11/06/2020 0:42	FALSE	19	TRUE	FALSE	NA	NA
3	RT @almacardi: Top story: @Hana_ElSayyed: 'power paint	NA	11/06/2020 0:41	FALSE	11	TRUE	FALSE	NA	NA
4	RT @DrJackiSmall: Our lives depends on it #ClimateChange https://t.co/mX6C2jzTCv	NA	11/06/2020 0:41	FALSE	2	TRUE	FALSE	NA	NA
5	Minister Champagne spoke to Foreign Minister Casten Nemra of the #MarshallIslands to	NA	11/06/2020 0:41	TRUE	1	FALSE	FALSE	NA	NA
6	Outside between a full afternoon and evening of events. The wind is warm and strong.	NA	11/06/2020 0:41	TRUE	0	FALSE	FALSE	NA	NA
7	RT @GerberKawasaki: Dear @elonmusk and @Tesla - your sustainability report was ama	NA	11/06/2020 0:41	FALSE	313	TRUE	FALSE	NA	NA
8	The world knew, in a sense, this was coming. COVID-19 is Disease X, according to ANZ ch	NA	11/06/2020 0:40	TRUE	0	FALSE	FALSE	NA	NA
9	#2018 #Conference - #author #AndrewKimbrell; has been a leading proponent of regene	NA	11/06/2020 0:40	TRUE	0	FALSE	FALSE	NA	NA
10	RT @LowyInstitute: Leadership from big-city mayors, CEOs, and others on climate chang	NA	11/06/2020 0:38	FALSE	1	TRUE	FALSE	NA	NA
11	RT @saveearth1928: A TOTALLY #TRANSPARENT TOTALLY #EDUCATION & #EQUALITY	NA	11/06/2020 0:38	FALSE	1	TRUE	FALSE	NA	NA
12	RT @AnikaMolesworth: We must transition quickly & effectively away from	NA	11/06/2020 0:37	FALSE	19	TRUE	FALSE	NA	NA
13	RT @PercievedLogic: A TOTALLY #TRANSPARENT TOTALLY #EDUCATION & #EQUALIT	NA	11/06/2020 0:37	FALSE	1	TRUE	FALSE	NA	NA
14	RT @arikring: @cberrl @DawnRoseTurner @C37H42Cl2N2O6 @RuthPtn @ZacharyPBeasl	NA	11/06/2020 0:37	FALSE	357	TRUE	FALSE	NA	NA
15	RT @AnikaMolesworth: We must transition quickly & effectively away from	NA	11/06/2020 0:36	FALSE	19	TRUE	FALSE	NA	NA
16	RT @blairpalese: The #climatechange induced #bushfire season is on in the northern he	NA	11/06/2020 0:36	FALSE	1	TRUE	FALSE	NA	NA
17	RT @UNinIndia: Though it can be challenging to think beyond immediate recovery amids	NA	11/06/2020 0:36	FALSE	2	TRUE	FALSE	NA	NA
18	RT @alvinfoo: Iceberg falls off the cliff, extremely rare capture clip. #ClimateChange htt	NA	11/06/2020 0:36	FALSE	62	TRUE	FALSE	NA	NA
19	RT @AnikaMolesworth: We must transition quickly & effectively away from	NA	11/06/2020 0:36	FALSE	19	TRUE	FALSE	NA	NA
20	RT @UNinIndia: Over 100.000 people around the world think that #ClimateChange and e	NA	11/06/2020 0:36	FALSE	4	TRUE	FALSE	NA	NA

FIGURE 21.3
Raw sample Tweets.

```
#twiits = searchTwitter("#ClimateChange ", n = 2000, lang="en",
locale="Nigeria") ClimateChange<- searchTwitter("#ClimateChange", n=2000,
lang = "en", locale = "Nigeria") #the funcion twListToDF takes a list of
objects from a single
#twitteR class and return a data frame version of the members
ClimateDAta<- twListToDF(ClimateChange)
ClimateDAta
write.csv(ClimateDAta, "climateChange.csv")#this writes the dataset to disk
```

The tweets containing the #ClimateChange data frame, thus, show that 2,000 elements were harvested. This is then converted to a data frame ClimateDAta of 2,000 cases with 16 variables. Part of the output is as shown in Figure 21.3.

21.10 Retrieving Data from a Given Web Page

The following web page consists of a dataset of a 750×5 array of prison record; this consists of 750 cases with 5 attributes. The following R code, thus, scrapes this dataset from the web page into the object, christened URL as follows.

```
URL <-'https://raw.githubusercontent.com/guru99-edu/R-Programming/
master/prison.csv' #this is a single line from the previous line
myData <- read.csv(URL)[1:5]
myData            #to view/print the content retrieved from the
                   #internet/URL
str(myData)        #to check the structure of the dataset
head(myData, 5)  #to view the first five rows on top
tail (myData, 5) # to view the trailing/last five rows

#TO SAVE THE RETRIEVED DATASET TO A FILE ON YOUR COMPUTER
#SYNTAX, write.csv (source file, 'destination file')
write.csv (myData, "prison.csv")
```

21.11 Retrieving a Local CSV file

Since textual data can be stored in any file format such as CSV, TXT, JSON, RDA, PDF, HTML, or XML, data could be retrieved from a file of any type. Although R natively has support to read CSV and TXT, additional format-specific packages are required to process other file formats (Welbers, 2017). As this approach of working with various packages to access different file formats may be a herculean task, a convenient task to handle this is through a package called 'readtext' as per Welbers (2017). In this context, the 'read.csv' function will be employed to achieve the purpose of this section.

```
install.packages("readr")
 library(readr)
ClimateDAta <- read.csv ("climate.csv", header =TRUE, sep = ", ",
stringsAsFactors = FALSE)
```

Another option is to choose a given file from a local drive, this must of course support the file format of the function used, in this case, read.csv. The following R code could be used to load the file. As the code executes, it will display a dialog box allowing the user to navigate to any given file on the local drive(s).

```
ClimateDAta<- read.csv(file.choose(), sep = ", ", stringsAsFactors =
FALSE, header = TRUE)
```

21.12 Data Preparation

21.12.1 Data Cleaning

The next procedure is to select the required field (variable) from the dataset, which in this case is the tweets, represented by 'text'. The following procedure shows the R implementation to reap the tweets from the dataset.

```
climateText<- paste(ClimateDAta$text, collapse = " ")
climateText
```

Looking at the data harvested on climate change, it is very obvious that there is the need to clean up unnecessary characters and links associated with SM, which may not be useful for text mining analytics. This process is called data cleaning; there is a way to carry this out in R, which the following codes show how to do.

```
# The first stage is to install and load the required packages
install.packages("stringi") #install package
install.packages("tm")
library(stringi) #load package
library(tm)

#This section cleans the data of unnecessary characters
climateClean<- stri_replace_all(climateText, "", regex = "<.*?>")
climateClean<- stri_trim(climateClean)
#creating Vector source and Corpus
```

```
climateVSource<- VectorSource(climateClean)
ClimateCorpus<- VCorpus(climateVSource)

####################################################################
#Removal of urls, special characters and numbers
ClimateCorpus<- tm_map(ClimateCorpus, removePunctuation)
ClimateCorpus<- tm_map(ClimateCorpus, removeNumbers)
hashtagRemoval <- function(x) gsub("#\\S+", "", x)
HandleRemoval <- function(x) gsub("@\\S+", "", x)
shortWordRemoval <- function(x) gsub('\\b\\w{1,5}\\b',' ', x)
urlRemoval <- function(x) gsub("http:[[:alnum:]]*", "", x)
ClimateCorpus<- tm_map(ClimateCorpus, content_
transformer(hashtagRemoval)) ClimateCorpus<- tm_map(ClimateCorpus,
content_transformer(HandleRemoval)) ClimateCorpus<- tm_map(ClimateCorpus,
content_transformer(shortWordRemoval)) ClimateCorpus<- tm_
map(ClimateCorpus, content_transformer(urlRemoval))
ClimateCorpus<- tm_map(ClimateCorpus, removeWords, c("ClimateChange",
"httpstcoEKIZYxdhg", "\U0001f30f", "…"))
####################################################################

# TOKENIZATION AND STEMMING
install.packages("quanteda")
 library(quanteda)
ClimateCorpusChar<- corpus(ClimateCorpus) #reverting the treated corpus #
               #back to character
climateToken<- tokens(ClimateCorpusChar) #tokenise to unigram
climateToken<- tokens_tolower(climateToken)
climateToken <- tokens_wordstem(climateToken)
 stopWds<- stopwords("english")
climateTokenNoSTOP<- tokens_remove(climateToken, stopWds)#stopwords
removal
# Weighting using dfm
dtm<-dfm(climateTokenNoSTOP) dtm<- dfm_remove(dtm, c('#*', '@*'))
dtm<- dfm_remove(dtm, c('rt', 'amp'))
#THE FOLLOWING REMOVES UNWANTED WORDS IN THE CORPUS
dtm<- dfm_remove(dtm, c('climatechang',' climat', 'chang',' …'
,' "',' httpstcowpajmb',' gerberkawasaki',' geraldkutney',' 
pauledawson',' ayanaeliza',' elonmusk',' profstrachan',' gretathunberg',' 
thunberg',' energicaus',' -
'))
frequency <- colSums(dtm)
frequency <- sort(frequency, decreasing=TRUE) head(frequency)
dtm2<-dtm[, frequency>=4] dtm3<-dfm(dtm2)

#WORDCLOUD
textplot_wordcloud(dtm, min_size = 1.2, max_size = 4, min_count = 35,
max_words = 150, color
= "darkblue", font = NULL, rotation = 0.1)
```

21.13 Frequent Terms and *#climatechange* Discourse

One of the main features of text mining is the frequency terms, showing how frequently each term occurs in a given corpus. When terms appear frequently in a given corpus, they

FIGURE 21.4
Wordcloud showing frequent terms.

are likely to provide insights on the subjects of interest. This has been used widely in literature and remains a topical approach today since text mining in itself is a new area of NLP (Dayeen *et al.*, 2020).

In the context of tweets with the climate change hashtag, Figure 21.4 presents the word-cloud. This shows the most frequent related words used with the subject, which include 'sustain', 'impact', 'report', and 'environ'. The word 'sustain' was stemmed from words like 'sustainable', 'sustainability', 'sustained', and 'sustains'. Within the raw data, this reveals the agitation and advocacy of sustainable measures to mitigate the effect of climate change, which is important in Nigeria, where limited resources could be a barrier to appropriate measures. In the same vein, the impact of climate change on the 'environment' is also noted to be a matter of great concern. The wordcloud in Figure 21.4 further provides information about the factors contributing to environmental degradation as codified by 'environment'. Some of these are 'pollut', 'globalwarm', 'emiss', 'deforest', and 'burn'. Further inspection of the raw data shows that pollution from emissions, bush burning, fossil fuel, and deforestation characterizes the issues around climate change in Nigeria's space. Topic models have been designed and used in NLP, which has the propensity to help categorize related words under relevant features. Some of these models are the joint sentiment topic model (Lin and He, 2009) and latent Dirichlet allocation, instituted by Blei, Ng, and Jordan (2003).

21.14 Sentiment Analysis

This section describes the pattern of the tweets on climate change; this shows the polarity of the tweets as *positive*, *negative*, or *neutral* opinions. The procedure powering this underlying feature uses algorithms that combine text analyses, computational linguistics, and NLP to understand the sentiments of the words in the tweets collected. The following procedure is an R programming implementation to plot the sentiment visualizations presented in this work using the syuzhet package in R.

```
# GRAPH PLOTS
library(syuzhet)
s_v <- get_sentences(climateText)
s_v_sentiment <- get_sentiment (s_v, method="bing")
plot(s_v,
                type="h", #other option here is "l"
                "h" main="Climate Change Tweets",
                xlab="Narrow Time",
                ylab="Emotional Valence",
                col="blue")
```

21.15 The Trajectory of Climate Change Discourse

The polarity of the opinions of the respondents as revealed in Figure 21.5 suggests mixed opinions, which range between two equal ends of emotional valence. Some were optimistic that the effects of climate change vis-à-vis the present situation can be put under control, while others expressed fear that the effects may be disastrous. In a situation like this, it is important to compare what other countries of the world, including developing economies, emerging nations, and most especially advanced countries, had done. Adopting best practices for mitigating the effect of climate change seems to be the way forward for developing countries like Nigeria.

21.16 Conclusion

The Third Industrial Revolution was characterized by the advent of many Internet technologies. The Internet technologies themselves, thus, provided the platform by which the emerging technologies thrive. Today, our way of life has been perturbed by the use of SM and its related technologies. This has been beneficial to humankind, especially in the way we communicate, interact, and share information. Despite the ubiquitous benefits that this

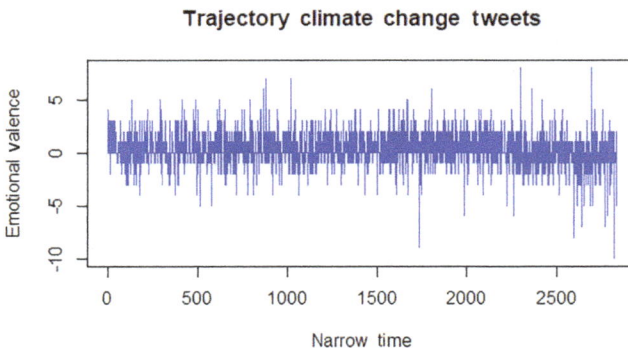

FIGURE 21.5
Trajectory of climate change tweets.

has brought to us, the challenge remains of the avalanche of data being generated daily by the use of these technologies. Improper management of this has been noted to have created bottlenecks in the way information is gathered and presented for decision-making in organizations. This is what necessitates a body of knowledge on IR in efforts to continually seek strategies and methods to retrieve useful information in the most effective way. This chapter, therefore, contributes to these efforts by using the NLP techniques via R implementation to mine the characteristics of useful information within the domain of climate change as it relates to the Nigerian cyberspace.

References

Adomavicius, G., Tuzhilin, A. (2005). Toward the next generation of recommender systems: a survey of the state-of-the-art and possible extensions. *IEEE Transactions on Knowledge and Data Engineering*, 17(6), 734–749.

Agerri, R. Artola, X., Beloki, Z., Rigau, G., Soroa, A. (2015). Big data for natural language processing: a streaming approach. *Knowledge-Based Systems*, 79, 36–42. https://doi.org/10.1016/j.knosys.2014.11.007.

Ahmed, W. (2017). Amplified messages: how hashtag activism and Twitter diplomacy converged at #Thi-sIsACoup – and won. In Veneti, A., Reilly, P., and Atanasova, D., eds. *Politics, Protest, Emotion: Interdisciplinary Perspectives. A Book of Blogs*, 109–114. Sheffield: University of Sheffield Information School.

Ahmed, W., Bath, P. (2015). The Ebola epidemic on twitter: challenges for health informatics. *Proceeding of 17th International Symposium on Health Information Management Research*, York, UK. Available on http://eprints.whiterose.ac.uk/87728/2/Abstract%2034.pdf.

Awoleye, O.M., Ojuloge, B., Ilori, M.O. (2014). Web application vulnerability assessment and policy direction towards a secure smart government. *Government Information Quarterly*, 31(1), S118–S125,

Baltas, A., Kanavos, A., Tsakalidis, A. (2016). An Apache Spark implementation for sentiment analysis on Twitter data. In *Proceedings of the International Workshop on Algorithmic Aspects of Cloud Computing (ALGOCLOUD)*, Aarhus, Denmark.

Blei, D., Ng, A.Y., Jordan, M.I. (2003). Latent Dirichlet allocation. *Journal of Machine Learning Research*, 3, 993–1022. Available on http://www.jmlr.org/papers/volume3/blei03a/blei03a.pdf.

Bose, R., Dey, R.K., Roy, S., Sarddar, D. (2019). Analyzing political sentiment using Twitter data. In *Information and Communication Technology for Intelligent Systems. Smart Innovation, Systems and Technologies*, 107.

Chae, B.K. (2015). Insights from hashtag #supplychain and Twitter analytics: Considering Twitter and Twitter data for supply chain practice and research. *International Journal of Production Economics*, 165, 247–259.

Dayeen, F.R., Sharma, A.S., Derrible, S. (2020). A text mining analysis of the climate change literature in industrial ecology. *Journal of Industrial Ecology*, 24(2). Available on https://doi.org/10.1111/jiec.12998.

Ferraiolo, D., Barkley, J., Kuhn, R. (1999). A role-based access control model and reference implementation within a corporate Internet. *ACM Transactions on Information and System Security*, 2(1), 34–64.

Glen, D. (2008). Opinion piece: social media: should marketers engage and how can it be done effectively? *Journal of Direct, Data and Digital Marketing Practice*, 9, 274–277.

Kaplan, A., Haenlein, M. (2010). Users of the world, unite! The challenges and opportunities of social media. *Business Horizons*, 53(1), 59–68.

Kim, Y., Nordgren, R., Emery, S. (2020). The story of Goldilocks and three Twitter's APIs: a pilot study on Twitter data sources and disclosure. *International Journal of Environmental Research and Public Health*, 17, 864. https://doi.org/10.3390/ijerph17030864.

Kumar, M., Bala, A. (2016). Analyzing Twitter sentiments through Big Data. *3rd International Conference on Computing for Sustainable Global Development (INDIACom'16)*. IEEE, New Delhi.

Lang, K. (1995). NewsWeeder: learning to filter Netnews. *Proceedings of ICML-95, 12th International Conference on Machine Learning*, Tahoe, CA, pp. 331–339.

Lin, C., He, Y. (2009). Joint sentiment/topic model for sentiment analysis CIKM '09: *Proceedings of the 18th ACM Conference on Information and Knowledge Management*, November 2009, pp. 375–384. https://doi.org/10.1145/1645953.1646003.

McMurdie, P.J., Holmes, S. (2013). phyloseq: an R package for reproducible interactive analysis and graphics of microbiome census data. *PLoS One*, 8(4), e61217. https://doi.org/10.1371/journal.pone.0061217.

Oltulu, P., Mannan, R., Gardner, J.M. (2018). Effective use of Twitter and Facebook in pathology practice. *Human Pathology*, 73, 128–143.

Onyijen, O.H., Awoleye, O.M., Olaposi, T.O. (2019). Effectiveness of social media platforms for product marketing in Southwestern Nigeria: a firm-level analysis. *International Journal of Development and Management Review*, 14(1), 175–192.

Oren, N. (2002). Reexamining tf.idf based information retrieval with genetic programming. *Proceedings of the 2002 Annual Research Conference of the South African Institute of Computer Scientists and Information Technologists on Enablement through Technology (SAICSIT)*, pp. 224–234.

Pant, G., Srinivasan, P., Menczer, F. (2004). 'Crawling the Web', *Web Dynamics: Adapting to Change in Content, Size, Topology and Use*, pp. 153–178. Berlin: Springer-Verlag.

Polettini, N. (2004). The vector space model in information retrieval-term weighting problem. *Entropy*, 1–9. Available on http://citeseerx.ist.psu.edu/viewdoc/download?doi= 10.1.1.104.3479&rep=rep1&type=pdf.

Procter, R., Vis, F., Voss, A. (2013). Reading the riots on Twitter: methodological innovation for the analysis of big data. *International Journal of Social Research Methodology*, 16(3), 197–214.

Salton, G. (1989). *Automatic Text Processing: The Transformation, Analysis, and Retrieval of Information by Computer*. Boston, MA: Addison Wesley.

Shkapenyuk, V., Suel, T. 2002. Design and implementation of a high-performance distributed web crawler. *Proceedings of the International Conference on Data Engineering*. Available on citeseer.ist.psu.edu/shkapenyuk02design.html.458,530,531.

Takahashi, B., Tandoc, E.C., Carmichael, C. (2015). Communicating on Twitter during a disaster: an analysis of Tweets during Typhoon Haiyan in the Philippines. *Computers in Human Behavior*, 50, 392–398.

Vitolo, C., Elkhatib, Y., Reusser, D., Macleod, C.J.A., Buytaert, W. (2015). Web technologies for environmental big data. *Environmental Modelling & Software*, 63, 185–198, https://doi.org/10.1016/j.envsoft.2014.10.007.

22

Control Charts and Capability Analysis for Statistical Process Control

Arun Kumar Sinha and Richa Vatsa
Central University of South Bihar (CUSB)

CONTENTS

DOI: 10.1201/9781003261148-26

22.1 Introduction

This chapter has four main sections. Section 22.2 begins with the basics of control charts, which also includes the justifications of three-sigma limits. A control chart is a statistical tool to detect whether a process is under statistical control. Control charts for statistical process control (SPC) are described in Section 22.3. As the appropriateness of a control chart depends on the availability of the data set, the types of data are explained. This helps determine the suitability of statistical theory for a desired control chart. Section 22.4.1 describes variable control charts with such details as the estimators of parameters, because standard values are usually not easily available. Sections 22.4.5 and 22.4.7 contain attribute control charts, including p, c, and u charts. Illustrations of all control charts are added. In Section 22.5, the capability analysis of the process control is explained with an illustration. The current scenarios of applications of SPC in developing countries are discussed in Section 22.6.

22.2 Basics of Control Chart

A control chart is a statistical technique to detect the presence of assignable causes of variation. In general, a control chart has three horizontal lines that include (1) a central (middle) line (CL) to show the desired standard, (2) an upper control limit (UCL), just above the CL, and (3) a lower control limit (LCL), just below the CL. The CL is always a horizontal line, while UCL and LCL may or may not be horizontal ones. However, the distance between the CL and UCL is the same as that of between CL and LCL. If t denotes a statistic of the characteristic of interest, we have

$$UCL = E(t) + 3SE(t), \ CL = E(t) \text{ and } LCL = E(t) - 3SE(t),$$

where $E(t)$ denotes the expected value of t, while $SE(t)$ represents the standard error of t.

This is referred to as the control chart based on three-sigma limits because $3SE(t)$ is added to $E(t)$ for the UCL and subtracted from $E(t)$ for the LCL. To draw a control chart, sample number is plotted on the X-axis and the values of t on the Y-axis. Further, suppose there are k samples; t is then computed for each sample, which is plotted on the Y-axis and the k sample numbers are on the X-axis. For each value of k, we have a separate value of t, and these are plotted on a two-dimensional graph.

22.2.1 Justification of Three-Sigma Limits

The control charts are based on three-sigma limits. First, we consider a normal population. In this situation, the probability is 0.0027 that a point will fall outside the control limits, that is, above the UCL or below the LCL, even though the process is in control. This is known as the probability of committing type-I error. In other words, this is the probability of rejecting a correct hypothesis. Similarly, if the process goes out of control, there is a probability greater than zero that a point will lie within the control limits. This is referred to as the probability of type-II error. This denotes the probability of accepting a wrong hypothesis. These two errors are related to each other. A decrease in one results in increase in the other and vice versa.

Under three-sigma limits, the probability of type-I error is 0.0027. If two-sigma limits are considered, the probability of type-I error is increased to 0.0455, because the area between −2 and 2 values of standard normal variate is 0.9545 and the value of type-I error $(\alpha)=1-0.9545=0.0455$, but the probability of type-II error is decreased. If we consider four-sigma limits, the probability of type-I error $(\alpha)=1-$ area between −4 and 4 values of standard normal variates $=1-0.9980=0.0020$.

On comparing the probabilities of type-I errors under two-sigma, three-sigma, and four-sigma, limits are obtained as 0.0455, 0.0027, and 0.0020, respectively. As the probabilities of type-I errors decrease, the probabilities of type-II errors increase, so we accept the three-sigma limits as a compromise between two-sigma and four-sigma limits. Considering these facts, the concept of three-sigma limits has been observed to function reasonably well for all practical situations.

For a non-normal population, we use Tchebycheff's inequality for $k>0$, because this holds for any distribution. It is given below:

$$\text{Prob}\left[|t-E(t)|<k\sigma\right]\geq 1-1/k^2. \text{ On entering } k=3 \text{ we get}$$

$$\text{Prob}\left[|t-\mu_t|<3\sigma_t\right]\geq 1-1/9=0.9 \left(\text{approximately}\right),$$

where $E(t)=\mu_t$ and σ_t represents the standard deviation of t.

This is a fairly high probability for practical purposes. This, in turn, suggests that even for non-normal populations, three-sigma limits are almost universally used because they have been found to be the most acceptable empirically, as they provide excellent protection against type-I and type-II errors.

The central limit theorem states that the mean (arithmetic mean) asymptotically follows a normal distribution. This, in turn, supports the use of the normal distribution in the construction of control charts. More specifically, if $x_1, x_2, \ldots, x_n$ are independent random variables with mean μ_i and variance σ_i^2 and if $y=x_1+x_2+\ldots+x_n$, the distribution

$$\frac{Y-\sum_{i=1}^{n}\mu_i}{\sum_{i=1}^{n}\sigma_i^2}=\frac{\bar{y}-\bar{\mu}}{\sqrt{\sum_{i=1}^{n}\frac{\sigma_i^2}{n^2}}}$$

approaches the $N(0, 1)$ distribution as n approaches infinity.

22.3 Control Charts for Statistical Process Control

A control chart is a graphical statistical technique to examine whether a process is under statistical control. If a point exceeds the UCL or falls below the LCL, this suggests that the process is out of statistical control. These scenarios, in turn, indicate the presence of one or more special (assignable) causes of variations, which have led the process out of statistical control. Though this statistical tool was initially developed to check the current status of a manufacturing process by Shewhart (1931), with the passage of time this technique has become very popular for examining the process of service sectors as well.

A process is said to be under statistical control if all points (statistics) lie within the UCL and LCL, but if a point exceeds UCL or falls below the LCL, the process is said to be out of statistical control. In this situation, all points outside the control limits are removed from the given data set and fresh control limits with the remaining observations are computed. Again, the remaining points are plotted with the new control limits. The process is continued until all points lie within the control limits. This is referred to as the final control chart, while the first control chart is known as the trial control chart.

First we discuss the types of data that we encounter in real-life situations, and then the appropriate control charts for each type of data sets are explained, pointing out their statistical backgrounds.

22.4 Types of Data and Control Charts

We usually come across the following three types of data in both the manufacturing and service sectors.

   i. **Data**: Variable or measurable data, **Charts**: $\bar{x}$, R, and sigma charts
  ii. **Data**: Attribute data (fraction defective or fraction nonconforming), **Chart**: p chart
 iii. **Data**: Number of count of defects or nonconformities (defects), **Chart**: c chart

Note that some experts of statistical quality control consider only two types of data, measurable and attribute, but attribute data consist of the fraction nonconforming (ii) and nonconformities (iii). See Montgomery (2009) and Mitra (2008) for more details.

22.4.1 Variables Control Charts

For the above data, two types of control charts are used, which are referred to as control charts for (1) process averages and (2) process variability. This means that one must decide one's preference to examine whether a process is under control with respect to process average or process variability. In general, we first examine whether a process is under control with respect to process variability, and once the process is found to be under statistical control then the status of process average is examined.

For process average, we have two types of $\bar{x}$ charts, those

   i. involving the range R, and
  ii. involving the standard deviation σ.

For process variability, we have two charts, the

   i. R chart and
  ii. Sigma chart.

22.4.2 Data Structure for $\bar{x}$ Control Chart

Suppose we have k samples of equal size n from a population. The data set is given below.

Sample No.	Observations	Mean	Range	S	s
1	$x_{11}, x_{12}, \dots, x_{1n}$	$\bar{x}_1$	R_1	S_1	s_1
2	$x_{21}, x_{22}, \dots, x_{2n}$	$\bar{x}_2$	R_2	S_2	s_2
.					
.					
.					
i	$x_{i1}, x_{i2}, \dots, x_{in}$	$\bar{x}_i$	R_i	S_i	s_i
.	.	.	.		
.	.	.	.		
.	.	.	.		
K	$x_{k1}, x_{k2}, \dots, x_{kn}$	$\bar{x}_k$	R_k	S_k	s_k

Let x_{ij} denote the jth measurement of the ith sample, $j=1, 2, \dots, n$ and $i=1, 2, \dots, k$. The mean of the ith sample is denoted by $\bar{x}_i$ and it is computed as shown below.

$$\bar{x}_i = \frac{\sum\limits_{1}^{n} x_{ij}}{n}$$

The range of the ith sample is represented by R_i and is obtained by taking the difference between its highest and the lowest values of the measurements.

$$R_i = \text{Max}(x_{ij}) - \text{Min}(x_{ij}) \cdot S_i = \sqrt{S_i^2} \quad \text{where} \quad S_i^2 = \frac{\sum\limits_{1}^{n}(x_i - \bar{x})^2}{n}. \quad \text{Similarly,} \quad s_i = \sqrt{S_i^2} \quad \text{and}$$

$$S_i^2 = \frac{\sum\limits_{1}^{n}(x_i - \bar{x})^2}{n-1}.$$

Suppose μ denotes the population mean and σ_2 represents the population variance. Then $\text{var}(\bar{x}) = \sigma^2/n$ and $SE(\bar{x}) = \sigma/\sqrt{n}$, where SE stands for standard error.

22.4.3 Three-Sigma Control Limits for the $\bar{x}_i$ Control Chart

$$\text{Upper Control Limit (UCL)} = E(\bar{x}) + 3\, SE(\bar{x}) = \mu + 3\sigma/\sqrt{n}$$

$$\text{Central Line (CL)} = E(\bar{x}) + \mu$$

$$\text{Lower Control Limit (LCL)} = E(\bar{x}) - 3\, SE(\bar{x}) = \mu - 3\sigma/\sqrt{n}$$

Thus, we have

$$\text{UCL} = \mu + A\sigma$$

$$\text{CL} = \mu$$

$$\mathrm{LCL} = \mu - A\sigma$$

where $A = 3\sigma/\sqrt{n}$.

Note that μ and σ are parameters (standards), which are usually not known. Thus, they are, in turn, replaced by their estimators.

Estimator of μ

μ is estimated by $\bar{x}_i$, which is the grand mean of the k means. It is given by

$$\bar{x}_i = (\bar{x}_1 + \bar{x}_2 + \ldots + \bar{x}_k)/k$$

where $\bar{x}_i$ is the mean of the ith sample.

Estimators of σ

i. σ is estimated by $\bar{S}/c_2$ and $\bar{S}$ is defined as

$$\bar{S} = (S_1 + S_2 + \ldots + S_k)/k$$

and C_2 is a constant depending on the sample size n.
   Using these estimators, we obtain the following control limits for the variables.

$$\mathrm{UCL} = \bar{\bar{x}} + 3\bar{S}/\left(c_2\sqrt{n}\right) = \bar{\bar{x}} + A_1\bar{S}$$

$$\mathrm{CL} = \bar{\bar{x}}$$

$$\mathrm{LCL} = \bar{\bar{x}} + 3\bar{S}/\left(c_2\sqrt{n}\right) = \bar{\bar{x}} - A_1\bar{S}$$

where $A_1 = 3/\left(c_2\sqrt{n}\right)$
ii. σ is also estimated by $\bar{R}/d_2$ where

$$\bar{R} = (R_1 + R_2 + \ldots + R_k)/k$$

and d_2 is a constant that depends on the sample size n.
   With these estimators, we obtain the following control limits for variables.

$$\mathrm{UCL} = \bar{\bar{x}} + 3\bar{R}/\left(d_2\sqrt{n}\right) = \bar{\bar{x}} + A_2\bar{R}s$$

$$\mathrm{CL} = \bar{\bar{x}}$$

$$\mathrm{LCL} = \bar{\bar{x}} - 3\bar{R}/\left(d_2\sqrt{n}\right) = \bar{\bar{x}} - A_2\bar{R}$$

where $A_2 = 3/\left(d_2\sqrt{n}\right)$
iii. When σ is estimated by $\bar{s}/c_4$, where $\bar{s}$ denotes the mean of $s_1, s_2, \ldots, s_k$ and c_4 is a constant that depends on the sample size n, we have the following control limits.

$$UCL = \bar{\bar{x}} + 3\bar{s}/(c_4\sqrt{n}) = \bar{\bar{x}} + A_3\bar{S}$$

$$CL = \bar{\bar{x}}$$

$$UCL = \bar{\bar{x}} - 3\bar{s}/(c_4\sqrt{n}) = \bar{\bar{x}} - A_3\bar{S}$$

where $A_3 = 3/(c_4\sqrt{n})$

To detect the lack in the process variability, we use R and σ charts. To obtain the control limits of the R chart, we proceed as follows.

$$UCL = E(R) + 3\ SE(R) = d_2\sigma + 3d_3\sigma = (d_2 + 3d_3)\sigma = D_2\sigma$$

$$CL = E(R) = d_2\sigma$$

$$LCL = E(R) - 3SE(R) = d_2\sigma - 3d_3\sigma = (d_2 - 3d_3)\sigma = D_1\sigma$$

As σ is also estimated by $\bar{R}/d_2$, we obtain

$$UCL = (1 + 3d_3/d_2)\bar{R} = D_4\bar{R}$$

$$CL = \bar{R}$$

$$LCL = (1 - 3d_3/d_2)\bar{R} = D_3\bar{R}$$

The values of d_2, d_3, D_1, D_2, D_3, and D_4 are available for values of $n=2, 3, \dots, 25$ or smaller values of n in almost all standard textbooks of statistics that contain a chapter on statistical quality control. However, we have seen that

$$D_1 = (d_2 - 3d_3),\ D_2 = (d_2 + 3d_3),\ D_3 = (1 - 3\ d_3/d_2),\ D_4 = (1 + 3\ d_3/d_2).$$

See Montgomery (2009), Mitra (2008), and Bowker and Lieberman (1972) for more detailed discussions.

22.4.4 Illustrations of *R* and $\bar{x}$ Control Charts

To study the aspect of service quality, regarding the amount of time it takes to deliver luggage (as measured from the time the guest completes check-in procedures to the time the luggage arrives in the guest's room), data were recorded over a 4-week period. Subgroups of five deliveries were selected from the evening shift on each day for analysis. The data set given below summarizes the result for all 28 days. The hotel management has instituted a policy that 99% of all luggage deliveries must be completed in 14 minutes or less.

Day	Luggage Delivery Times in Minutes				
1	6.7	11.7	9.7	7.5	7.8
2	7.6	11.4	9.0	8.4	9.2
3	9.5	8.9	9.9	8.7	10.7
4	9.8	13.2	6.9	9.3	9.4

(Continued)

Day	Luggage Delivery Times in Minutes				
5	11.0	9.9	11.3	11.6	8.5
6	8.3	8.4	9.7	9.8	7.1
7	9.4	9.3	8.2	7.1	6.1
8	11.2	9.8	10.5	9.0	9.7
9	10.0	10.7	9.0	8.2	11.0
10	8.6	5.8	8.7	9.5	11.4
11	10.7	8.6	9.1	10.9	8.6
12	10.8	8.3	10.6	10.3	10.0
13	9.5	10.5	7.0	8.6	10.1
14	12.9	8.9	8.1	9.0	7.6
15	7.8	9.0	12.2	9.1	11.7
16	11.1	9.9	8.8	5.5	9.5
17	9.2	9.7	12.3	8.1	8.5
18	9.0	8.1	10.2	9.7	8.4
19	9.9	10.1	8.9	9.6	7.1
20	10.7	9.8	10.2	8.0	10.2
21	9.0	10.0	9.6	10.6	9.0
22	10.7	9.8	9.4	7.0	8.9
23	10.2	10.5	9.5	12.2	9.1
24	10.0	11.1	9.5	8.8	9.9
25	9.6	8.8	11.4	12.2	9.3
26	8.2	7.9	8.4	9.5	9.2
27	7.1	11.1	10.8	11.0	10.2
28	11.1	6.6	12.0	11.5	9.7

Source: Levine et al. (2016). The data set was collected under *Analyze* phase of the Measure phase of the Six Sigma Project of Beachcomber Hotel.

We need

   i. to draw charts to examine the current status of the hotel service quality.

   ii. to examine whether the process is capable of meeting the 99% goal set forth by the hotel management.

   iii. to calculate the CPU for measuring the process performance.

Solution:

   i. To investigate the current status of hotel service quality, we need to draw R and X bar charts first. First the R chart is to be drawn to determine whether the process is in control with respect to the process variability. Then whether the process is in control with respect to the process average is examined. The control limits for the R chart are given below.

$$UCL = D_4\bar{R}$$

$$CL = \bar{R}$$

$$LCL = D_3\bar{R}$$

We need to calculate the range R for each day of luggage delivery time and then $\bar{R}$, the arithmetic mean of all 28 R values, is obtained. The values of $D_3=0.000$ and $D_4=2.114$ for $n=5$ were noted from a standard table of constants. Here, $n=5$ because each day 5 deliveries of luggage times were recorded.

The control limits require the arithmetic mean of ranges of the deliveries of luggage times for all 28 days, $\bar{R}$, but for drawing the R chart the separate value of R for each day is required. The control limits of x bar are given below.

$$UCL = \bar{\bar{x}} + A_2\bar{R}$$

$$CL = \bar{\bar{x}}$$

$$LCL = \bar{\bar{x}} - A_2\bar{R}$$

These require $\bar{\bar{x}}$, the arithmetic mean of all means of the delivery times for 28 days together, but drawing the x bar chart requires the arithmetic mean of delivery times for each day separately. The drawing of both charts requires the mean and range of delivery of luggage time of each day. These are shown in Table 22.1.

TABLE 22.1

Mean and Range of the Delivery of Luggage Times for all 28 days

Day	Mean	Range
1	8.68	5.0
2	9.12	3.8
3	9.54	2.0
4	9.72	6.3
5	10.46	3.1
6	8.66	2.7
7	8.02	3.3
8	10.04	2.2
9	9.78	2.8
10	8.80	5.6
11	9.58	2.3
12	10.00	2.5
13	9.14	3.5
14	9.30	5.3
15	9.96	4.4
16	8.96	5.6
17	9.56	4.2
18	9.08	2.1
19	9.12	3.0
20	9.78	2.7
21	9.64	1.6
22	9.16	3.7
23	10.30	3.1
24	9.86	2.3
25	10.26	3.4
26	8.64	1.6
27	10.04	4.0
28	10.18	5.4

22.4.5 The R Chart

We obtain $\bar{R} = (R_1 + R_2 + \ldots + R_k)/k = 3.482$

The control limits for the R chart are obtained below.

$$\text{UCL} = D_4\bar{R} = 2.114 * 3.482 = 7.361, \ \text{CL} = 3.482, \ \text{and} \ \text{LCL} = D_3\bar{R} = 0$$

An R chart for luggage delivery times is depicted below using Minitab 19.

Figure 22.1 does not indicate any range value outside the control limits or any clear patterns. This means that the process is in statistical control, free from any special cause of variation.

22.4.6 The $\bar{x}$ Chart

As the R chart shows that the process is in control with respect to the process variability, we next plot the $\bar{x}$ chart to examine the current status of the process with respect to the process average. Figure 22.2 displays the $\bar{x}$ chart for the luggage delivery times.

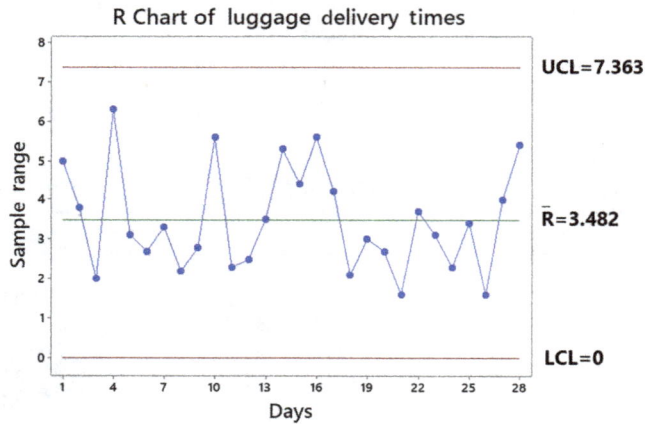

FIGURE 22.1
R chart for luggage delivery times.

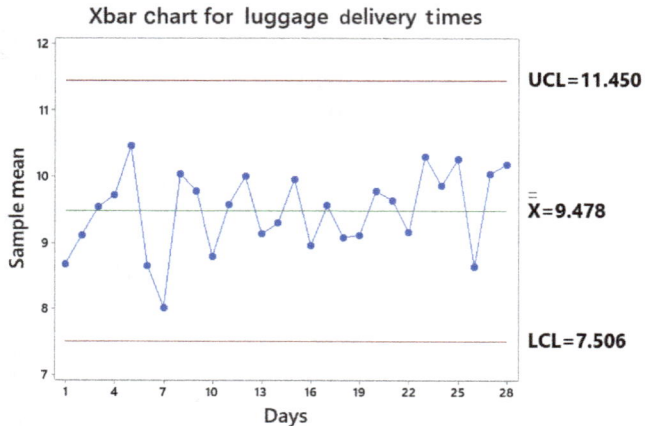

FIGURE 22.2
$\bar{x}$ chart for the luggage delivery times.

For the control limits of $\bar{x}$ chart, we have $\bar{\bar{x}} = 265.38/28 = 9.478$, $A_2 = 0.577$, $\bar{R} = 3.482$. Finally, we compute

$$UCL = 9.478 + 0.577 * 3.482 = 11.487, CL = 9.478, LCL = 9.478 - 0.577 * 3.482 = 7.469.$$

The control chart is drawn using Minitab 19.

Figure 22.2 does not disclose an average value of the luggage delivery time for any of the 28 days outside the control limits. This, in turn, indicates that the process is in statistical control with respect to the process average. As both the R and $\bar{x}$ charts are in control, we conclude that the luggage delivery process is in a state of statistical control.

The rest of the illustrations, (ii) and (iii), are explained under the illustration of Capability Analysis of Section 22.5.

22.4.7 Control Chart for Standard Deviation

Neither R nor σ has a normal distribution, though these functions are random variables. The probability distribution of σ is related to the χ^2 distribution and the distribution of R is approximately related to the χ^2 distribution. Note that almost all of the probability distributions are contained within the expected value of these variables ± 3 standard error of the variables. For this reason, it is necessary to calculate the expected value and its standard deviation/standard error for a control chart.

If σ^2 is the unknown variance of a population distribution, its unbiased estimator is the sample variance s^2, which is defined by $\sum_i^n (x_i - \bar{x})^2 / n - 1$.

Note that the sample standard deviation s is not an unbiased estimator of σ. We have $E(s) = c_4 \sigma$ and $SE(s) = \sigma \left(1 - c_4^2\right)^{1/2}$, where c_4 is a constant that depends on n.

Now, $USL = E(s) + 3SE(s) = c_4 \sigma + 3\sigma \left(1 - c_4^2\right)^{1/2} = B_6 \sigma$ where $B_6 = \left(c_4^2 + 3\left(1 - c_4^2\right)^{1/2}\right)$ $CL = c_4 \sigma$ and 1

$$LCL = E(s) - 3SE(s) = c_4 \sigma - 3\sigma \left(1 - c_4^2\right)^{1/2} = B_5 \sigma, \quad \text{where } B_5 = \left(c_4 - 3\left(1 - c_4^2\right)^{1/2}\right).$$

This works when σ is available. If σ is not available, it is estimated by $\bar{s}/c_4$, where $\bar{s} = (s_1 + s_2 + \ldots + sk)/k$. Now we have the following control limits: $UCL = B_4 \bar{s}$, where $B_4 = B_6/c_4$

$$CL = \bar{S}$$

$LCL = B_3 \bar{s}$, where $B_3 = B_5/c_4$

The values of B_3, B_4, B_5, B_6, and c_4 are given for various values of n. For more details, see Montgomery (2009) and Mitra (2008).

22.4.8 Control Charts for Attributes

p Chart or control chart for Fraction Defective (Proportion Nonconforming).

For the p chart, characteristics are observed by classifying the items into two classes or groups, usually defective or nondefective. We use the binomial distribution to obtain the mean and standard deviation of the proportion of defective items. Suppose P denotes the probability of an item being defective, while $(1 - P)$ shows the probability of an item being nondefective.

The probability of obtaining exactly d defectives in a sample of size n items inspected is given by $P_d = \begin{pmatrix} n \\ d \end{pmatrix} P^d (1-P)^{n-d}, \quad d = 0, 1, \ldots, n.$

Thus, we have $E(d) = np$ and $\mathrm{var}(d) = nP(1-P)$. Here, P denotes the proportion of defective items in the population, and it is usually unknown.

Let us suppose that the proportion of defective items in the sample of size n is given by $p = \dfrac{d}{n}.$

This means that d defective items were found out of n items inspected.

Then we have $E\left(\dfrac{d}{n}\right) = P$ and $\mathrm{var}\left(\dfrac{d}{n}\right) = \dfrac{P(1-P)}{n}$. Thus, the control limits are given below.

$$\mathrm{UCL} = P + 3\sqrt{\frac{P(1-P)}{n}}$$

$$\mathrm{CL} = P$$

$$\mathrm{LCL} = P - 3\sqrt{\frac{P(1-P)}{n}}$$

As P is usually unknown, it is estimated by $\bar{p}$, where
$\bar{p} = (d_1 + d_2 + \ldots + d_k)/nk$, where d_i defective items were found in the ith sample. Thus, on replacing P by $\bar{p}$ we obtain the following control limits.

$$\mathrm{UCL} = \bar{p} + 3\sqrt{\frac{\bar{p}(1-\bar{p})}{n}}$$

$$\mathrm{CL} = \bar{p}$$

$$\mathrm{LCL} = \bar{p} - 3\sqrt{\frac{\bar{p}(1-\bar{p})}{n}}$$

22.4.9 Case of Variable Sample Sizes

When the samples are not of the same sizes, the control limits are modified as shown below.

$$\mathrm{UCL} = \bar{p} + 3\sqrt{\frac{\bar{p}(1-\bar{p})}{n_i}}$$

$$\mathrm{CL} = \bar{p}$$

$$\mathrm{LCL} = \bar{p} - 3\sqrt{\frac{\bar{p}(1-\bar{p})}{n_i}}$$

Note that the above control limits have n_i in place of n. These will result in zigzag UCL and LCL instead of straight lines for a fixed value of the sample size, n. But CL will always be a straight line. Here,

$$\bar{p} = (d_1 + d_2 + \ldots + d_k)/(n_1 + n_2 + \ldots + n_k)$$

where d_i represents the number of defective items in the ith sample of size n_i, $i = 1, 2, \ldots, k$. Another approach is to use the average sample size as $\bar{n}$, where $\bar{n}$ denotes the average of n_1, $n_2, \ldots, n_k$. This leads to an approximate set of control limits. See Montgomery (2009) for a more detailed discussion of this issue.

22.4.10 Percentage Chart

In this case, UCL, CL, and LCL are multiplied by 100. Then we have

$$UCL = 100 \left(\bar{p} + 3\sqrt{\frac{\bar{p}(1 - \bar{p})}{n}} \right)$$

$$CL = 100\bar{p}$$

$$LCL = 100 \left(\bar{p} - 3\sqrt{\frac{\bar{p}(1 - \bar{p})}{n}} \right)$$

22.4.11 np or d Chart (No. of Defective Chart)

In this case, UCL, CL, and LCL are multiplied by n because $p = d/n$ or $d = np$. We then have

$$UCL = n\bar{p} + 3\sqrt{n\bar{p}(1 - \bar{p})}$$

$$CL = n\bar{p}$$

$$CL = n\bar{p} - 3\sqrt{n\bar{p}(1 - \bar{p})}$$

In case we obtain a negative value of LCL, we consider it as zero. See Montgomery (2009), Mitra (2008), and Bowker and Lieberman (1972) for more details.

22.4.12 Illustration of p Chart

For 32 days, 500 film canisters were sampled and inspected. Table 22.2 lists the number of defective canisters (the nonconforming items) for each day (the subgroup). Construct a p chart and examine the state of statistical control.

Solution:
   Here $n = 500$,
   $\bar{p}$ = Total number of nonconforming items/Total items inspected = 761/500*32 = 761/16,000 = 0.0476.
   UCL = 0.0761
   CL = 0.0476
   LCL = 0.0190
   As all points lie between UCL and LCL, we conclude that the process is in a state of statistical control (Figure 22.3).

TABLE 22.2

Number of Defective Canisters Per Day

Day	Number Nonconforming	Day	Number Nonconforming
1	26	17	23
2	25	18	19
3	23	19	18
4	24	20	27
5	26	21	28
6	20	22	24
7	21	23	26
8	27	24	23
9	23	25	27
10	25	26	28
11	22	27	24
12	26	28	22
13	25	29	20
14	29	30	25
15	20	31	27
16	19	32	19

Source: Levine et al. (2016).

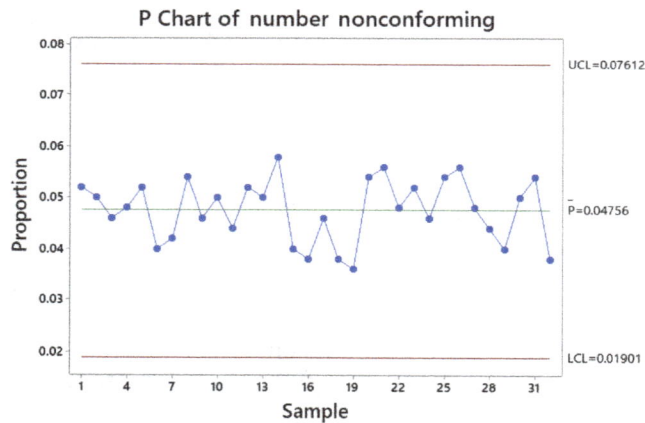

FIGURE 22.3
p chart of number nonconforming.

22.4.13 Control Charts for Nonconformities (Defects) or c Chart

The failure to meet a specification at a point results in a defect or nonconformity. An item or a product having at least one defect or nonconformity is referred to as defective or nonconforming item. For this, a p chart is drawn to examine whether a process is in a state of statistical control with respect to fraction nonconforming items. Conversely, a c chart is plotted for defects or nonconformities per unit. The unit considered may be a single item or a group of items, part of an item, etc. There are numerous opportunities for defects to occur on an item, but the probability of the occurrence of a defect is very small. For these reasons, the statistical theory of the c chart is based on the Poisson distribution.

22.4.14 Statistical Theory of c Chart

The probability of locating c defects on an item where the number of defects c follows the Poisson distribution is given by

$$p(x) = e^{-c}c^x/x!, \quad x = 0, 1, 2, \ldots$$

where x denotes the number of nonconformities and $c > 0$ is the parameter of the Poisson distribution.

As mean = variance in this distribution, mean = c and variance = c. Thus, the control limits for nonconformities or c chart with standard given are given below.

$$\text{UCL} = c + 3\sqrt{c}$$

$$\text{CL} = c$$

$$\text{LCL} = c - 3\sqrt{c}$$

As c, the parameter of the Poisson distribution, is not known, it is estimated by the average value of the observed values of the nonconformities. Let us represent it by $\bar{c}$. Thus, the control limits of the c chart with the estimated value of c are mentioned below.

$$\text{UCL} = \bar{c} + 3\sqrt{\bar{c}}$$

$$\text{CL} = \bar{c}$$

$$\text{LCL} = \bar{c} - 3\sqrt{\bar{c}}$$

22.4.15 Control Chart for Average Number of Nonconformities per Unit or u Chart

Suppose there are x nonconformities in sample of n inspection units. Then the average number of nonconformities per inspection unit n is given by $\frac{x}{n}$. Using these values of u, we obtain $\bar{u} = (u_1 + u_2 + \ldots + u_k)/k$ and $\text{var}(\bar{u}) \frac{\bar{u}}{n}$

$$\text{UCL} = \bar{u} + 3\sqrt{\frac{\bar{u}}{n}}$$

$$\text{CL} = \bar{u}$$

$$\text{LCL} = \bar{u} - 3\sqrt{\frac{\bar{u}}{n}}$$

This per-unit chart is referred to as the control chart for the average number of nonconformities per unit or u chart.

22.4.16 Procedures with Variable Sample Size

If the sample size varies from sample to sample, then n is replaced by n_i or $\bar{n}$ where $\bar{n} = (n_1 + n_2 + \ldots + n_k)/k$. See Montgomery (2009) and Mitra (2008) for more details.

TABLE 22.3

Number of Nonconformities Per Unit

Time	Nonconformities per Unit	Time	Nonconformities per Unit
1	25	6	15
2	11	7	12
3	10	8	10
4	11	9	9
5	6	10	6

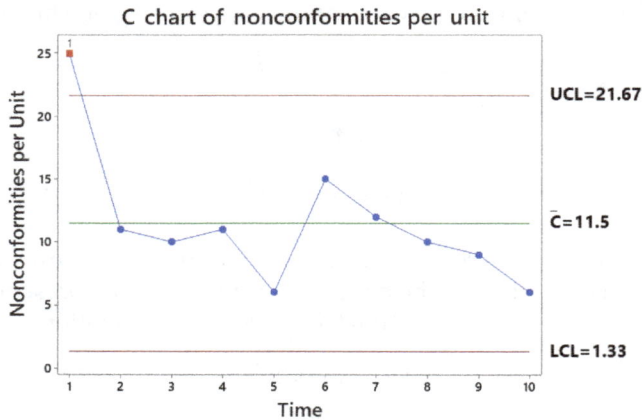

FIGURE 22.4
c chart of nonconformities per unit.

22.4.17 Illustration of c Chart

The following data were collected on the number of nonconformities per unit for 10 time periods (Table 22.3).

Construct the appropriate control chart and determine UCL, CL, and LCL. What is the state of statistical quality control?

Solution:
We get $\bar{u} = 11.5$

UCL $= 11.5 + 3 \, (11.5)^{1/2} = 11.53 + 10.1735 = 21.6735$

CL $= 11.5$

LCL $= 11.5 - 3 \, (11.5)^{1/2} = 11.53 - 10.1735 = 1.3265$

The control chart is shown in Figure 22.4. The c control chart was drawn using Minitab 19. The chart shows UCL=21.67, CL=11.5, and LCL=1.33.

At time 1, the observation is above UCL. This reveals that the process is not in statistical control because of the presence of some special causes of variation.

22.5 Capability Analysis

To analyze the capability process, we estimate the percentage of products or services that are within specifications. A process has to be under statistical control for this purpose because an out-of-control process does not allow for a reliable prediction of its capability.

Assuming that the process is in control and X is approximately normal, we can estimate the probability of a process outcome being within specifications.

22.5.1 Estimating the Process Capability

If the lower specification limit and upper specification limit are represented by LSL and USL, respectively, then

$$\text{Prob}(\text{outcome within specifications}) = \text{Prob}(\text{LSL} < X < \text{USL})$$

$$= \text{Prob}\left[(\text{LSL} - \bar{\bar{x}})/\bar{R}/d_2 < Z < (\text{USL} - \bar{\bar{x}})/\bar{R}/d_2\right]$$

where $Z = (X - \bar{\bar{x}})/\bar{R}/d_2$

22.5.2 Special Cases

22.5.2.1 Case I

For a characteristic with only an USL

$$\text{Prob}(\text{outcome within only upper specification}) = \text{Prob}(X < \text{USL})$$

$$= \text{Prob}\left[Z < (\text{USL} - \bar{\bar{x}})/\bar{R}/d_2\right]$$

$$= \text{Prob}\left[(X - \bar{\bar{x}})/\bar{R}/d_2 < (\text{USL} - \bar{\bar{x}})/\bar{R}/d_2\right]$$

22.5.2.2 Case II

For a characteristic with only an LSL

$$\text{Prob}(\text{outcome within only lower specification}) = \text{Prob}(\text{LSL} < X)$$

$$= \text{Prob}\left[(\text{LSL} - \bar{\bar{x}})/\bar{R}/d_2 < Z\right]$$

$$= \text{Prob}\left[(\text{LSL} - \bar{\bar{x}})/\bar{R}/d_2 < (X - \bar{\bar{x}})/\bar{R}/d_2\right]$$

22.5.3 Capability Indices

The capability indices are used to interpret the capability of a process. A capability index is defined as an aggregate measure of a process's ability to meet specification limits or customer requirements/satisfaction. Larger values are interpreted as indicating that the process is more capable of meeting customer satisfaction. Considered the most commonly used index, it is denoted by C_p and defined below.

$$C_p = (\text{USL} - \text{LSL})/6(\bar{R}/d_2) = \text{Specification spread}/\text{Process spread}$$

The difference between USL and LSL shows the *specification spread*. The denominator represents six times the standard deviation, which is estimated by $\bar{R}/d_2$. It is a six-standard-deviation spread in the data (the mean ± 3 standard deviation), referred to as the *process spread*. For a normal distribution, this area is $0-9973$ or 99.73%. Ideally, the process spread should be small in comparison to the specification spread so that majority of the process output falls within the specification limits. Considering this fact, it is said that the *larger the value of* C_p, *the better the capability of the process.*

22.5.4 Limitations of C_p

This index has serious limitations because it does not involve the current average of the output. It is a measure of process potential, not of actual performance. There are three possible values of C_p.

 i. In the case of $C_p=1$, we could hope that approximately 99.73% of outcomes would be inside the specification limits, if the process mean is located at the center.

 ii. $C_p>1$ shows that a process has potential of having more than 99.73% of its outputs within specifications.

 iii. $C_p<1$ reveals that the process is not very capable of meeting customer requirements. In the case of a perfectly centered process mean, less than 99.73% of the outcomes would be within specifications.

Earlier, most companies used to use $C_p \geq 1$. Now the situations have changed drastically with improved quality consciousness. Most companies adopting Six Sigma management now require $C_p = 1.33$, 1.5, or 2.0 for companies. See Levine et al. (2016) for more detailed discussions.

22.5.5 CPL, CPU, and C_{pk}

A capability index C_p is an aggregate measure of a process's ability (potential) to meet specification limits or customers' requirements. This formula suffers from an inherent weakness because it does not consider the average of the process outcomes. As an improvement over the capability index C_p, newer measures of a process capability in terms of actual performance are available, which include CPL, CPU, and C_{pk}. CPL and CPU are defined below.

$$CPL = \left(\bar{X} - LSL\right)/3\left(\bar{R}/d_2\right)$$

$$CPU = \left(USL - \bar{X}\right)/3\left(\bar{R}/d_2\right)$$

A value of CPL (or CPU) equal to 1.0 shows that the process mean is three standard deviations away from the LSL (or USL). Note that with only an LSL, the CPL measures process performance. Similarly, with only a USL, the CPU measures the process performance. For both capability indices, larger values of the index show a greater capability of the process.

Further, the capability index C_{pk} measures the actual performance for quality characteristic with two-sided specification limits. It is defined below.

$$C_{pk} = MIN\left[CPL, \ CPU\right]$$

This states that C_{pk} is equal to the minimum of the two indices, CPL and CPU. If the characteristic is normally distributed, then a value of 1 suggests that at least 99.73% of the current output is within specification limits. Like other indices, we always look for larger values of C_{pk}.

22.5.6 Illustration of Capability Analysis

See (ii) and (iii) parts of the illustration of $\bar{x}$ and R charts.

(ii) We have seen in solution (i) that the hotel luggage delivery process was in a state of statistical control. It is further mentioned that the hotel management has instituted a policy that 99% of the luggage deliveries must be completed in 14 minutes or less.

Further, it is known that

$n = 5, \bar{\bar{x}} = 9.478, \bar{R} = 3.482$ and $d_2 = 2.326$

$$\text{Prob (outcome within only upper specification)} = \text{Prob}\left[Z < \left(\text{USL} - \bar{\bar{x}}\right)/\bar{R}/d_2\right]$$

$$= \text{Prob}\left[Z < (14 - 9.478)/(3.482/2.326)\right]$$

$$= \text{Prob}\left[Z < 3.02\right]$$

$$= 0.99874$$

We evaluate from the normal area table that Prob $[Z < 3.02] = 0.99874$.

This suggests that 99.874% of the luggage deliveries will be made within the specified time. We, therefore, conclude that the process is capable of meeting the goal of 99% set forth by the hotel management.

(iii) As CPU $= \left(\text{USL} - \bar{\bar{x}}\right)/3\left(\bar{R}/d_2\right)$, on entering USL $= 14, \bar{\bar{x}} = 9.478, \bar{R} = 3.482, d_2 = 2.326$, we obtain CPU $= 1.01$. This indicates that the USL is slightly more than three standard deviations above the mean.

22.6 Scenarios in Developing Countries

SPC techniques played a very important role in helping rebuild the economy of Japan after the Second World War under the supervision of W. E. Deming, a US expert on statistical quality control and improvement. In developing countries, SPC techniques can help improve economic conditions tremendously if employed properly because of their cheap labor and natural resources. Madanhire and Mbohwa (2016) described the applications of SPC in Zimbabwean manufacturing industries without mentioning a data set. They also discuss the associated problems with manufacturing industries there, which could be minimized. Abtew et al. (2018) used the data set of a garment-producing company known as Silver Spark Apparel Limited, a Raymond Group Company, India, as an illustration, but the data set used is not given in their paper. There are good garment companies in Ethiopia, China, India, and other countries, and other industries in developing countries that could contribute considerably to the incomes of their countries using SPC techniques. These countries could follow the example of Japan to achieve their goals.

Acknowledgments

We express our deep appreciation to the learned reviewer for his constructive suggestions and to Professor O. Olawale Awe, Editor, for his interest and patience while the paper was being prepared. These helped improve the content of the paper considerably.

References

Abtew, M. A., Kropi, S., Hong, Y., and Pu, L. (2018). Implementation of statistical process control (SPC) in the sewing section of garment industry for quality improvement. *AUTEX Research Journal*, 18 (2), 160–172.

Bowker, A. H. and Lieberman, G. J. (1972). *Engineering Statistics*, 2nd ed. Prentice-Hall, Englewood Cliffs, NJ.

Levine, D. M., Stephen, D. F., and Szabat, K. A. (2016). *Statistics for Managers: Using Microsoft Excel*, 7th ed. Pearson India Education Services Pvt., Ltd, Noida.

Madanhire, I. and Mbohwa, C. (2016). Application of statistical process control in SPC manufacturing industry in a developing country. *Procedia CIRP*, 40, 580–583 (Available online at www.sciencedirect.com).

Mitra, A. (2008). *Fundamentals of Quality Control and Improvement*, 3rd ed. Wiley, Hoboken, NJ. Authorized reprint in 2014 by Wiley India (P) Ltd.

Montgomery, D. C. (2009). *Statistical Quality Control: A Modern Introduction*, 6th ed. Wiley, Hoboken, NJ. Authorized reprint in 2011 by Wiley India (P) Ltd.

Shewhart, W. A. (1931). *Economic Control of Quality of Manufactured Product*. D. Van Nostrand Co., New York.

23

Determination of Sample Size and Errors

Haftom Temesgen Abebe

Mekelle University

CONTENTS

DOI: 10.1201/9781003261148-27

23.1 Introduction

One of the most frequent problems in planning a statistical investigation is the determination of the appropriate sample size. Sample size is the number of participants or observations included in a survey or study. It is called a sample because it only represents part of the population (or target population).

In planning a survey, the stage is always reached of the determination of the sample size. Among the questions that an investigator should ask when planning a survey is "How large a sample do I need?" The answer depends on the aims, nature, and scope of the study and on the expected result, all of which should be carefully considered at the planning stage. One may also ask why sample size is so important. The answer to this is that an appropriate sample size is required to ensure the validity and accuracy of the results. The larger the sample size, the more reliable the results are, but a larger sample size needs longer time and more money. In contrast, using too few subjects cannot adequately address the research question. Thus, we may fail to detect important effects or may estimate the effects too imprecisely. The feasible sample size is also determined by the availability of resources: time, manpower, transport, available facility, and money.

Studies that have either an excessively large sample size or an inadequate sample size are both wasteful of the time of participants and investigators, the resources to conduct the assessments, and analytical efforts. Therefore, studies should be designed to include a sufficient number of participants to adequately address the research question. Moreover, having the proper sample size is crucial for obtaining statistically significant results.

There are numerous approaches to determining the sample size. One method is to use the entire population (census). This method is very expensive in terms of money and time, which makes it impossible for large populations. Another approach is to use the same sample size as those of similar studies to the one you plan. This approach is not recommended, as you may run the risk of repeating errors that were made in determining the sample size for another study. A third approach is to rely on published tables of the sample size for a given set of criteria, and the fourth approach is to use a formula to determine the sample size. In this chapter, we focus on formulas.

23.2 Sample Size Determination Using Formulas

Here we introduce formulas that can be used to estimate the sample size needed to produce a confidence interval estimate with a specified margin of error (precision) or to ensure a test of hypothesis with a high probability of detecting a meaningful difference in the parameters.

In estimating a certain characteristic of a population, sample size calculations are important to ensure that estimates are obtained with required precision or confidence. The sample should be sufficient to represent the characteristics of interest of the study population. In order to decide the type of formula, ask yourself: Do you want to learn about a mean or proportion? A difference in means or proportions? Do you want to estimate a population parameter with a given precision, or do you want to test with a given power? What type of sample(s) will you be working with, a single group, two or more independent groups, or matched pairs?

23.3 Sample Size Calculation Based on Confidence Intervals

23.3.1 Sample Size Determination for Estimating a Single Population Mean

In studies where the objective is to estimate the mean of a continuous outcome variable in a single population, the confidence interval (CI) for the mean population, μ is

$$\mu \in \left(\bar{X} \pm d\right) \text{ where } d = Z_{1-\alpha/2}\frac{\sigma}{\sqrt{n}}, \tag{23.1}$$

where d ($= e$ in some text books) is the margin of error/acceptable sample error or absolute precision, which is half of the width of the CI, $\bar{X}$ is the sample mean, n is the sample size, $Z_{1-\alpha/2}$ is the critical value for a given confidence level, and σ is the expected standard deviation (SD) of the variable to be studied. The formula for determining sample size for a single population mean based on the CI is given by

$$n = \frac{(Z_1 - \alpha/2)\sigma^2}{d^2}, \tag{23.2}$$

where n is the minimum sample size.

23.3.1.1 Practical Considerations in Sample Size Determination for a Single Population Mean

How are we to determine the sampling error and confidence level? Researchers or investigators should work with managers to make these decisions. How much error is the manager willing to tolerate (less error means more accuracy)? The convention is +5%. The more important the decision, the less the acceptable level of the sampling error should be, and the higher the desired confidence level, the larger the sample size required. The convention is a 95% confidence level ($Z = 1.96$). The more important the decision, the more likely the researcher will want more confidence. Moreover, most of the time the population SD σ is unknown, and in that case we have to substitute an estimated value sample SD (s) to obtain an approximate sample size. The SD of the sample, s, may be used as a measure of variance and can be estimated from a previous (similar) study or a pilot study conducted.

Example 23.1

Suppose that for a certain group of cancer patients, we are interested in estimating the mean age at diagnosis. We would like a 95% CI of 5 years wide. If the population SD is 12 years, how large should our sample be?

Solution:

Given: 95% CI ($Z = 1.96$), $\sigma = 12$ and $d = 5/2 = 2.5$. Substituting these into equation (23.2) yields:

$$n = \frac{(1.96)^2 \, (12)^2}{(2.5)^2} = 88.5 \approx 89$$

Suppose the margin error is 1 ($d = 1$), then the sample size increases

$$n = \frac{(1.96)^2 \, (12)^2}{(1)^2} = 554$$

Example 23.2

A hospital director wishes to estimate the mean weight of babies born in the hospital. How large a sample of birth records should be taken if she/he wants a 95% CI of 0.5 wide? Assume that a reasonable estimate of σ is 2.

Solution:

Given: 95% CI ($Z=1.96$), $\sigma=12$, and $d=0.5/2=0.25$. Substituting these into equation (23.2) yields:

$$n = \frac{(1.96)^2\, 2^2}{(0.25)^2} = 246 \text{ birth records}$$

Example 23.3

Find the minimum sample size needed to estimate the drop in heart rate (μ) for a new study using a higher dose than the standard one. We require that the two-sided 95% CI for μ be no wider than 5 beats per minute and the sample SD, s, for change in heart rate equals 10 beats per minute.

Solution:

Given: 95% CI ($Z=1.96$), $s=10$, and $d=5/2=25$. Substituting these into equation (23.2) yields:

$$n = \frac{(1.96)^2\, 10^2}{(2.5)^2} = 62 \text{ patients}$$

The minimum sample size needed is 62 patients satisfying the criteria.

23.3.2 Sample Size Determination for Estimating a Single Population Proportion

In studies where the objective is to estimate the proportion of a dichotomous outcome variable in a single population, the CI for the proportion population, p, we have

$$p \in (\hat{p} \pm d), \text{ where } d = Z_{1-\alpha/2}\sqrt{\frac{\hat{p}(1-\hat{p})}{n}} \tag{23.3}$$

where $\hat{p}$ is the sample proportion, n is the sample size, d is margin of error, and $Z_{1-\alpha/2}$ is the appropriate value from the standard normal distribution for the desired confidence level. For large populations, the formula for the sample size for a single population proportion based on CI is given by

$$n = \frac{(Z_{1-\alpha/2})^2\, p(1-p)}{d^2} \tag{23.4}$$

where n is the minimum sample size required for a very large population ($\geq 10{,}000$) and p is the expected proportion of the event to be studied (to be estimated based on findings of previous studies).

23.3.2.1 Finite Population Correction for Proportions

If the population is small (<10,000), then the sample size can be reduced slightly. This is because a given sample size provides proportionately more information for a small population than for a large population. The sample size (n) can be adjusted using the formula for the finite population factor:

$$\text{Corrected sample size} = \frac{n}{1 + \frac{(n-1)}{N}} \tag{23.5}$$

where n is the noncorrected sample size (23.4) and N is the population size. Depending on the nature of the study, a 10%–15% contingency should be added.

23.3.2.2 Practical Considerations in Sample Size Determination for a Single Population Proportion

How should we estimate the variability (p and $q=1-p$ shares) in the population? We do not know the variability in the proportion that will adopt the practice. Thus, we estimate the variability from the results of previous studies or conduct a pilot study, or expect the worst case $p=0.5$ (maximum variability).

Example 23.4

Suppose that you are interested in knowing the proportion of infants who breastfeed > 18 months of age in a rural area. Suppose that in a similar area, the proportion (p) of breastfed infants was found to be 0.20. What sample size is required to estimate the true proportion within ±3% points with 95% confidence?

Solution:

Given: 95% CI ($Z=1.96$), $p=0.2$, $q=1-0.2=0.8$, and $d=0.03$

$$n = \frac{(Z_{1-\alpha/2})^2 \, p(1-p)}{d^2} = \frac{(1.96)^2 \, 0.2(0.8)}{0.03^2} = 683$$

Example 23.5

A researcher is interested in determining the prevalence of family planning use in Mekelle city. A previous study indicates the prevalence is around 55%. If the researcher accepts a 95% CI and 5% of margin of error, what number of women of reproductive age should be included in his study?

Solution:

Given 95% CI ($Z=1.96$), $p=0.55$, $q=1-0.55=0.45$, and $d=0.05$,

$$n = \frac{(Z_{\alpha/2})^2 \, p(1-p)}{d^2} = \frac{(1.96)^2 \, 0.55(0.45)}{0.05^2} = 381$$

23.3.3 Sample Size Determination for Estimating the Difference between Two Population Means

In studies where the objective is to estimate the difference in means between two independent populations, the CI for the population mean difference $\mu_1 - \mu_2$ is given by

$$\left(\bar{X}_1 - \bar{X}_2\right) \pm d, \text{ where } d = Z_{1-\alpha/2}\sqrt{\left(\frac{\sigma_1^2}{n_1} + \frac{\sigma_2^2}{n_2}\right)}, \tag{23.6}$$

where $\bar{X}_1$ and $\bar{X}_2$ are the sample means. The formula for determining the sample sizes required in each comparison group assuming equal variability $\sigma_1^2 = \sigma_2^2$ and equal sample sizes $n_1 = n_2 = n$ is given below:

$$n_i = 2\left(\frac{Z_{1-\alpha/2}\sigma}{d}\right)^2 = \frac{2Z_{1-\alpha/2}^2\sigma^2}{d^2}, \tag{23.7}$$

where n_i is the sample size required in each group ($i=1,2$), Z is the value from the standard normal distribution reflecting the confidence level that will be used, and d is the desired margin of error. σ again reflects the population SD of the outcome variable.

Note that in (23.7), σ is the common population SD, and we have assumed equal sample sizes. If σ is unknown, we substitute an estimated value s to get an approximate sample size. The sample sizes obtained using this formula are usually approximate because we have to substitute an estimated value of σ, the common population SD. This estimate will probably be based on an educated guess from a previous study or on the range of population values.

In most cases the common SD σ is unknown, then we used S_p, the pooled estimate of the common SD, as a measure of variability in the outcome (based on pooling the data), where S_p is computed as follows:

$$S_p = \sqrt{\frac{(n_1-1)S_1^2 + (n_2-1)S_2^2}{(n_1+n_2-2)}} \tag{23.8}$$

If data are available on variability of the outcome in each comparison group, then the pooled estimate of the common SD, S_p can be computed and used in the sample size formula. However, it is more often the case that data on the variability of the outcome are available only from one group, often the untreated (e.g., placebo control) or unexposed group. The known SD of the outcome variable measured in the placebo, control, or unexposed group can be used for both S_1 and S_2 in the formula shown above.

A sample size calculation can also be done using the formulas shown above when $n_1 \neq n_2$. In this situation, we let n_2 be some multiple r of n_1; then we substitute $(r+1)/r$ for 2 in the sample size formulas (23.7). After solving for n_1, we substitute $n_2 = rn_1$.

$$n_1 = \frac{r+1}{r}\left(\frac{Z_{1-\alpha/2}\sigma}{d}\right)^2 = \left(\frac{r+1}{r}\right)\frac{Z_{1-\alpha/2}^2\sigma^2}{d^2}, \tag{23.9}$$

where n_1 is the sample size of the smaller group, r is the ratio of the larger group to the smaller group, and σ is the population SD, assumed known. If σ is unknown, we use S_p, the pooled estimate of the common SD (23.8), as a measure of variability in the outcome.

Example 23.6

An investigator wants to plan a clinical trial to evaluate the efficacy of a new drug designed to increase high-density lipoproteins (HDL) cholesterol (the "good" form of cholesterol). The plan is to enroll participants and randomly assign them to receive either the new drug or a placebo. HDL cholesterol will be measured in each participant after 12 weeks on the assigned treatment. Based on prior experience with similar trials, the investigator expects that 10% of all participants will be lost to follow-up or will drop out of the study over 12 weeks. A 95% CI will be estimated to quantify the difference in mean HDL levels between patients taking the new drug as compared to placebo. The investigator would like the margin of error to be no more than 3 units. How many patients should be recruited into the study? Use the SD of HDL cholesterol is 17.1.

Solution:

Given: 95% CI ($Z=1.96$), $d=3$, and $\sigma=17.1$. We will use these as inputs to compute the sample sizes as follows:

$$n_i = 2\left(\frac{Z_{1-\alpha/2}\sigma}{d}\right)^2 = 2\left(\frac{1.96(17.1)^2}{3}\right) = 250$$

Samples of size $n_1=250$ and $n_2=250$ will ensure that the 95% CI for the difference in mean HDL levels will have a margin of error of no more than 3 units. Note that these sample sizes refer to the numbers of participants with complete data. The investigators hypothesized a 10% attrition (or drop-out) rate (in both groups). In order to ensure that the total sample size of 500 is available at 12 weeks, the investigator needs to recruit more participants to allow for attrition.

(The number to enroll) * (% retained)=desired sample size
The number to enroll=desired sample size/(% retained)
The number to enroll=500/0.90=556.

If a 10% attrition rate is anticipated, the investigators should enroll 556 participants. This will ensure 500 participants with complete data at the end of the trial. That is, final sample sizes $n_1=278$ and $n_2=278$ will be needed to ensure that the 95% CI for the difference in mean HDL levels will have a margin of error of no more than 3 units.

Example 23.7

An investigator wants to compare two diet programs in children who are obese. One diet is a low-fat diet, and the other is a low-carbohydrate diet. The plan is to enroll children and weigh them at the start of the study. Each child will then be randomly assigned to either the low-fat or the low-carbohydrate diet. Each child will follow the assigned diet for 8 weeks, at which time they will again be weighed. The number of pounds lost will be computed for each child. Based on data reported from diet trials in adults, the investigator expects that 20% of all children will not complete the study. A 95% CI will be estimated to quantify the difference in weight loss between the two diets and the investigator would like the margin of error to be no more than 3 pounds. How many children should be recruited into the study?

The sample sizes are computed as follows

$$n_i = 2\left(\frac{Z_{1-\alpha/2}\sigma}{d}\right)^2 = \frac{2Z_{1-\alpha/2}^2\sigma^2}{d^2}$$

Again, the issue is determining the variability in the outcome of interest (σ), here the SD in pounds lost over 8 weeks. To plan this study, investigators use data from a published study in adults. Suppose one such study compared the same diets in adults with 100 participants in each diet group and reported an SD in weight loss over 8 weeks on a low-fat diet of 8.4 pounds and an SD in weight loss over 8 weeks on a low-carbohydrate diet of 7.7 pounds. These data can be used to estimate the common SD in weight loss as follows:

$$S_p = \sqrt{\frac{(n_1-1)S_1^2+(n_2-1)S_2^2}{(n_1+n_2-2)}} = \sqrt{\frac{(100-1)8.4^2+(100-1)7.7^2}{(100+100-2)}} = 8.1$$

We now use this value and the other inputs to compute the sample sizes:

$$n_i = 2\left(\frac{Z_{1-\alpha/2}\sigma}{d}\right)^2 = 2\left(\frac{1.96(8.1)}{3}\right)^2 = 56$$

Samples of size $n_1=56$ and $n_2=56$ will ensure that the 95% CI for the difference in weight loss between diets will have a margin of error of no more than 3 pounds. Again, these sample sizes refer to the numbers of children with complete data. The investigators anticipate a 20% attrition rate. In order to ensure that the total sample size of 112 is available at 8 weeks, the investigator needs to recruit more participants to allow for attrition. Thus,

(The number to enroll) * (% retained)=desired sample size
The number to enroll=desired sample size/(% retained)
The number to enroll=112/0.80=140.

23.3.4 Sample Size Determination for Comparing Two Means from Paired Samples

Sample sizes for estimating μ_d and conducting a statistical test for μ_d based on paired data (differences) are found using the formulas seen as before. The only change is that we are working with a single sample of differences rather than a single sample.

In studies where the objective is to estimate the mean difference of a continuous outcome based on paired data, the CI for the population mean difference $\mu_d=\mu_1-\mu_2$ is given by

$$X_d \pm d, \text{ where } d = Z_{1-\alpha/2}\sqrt{\frac{\sigma_d^2}{n}}, \tag{23.10}$$

where $\overline{X}_d$ is the sample mean for the paired difference. The sample size formula required for a $(1-\alpha)100\%$ CI for μd is given by

$$n = \left(\frac{Z_{1-\alpha/2}\sigma_d}{d}\right)^2, \tag{23.11}$$

where n is the minimum sample size, $Z_{1-\alpha/2}$ is the value from the standard normal distribution reflecting the confidence level that will be used (e.g., $Z=1.96$ for 95%), d is the desired margin of error, and σ_d is the SD of the difference scores (e.g., the difference based on measurements over time or the difference between matched pairs). If σ_d is unknown, substitute an estimated value SD to obtain approximate sample size.

23.3.5 Sample Size Determination for Estimation of the Difference between Two Population Proportions

In studies where the objective is to estimate the difference in proportions between two independent populations (i.e., to estimate the risk difference), the CI for the population proportion difference $P_1 - P_2$ is given by $(p_1 - p_2) \pm d$, where $d = Z_{1-\alpha/2} \sqrt{\left(\dfrac{p_1(1-p_1)}{n_1} + \dfrac{p_2(1-p_2)}{n_2} \right)}$ and p_1 and p_2 are the sample proportions of success in each group. The formula for determining the sample sizes required in each comparison group based on CI is

$$n_i = \left(p_1(1-p_1) + p_2(1-p_2) \right)\left(\frac{Z_{1-\alpha/2}}{d} \right)^2, \tag{23.12}$$

where n_i is the sample size required in each group ($i=1,2$), $Z_{1-\alpha/2}$ is the value from the standard normal distribution reflecting the confidence level that will be used (e.g., $Z_{1-\alpha/2} = 1.96$ for 95%), and d is the desired margin of error. In order to determine the sample size, we need approximate values of p_1 and p_2. If there is no information available to approximate p_1 and p_2, 0.5 can be used to generate the most conservative, or largest, sample sizes.

Similar to the situation for the difference in means between two independent populations above, it may be the case that data are available on the proportion of successes in one group, usually the untreated or unexposed group. If so, the known proportion can be used for both p_1 and p_2 in the formula shown above.

The formula shown above (23.12) generates sample size estimates for samples of equal size. If a study is planned where different numbers of patients/subjects will be assigned or different numbers of patients will make up the comparison groups, then alternative formulas can be used. A sample size formula can also be derived for $n_1 \neq n_2$. In this situation, we let n_2 be some multiple r of n_1; then after solving for n_1, we substitute $n_2 = rn_1$. The general sample size formula is

$$n_1 = \frac{r+1}{r}\left(\bar{p}(1-\bar{p}) \right)\left(\frac{Z_{1-\alpha/2}}{d} \right)^2, \tag{23.13}$$

where n_1 is the sample size of a small group, r is the ratio of the larger group to the smaller group, and $\bar{p}$ is the overall proportion based on pooling the data from the two comparison groups [it can be computed by taking the mean of the proportions in the two comparison groups, and $\bar{p}(1-\bar{p})$ is a measure of variability, similar to SD].

Example 23.8

An investigator wants to estimate the impact of smoking during pregnancy on premature delivery. Normal pregnancies last approximately 40 weeks and premature deliveries are those that occur before 37 weeks. The 2005 National Vital Statistics report indicates that approximately 12% of infants are born prematurely in the United States (National Center for Health Statistics, 2005). The investigator plans to collect data through medical record review and to generate a 95% CI for the difference in proportions of infants born prematurely to women who smoked during pregnancy compared to those who did not. How many women should be enrolled in the study to ensure that the 95% CI for the difference in proportions has a margin of error of no more than 4%?

Solution:

The sample sizes (i.e., numbers of women who smoked and did not smoke during pregnancy) can be computed using the formula shown above (23.12). National data suggest that 12% of infants are born prematurely. We will use that estimate for both groups in the sample size computation.

$$n_i = \left(p_1 (1 - p_1) + p_2 (1 - p_2) \right) \left(\frac{Z_{\alpha/2}}{d} \right)^2 = (0.12(1 - 0.12) + 0.12(1 - 0.12)) \left(\frac{1.96}{0.04} \right)^2 = 508$$

Samples of size $n_1 = 508$ women who smoked during pregnancy and $n_2 = 508$ women who did not smoke during pregnancy will ensure that the 95% CI for the difference in proportions who deliver prematurely will have a margin of error of no more than 4%.

23.3.6 Sample Size Calculation Based on Hypothesis Testing

In planning an experiment to compare two treatments, the following methods are employed to estimate the sample size needed. The investigators first decide whether a value Δ that represents the size of the difference between the true effects of the treatment, i.e., $\Delta = \mu_1 - \mu_2$, that is regarded as important. As long as the true difference is as large as Δ, the experimenter would like the experiment to have a high probability of showing a statistically significant difference between the treatment means. That is, the investigator wants the power of the test to be high when the difference is Δ. In other words, the investigator wants a large P(*reject H_0/H_0 is false*)=p =$1-\beta$. The power of the test $(1-\beta)$ is the probability that the null hypothesis is rejected when it is false. The power of the test is higher for alternatives farther away from the null.

In designing studies, most people consider power of 80% or 90% (just as we generally use 95% as the confidence level for CI estimates). The inputs for the sample size determination formulas based on hypothesis testing include the level of significance, the desired power, and the effect size. The effect size is selected to represent a practically important difference or a clinically meaningful difference in the parameter of interest.

The formulas we present below yield the minimum sample size needed to ensure that the test of hypothesis will have a specified probability of rejecting the null hypothesis when it is false (i.e., a specified power). In planning studies, investigators should account for attrition or loss to follow-up.

23.3.7 Sample Size Determination for One Sample, Continuous Outcome

In studies where the objective is to perform a test of hypothesis comparing the mean of a continuous outcome variable in a single population to a known mean, the hypotheses of interest are

H_0: $\mu = \mu_0$ against

H_1: $\mu \neq \mu_0$

H_1: $\mu < \mu_0$

H_1: $\mu > \mu_0$, where μ_0 is a known mean (e.g., a historical control).

The formula for determining the sample size to ensure that the test has a specified power is given below:

$$\text{Two-sided test}: n = \left(\frac{Z_{1-\alpha/2} + Z_{1-\beta}}{ES} \right)^2, \text{ where Effect size} = ES = \frac{|\mu_1 - \mu_0|}{\sigma} \qquad (23.14)$$

$$\text{One-sided test}: n = \left(\frac{Z_{\alpha} + Z_{1-\beta}}{ES} \right)^2, \text{ where Effect size} = ES = \frac{|\mu_1 - \mu_0|}{\sigma}, \qquad (23.15)$$

where μ_0 is the mean under null hypothesis H_0, μ_1 is the mean under alternative hypothesis H_1, σ is the population SD of the outcome of interest, α is the selected level of significance, and $Z_{1-\alpha/2}$ is the value from the standard normal distribution that holds $1-\alpha/2$ below it; $1-\beta$ is the selected power, while $Z_{1-\beta}$ is the value from the standard normal distribution that holds $1-\beta$ below it. The numerator of the effect size, the absolute value of the difference in means $|\mu_1 - \mu_0|$, represents what is considered a practically important difference in means or a clinically meaningful difference.

Similar to the issue of planning studies to estimate CIs, it can sometimes be difficult to estimate the SD σ. In sample size determinations, researchers often use a value for the SD from a previous study or a study performed in a different but comparable population.

Example 23.9

An investigator hypothesizes that in people free of diabetes, fasting blood glucose, a risk factor for coronary heart disease, is higher in those who drink at least 2 cups of coffee per day. A cross-sectional study is planned to assess the mean fasting blood glucose levels in people who drink at least two cups of coffee per day. The mean fasting blood glucose level in people free of diabetes is reported as 95.0 mg/dL with a SD of 9.8 mg/dL. If the mean blood glucose level in people who drink at least 2 cups of coffee per day is 100 mg/dL, this would be important clinically. Based on prior experience, the investigator expects that 10% of the participants will fail to fast or will refuse to follow the study protocol. How many patients should be enrolled in the study to ensure that the power of the test is 80% to detect this difference? A two-sided test will be used with a 5% level of significance.

Solution:

Given: $\mu_0 = 95$, $\mu_1 = 100$, $\sigma = 9.8$, $Z_{0.95} = 1.96$ and $Z_{0.80} = 0.84$
To compute the sample size we substitute these values in equation (23.14):

$$n = \left(\frac{Z_{1-\alpha/2} + Z_{1-\beta}}{ES} \right)^2 = \sigma^2 \left(\frac{Z_{1-\alpha/2} + Z_{1-\beta}}{\mu_1 - \mu_0} \right) = 9.8^2 \left(\frac{1.96 + 0.84}{100 - 95} \right) = 31$$

(The number to enroll) * (% retained) = desired sample size
The number to enroll = desired sample size/(% retained)
The number to enrolled = 31/0.90 = 35.

Therefore, a sample of size $n = 35$ will ensure that a two-sided test with $\alpha = 0.05$ has 80% power to detect a 5 mg/dL difference in mean fasting blood glucose levels.

23.3.8 Sample Size Determination for One Sample, Dichotomous Outcome

In studies where the objective is to perform a test of a hypothesis comparing the proportion of successes in a dichotomous outcome variable in a single population to a known proportion, the hypotheses of interest are

$H_0: p=p_0$ against

$H_1: p \neq p_0$

$H_1: p < p_0$

$H_1: p > p_0$,

where p_0 is the known proportion (e.g., a historical control, as from a previous study). The formula for determining the sample size to ensure that the test has a specified power is given below:

$$\text{Two-sided test}: n = \left(\frac{Z_{1-\alpha/2} + Z_{1-\beta}}{ES} \right)^2, \text{ where Effect size} = ES = \frac{p_1 - p_0}{\sqrt{p_1(1-p_1)}} \qquad (23.16)$$

$$\text{One-sided test}: n = \left(\frac{Z_{\alpha} + Z_{1-\beta}}{ES} \right)^2, \text{ where Effect size} = ES = \frac{p_1 - p_0}{\sqrt{p_1(1-p_1)}}, \qquad (23.17)$$

where p_0 is the proportion under the null hypothesis H_0 and p_1 is the proportion under the alternative hypothesis H_1. The numerator of the effect size, the absolute value of the difference in proportions $|p_1 - p_0|$, again represents what is considered a clinically meaningful or practically important difference in proportions.

Example 23.10

A recent report from the Framingham Heart Study indicated that 26% of people free of cardiovascular disease had elevated LDL cholesterol levels, defined as LDL > 159 mg/dL (Rutter et al., 2004). An investigator hypothesizes that a higher proportion of patients with a history of cardiovascular disease will have elevated LDL cholesterol. How many patients should be studied to ensure that the power of the test is 90% to detect a 5% difference in the proportion with elevated LDL cholesterol? A two-sided test will be used with a 5% level of significance.

Solution:

Given: $p_1 - p_0 = 0.05$, $p_0 = 0.26$, $Z_{0.95} = 1.96$, and $Z_{0.90} = 1.28$.
    To compute the sample size we substitute these values in equation (23.16)

$$n = \left(\frac{Z_{1-\alpha/2} + Z_{1-\beta}}{ES} \right)^2 = p_0(1-p_0) \left(\frac{Z_{1-\alpha/2} + Z_{1-\beta}}{p_1 - p_0} \right)^2 = 0.26(1-0.26) \left(\frac{1.96 + 1.28}{0.05} \right)^2 = 869$$

A sample of size $n=869$ will ensure that a two-sided test with $\alpha=0.05$ has 90% power to detect a 5% difference in the proportion of patients with a history of cardiovascular disease who have an elevated LDL cholesterol level.

23.3.9 Sample Size Determination for Two Independent Samples, Continuous Outcome

In studies where the objective is to perform a test of hypothesis comparing the means of a continuous outcome variable in two independent populations, the hypotheses of interest for one- and two-sided tests are

$H_0: \mu_1 - \mu_2 = 0$ against
$H_1: \mu_1 - \mu_2 \neq 0 = \Delta$
$H_1: \mu_1 - \mu_2 > 0 = \Delta$
$H_1: \mu_1 - \mu_2 < 0 = \Delta,$

where μ_1 and μ_2 are the means in the two comparison populations and Δ is the difference in means between the two groups expected under the alternative hypothesis, H_1.

Here we assume that the population SDs σ for independent samples is known. We also assume that $\sigma_1 = \sigma_2 = \sigma$. If σ is unknown, we substitute an estimated value to obtain an approximate sample size. Then, the formula for determining the sample sizes (assuming equal-sized groups) to ensure that the test has a specified power is

$$\text{Two-sided test}: n_i = \left(\frac{Z_{1-\alpha/2} + Z_{1-\beta}}{\text{ES}} \right)^2, \text{ where Effect size} = \text{ES} = \frac{|\mu_1 - \mu_2|}{\sigma} = \frac{|\Delta|}{\sigma} \quad (23.18)$$

$$\text{One-sided test}: n_i = \left(\frac{Z_\alpha + Z_{1-\beta}}{\text{ES}} \right)^2, \text{ where Effect size} = \text{ES} = \frac{|\mu_1 - \mu_2|}{\sigma} = \frac{|\Delta|}{\sigma}, \quad (23.19)$$

where n_i is the minimum sample size required in each group ($i = 1, 2$; assuming equal-sized groups), α is the selected level of significance and $Z_{1-\alpha/2}$ is the value from the standard normal distribution holding $1 - \alpha/2$ below it, and $1 - \beta$ is the selected power and $Z_{1-\beta}$ is the value from the standard normal distribution holding $1 - \beta$ below it.

Note also that the formula shown above generates sample size estimates for samples of equal size. If a study is planned where different numbers of subjects/participants will be assigned or different numbers of patients/participants will comprise the comparison groups, then alternative formulas can be used.

The sample size calculation can also be derived for $n_1 \neq n_2$. In this situation: we let n_2 be some multiple r (e.g., $r = 0.5$) of n_1; then we substitute $(r+1)/r$ for 2 in the sample size formulas (23.18 and 23.19). After solving for n_1, we substitute $n_2 = rn_1$.

$$\text{Two-sided test}: n_1 = \left(\frac{r+1}{r} \right) \sigma^2 \left(\frac{Z_{1-\alpha/2} + Z_{1-\beta}}{\Delta} \right)^2 \quad (23.20)$$

$$\text{One-sided test}: n_1 = \left(\frac{r+1}{r} \right) \sigma^2 \left(\frac{Z_\alpha + Z_{1-\beta}}{\Delta} \right)^2, \quad (23.21)$$

where n_1 is the sample size of small group, r is the ratio of the larger group to the smaller group, and σ is the population SD and assuming known. If σ is unknown, we used S_p, the pooled estimate of the common SD (23.8), as a measure of variability in the outcome.

Example 23.11

What is the sample size needed to find a mean blood pressure difference of 5 mmHg significant at the 0.05 level (2-sided), assuming SD = 10 mmHg, 80% power, and equal group sizes?

Solution:

Given: Δ=mean difference=5, $Z_{1-0.05/2}$=1.96, $Z_{0.8}$=0.84, SD=10, and r=1. To compute the sample sizes we substitute these values in equation (23.18):

$$n_i = 2\left[\frac{\left(Z_{1-\alpha/2}+Z_\beta\right)\times\sigma}{\Delta}\right]^2 = 2\left[\frac{(1.96+0.84)\times10}{5}\right]^2 = 67.72 = 63 \text{ per group}$$

Samples of sizes n_1=63 and n_2=63 will be needed to ensure a 95% CI with 80% power and with a difference in means of 5 mmHg and SD=10 mmHg. 126 total subjects needed.

Example 23.12

What is the sample size needed to find a mean blood pressure difference of 5 mmHg significant at the 0.05 level (2-sided), assuming SD=10 mmHg, 90% power, and equal group sizes?

This time, let us increase our power requirement to 90% (Z_β=1.28). What happens to the sample size needed?

Solution:

Given: Δ=mean difference =5, $Z_{1-0.05/2}$=1.96, $Z_{0.9}$=1.28, SD=10, and r=1.

$$n = 2\left[\frac{\left(Z_{1-\alpha/2}+Z_\beta\right)\times\sigma}{\Delta}\right]^2 = 2\left[\frac{(1.96+1.28)\times10}{5}\right]^2 = 83.98 = 84 \text{ per group}$$

Samples of sizes n_1=84 and n_2=84 will be needed to ensure a 95% CI with 90% power and with a difference in means of 5 mmHg and SD=10 mmHg. 168 total subjects needed.

23.3.10 Sample Size Determination for Comparing Two Means from Paired Samples

In studies where the objective is to conduct a test of a hypothesis on the mean difference in a continuous outcome variable μd based on paired data (differences), the hypotheses of interest for one- and two-sided tests are

H_0: μ_d=0 against
H_1: μ_d≠0=Δ
H_1: μ_d>0=Δ
H_1: μ_d<0=Δ,

where μ_d is the mean difference in the population and Δ is the difference in means between the two groups expected under the alternative hypothesis, H_1.

Sample sizes for estimating μ_d and conducting a statistical test for μ_d based on paired data are found using the formulas seen before. The only change is that we are working with a single sample of differences rather than a single sample of x values. For convenience, the appropriate formula for determining the sample size to ensure that the test has a specified power is given below:

$$\text{Two}-\text{sided test}: n = \left(\frac{Z_{1-\alpha/2}+Z_{1-\beta}}{\text{ES}}\right)^2, \text{ where Effect size} = \text{ES} = \frac{\Delta}{\sigma_d} \qquad (23.22)$$

$$\text{One-sided test}: n = \left(\frac{Z_\alpha + Z_{1-\beta}}{ES}\right)^2, \text{ where Effect size} = ES = \frac{\Delta}{\sigma_d}, \quad (23.23)$$

where n is the sample size, σ_d is the SD of the within-pair difference, and Δ is the clinically meaningful difference.

Example 23.13

An investigator wants to evaluate the efficacy of an acupuncture treatment for reducing pain in patients with chronic migraine headaches. The plan is to enroll patients who suffer from migraine headaches. Each will be asked to rate the severity of the pain they experience with their next migraine before any treatment is administered. Pain will be recorded on a scale of 1–100 with higher scores indicative of more severe pain. Each patient will then undergo the acupuncture treatment. On their next migraine (post-treatment), each patient will again be asked to rate the severity of the pain. The difference in pain will be computed for each patient. A two-sided test of hypothesis will be conducted at $\alpha=0.05$, to assess whether there is a statistically significant difference in pain scores before and after treatment. How many patients should be involved in the study to ensure that the test has 80% power to detect a difference of 10 units on the pain scale? Assume that the SD in the difference scores is approximately 20 units.

Solution:

Given: $\Delta = \mu_d = 10$, $Z_{1-0.05/2} = 1.96$, $Z_{0.8} = 0.84$, and $\sigma_d = 20$. Then, substituting these values to compute the sample size (23.22)

$$n = \sigma_d^2\left(\frac{Z_{1-\alpha/2} + Z_{1-\beta}}{\Delta}\right)^2 = 20^2\left(\frac{1.96 + 0.84}{10}\right)^2 = 32$$

A sample of size $n=32$ patients with migraine will ensure that a two-sided test with $\alpha=0.05$ has 80% power to detect a mean difference of 10 points in pain before and after treatment, assuming that all 32 patients complete the treatment.

23.3.11 Sample Size Determination for Two Independent Samples, Dichotomous Outcomes

In studies where the objective is to perform a test of hypothesis comparing the proportions of successes in two independent populations, the hypotheses of interest are

$H_0: p_1 - p_2 = 0$ against
$H_1: p_1 - p_2 \neq 0 = \Delta$
$H_1: p_1 - p_2 > 0 = \Delta$
$H_1: p_1 - p_2 < 0 = \Delta$,

where p_1 and p_2 are the proportions in the two comparison populations and Δ is the difference in proportions between the two groups expected under the alternative hypothesis, H_1. The formula for determining the sample sizes to ensure that the test has a specified power is given below:

$$\text{Two-sided test}: n_i = 2\left(\frac{Z_{1-\alpha/2} + Z_{1-\beta}}{ES}\right)^2, \text{ where Effect size} = ES = \frac{\Delta}{\sqrt{\overline{p}(1-\overline{p})}} = \frac{p_1 - p_2}{\sqrt{\overline{p}(1-\overline{p})}} \quad (23.24)$$

$$\text{One} - \text{sided test}: n_i = 2\left(\frac{Z_\alpha + Z_{1-\beta}}{\text{ES}}\right)^2, \text{ where Effect size} = \text{ES} = \frac{\Delta}{\sqrt{\overline{p}(1-\overline{p})}} = \frac{p_1 - p_2}{\sqrt{\overline{p}(1-\overline{p})}}, \qquad (23.25)$$

where n_i is the sample size in each group (assuming equal-sized groups, $i=1,2$), Z_β represents the desired power, $Z_{1-\alpha/2}$ represents the desired level of statistical significance, $p_1 - p_2$ is the difference in proportions between the two groups expected under the alternative hypothesis, H_1, and $\overline{p}$ is the overall proportion based on pooling the data from the two comparison groups (it can be computed by taking the mean of the proportions in the two comparison groups, assuming that the groups will be of approximately equal size and $\overline{p}(1-\overline{p})$ is a measure of variability similar to the SD).

Sample size calculation can also be done when $n_1 \neq n_2$. In this situation, we let n_2 be some multiple r of n_1; then we substitute $(r+1)/r$ for 2 in the sample size formulas (23.24 and 23.26). After solving for n_1, we substitute $n_2 = rn_1$. The general sample size formula is

$$\text{Two} - \text{sided test}: n_1 = \frac{r+1}{r}\left(\frac{Z_{1-\alpha/2} + Z_{1-\beta}}{\text{ES}}\right)^2, \text{ where Effect size} = \text{ES} = \frac{\Delta}{\sqrt{\overline{p}(1-\overline{p})}} = \frac{p_1 - p_2}{\sqrt{\overline{p}(1-\overline{p})}}$$

$$(23.26)$$

$$\text{One} - \text{sided test}: n_1 = \frac{r+1}{r}\left(\frac{Z_\alpha + Z_{1-\beta}}{\text{ES}}\right)^2, \text{ where Effect size} = \text{ES} = \frac{\Delta}{\sqrt{\overline{p}(1-\overline{p})}} = \frac{p_1 - p_2}{\sqrt{\overline{p}(1-\overline{p})}},$$

$$(23.27)$$

where n_1 is the sample size of the small group and r is the ratio of the larger group to the smaller group.

Example 23.14

I am going to run a case-control study to determine if pancreatic cancer is linked to drinking coffee. If I want 80% power to detect a 10% difference in the proportion of coffee drinkers among cases vs. controls (if coffee drinking and pancreatic cancer are linked, we would expect that a higher proportion of cases would be coffee drinkers than controls), how many cases and controls should I sample? About half the population drinks coffee and a two-sided test of hypothesis will be conducted at $\alpha=0.05$.

Solution:

Given: $\Delta = p_1 - p_2 = 0.1$, $Z_{1-0.05/2} = 1.96$, $Z_{0.8} = 0.84$, $\overline{p} = 0.5$, and $r = 1$. Then, substituting these values to compute the sample size (23.26)

$$n_i = 2\overline{p}(1-\overline{p})\left(\frac{Z_{1-\alpha/2} + Z_{1-\beta}}{p_1 - p_2}\right)^2 = 0.5\left(\frac{1.96 + 0.84}{0.1}\right)^2 = 392$$

Therefore, we would take 392 cases and 392 controls to have 80% power. The total sample is 784.

Example 23.15

How many total cases and controls would I have to sample to get 80% power for the same study, if I sample 2 controls for every case? Ask yourself, what changes here? In this case r

$$n_1 = \frac{r+1}{r}\bar{p}(1-\bar{p})\left(\frac{Z_{1-\alpha/2}+Z_{1-\beta}}{p_1-p_2}\right)^2 = 0.375\left(\frac{1.96+0.84}{0.1}\right)^2 = 294$$

Therefore, we need 294 cases and $2\times294=588$ controls. The total sample size is 882.

Note you get the best power for the lowest sample size if you keep both groups equal (882>784). You would only want to make groups unequal if there was an obvious difference in the cost or ease of collecting data on one group. For example, cases of pancreatic cancer are rare and take time to find.

23.3.12 Sample Size Determination for Comparing Two Proportions from Paired Samples

In studies where the objective is to conducting a statistical test for proportion difference based on paired data, the hypotheses of interest for one- and two-sided tests are

H_0: $p_1-p_2=0$ against
H_1: $p_1-p_2 \neq 0=\Delta$
H_1: $p_1-p_2>0=\Delta$
H_1: $p_1-p_2<0=\Delta$,

where p_1 and p_2 are the proportions based on paired data and Δ is the difference in proportions between the two groups expected under the alternative hypothesis, H_1.

The sample size for the paired data in proportion is

$$\text{Two}-\text{sided test}: n = \left(\frac{Z_{1-\alpha/2}+Z_{1-\beta}}{ES}\right)^2, \text{ where Effect size} = ES = \frac{\Delta}{\sqrt{\bar{p}(1-\bar{p})}} = \frac{p_1-p_2}{\sqrt{\bar{p}(1-\bar{p})}} \quad (23.28)$$

$$\text{One}-\text{sided test}: n = \left(\frac{Z_{\alpha}+Z_{1-\beta}}{ES}\right)^2, \text{ where Effect size} = ES = \frac{\Delta}{\sqrt{\bar{p}(1-\bar{p})}} = \frac{p_1-p_2}{\sqrt{\bar{p}(1-\bar{p})}}, \quad (23.29)$$

where n is the minimum sample size and $\Delta=p_1-p_2$ is the clinically meaningful difference in dependent proportions expected under the alternative hypothesis H_1.

23.4 Estimates of Sampling Errors

The estimates from a sample survey can be affected by two types of errors: sampling errors and nonsampling errors. Nonsampling errors are the results of mistakes made in implementing data collection and data processing, such as misunderstanding the questions by either the interviewer or the respondent, and data entry errors. Nonsampling errors are difficult to avoid and evaluate statistically.

On the other hand, sampling errors can be evaluated statistically. Sampling error is the error caused by observing a sample instead of the whole population. The sample of respondents selected from a survey is only one of many samples that could have been selected from the same population using the same design and the expected size. Each of those samples may yield results that differ somewhat from the results of the actual sample selected. Sampling errors are a measure of the variability between all possible samples and represent the approximate amount of variance that we can expect if we ran the same poll with a different sample. Even though the degree of variability is not known exactly, it can be estimated statistically from the survey results.

Sampling error is usually measured in terms of the standard error for a particular statistic (such as percentage or mean), which is the square root of the variance. The standard error can be used to calculate CIs within which the true value for the population can reasonably be assumed to fall.

23.5 Conclusion

The determination of the appropriate sample size is a critical component in study design. Determining the appropriate design of a study is more essential than the statistical analysis. A poorly designed study can never be improved, while a poorly analyzed study can be reanalyzed. Therefore, in estimating a certain characteristic of a population, sample size calculations are essential to ensure that estimates are obtained with required precision or confidence. Moreover, having the proper sample size is crucial for finding a statistically significant result.

References

National Center for Health Statistics. *Health, United States, 2005 with Chartbook on Trends in the Health of Americans.* Hyattsville, MD: US Government Printing Office; 2005.

Rutter MK, Meigs JB, Sullivan LM, D'Agostino RB, Wilson PW. C-reactive protein, the metabolic syndrome and prediction of cardiovascular events in the Framingham Offspring Study. *Circulation.* 2004;110:380–385.

Wechsler H, Lee JE, Kuo M, Lee H. College binge drinking in the 1990s: A continuing problem. Results of the Harvard School of Public Health 1999 College Alcohol Study. *Journal of American College Health,* 2000;48:199–210.

Part 5

New Approaches to Statistical Learning in Developing Countries

24

Active and Agnostic: A Multidisciplinary Approach to Statistical Learning

Leonardo César Teonácio Bezerra

Federal University of Rio Grande do Norte

CONTENTS

24.1 Introduction

Statistical learning (SL) is a multidisciplinary field in which professionals with a solid background in statistics are expected to develop models to be used by professionals with a deep understanding of its application domain. The former are generally referred to as *data scientists*, whereas the latter are known as *business analysts* (Provost and Fawcett, 2013). These multidisciplinary teams are present in most *business intelligence* (BI) departments of competitive companies. Thus, a proper understanding of both model and application domains is critical for an effective team. In this work, we describe a learning approach to SL that is intended to better meet that requirement. Specifically, our goal is to promote the inclusion of students from backgrounds other than STEM (science, technology, engineering, and math) who generally do not take part in SL courses, and thus better leverage the benefits of multidisciplinarity within the courses.

To achieve this goal, we focus on two major factors we believe that have deterred non-STEM students from participating in traditional SL courses. The first factor is the lack of interaction, both among students and between students and professors. Specifically, STEM-related courses frequently rely heavily on keynotes and assignments, where large audiences watch the course lead professor lecture and refer to teaching assistants for guidance on the assignments (Freeman et al., 2014). In developing countries, a modified version of this model is employed, as teaching assistant resources are far more limited. Socioeconomic reality renders this passive learning approach even less nurturing of human interaction, and students from fields in which active learning approaches are adopted tend to have difficulty adapting to this formal structure.

DOI: 10.1201/9781003261148-29

The second factor that likely keeps non-STEM students out of SL courses is the need for above-average information technology (IT) skills, particularly computational thinking (CT). In developing countries, the discussion on the inclusion of CT courses in secondary education is still incipient (MEC, 2017; SBC, 2017), and even university-level courses are often only available in STEM-related curricula. More importantly, a large percentage of students that attend STEM-related courses fail, and CT courses are no exception to this phenomenon (Freeman et al., 2014). This educational reality renders SL courses even less accessible to non-STEM students, whose courses often make very limited use of IT resources.

In the proposed approach, students are assigned to groups to organize tutorials and develop projects that actively promote interaction and leverage multidisciplinarity. While tutorials encourage students to actively discuss relevant concepts in terms that are simpler and easier for them, projects allow them to investigate topics that they deem most interesting. Furthermore, the educational objects adopted combine both graphical user interface (GUI) and command-line interface (CLI) tools to increase the digital accessibility of SL courses. The former, such as Spreadsheets and Orange3, present a smooth learning curve and are particularly effective for the interactive exploration of data visualization and unsupervised learning concepts. The latter, such as Colaboratory and Jupyter, render powerful SL libraries such as Pandas, Scikit-learn, Keras, Seaborn, and Plotly accessible to non-STEM students. More importantly, educational objects based on CLI tools are designed to use only high-level programming concepts, such as names, values, and methods. Indeed, even basic structured programming concepts such as decision and repetition are vetted.

To leverage multidisciplinarity, group composition is based on student segmentation, which is done on the fly on the first day of class with data that students help collect and analyze. In addition, segmentation also facilitates the agnostic aspect of the approach, balancing the IT-skill levels of the group. More recently, feedback provided by students has led to a variant of this approach in which the group composition for tutorial presentations differs from the group composition for project development. This further promotes interaction among students.

In the remainder of this work, we will further ground and detail the methodology adopted (Section 24.2) and describe a pilot project ongoing in Brazil (Section 24.3). So far, five classes from three different courses composed of approximately 120 students have been executed, and the initiatives adopted to promote the inclusion of non-STEM students have shown benefits across the students enrolled in the courses. Among the most relevant, several deliverables have been made publicly available to the community and/or used for job interviews. We conclude in Section 24.4 discussing future work directions based on the results and challenges observed so far.

24.2 An Active and Agnostic Approach to Multidisciplinary SL

The fundamental principles of the approach we discuss in this work are active learning and IT-agnostic education. Next, we discuss each of these topics, detailing the initiatives to promote them and the context in which they are situated. Furthermore, student segmentation is adopted to enhance the benefits of these principles and is detailed at the end of this section.

24.2.1 Active Versus Passive Learning

Several definitions of active learning have been proposed by various authors. In this work, we use a definition proposed by Cochran (2015), which is an amalgam of definitions proposed by other authors:

> Instructional strategies and exercises designed to engage students through participation in exercises that involve them in higher-order thinking tasks such as analysis, synthesis, and evaluation of course material.

Lectures can effectively transfer information or provide for the demonstration of simple concepts, but they are passive from the student perspective (Hartley and Cameron, 1967; McLeish, 1968). Instructors who utilize active learning exercises and strategies in their classrooms are generally motivated by the beliefs that (1) learning is a naturally active process and (2) understanding of a complex concept is most effectively fostered through student interaction and engagement with other students, the instructor, and the subject matter. This may appear to many instructors to be at odds with how they learn; however, most students differ greatly from their instructor in their aptitude for the subject, their interest in the subject, and their learning modality. An instructor of any relatively technical discipline must avoid the common trap of assuming that s/he and her/his students' learning styles are similar.

Russell et al. (1984) found that students tend to retain more information presented in lectures when the density of new information is relatively low. These authors propose that no more than half of any class meeting be devoted to coverage of new concepts or material. McKeachie (1999) asserts that holding a student audience's attention throughout a long lecture is challenging for most instructors. This could be explained, at least in part, by the results of Hartley and Davies (1978), who find that a typical student's level of attention diminishes rapidly after the first 10 minutes of a typical lecture. These results indicate that overreliance on lectures may quickly create a stagnant learning environment that fails to engage students or challenge them to think critically about or question concepts. Russell et al. (1984) reported that students achieve only a superficial understanding of the concepts and ideas discussed in such learning environments.

Chilcoat (1989) reviewed several research studies that report that changes in the classroom environment during a lecture can reinvigorate student interest and engagement, enabling the instructor to recapture her/his students' attention. Several studies (Cashin, 1985; Brown and Atkins, 1988; Campbell and Smith, 1995; Cochran, 2001a, 2001b) have extended this finding. In related work, Chickering and Gamson (1987) summarized this perspective on learning and education as follows:

> Learning is not a spectator sport. Students do not learn much just by sitting in class listening to teachers, memorizing prepackaged assignments, and spitting out answers. They must talk about what they are learning, write about it, relate it to past experiences, apply it to their daily lives. They must make what they learn part of themselves.

A recent meta-analysis of active learning in STEM publications has discussed the benefits of these approaches in comparison to traditional lecturing (Freeman et al., 2014). Among the most important findings are that benefits are (1) seen for all STEM disciplines, class sizes, course types, and levels; (2) higher in small classes and when students are evaluated through concept inventories; and (3) more observable regarding a reduction in fail rate than on overall scores.

Active learning has gained popularity in some STEM disciplines over the past two decades. Faculty teaching courses in applied mathematics, particularly operations

Skills	Unit 01: the data science process	Project domains
Understanding pipelines o Preparation o Feature engineering o Estimators **Evaluating pipelines** o Sampling o Hyperparameters o Metrics o Comparison **Applying pipelines** o Domains o Tasks	**Tutorial: pandas-zero (2,0pt)** o [10/03] Dataframes as databases (G01) o [10/03] Data analysis and presentation (G02) o [12/03] ETL (G03) o [17/03] Combining information from multiple bases (G04) o [17/03] Merging information from multiple bases (G05) o [19/03] Visualizing and identifying distributions (G06) o [24/03] Relations between features (G07) o [24/03] Visually interacting with data (G08) **Project: domain, goals, data, and EDA (3,0pt)** o [26/03] Groups 01-04 o [31/03] Groups 05-08 **Assessment and conclusion (5,0pt)** o [02/04] Assessment o [07/04] Feedback, discussion, and Unit 02 assignments	• Agroindustry • Bioinformatics • Climate and environment • Cybersecurity • Education • Energy • Financial market • Food and poverty • Fashion • Industry • Health • Law • Robotics • Social networks • Sports

FIGURE 24.1

A machine learning course following the proposed approach. Left: course curriculum- to-skill mapping. Center: schedule for the first unit, where tutorials and projects alternate. Right: generator topics for project development. Prior to Unit 01, students are subject to student segmentation in a practical introduction to data science and machine learning using GUI applications.

research, probability, and statistics, have seen great success with active learning. For a broad discussion, see Rosenthal (1995) or Cochran (2005a, 2009b, 2012). For example, from statistics, see Gnanadesikan, et al. (1997), Kvam (2000), Cochran (2002), Tougaw (2005), or Carlson and Winquist (2011). For results from probability, see Adair et al. (2018), Cochran (2001a, 2004, 2009a), or Keeler and Steinhorst (2001). For results on active learning and operations research, see Liebman (1994, 1998), Cochran (2005b, 2015), or Leseure (2019).

In Brazil, active learning is most iconically represented by Paulo Freire's *Pedagogy of the Oppressed* (Freire, 1972). In summary, Freire discussed (1) education as a central political actor in society, either promoting conformity or emancipation, and (2) the role of the student (background) in education, contrasting a banking model of education (passive) with a problem-posing approach (active). These two core concepts have led to a significant dichotomy in Brazil, being central to non-STEM educational theory but marginal to STEM education.

In an attempt to bridge these worlds, we propose that students act as the central executors of the SL courses. Specifically, each course is based on tutorial hosting and project development, which students are responsible for. *A priori*, the course curriculum is mapped to the skills the students are expected to develop during the course (Figure 24.1, left). The course schedule then alternates tutorial and project weeks, as described in Figure 24.1 (center). While the tutorial assignment is predefined, students are free to choose among a list of generator topics. Next, we further detail tutorial and project dynamics.

Tutorial-based learning: During tutorial weeks, each class comprises one or more tutorials that students must host, preassigned as a function of the course curriculum (Figure 24.1, center). Tutorials must be interactive and combine lectures, discussions, and practice, but students are free to determine how to accomplish this. They are encouraged to produce reusable tutorials, make them available online, and submit them to meetups and events to which they want to contribute. Given the reusability of the tutorials, students may choose to refine tutorials from past courses, creating novel educational objects to enrich the learning process.

Problem-based learning: During project weeks, students must propose and develop a project aligned with one of the generator topics of the course (an example is given in Figure 24.1, right), as well as with the United Nations goals from the 2030 Agenda for Sustainable Development (UN, 2015). In every other aspect, students are free to originally conceive and develop the project. Classes are used for progress reporting and discussion. Specifically, students must describe how the tutorials have helped in the development of the project and get feedback from other students. Since proposing a project at the beginning of the course can be challenging for students, they are encouraged to revisit their project definitions, goals, and schedules throughout the course. More importantly, students are also encouraged to publicize their projects for networking with other students and/or job positions.

24.2.2 IT-agnostic Learning

Given the role of IT in most SL tools, traditional courses require above-average IT skills. Although GUI applications help mitigate this need, some advanced topics in SL are typically confined to programming language libraries. In general, courses that address these libraries require students to be proficient in CT, particularly computer programming. Furthermore, installing those libraries and running algorithms often requires knowledge of operating systems and computer architecture.

In Brazil, access to computing devices is highly imbalanced, and the imbalance is even more dramatic for desktops (IBGE, 2012). Even the private education system is only beginning to integrate CT into its curricula, since the recently proposed national curriculum is still discussing this inclusion (MEC, 2017; SBC, 2017). As a result, the students from the general public who reach postsecondary education cannot be expected to succeed in traditional SL courses as they currently stand. This issue is partially mitigated in STEM curricula, but the introductory CT courses they include are no exception to the high level of failure seen in other STEM courses (Freeman et al., 2014).

To render SL courses accessible to non-STEM students, we propose to (1) adopt GUI applications for their smooth learning curve and (2) produce educational objects on CLI tools devised for an audience that has no background on CT. Furthermore, we adopt Colab as the CLI platform for its interactiveness and abstraction of the more advanced concepts required to install advanced SL libraries.

GUI applications: Introductory concepts such as tabular data, data collection, exploratory data analysis (EDA), and data visualization are easily presented using applications such as Google Forms and Spreadsheets. Form is particularly suited for data collection and basic EDA, whereas Spreadsheets renders simple concepts such as descriptive statistics and data aggregation through pivot tables. In turn, Orange3 furnishes an intuitive workflow perspective of SL, providing powerful data visualization and machine learning widgets. Furthermore, add-ons for Orange3 enable its application to specific domains, such as natural language processing, computer vision, time series analysis, and bioinformatics.

Python libraries: Advanced concepts such as data cleaning and transformations are available through high-level methods in the Pandas library. In addition, libraries such as Seaborn (static) and Plotly (interactive) provide high-level, one-line methods for data visualization. These three libraries are covered in the pandas-zero course available in GitHub,[1] prepared with the help of IT students. More advanced topics such as data preparation

---

[1] https://github.com/leobezerra/pandas-zero.

and machine learning are provided by the Scikit-learn library, which is not as high level as Pandas. Yet, the scikit-zero course[2] was designed as a follow-up on pandas-zero and revisits Scikit-learn to render it accessible to non-STEM students. Finally, our goal is to also provide a course on deep learning (DL) using the Keras library, which provides a high-level approach and has been adopted by many universities for use in introductory DL courses.[3]

Colab as an interactive platform: Interpreted programming languages enable interactive programming through REPL (read–eval–print–loop) tools. In the Python community, Jupyter notebooks are popular REPL tools due to their user-friendly, experience-rich web interface. Colab is a cloud-based REPL inspired in Jupyter notebooks through which Google offers access to its infrastructure. In addition to providing the most relevant SL Python libraries preinstalled, one can also run computationally expensive algorithms such as DL using specialized hardware in a transparent way. User experience is enriched further through the integration with Google Drive and notebook sharing (though concurrent editing is still a work in progress).

24.2.3 Student Segmentation

Though the principles discussed above have been designed to foster the inclusion of non-STEM students, they would likely not be as effective if students were assigned to groups randomly or based on their social relationships. Instead, we adopt student segmentation to ensure that each group comprises distinct backgrounds and IT proficiency levels. This segmentation is done on the fly on the first day of class using data that students help collect and analyze. In addition to improving the efficacy of the principles behind this approach, on-the-fly segmentation introduces students to the data science process as a whole in a practical and personal way.

More recent feedback provided by students led to a variant of this approach in which each student is assigned to two groups (one for tutorial presentation and another for project development). The goal is to maximize experience interchange between students, as tutorial and project experience are seen as complementary. In its current version, group assignment is performed at random so long as the groups comprise different profiles as identified during segmentation. No specific algorithm is adopted to minimize the overlap between groups, as random assignment already provides good results. Students who miss the first class are classified by their profiles and assigned to groups for which they would not compromise the existing balance nor increase the overlap between tutorial and project group composition.

24.3 A Pilot Project in Brazil

Over the past year, the approach proposed in this chapter has been evaluated at Universidade Federal do Rio Grande do Norte, in Natal, Brazil. Table 24.1 details the courses held so far,

---

[2] https://github.com/leobezerra/scikit-zero.
[3] Though even basic programming concepts such as structured programming are not required in all repositories mentioned, students who still struggle with Python syntax are referred to the python-zero course also available in the same GitHub account, where the first two tutorials cover most of the topics that improve syntax understanding.

TABLE 24.1

Courses Held in a Pilot Project at Federal University of Rio Grande do Norte Over 2019–2020

Level	Curriculum	Course	Number of Classes	Total Students	Non-IT Students
Undergraduate	Information technology	Data science	1	36	5
Undergraduate	Information technology	Machine learning	1	39	9
Graduate	Information technology	Data mining	3	42	9

which have included approximately 120 students. About 20% of these students did not present a strong IT background but succeeded on the courses. Our brief discussion of the most relevant outcomes of the courses follows.

Interactive tutorials: The pandas-zero and scikit-zero tutorials are a direct product of these courses. Specifically, students produce the initial version of each notebook, which is later reviewed and made available on GitHub. A secondary benefit of this process is that many students are able to contribute to an open-source project for the first time.

Tutorials for social media: The ds-zero repository[4] gathers Medium posts produced by students about SL topics and projects they developed. Different from the interactive tutorials, the goal of these posts is to foster science communication and data storytelling.

Social impact: Tutorials and projects have led students to succeed in job applications and participate in community events. Moreover, during the COVID-19 pandemic, tutorials were used for live broadcastings and to help in data science projects.

In addition to the positive outcomes described above, it is important to discuss the most meaningful challenges observed.

Adaptation to active methodologies: Given their history with passive methodologies, some students find it difficult to integrate into active approaches. This was more evident for younger students who were used to sitting back and listening, focusing only on exams. By contrast, more mature students stand out in the courses given what their background brings to the classes.

Social unease: STEM students generally struggle with social interactions. In addition to public speaking, the course requires the use of soft skills such as teamwork and conflict resolution, which IT students find particularly difficult. Furthermore, student segmentation can lead to groups in which students did not know each other previously. Though typical in the market, this is not as often in academia and may worry some students.

Privacy concerns: Older students are often unfamiliar with open-source initiatives and feel uncomfortable sharing the outcomes of their work publicly. Indeed, one class comprising mostly students who had been on the market for over a decade produced some of the best tutorials, but these students generally did not feel confident enough to make them public.

Curriculum limitations: Though the methodology described here focuses on the inclusion of non- STEM students, traditional academic curricula often include course prerequisites that limit this benefit. Indeed, this pilot project was only feasible due to special programs created by the university to let students enroll in the courses without fulfilling those prerequisites. Even so, the percentage of students allowed to do so was limited since it was a pilot project.

---

[4] https://github.com/leobezerra/ds-zero.

24.4 Conclusion

SL is a field in which multidisciplinarity is deeply rooted, leading to company teams composed of members who have contrasting and complementary backgrounds. In contrast to this market pattern, academia has generally produced silos in which courses focus either on model developing or on business insights, with a clear distinction between the audiences that participate in either. As most undergraduate curricula include at least some introductory courses on statistics, students from degree programs other than STEM should be expected to participate in SL courses. Furthermore, when the audience in a[5] course is mixed with respect to STEM background and preparation, it should be expected that the outcomes of the courses would represent the combined backgrounds of those students.

This work has described an approach to leverage multidisciplinarity in SL courses, promoting the inclusion of non-STEM students through an active and IT-agnostic methodology. These principles are further bolstered by student segmentation, which helps mix the different backgrounds students present and balance the unequal IT skills of the members. An ongoing pilot project in Brazil is demonstrating the benefits of this approach and identifying challenges for future work to focus on.

Topic-wise, we plan on devising a course on DL based on Keras, with the main goal of producing a Keras-zero tutorial repository following the same principles of pandas-zero and scikit- zero. Challenge-wise, we plan on hosting an outreach program that would overcome prerequisite issues and expand the courses to people in the market who could also help enrich student interaction. Finally, we plan on collaborating with the pedagogical and psychological units of the university to help develop soft skills in STEM students, helping them integrate more easily into the proposed approach.

References

Adair, D., Jaeger, M., and Price, O.M. 2018. Promoting Active Learning when Teaching Introductory Statistics and Probability Using a Portfolio Curriculum Approach, *International Journal of Higher Education*, 7(2), 175–188.

Brown, G. and Atkins, M. 1988. *Effective Teaching in Higher Education*. Methune, London.

Campbell, W.E. and Smith, K.A. (eds.). 1995. *New Paradigms for College Teaching*. Jossey-Bass Publishers, San Francisco CA.

Carlson, K.A. and Winquist, J.R. 2011. Evaluating an Active Learning Approach to Teaching Introductory Statistics: A Classroom Workbook Approach. *Journal of Statistics Education*, **19**(1).

Cashin, W.E. 1985. *Improving Lectures*. IDEA Paper No. 14. Kansas State University Center for Faculty Evaluation and Development, Manhattan, KS.

Chickering, A.W. and Gamson, Z.F. 1987. Seven Principles for Good Practice. *AAHE Bulletin*, **39**, 3–7.

Chilcoat, G.W. 1989. Instructional Behaviors for Clearer Presentations in the Classroom. *Instructional Science*, **18**, 289–314.

Cochran, J.J. 2001a. Probability, Stats & Playing Games. *ORMS Today*, **28**(2), 14.

Cochran, J.J. 2001b. Who Wants To Be A Millionaire®: The Classroom Edition. *INFORMS Transactions on Education*, **1**(3), 112–116.

---

[5] Though the ultimate goal of the proposed approach is to reach non-STEM students, current limitations of the university still restrict access from those students to the courses offered so far. For this reason, we group non-STEM students with STEM students without a strong IT background in this analysis.

Cochran, J.J. 2002. Data Management, Exploratory Data Analysis, and Regression Analysis with 1969–2000 Major League Baseball Attendance Data, *The Journal of Statistics Education*, **10**(2).

Cochran, J.J. 2004. Bowie Kuhn's Worst Nightmare: An Integer Programming & Simpson's Paradox Case, *INFORMS Transactions on Education*, **5**(1), 18–36.

Cochran, J.J. 2005a. Can You *Really* Learn Basic Probability by Playing Sports Board Games? *The American Statistician*, **59**(3), 266–272.

Cochran, J.J. 2005b. Active Learning for Quantitative Courses, Chapter 9 of *TutORials*, J.C. Smith, ed., INFORMS.

Cochran, J.J. 2009a. All of Britain Must Be Stoned! An Effective Introductory Probability Case. *INFORMS Transactions on Education*, **10**(2), 62–64.

Cochran, J.J. 2009b. Pedagogy in Operations Research: Where Have We Been, Where Are We Now, and Where Should We Go? *ORiON*, **25**(2), 161–184.

Cochran, J.J. 2012. You Want Them to Remember? Then Make It Memorable! *European Journal of Operational Research*, **219**(3), 659–670.

Cochran, J.J. 2015. Extending "Lego® My Simplex", *INFORMS Transactions on Education*, **15**(3), 224–231.

Freeman, S., Eddy, S.L., McDonough, M., Smith, M.K., Okoroafor, N., Jordt, H., and Wenderoth, M.P. (2014). Active Learning Increases Student Performance in Science, Engineering, and Mathematics. *Proceedings of the National Academy of Sciences*, **111**(23), 8410–8415.

Freire, P. (1972). *Pedagogy of the Oppressed*. 1968. Trans. M. B. Ramos. Herder, New York.

Gnanadesikan, M., Scheaffer, R.L., Watkins, A.E., and Witmer, J.A. 1997. An Activity-Based Statistics Course. *Journal of Statistics Education*, **5**(2).

Hartley, J. and Cameron, A. 1967. Some Observations on the Efficiency of Lecturing. *Educational Review*, **20**, 30–37.

Hartley, J. and Davies, I.K. 1978. Note-Taking: A Critical Review. *Programmed Learning and Educational Technology*, **15**, 207–224.

Instituto Brasileiro de Geografia e Estatística. 2012. Pesquisa nacional por amostra de domicílios contínua – PNAD contínua. https://www.ibge.gov.br/estatisticas/sociais/populacao/9171-pesquisa-nacional-por-amostra-de-domicilios-continua-mensal.html?=&t=o-que-e.

Keeler, C. and Steinhorst, K. 2001. A New Approach to Learning Probability in the First Statistics Course. *Journal of Statistics Education*, **9**(3).

Kvam, P.H. 2000. The Effect of Active Learning Methods on Student Retention in Engineering Statistics. *The American Statistician*, **54**(2), 136–140.

Leseure, M. 2019. Teaching Operations Planning at the Undergraduate Level. *Sage Open*, **9**, 2158244019855854.

Liebman, J.S. 1994. New Approaches in Operations Research Education. *International Transactions in Operational Research*, **1**(2), 189–196.

Liebman, J.S. 1998. Teaching Operations Research: Lessons from Cognitive Psychology. *Interfaces*, **28**(2), 104–110.

McKeachie, W.J. 1999. *McKeachie's Teaching Tips*, 10th ed. Houghton Mifflin, Boston, MA.

McLeish, J. 1968. *The Lecture Method*. Cambridge Monographs on Teaching Methods. Cambridge Institute of Education, Cambridge MA.

Ministério da Educação e Cultura. 2017. Base Nacional Comum Curricular. http://basenacionalcomum.mec.gov.br/.

Provost, F. and Fawcett, T. 2013. *Data Science for Business: What You Need to Know about Data Mining and Data-Analytic Thinking*. O'Reilly Media, Inc., Sebastopol, CA.

Rosenthal, J.S. 1995. Active-Learning Strategies in Advanced Mathematics Classes. *Studies in Higher Education*, **20**, 223–228.

Russell, I.J., Hendricson, W.D., and Herbert, R.J. 1984. Effects of Lecture Information Density on Medical Student Achievement. *Journal of Medical Education*, **59**, 881–889.

Sociedade Brasileira de Computação. 2017. Nota técnica sobre a BNCC (Ensino médio e fundamental). Technical Report. https://www.sbc.org.br/institucional-3/cartas-abertas/summary/93-cartas-abertas/1197-nota-tecnica-sobre-a-bncc-ensino-medio-e-fundamental.

Tougaw, D. 2005. Integration of Active Learning Exercises into a Course on Probability and Statistics. *Computer Science*, **1**(1).

United Nations. 2015. Transforming Our World: The 2030 Agenda for Sustainable Development. https://sustainabledevelopment.un.org/post2015/transformingourworld.

25

Modernizing the Curricula of Statistics Courses through Statistical Learning

Marcus Alexandre Nunes

Federal University of Rio Grande do Norte

CONTENTS

25.1 Introduction

Statistics is an evolving field. The past 20 years have shown how this discipline changed. While new statistical models have appeared, computers became faster and cheaper, allowing professionals to apply these new methods with little trouble. Moreover, more data are generated, collected, and made accessible. Therefore, everyone with Internet access can download new data and analyze it independently. While traditional concepts such as inference, aggregation, likelihood, and experimental design are still valid, there are real-world problems that cannot be addressed in the same fashion as before.

Data visualization, for example, is now much more developed than it was 10 years ago. Consideration of concepts such as color palettes, font options, and different plot types is more commonplace. There are many free statistical software options available for every operational system. New plot types are becoming more popular on a daily basis. Dashboards are simpler to build and update. With this in mind, we believe that universities' undergraduate curricula should be adapted to this new world.

In 2014, the American Statistical Association (ASA) published its guidelines on the undergraduate curriculum of statistics. These guidelines can be summarized in four key points:

1. Increased importance of data science;

2. Real applications;

3. More diverse models and approaches;

4. Ability to communicate.

DOI: 10.1201/9781003261148-30

In this chapter, we report how these guidelines have been applied in a course called *Introduction to Big Data Modeling*, offered since 2015 at the Federal University of Rio Grande do Norte, Brazil. We also report on students' impressions of the discipline as well as some of the tasks that were proposed for them during that time.

Introduction to Big Data Modeling is offered regularly as an elective course to second-year students. Its prerequisites are basic statistical inference (*t*-test, ANOVA, simple linear regression) and R programming. It is not a course in which deep mathematical understanding is requested. Few proofs are presented during the course, and the mathematical requirements are equivalent to a statistics 300-level course.

This chapter is structured as follows: in Section 25.2, we show how the course is structured in relation to the ASA Guidelines. In Section 25.3, we present in detail the web-scraping module of the course, showing how we teach our students to collect, process, and prepare real-world data for analysis. We use Section 25.4 to discuss the results of this course and our future plans for it.

25.2 Relation between Introduction to Big Data Modeling and ASA Guidelines

25.2.1 Increased Importance of Data Science

The impact of data science has been increasing during the past years (Donoho, 2017). One simple Google search for the terms "data science" shows how the interest for this term has increased tenfold during the 2010s, as we can see below (Figure 25.1).

But what is data science? There are many available definitions. We understand data science to be an interdisciplinary field in which statistical and computer science techniques are merged. In our experience, the best results in modeling are obtained by combining these two techniques: the statistical base and the computational power.

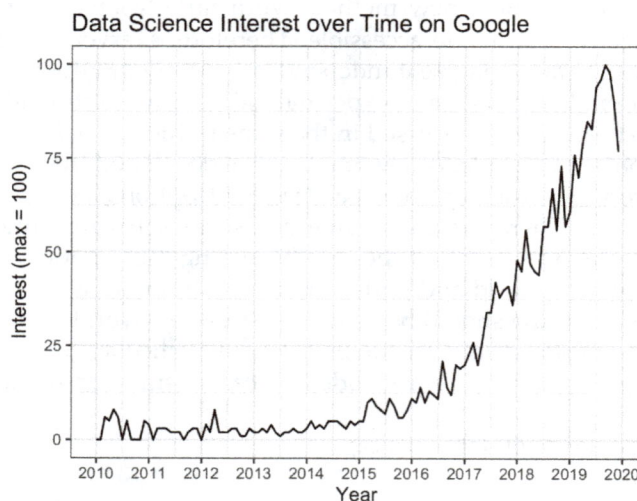

FIGURE 25.1
Search volume for "data science" in Google from 2010 to 2020. (Data obtained from Google Trends.)

Introduction to Big Data Modeling takes this approach to teach students how to use computational power to solve problems that are not solvable with traditional statistical methods. For example, it is easier to fit a random forest model instead of looking for a parametric distribution that works well with a generalized linear model.

Data science is concerned with data visualization. Many different types of charts are used to generate insights and ideas for analysis, data cleaning, and feature engineering. Our course places great emphasis on data visualization because it is important to show the students how good data visualization is the first step in a successful data analysis.

In addition to the graphical representation of information, this course shows the students how to fit predictive models using machine learning methods. After all, data science is known for the intense use of machine learning and computational resources in the data modeling step. Methods such as cross-validation, random forest, support vector machines, and others demand assets that cannot always be fulfilled by spreadsheets. Therefore, a programming language is needed.

Our programming language of choice is R. While Python is the most popular choice for machine learning modeling, we believe there is no fundamental difference between them. Since R and Python are multiparadigm object-oriented programming languages with many functions written for data science, they have similar performance in this field. We prefer to use R in the course because during the fourth semester, the students have already have been trained in R. Therefore, we can start the data visualization and modeling part at the onset of the course, with no need to teach students how to load data, create loops, or apply functions to objects.

25.2.2 Real Applications

One of the pillars of our *Introduction to Big Data Modeling* is the use of real datasets. We understand that the students are more motivated when they see data collected from the real world (Hicks and Irizarry, 2016). We use simple datasets, such as Fisher's Iris dataset, but we also use very complex data, such as the players' attributes from the electronic game FIFA Soccer.

However, there is a spectrum of complexity in the datasets used in this course. One of the first datasets analyzed in the course is Fisher's Iris dataset, a small but interesting collection of flower measurements. Despite its 150 observations in 5 columns, it is possible to run classification and regression algorithms and achieve fast and interesting results. It is interesting to start with this dataset because it does not need to be cleaned or preprocessed, allowing the students to immediately start employing their knowledge in a data science application.

As the course advances, the datasets become more complex. They need to be cleaned, removing observations or grouping them. Categorical variables must be converted to numerical and numerical variables need to be standardized. The students need to learn how to deal with difficult data and we believe the best way to learn the desired skills is by practicing and applying new concepts little by little.

The Internet is full of great sources with interesting datasets. Many federal governments around the world make some of theirs freely available. The US Government's open data can be found at https://www.data.gov/. The Brazilian Institute of Geography and Statistics hosts census results at https://downloads.ibge.gov.br/ (in Portuguese).

There are nongovernmental data sources that provide high-quality data for the classes. Thatmost used in this course is the UCI Machine Learning Repository (https://archive.ics.uci.edu/ml/index.php), maintained by the University of California, Irvine. It has many of the standard datasets used as benchmarks for machine learning algorithms.

More recently, Kaggle (https://www.kaggle.com/datasets) has emerged as one of the most popular places for hosting datasets and machine learning competitions. As of May 2020, Kaggle lists more than 37,000 datasets available for download and free to use.

Besides the datasets provided by the instructor, the students have to work on a project of their own, analyzing a dataset chosen by themselves. The only restriction they have to follow is to use a dataset not analyzed before during the course.

This approach has yielded promising results. When the students are free to choose anything, they tend to focus on subjects in which they have a personal interest. Our students have analyzed data on songs available in Spotify, UFO sightings, evolution of the Human Development Index (HDI) in Brazil in the past three decades, and many other topics.

25.2.3 More Diverse Models and Approaches

While many courses in undergraduate level choose to show fewer modeling techniques to the students, proving results and going deep on the math behind them, in our course, we go in the opposite direction. We prefer to present models on a user level, focusing on the model's strengths and limitations. The students are required to intuitively know how the algorithms work, but the majority of the mathematical proofs are omitted. The main result we want from this course is that the students learn when the algorithms work and when they do not.

The list of models taught in the course has changed over time. In its first iteration, one of the classification techniques the students had to learn was linear discriminant analysis (LDA). However, as the course evolved, we decided that the students were working with too many linear classification methods. We changed to random forests, a more general method capable of dealing with nonlinear problems.

Introduction to Big Data Modeling is a 15-week course offered during our spring semester. It lasts for one semester, but that has proven to be sufficient to show the students the main ideas behind validation techniques, data preprocessing, dimension reduction, and many diverse models. These were the subjects covered the last time *Introduction to Big Data Modeling* was offered:

- Data preparation
- *k*-means
- Hierarchical clustering
- Principal components analysis
- Data acquisition
- Cross-validation
- *K* nearest neighbor
- Support vector machine
- Classification and regression trees
- Random forests
- Model ensemble

We spend roughly one week on each topic. Half of the time the students have expository lectures, where they learn the basics behind the methods they will learn, and half of the time they practice what they learn, applying their knowledge to solve real-world problems.

25.2.4 Ability to Communicate

The course also has an important communication component. In addition to the report the students need to write, their final project must be presented in front of an audience. Each group has 15–20 minutes to show their results to the other students in a slide presentation. Usually, the presentation is based on slide presentations. However, some students like to experiment and build dashboards to present their results.

The presentations are not graded only for their content, but also for their delivery. The students must demonstrate that they know the subject they are talking about. It is important for them to explain the dataset(s) they are using, how the data were obtained, what modeling worked best, what has failed, and so on.

In addition, the students are encouraged to ask questions of their peers, creating a respectful environment where everybody learns from each other.

25.3 Case Study

The case study we present here is the project the students have to complete on the web-scraping module. This is the fifth module of the course. By this point, students already know how to prepare the data for analysis (cleaning, transforming, and removing outliers) and have experience with three multivariate analysis methods: principal component analysis, *k*-means, and hierarchical clustering.

We believe the best way to engross the students in statistics is to engage them with projects to which they can relate. The project we propose for web scraping is to automatically extract data from Brazilian cities using Wikipedia in Portuguese. While the data are in Portuguese, it is very easy to translate those into English and understand what is happening in this project.

For this course, our programming language of choice is R. According to our knowledge, R is one the most used (if not the most used) languages in statistics departments around the world. Besides that, R is free, so money for software is not a bottleneck for the students. Other languages, such as python and Julia, could be used as well.

First, the students need to load the packages needed for the analysis. For this particular project, there are five packages needed:

- rvest: downloads web pages and extract the information we are interested in;
- dplyr: data manipulation, such as filtering and merging;
- ggplot2: used for plots;
- stringr: string cleaning;
- scales: formats plot labels.

Each package can be installed using the command install.packages(package), where package stands for each one of the five packages listed above. After the installation, they need to be loaded into R memory:

```
library(rvest)
library(dplyr)
library(ggplot2)
```

```
theme_set(theme_bw()) # set plot style
library(stringr)
library(scales)
```

The first step in the analysis is to define the web page address we want to download. We are interested in the link

https://pt.wikipedia.org/wiki/Lista_de_municípios_do_Brasil_por_ população, which has a table with Brazilian population data per city. In the code below we inform it to R and download its contents using the function read_html.

```
url <- "https://pt.wikipedia.org/wiki/
Lista_de_munic%C3%ADpios_do_Brasil_por_popula%C3
%A7%C3%A3o"
population <- url %>%
  read_html()
```

After downloading the web page, we need to extract the information we want from it. In this case, we will use the function html_table to obtain the table already present.

```
population <- population %>%
 html_table(fill=TRUE)

population <- population[[1]]
```

Since its column names are in Portuguese, we will translate them and check the result:

```
names(population) <- c("Position", "IBGE.Code", "City", "State",
"Population")
```

```
head(population)
```

##	Position	IBGE.Code	City	State	Population
## 1	1°	3550308	São Paulo	São Paulo	12252023
## 2	2°	3304557	Rio de Janeiro	Rio de Janeiro	6718903
## 3	3°	5300108	Brasília	Distrito Federal	3015268
## 4	4°	2927408	Salvador	Bahia	2872347
## 5	5°	2304400	Fortaleza	Ceará	2669342
## 6	6°	3106200	Belo Horizonte	Minas Gerais	2512070

After the population data are obtained, we proceed in a similar fashion to obtain the city areas in square kilometers:

```
url <- "https://pt.wikipedia.org/wiki/Lista_de_munic%C3%ADpios_
brasileiros_por_%C3%A1re a_decrescente"
```

```
area <- url %>%
  read_html()
```

```
area <- area %>%
  html_table(fill=TRUE)
```

```
area <- area[[1]]
```

```
names(area) <- c("Position", "City", "IBGE.Code", "State", "Area")

head(area)
```

##	Position	City	IBGE.Code	State	Area
## 1	1	Altamira	1500602	Pará	159 695,938
## 2	2	Barcelos	1300409	Amazonas	122 475,728
## 3	3	São Gabriel da Cachoeira	1303809	Amazonas	109 184,896
## 4	4	Oriximiná	1505304	Pará	170 602,992
## 5	5	Tapauá	1304104	Amazonas	89 324,259
## 6	6	São Félix do Xingu	1507300	Pará	84 212,426

Notice that both datasets have a column called IBGE.Code. IBGE stands for Instituto Brasileiro de Geografia e Estatística (English: Brazilian Institute of Geography and Statistics). Every city in Brazil has a unique IBGE code associated with it. This column is important to handle city misnaming when we join the datasets. The following code makes a new dataset with all the data we have so far, organizing the information according to the IBGE code:

```
brazil <- left join(population, area, by = "IBGE.Code")

head(brazil)
```

##	Position.x	IBGE.Code	City.x	State.x	Population	Position.y
## 1	1	3550308	São Paulo	São Paulo	12252023	966
## 2	2	3304557	Rio de Janeiro	Rio de Janeiro	6718903	1236
## 3	3	5300108	Brasília	Distrito Federal	3015268	259
## 4	4	2927408	Salvador	Bahia	2872347	1903
## 5	5	2304400	Fortaleza	Ceará	2669342	3336
## 6	6	3106200	Belo Horizonte	Minas Gerais	2512070	3215

##	City.y	State.y	Area
## 1	São Paulo	São Paulo	1 522, 986
## 2	Rio de Janeiro	Rio de Janeiro	1 182,296
## 3	Brasília	Distrito Federal	5 801,937
## 4	Salvador	Bahia	706,799
## 5	Fortaleza	Ceara	313,140
## 6	Belo Horizonte	Minas Gerais	330,954

Notice that some columns have duplicate data, such as City.x and City.y. It is redundant to have both in the same dataset. We remove the columns we do not want in the final dataset and rename them accordingly:

```
brazil <- brazil %>%
  select(City.x, State.x, Area, Population)

names(brazil) <- c("City", "State", "Area", "Population")

head(brazil)
```

##	City	State	Area	Population
## 1	São Paulo	São Paulo	1 522,986	12252023
## 2	Rio de Janeiro	Rio de Janeiro	1 182,296	6718903
## 3	Brasília	Distrito Federal	5 801,937	3015268
## 4	Salvador	Bahia	706,799	2872347
## 5	Fortaleza	Ceará	313,140	2669342
## 6	Belo Horizonte	Minas Gerais	330,954	2512070

However, area is not ready for analysis, as there are spaces and decimal separators in it. This will make R understand area not as numbers, but as strings. Therefore, we need to clean these values and transform them into a numeric variable.

```
brazil <- brazil %>%
  # Area transformation
  mutate(Area = str_replace(Area, "[[:space:]]", "")) %>%
  mutate(Area = str_replace(Area, ", ", ".")) %>%
  mutate(Area = as.numeric(Area))

head(brazil)
```

##	City	State	Area	Population
## 1	São Paulo	São Paulo	1 522,986	12252023
## 2	Rio de Janeiro	Rio de Janeiro	1 182,296	6718903
## 3	Brasília	Distrito Federal	5 801,937	3015268
## 4	Salvador	Bahia	706,799	2872347
## 5	Fortaleza	Ceará	313,140	2669342
## 6	Belo Horizonte	Minas Gerais	330,954	2512070

Now the dataset is ready for analysis. One of the tasks proposed to the students is to check if there is a linear relationship between city area and population. After all, it is expected that the larger the area of the city, the more people will live there. However, the plot does not reflect this (Figure 25.2):

```
ggplot(brazil, aes(x=Area, y=Population)) +
  geom_point() +
  labs(x="Area (km^2)", y="Population")
```

Most students notice that the reason behind this behavior is the population density. Hence, they are asked to look for the cities' densities and check if they are unequal across the dataset:

```
brazil <- brazil %>%
  mutate(Density = Population/Area)

brazil %>%
  arrange(desc(Density)) %>%
  head(5)
```

##	City	State	Area	Population	Density
## 1	Taboão da Serra	São Paulo	20.478	289664	14145.13
## 2	Diadema	São Paulo	30.650	423884	13829.82
## 3	São João de Meriti	Rio de Janeiro	34.838	472406	13560.08
## 4	Carapicuíba	São Paulo	34.967	400927	11465.87
## 5	Osasco	São Paulo	64.935	698418	10755.65

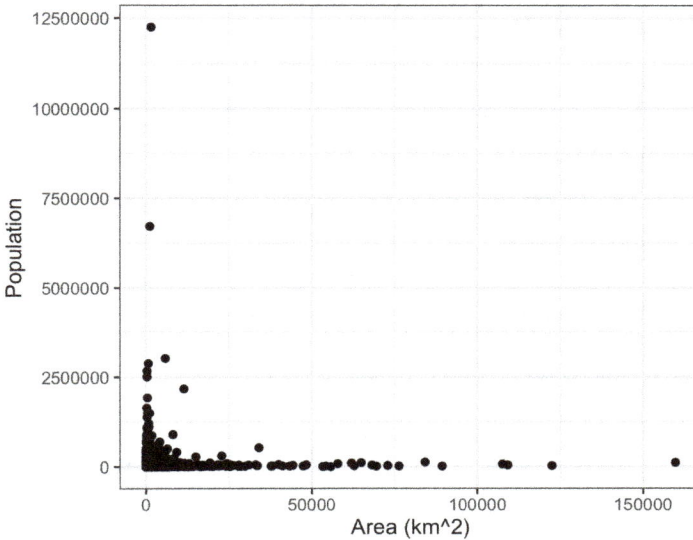

FIGURE 25.2
Relationship between population and area for the Brazilian cities.

```
brazil %>%
  arrange(desc(Density)) %>%
  tail(5)
```

##	City	State	Area	Population	Density
## 5566	Ipiranga do Norte	Mato Grosso	NA	7667	NA
## 5567	Itanhangá	Mato Grosso	NA	6737	NA
## 5568	Paraíso das Águas	Mato Grosso do Sul	NA	5555	NA
## 5569	Pinto Bandeira	Rio Grande do Sul	NA	3003	NA
## 5570	Aroeiras do Itaim	Piauí	NA	2551	NA

Many other tasks are presented to the students regarding to this specific dataset. These problems are made to challenge them, in order to make them relate different subjects and apply different statistical methods.

Note that in this chapter, we only deal with obtaining data through web scraping. As the course progresses, new topics are introduced to students. As they are cumulative issues, each new problem involves understanding what is being addressed and how it can be related to the issues addressed up to that point.

25.4 Final Remarks

We presented some considerations for our course *Introduction to Big Data Modeling*. As we stated previously, this course follows the ASA Guidelines for undergraduate programs in statistics.

Student evaluations indicate students are satisfied with this course contents. "The content of this course is very interesting and important for students who have worked with data science," "Excellent course! It covers a lot of knowledge in R programming," and "New discipline, but recommended for everyone, since it is a current and well-spoken subject worldwide" are some of the testimonials we have received so far.

We believe *Introduction to Big Data Modeling* is well implemented at our university. It was first offered in 2015 for students in the Statistics Department; 2019 was the first year the course was offered for the students enrolled in the Actuarial Science Department, and this met with huge success as well.

Our future plans for this course include expanding it from a one-semester to a two-semester course. Its first part will remain the same, but we plan to add more advanced topics in its second part. For example, time series are not covered in the present course. More recent methods, such as deep learning, are not considered either. Therefore, an expansion could be a good idea, since it would expose the students to more techniques and enable them to analyze a wider range of data.

This course is on par with the most recent introductory machine learning courses around the world. Since all data and software we use are free, budget is not a concern. However, some investment of time is necessary, since most professors are not used to these new learning approaches.

References

Curriculum Guidelines for Undergraduate Programs in Statistical Science. http://www.amstat.org/education/pdfs/guidelines2014-11-15.pdf.

Donoho, D. 2017. "50 years of data science." *Journal of Computational and Graphical Statistics* 26 (4): 745–766.

Grolemund, G., and H. Wickham. 2011. "Dates and times made easy with lubridate." *Journal of Statistical Software* 40 (3): 1–25. http://www.jstatsoft.org/v40/i03/.

Healy, K. J. 2018. *Data Visualization: A Practical Introduction*. Princeton University Press. https://books.google.com.br/books?id=o6BYtgEACAAJ.

Hicks, S. C. and R. A. Irizarry. 2016. "A guide to teaching data science." *The American Statistician* 72 (4): 382–391.

Lazar, N. A., J. Reeves, and C. Franklin. 2011. "A capstone course for undergraduate statistics majors." *The American Statistician* 65 (3): 183–189. https://doi.org/10.1198/tast.2011.10240.

Massicotte, P., and D. Eddelbuettel. 2020. *GtrendsR: Perform and Display Google Trends Queries*. https://CRAN.R-project.org/package=gtrendsR.

Wagaman, A. 2016. "Meeting student needs for multivariate data analysis: A case study in teaching an undergraduate multivariate data analysis course." *The American Statistician* 70 (4): 405–412. https://doi.org/10.1080/00031305.2016.1201005.

Wickham, H. 2016. *Ggplot2: Elegant Graphics for Data Analysis*. Springer-Verlag, New York. https://ggplot2.tidyverse.org.

Wickham, H. 2019a. *Rvest: Easily Harvest (Scrape) Web Pages.* https://CRAN.R-project.org/package=rvest.

Wickham, H. 2019b. *Stringr: Simple, Consistent Wrappers for Common String Operations.* https://CRAN.R-project.org/package=stringr.

Wickham, H., R. François, L. Henry, and K. Müller. 2020. *Dplyr: A Grammar of Data Manipulation.* https://CRAN.R-project.org/package=dplyr.

Wickham, H., and D. Seidel. 2020. *Scales: Scale Functions for Visualization.* https://CRAN.R-project.org/package=scales.

Wickham, H., and G. Grolemund. 2017. *R for Data Science.* O'Reilly Media. https://books.google.com.br/books?id=-7RhvgAACAAJ.

26

The Ten Most Similar Players: How to Use Statistics to Find the Best Soccer Players for Your Team

Thiago Valentim Marques

Federal Institute of Rio Grande do Norte

CONTENTS

26.1 Introduction

The market for electronic games, such as video and computer games, is continually expanding since they are products that increasingly rely on technological advances (Baltezarevic et al., 2018). Their great heterogeneity, from games that require an eminently sensory-motor activity to those that demand great strategic skills and understanding from the player, along with their close relationship with Internet use (since a large part of the existing games can be played online) also explains this expansion (Suzuki et al., 2009). In the Brazilian case, the Game Brazil 2018 Survey estimated that 75.5% of Brazilians play electronic games, regardless of the platform (Brasil, 2018).

Pro Evolution Soccer (PES) is a football simulation game (eFootball) developed by Konami and released annually since 2001, available for Microsoft Windows, PlayStation 3, PlayStation 4, Xbox 360, and Xbox One. Thanks to its advancement and realism, PES is one of the most played eFootball games in the world. In Brazil, many championships of this game involve the hiring of players. Players with the best skills have the highest prices. The game presents an option to search for players that uses filters, combining various attributes according to the player profile that the user wants. For example, the user can search for players who have a shot accuracy above 80 and a dribble above 75 (attribute between 0 and 100). However, there is no option to search for players similar to a user-set player.

Based on the evidence that electronic games are the gateway to the world of information and communication technologies (Belli and Raventós, 2008) and PES is one of the most practiced e-sports by Brazilian gamers (Brasil, 2018), the goal of the final project of the

DOI: 10.1201/9781003261148-31

course Introduction to Big Data Modeling was to create an application in shiny (system for developing web applications using R software) capable of searching for similar players using the following filters: name of the player to be compared, overall, height, position, and preferred foot. To accomplish this, the following specific objectives were proposed: (1) perform a descriptive analysis of the data, (2) build a radar chart of the main attributes of the players, (3) use clustering to group players by similarities, and (4) build an interactive page on the Internet to present the results.

26.2 Material and Methods

26.2.1 The dataset

The PES Data Base site (http://www.pesdb.net) presents the PES database since the 2011 version. Since there is no option to download this information directly, the dataset was obtained via web scraping (**rvest** package) by an algorithm developed in software R version 3.4.4 (R CORE TEAM, 2018). The dataset used in this work is comprised of 39 attributes for 11,158 players registered in PES 2019 through the last game update for October 2019. The attributes were the following (Table 26.1).

26.2.2 Methods

Hierarchical clustering was used to identify the players most similar to a given player. We considered each player as a vector in $\mathbb{R}^n$ space, where n is the number of player attributes. Hence, we defined the similarity between the players as the Euclidean distance between them, defined as follows:

Definition: Let $x, y \in \mathbb{R}^n$ be such that $x = (x_1, \ldots, x_n)$ and $y = (y_1, \ldots, y_n)$. The Euclidean distance between x and y, denoted by $d(x, y)$, is given by:

$$d(x, y) = \sqrt{(x_1 - y_1)^2 + \ldots + (x_n - y_n)^2} = \left[\sum_{i=1}^{n} (x_i - y_i)^2 \right]^{\frac{1}{2}}.$$

TABLE 26.1

Available Attributes of the Soccer Players Registered in PES 2019

Player Name	Squad Number	Team Name	League	Nationality
Region	Height	Weight	Age	Foot
Position	Attacking prowess	Ball control	Dribbling	Low pass
Lofted pass	Finishing	Place kicking	Swerve	Header
Defensive prowess	Ball Winning	Kicking Power	Speed	Explosive power
Unwavering balance	Physical contact	Jump	Goalkeeping	Gk catch
Clearing	Reflexes	Coverage	Stamina	Weak foot usage
Weak foot accuracy	Form injury	Resistance	Overall rating	

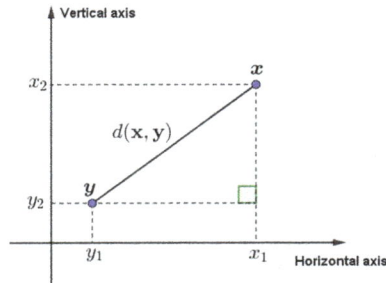

FIGURE 26.1

Geometric interpretation of the Euclidean distance between the vectors $x = (x_1, x_2)$ and $y = (y_1, y_2)$.

In particular, the geometric interpretation of the Euclidean distance between $x = (x_1, x_2)$ and $y = (y_1, y_2)$ can be seen in Figure 26.1. Note that the vectors x and y are represented by points in $\mathbb{R}^2$.

One can observe that a right triangle was formed when observing the projections of points x and y on the horizontal and vertical axes. Thus, by the Pythagorean Theorem, we have:

$$\left[d(x,y) \right]^2 = \left(x_1 - y_1 \right)^2 + \left(x_2 - y_2 \right)^2 \Rightarrow d(x,y) = \sqrt{\left(x_1 - y_1 \right)^2 + \left(x_2 - y_2 \right)^2}.$$

or

$$d(x,y) = \left[\sum_{i=1}^{2} \left(x_i - y_i \right)^2 \right]^{\frac{1}{2}}.$$

We used k-nearest neighbors to determine the group of ten players with whom a prefixed player was associated. This classifier searched within the dataset of 11,157 players (all players except the player chosen to be compared) to identify the ten players who have the shortest Euclidean distance. It is worth noting that other distances could be used as a measure of similarity (Minkowski distance, Manhattan distance, Canberra distance, Mahalanobis distance, among others), and although it is more sensitive to outliers, the greatest advantage of Euclidean distance is that its processing is faster than the others.

26.3 Results

26.3.1 Exploratory Data Analysis

Exploratory data analysis was performed to review the dataset variables and investigate possible correlations between them. For example, when selecting the variables height, weight, age, and overall (summary score of the player), it was possible to obtain a 4×4 chart with correlation and scatter and density plots using the **GGally** package's **ggpairs** function (Figure 26.2).

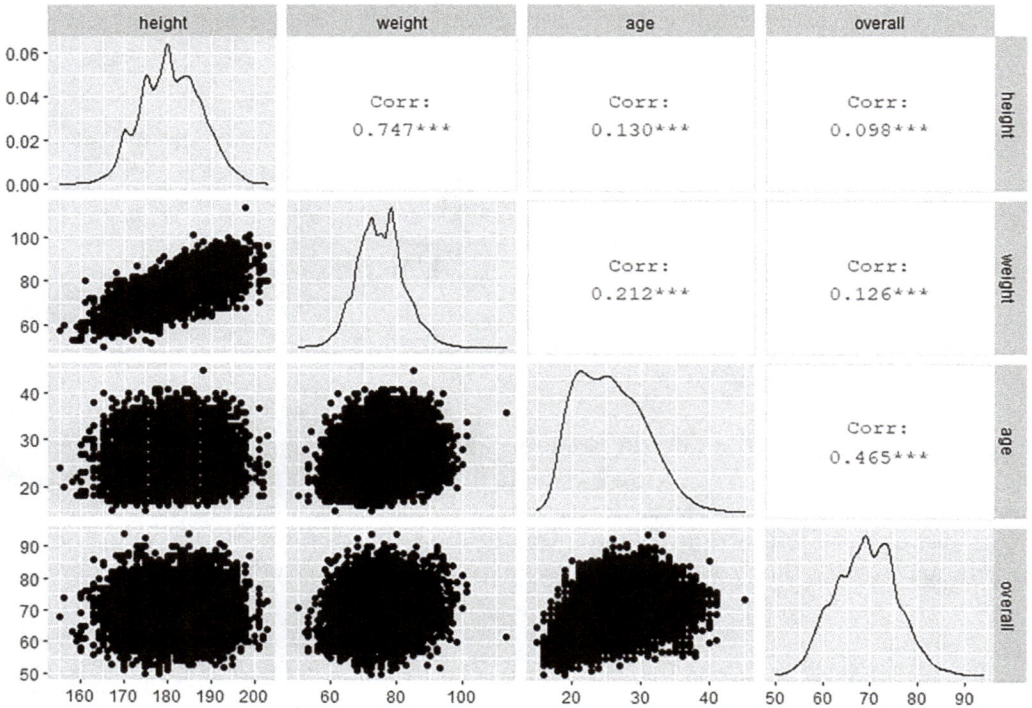

FIGURE 26.2
Scatter plots, densities, and correlations between height, weight, age, and overall variables.

Note that the height, weight, and overall variables have approximately symmetrical densities, while the age variable has an asymmetric density. Regarding the relationship between these variables, all correlations are significantly greater than zero (p-value < 0.001) and, according to the table proposed by Evans (1996) on the intensity of Pearson's correlation coefficient (r), the correlation between height and weight is strong ($r = 0.75$), the correlation between overall and age was moderate ($r = 0.47$), and all other correlations were weak.

Analyzing the histogram of the height variable (left panel of Figure 26.3), it was found that this variable fit a normal distribution of mean 180.8 cm and a standard deviation 6.8 cm reasonably well. This result is consistent with the assumption that the physical characteristics (e.g., height, weight, and foot size) of a population follow a normal distribution. Regarding the summary of the five numbers (right panel of Figure 26.3), this variable had a minimum of 155 cm, first quartile 176 cm, median 180 cm, average 180.8 cm, third quartile 186 cm, and maximum 203 cm. As a curiosity for the players, the smallest player in the game is Hiroto Nakagawa, a midfielder for Kashiwa Reysol of Japan, and the largest players are Costel Pantilimon (England's Sunderland goalkeeper) and Mena Qvist (left midfielder for Horsens from Denmark).

26.3.2 Web App

Shiny is a system for developing web applications using R, an R package (**shiny**), and a web server (shiny-server). The Shiny app is generated by a single script called **app.R**.

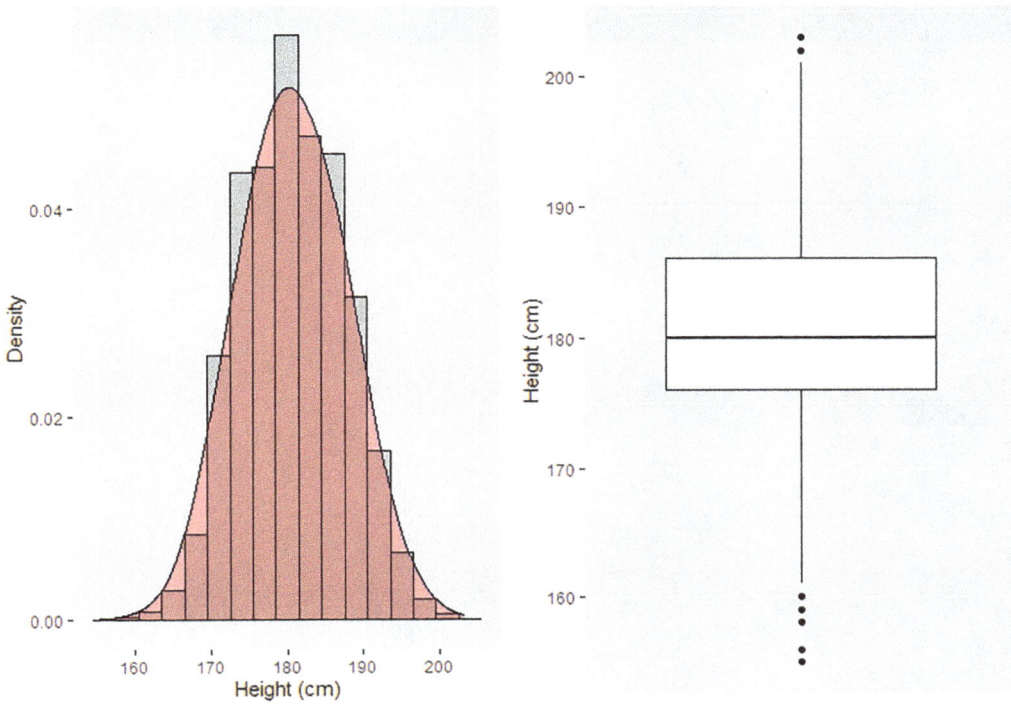

FIGURE 26.3
Histogram (left panel) and boxplot (right panel) of the height variable.

This script has three components: an object with the user interface (user side); a server() (server-side) function; and a call to the shinyApp() function. Based on this and using the **shiny, shinydashboard, shinythemes, plotly, shinycssloaders, tidyverse, knitr, kableExtra, ggfortify, plotly**, and **FNN** packages, the application "The has most similar players – Pro Evolution Soccer 2019" was developed. The ready-made application interface, containing the "Graphic," "About," and "Developers" tabs can be seen in Figure 26.4.

The application consists of selecting a player to be compared (the default is Cristiano Ronaldo). The user can then select the filters: overall (between 0 and 100), height (between 155 and 203 cm), position, and preferred foot. It is worth mentioning that the overall is a summary measure of a player's score. For example, Cristiano Ronaldo, a Juventus player from Italy (PM Black White because the game does not have the copyright to use the club's name), and Lionel Messi, of Barcelona from Spain, have an overall equal to 94 and are considered the best PES2019 players. On the other hand, the worst players in the game have an overall 50. In short, this attribute is strongly linked to the monetary value of the players. However, if a gamer wants to hire players with a high overall, he will have to pay accordingly.

The player's height can be filtered according to the gamer's playing style. For example, if the user wants a very high defense to avoid goals in aerial moves, then he will probably want to get taller players. The position to be chosen can be picked according to the intention of the gamer. For example, although the player being compared plays on the left, you can try to get a player with the same characteristics playing on the right. According to their

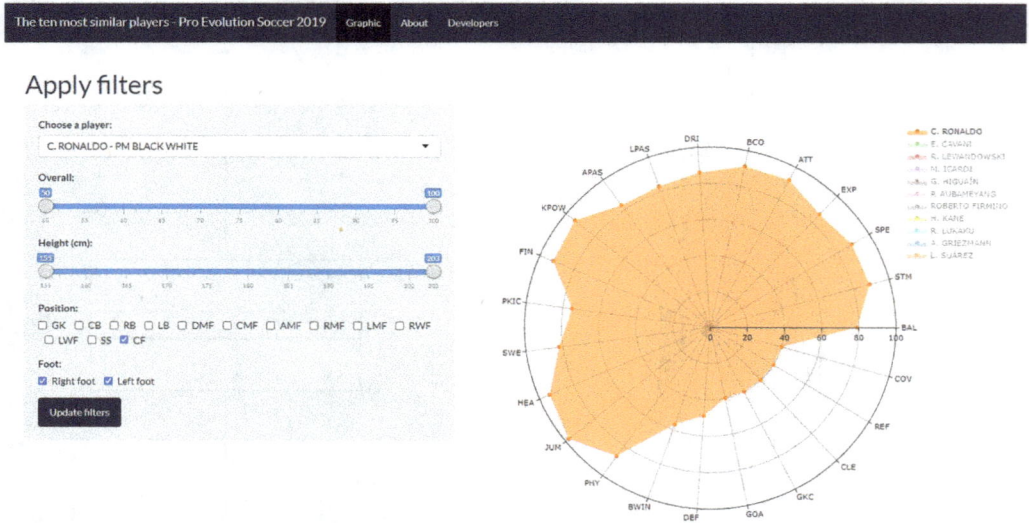

To visualize the graph of the player, click the icon at side of names in the graphic legend. It is worth noting that graphics will be overlapped.

FIGURE 26.4
Application interface "The ten most similar players – Pro Evolution Soccer 2019."

preference, the user can choose right-handed, left-handed, or both. It is worth mentioning that many players like to play with attackers or defenders with their feet changed in relation to their original position. For example, Arjen Robben is a southpaw and became known in football for his moves on the right side of the attack.

The graph chosen for the user to view the attributes of the players was the "radar chart." This graph consists of the display of multivariate data in the form of a two-dimensional graph, being quite informative and quite common in games whose characters have many attributes. For example, if you want the ten players most similar to Mohamed Salah, Liverpool England striker selecting the filters: overall, 75–85; height, from 180 to 190 cm; in the midfield positions on the left and right, attackers on the right and on the left and second attacker; in addition to being left-handed, the ten players that the application will return are shown in Figure 26.5.

26.4 Final Considerations

The creation of the application was highly relevant in the teaching and learning process of Data Science, since the last unit of the discipline is a project serving to put the acquired knowledge into practice. For example, we highlight the simplification of the presentation of data and complex results and the increase in the independence of the project's developers in seeking solutions using R programming, since the application is unprecedented in analyzing football data. It is noteworthy that in addition to the top score in the discipline, the application gained prominence in the competition results of the research "Shiny

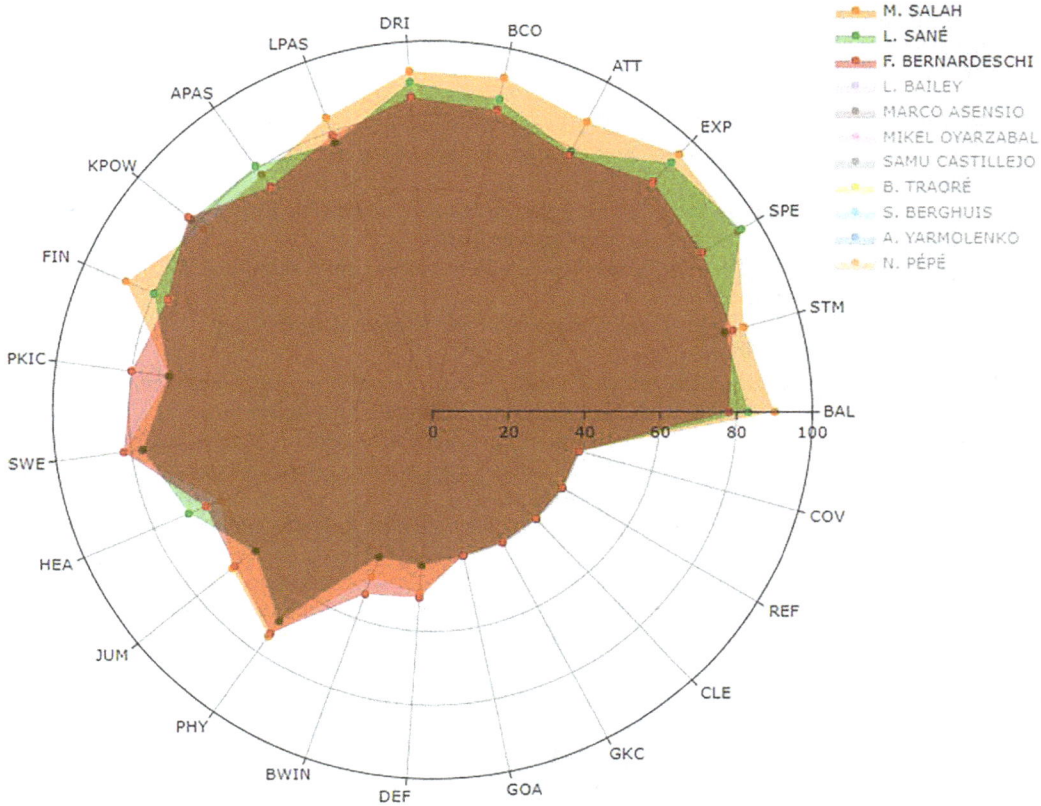

FIGURE 26.5
Radar chart of the ten players most similar to Mohamed Salah. The graphics by Manchester City's Leroy Sané and Juventus's Federico Bernardeschi are highlighted after being selected for viewing.

Contest" and today is part of the Shiny Gallery. In addition, an article on the largest television channel in the state was dedicated to the application.

References

Baltezarevic, Radoslav, Borivoje Baltezarevic, and Vesna Baltezarevic (2018). The video gaming industry: From play to revenue. *International Review*, 3–4, 71–76.

Belli, Simone, and Cristian López Raventós (2008). Breve historia de los videojuegos. *Athenea Digital. Revista de pensamiento e investigación social*, 14, 159–179.

Beygelzimer, Alina, Sham Kakadet, John Langford, Sunil Arya, David Mount, and Shengqiao Li (2018). *FNN: Fast nearest neighbor search algorithms and applications*. R package version 1.1.3. https://CRAN.R-project.org/package=FNN.

Chang, Winston (2018). *shinythemes: Themes for Shiny*. R package version 1.1.2. https://CRAN.R-project.org/package=shinythemes.

Chang, Winston, Joe Cheng, Joseph J. Allaire, Yihui Xie, and Jonathan McPherson (2018). *shiny: Web application framework for R*. R package version 1.4.0.2. https://CRAN.R-project.org/package=shiny.

Chang, Winston, and Barbara Borges Ribeiro (2018). *shinydashboard: Create dashboards with 'Shiny'*. R package version 0.7.1. https://CRAN.R-project.org/package=shinydashboard.

Evans, James D (1996). *Straightforward Statistics for the Behavioral Sciences*. Thomson Brooks/Cole Publishing Co, Pacific Grove, CA.

Horikoshi, Masaaki, and Yuan Tang (2016). *ggfortify: Data visualization tools for statistical analysis results*. https://CRAN.R-project.org/package=ggfortify.

Pesquisa Game Brasil (2018). *Pesquisa Game Brasil 18*. Recuperado em 6 setembro, 2018. https://pesquisagamebrasil.com.br.

R Core Team (2018). *R: A Language and Environment for Statistical Computing*. R Foundation for Statistical Computing, Vienna.

Sali, Andras, and Dean Attali (2018). *shinycssloaders: Add CSS loading animations to 'shiny' outputs*. R package version 0.3. https://CRAN.R-project.org/package=shinycssloaders.

Schloerke, Barret, Di Cook, Joseph Larmarange, Francois Briatte, Moritz Marbach, Edwin Thoen, Amos Elberg, and Jason Crowley (2018). *GGally: Extension to 'ggplot2'*. R package version 2.0.0. https://CRAN.R-project.org/package=GGally.

Sievert, Carson (2018). *Interactive Web-Based Data Visualization with R, Plotly, and Shiny*. Chapman and Hall/CRC, Boca Raton, FL.

Suzuki, Fernanda Tomie Icassati, Marcelo Vieira Matias, Maria Teresa Araujo Silva, and Maria Paula Magalhães Tavares de Oliveira (2009). O uso de videogames, jogos de computador e internet por uma amostra de universitários da Universidade de São Paulo. *Jornal Brasileiro de Psiquiatria*, 58(3), 162–168.

Wickham, Hadley (2018). *rvest: Easily Harvest (Scrape) web pages*. R package version 0.3.5. https://CRAN.R-project.org/package=rvest.

Wickham, Hadley, Mara Averick, Jennifer Bryan, Winston Chang, Lucy D'Agostino McGowan, Romain François, Garrett Grolemund, Alex Hayes, Lionel Henry, Jim Hester, Max Kuhn, Thomas Lin Pedersen, Evan Miller, Stephan Milton Bache, Kirill Müller, Jeroen Ooms, David Robinson, Dana Paige Seidel, Vitalie Spinu, Kohske Takahashi, Davis Vaughan, Claus Wilke, Kara Woo, and Hiroaki Yutani (2018). Welcome to the tidyverse. *Journal of Open Source Software*, 4(43), 1686. https://doi.org/10.21105/joss.01686.

Xie, Yihui (2015). *Dynamic Documents with R and Knitr*. 2nd edition. Chapman and Hall/CRC, Boca Raton, FL. ISBN 978-1498716963.

Zhu, Hao (2018). *kableExtra: Construct complex table with 'kable' and pipe syntax*. R package version 1.1.0. https://CRAN.R-project.org/package=kableExtra.

27

Teaching and Learning Statistics in Nigeria with the Aid of Computing and Survey Data Sets from International Organizations

Monday Osagie Adenomon

Nasarawa State University

CONTENTS

27.1 Introduction

This chapter discusses statistical education in Nigeria and proffers better ways to teach statistics in Nigeria through a program of Teaching and Learning Statistics with the Aid of Computing and Survey Datasets from International Organization to help graduate students in statistics relate theoretical statistics learned in the classroom to real-world problems across many disciplines and areas of application. The aim is to establish a platform to discuss the new framework, the TEAM (Teach, Examples with real-world data, Attitudinal influence, and Mentoring) framework, for teaching and learning statistics in developing countries such as Nigeria, which synchronizes the use of statistical computing and the use of survey data sets from international organization websites. This is consistent with the objectives of the international teaching effectiveness colloquium series established in 2005 (Cochran, 2006, 2007a, 2007b, 2008, 2009, 2010a, 2010b, 2013, 2015, 2018a, 2018b, 2020; Cochran and Cancela, 2007; Khumbah and Cochran, 2013a, 2013b; Sonomtseren et al., 2016) that has resulted in 20 events in 6 continents. Nations in which these events have been held include Uruguay, South Africa, Colombia, India, Argentina, Tanzania, Cameroon, Fiji, Uganda, Nepal, Kenya, Tunisia, Croatia, Cuba (twice), Mongolia, Moldova, Bulgaria, and Grenada.

DOI: 10.1201/9781003261148-32

27.2 Background of Statistics Education in Nigeria

Statistics can be defined as a science that formalizes the process of making inferences from observations to solve real-life problems in such fields as business, health, economics, agriculture, and security (Siegmund and Yakir, 2017). Also, statistics in the plural form is often used to refer to numerical or non-numerical facts or numbers (Oyejola and Adebayo, 2004). Also, Steel and Torrie (1980) define statistics as the science, pure and applied, of creating, developing, and analyzing techniques such that the uncertainty of inductive inferences may be evaluated.

The field of statistics had a long history in the world. According to its etymology in the *Oxford English Dictionary*, statistics was first applied to political science to address concerns about the facts of a state or community. Since then, statistical sciences have made extensive breakthroughs through their mathematical foundation, probability theory, and computing (Stigler, 1992). The history of statistics has the potential to help young statisticians to understand and appreciate the developments of the field of statistics and its usefulness in the present day (Fienberg, 1992), especially among students and tutors of statistics in developing countries such as Nigeria.

Statistics education at the university level first started at the University of Ibadan, Nigeria, in the period 1957–1973 (www.sci.ui.edu.ng/stathistory), but it still has not been accorded a place in policy formulation and national development of Nigeria (Agunloye, 2014). Ogum (1998) presented some reasons for the low level of statistical education in Nigeria. Adelodun and Awe (2013) stated that the major cause of low development in statistical education in Nigeria is a lack of understanding among educational stakeholders in Nigeria of statistics education as a separate discipline. At present, many universities in Nigeria run statistics programs at the B.Sc., M.Sc., and Ph.D. levels in Statistics, yet student enrollments are low and the mode of teaching statistics is poor, while Awe et al. (2015) stated that statistical capacity building is at a low level. Hence, the need for this chapter to proffer new approaches to teaching and learning statistics with the aid of statistical computing and real-world application.

- This chapter seeks to answer the following research questions:
- Do students of statistics require mentorship?
- Do students of statistics prefer computational statistics?
- Would students of statistics build a career in statistics after graduation?
- Can the application of Teaching and Learning Statistics with the Aid of Computing and Survey Datasets from International Organization improve statistical education in Nigeria?

These questions were the subject of a survey conducted in 2018 among final-year students of statistics at Nasarawa State University, Keffi, Nigeria.

27.3 Challenges of Statistical Teaching and Learning in Nigeria

Agunloye (2014) found that teaching statistics as part of mathematics has not helped the development of manpower training in statistics. Adelodun and Awe (2013) added that the adaptation of statistics education to national needs is inadequate, which has also led to unemployment among statistics graduates.

The survey results in Table 27.1 and Figure 27.1 show that few students applied for statistics in the university. The implication of this is that most secondary-school students do not apply to study statistics at the university, unlike courses such as computer science; this is because statistics is not a subject at the secondary-school level in Nigeria. For the Nigerian statistical community, the government needs to be informed of the need to include statistics as a subject at the secondary-school level so as to improve the awareness of statistics among secondary-school students in Nigeria, as this may increase the number of students that apply to university to study statistics. At the time of this survey, the total number of responses is 17, which represented the total number of final-year statistics students in the 2018/2019 Session, Nasarawa State University, Keffi, Nigeria. But the case in Nigerian universities is that few students originally apply to study statistics, while other students who are granted admission to statistics changed from the original course to which they applied.

Also, teaching and learning of statistics in Nigeria are still primitive, characterized by such factors as the lack of real-life applications, lack of computational applications, use of old curricula, lack of mentoring, lack of qualified lecturers and tutors, lack of creative ability, lack of computing facilities, lack of student interest, excessively theoretical teaching, and student frustrations with the lecturers.

These are the many problems affecting the Teaching and Learning of statistics in Nigeria, which needs to be reformed through the use of the aid of computing and survey data sets from international organizations and the use of a new framework called TEAM.

TABLE 27.1

Number of Students that Applied for Statistics among 400-Level Students, 2018/2019 Session in Statistics in Nasawara State University, Keffi, Nigeria in a Survey

Response	Frequency	Percent
No	10	58.82
Yes	7	41.18
Total	17	100.00

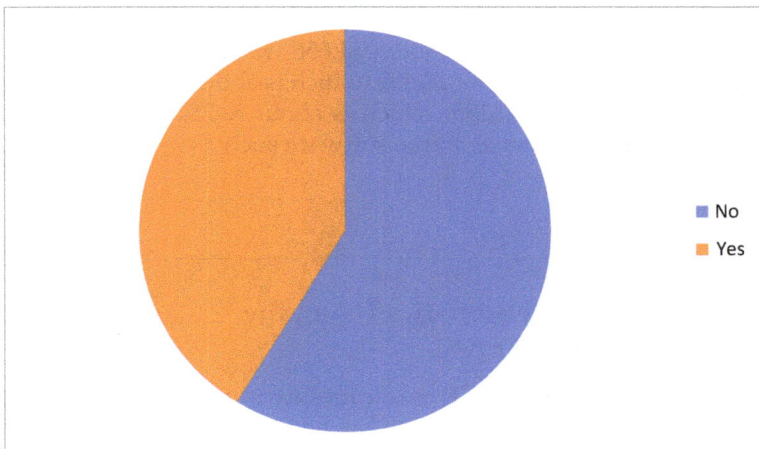

FIGURE 27.1
Number of 400-level students applying for statistics, 2018/2019 session.

27.4 Challenges and Cost of Statistical Software for Statistical Analysis

The section review some popular statistical software and their respective challenges and cost. The software packages are EView, MINITAB, and STATA.

EViews is a statistical and econometrics software package that can run on Windows. Econometrics deals with three types of data: cross-sectional data, time series data, and panel (longitudinal) data. EViews allows you to work with all three types of data. EViews is most commonly used for time series analysis in academics, business, and government, but it can be used easily with cross-sectional and/or panel data. EViews can deal with simple and complex statistical analyses. EView can be purchased from the company directly or on the open market, but EViews may not be suitable for big data analysis, and it is very expensive (Adenomon, 2019).

MINITAB is a statistical package developed at Pennsylvania State University by researchers Barbara F. Ryan, Thomas A Ryan, Jr., and Brian L Joiner in 1972. It began as a light version of OMNITAB, a statistical analysis program by NIST (National Institute of Standards and Technology; Wikipedia, 2017). The documentation for MINITAB was published in 1986. MINITAB is distributed by MINITAB Inc., a privately owned company headquartered in State College, Pennsylvania (Marques de Sá, 2007; Adenomon, 2018).

Stata was first created in 1985 by Stata Corp. It was originally written by William Gould, while the name "Stata" is an abbreviation of "statistics" and "data." It is used mostly in the fields of sociology, economics, political science, biomedical sciences, and epidemiology. STtata has four major versions, including Stata/MP for multiprocessor computers; Stata/SE for large databases; and Stata/IC, which is the standard numeric version, which supports any data of the sizes of any of the builds listed above. Stata capabilities include data management, statistical analysis, graphics, simulations, regression, and custom programming. It also possesses a system to disseminate user-written programs that can grow continuously. Lastly, Stata can import data in a variety of formats. This includes ASCII data formats such as CSV or data bank formats and spreadsheet formats, including various Excel formats (Hamilton, 2013; Bittmann, 2019; Adenomon, 2020).

The challenges of this software are that it is limited in application to complex analysis such as Bayesian Statistical techniques, spatial analysis, and big data analysis, and is very expensive for students and tutors in Nigeria and other developing countries. For instance, a single-user license of Eviews 11 costs about \$1,650 (which is about ₦660,000); a single-user license of MINITAB 19 costs about \$2,499 (which is about ₦999,600); and a single-user license of STATA 16 costs about \$1,195 (which is about ₦478,000). This software is very expensive, hence the need for the use of free software such as R.

27.5 Teaching and Learning Statistics Using the Framework Called the TEAM

Vance Smith (2019) proposed the ASCCR Frame for Learning Essential Collaboration Skills among statisticians and data scientists, which has had excellent results in developed countries. However, in Nigeria, and especially in North Central Nigeria, we teach and learn statistics using the TEAM framework, explained as follows:

T—Teach

The concept of Teach is very important in teaching and learning statistics in Nigeria as a developing country, because statistics are taught in secondary schools as part of mathematics and only at the senior secondary school level. Because statistics are not as well taught as mathematics, the awareness of statistics is very low, which is why students applying to study statistics are very few, as shown in Table 27.1. In this teaching method, we try to explain the concepts of statistics simply to enhance students' understanding and to make it appeal to their interests. Also, teaching students of statistics with computing techniques makes it more real to them than without computing techniques. Using this Teach method encourages the student to study statistics through interest.

Table 27.2 shows that 76.47% of the students have not experienced anything that discouraged them from studying statistics as the course of their choice. This means that a good teaching method would help encourage students to study statistics, keeping in mind that some students studying statistics did not originally apply to do so in the university. In addition, if students are encouraged to study statistics in the Nigerian university, that would lead to increased manpower in statistics in Nigeria. Although one cannot generalize this result from 17 responses.

E—Examples with Real-World Data

The goal of this chapter is to indicate our way of teaching and learning statistics with the aid of computing and survey data sets from international organizations to produce statistics graduates who can apply the theory of statistics learned in class to solve real-life problems, becoming more employable and entrepreneurial. For instance, teaching probability distribution with R software, we discovered students quickly learned the difference between probability distributions that are discrete from those that are continuous in nature. In Nigeria, North Central Nigeria in particular, we teach statistics using data obtained from the National Bureau of Statistics (NBS) such as the inflation rate and unemployment rate; data from the World Bank such as the malaria survey in Nigeria are also used in teaching, which helps students to apply statistical techniques learned in the classroom to solve real-life problem during their final-year project and also improve student collaborative skills (Awe et al., 2015).

In a survey in 2018 among final-year students of statistics at Nasarawa State University, Keffi, Nigeria, the question on the preferred area of statistics was answered as shown in Table 27.3.

From Table 27.3, we see that most students preferred computational and applied statistics. The implication is that theoretical statistics is very important in teaching statistics, but teaching and learning statistics in Nigeria would be better appreciated by statistics students if statistics is taught with examples using real-world data informing computational statistics or applied statistics, as is the case with our sample, corroborating Awe et al. (2015).

TABLE 27.2

Student Experience of Discouragement in the Field of Statistics

Response	Frequency	Percent
Yes	4	23.53
No	13	76.47
Total	17	100.00

TABLE 27.3

Area of Statistics Preferred by Students of Statistics

Response	Frequency	Percent	Cumulative
Computational statistics	7	41.18	41.18
Theoretical statistics	3	17.65	58.82
Applied statistics	7	41.18	100.00
Total	17	100.00	

TABLE 27.4

Willingness to Continue in the Field of Statistics after Graduation

Response	Frequency	Percent
Yes	14	82.35
No	3	17.65
Total	17	100.00

A—Attitudinal Influence

It was established in this chapter that statistical education in Nigeria is still very low. In many situations, students are forced to study statistics in university and polytechnics in Nigeria, so there is a need to influence the attitude of students in statistics to remain and continue in a career in statistics. To achieve this, we employ the use of orientation programs to expose to new students the possibility of becoming statisticians and data scientists. The use of statistical counseling is also used to influence the attitudes of students, in which lecturers used their personal stories and career progression to encourage the students to remain and study statistics, which is the course of the future.

Table 27.4 shows that 82.35% of the undergraduate students responded that they would continue in the field of statistics after graduation. From my personal experience, some undergraduate students of statistics may not want to continue a career in statistics because they felt the prospects of statistics to be low or because they originally did not apply to study statistics in the first instance. Some may want to study computer science or business administration after undergraduate studies in statistics. Attitudinal influence by tutors can help influence such students to continue a career in statistics after graduation because Nigeria needs more statisticians for better informed data-based decision-making.

M—Mentoring

According to the *Oxford Learner's Dictionary*, a mentor is an experienced person who advises and helps somebody with less experience for years. The place of mentoring in teaching and learning in statistics in developing countries such as Nigeria cannot be overemphasized because it improves statistical capacity building in Nigeria. Mentoring helps in teaching statistics, changing the attitude of the student toward statistical education in Nigeria, and also helping to produce students who can apply the theory of statistics learned in the classroom to solve a real-life problem.

According to our 2018 survey, the majority of the students (94.12%) want their lecturer to mentor them (see Table 27.5). Also, 100% of the final-year responded that mentoring by the lecturer can help improve the academic performance of students of statistics, as shown in Table 27.6.

TABLE 27.5

Students Desiring Mentorship

Response	Frequency	Percent
No	1	5.88
Yes	16	94.12
Total	17	100.00

TABLE 27.6

Role of Mentorship in Student Academic Performance

Response	Frequency	Percent
Yes	17	100.00
Total	17	100.00

Tables 27.5 and 27.6 show that most students of statistics in Nigeria desired mentorship, while all students responded that student mentoring by lecturers of statistics could enhance student academic performance in statistics. This means that the role of mentorship in teaching and learning statistics in Nigeria cannot be over-emphasized.

27.6 Overview of R Statistical Software and Some Free Statistical Software

R is a system for statistical analyses and graphics created by Ihaka and Gentleman (1996). R is both a software and a language, considered a dialect of the S language created by the AT&T Bell Laboratories for statistical computing and graphics that provides a wide variety of statistical methods (time series analysis, linear and nonlinear modeling, classical statistical tests, etc.) and graphical techniques, and is highly extensible. R is freely distributed under the terms of the GNU General Public License and the statistical and mathematical packages are downloaded and installed through the Internet (i.e., it is accessible online). Its development and distribution are carried out by several statisticians known as the R Development Core Team (2005). R is now widely used in academic research, education, and industry. It is constantly growing, with new versions of the core software released regularly and more than 2,600 packages available (Eubank & Kupresanin, 2011 this reference is almost 10 years old; please find a more recent one to compare with it and show how the number of R packages has grown over the years). The R language is, arguably, the de facto standard for statistical research purposes. There are now many books that detail its use (along with that of add-on packages) for the solution of data analysis problems.

In a broader sense, R is a very powerful functional language that merely happens to have built-in (and add-on) tools that perform some of the standard (and not so standard) statistical calculations with data.

According to Paradis (2005), the major advantages of R are the following: (1) it has many functions and packages for statistical analyses and graphics, (2) it allows the user to program loops to successively analyze several data sets, and (3) it is possible in R to combine in a single program different statistical functions to perform more complex analyses.

R is especially powerful for data manipulation, calculations, and plots. Its features include

i. An integrated and very well-conceived documentation system.
ii. Efficient procedures for data treatment and storage.
iii. A vast and coherent collection of statistical procedures for data analysis.
iv. A suite of operators for calculations on tables, especially matrices.
v. Advanced graphical capabilities.
vi. A simple and efficient programming language, including conditions, loops, recursion, and input-output possibilities (Figure 27.2).

Example of R packages to analyze survey data sets are as follows:
haven; readr; readxl; tydyverse; gmodels; tibble; kableExtra; Amelia; panelr.

haven: The haven package provides functions for importing data from SAS, SPSS, and Stata file formats with the following functions: read_sas(), read_sav, and read_dta(). The functionality is similar to that available in the base R, but the haven package is often faster.

readr: The **readr** package provides new functions for importing tabular data into **R**. Specifically, the functions read_table(), read_csv(), read_delim() are intended to be fast (around 10 times faster) replacements for the base **R** functions.

readxl: It used to be that the most reliable way to get data from Excel into **R** was to first save it as a tab (or comma) delimited text file. The easiest way to import tabular data from xls and xlsx formats files now is to use the **readxl** package. Importantly, it has no external dependencies, so is very straightforward to install and use on all platforms.

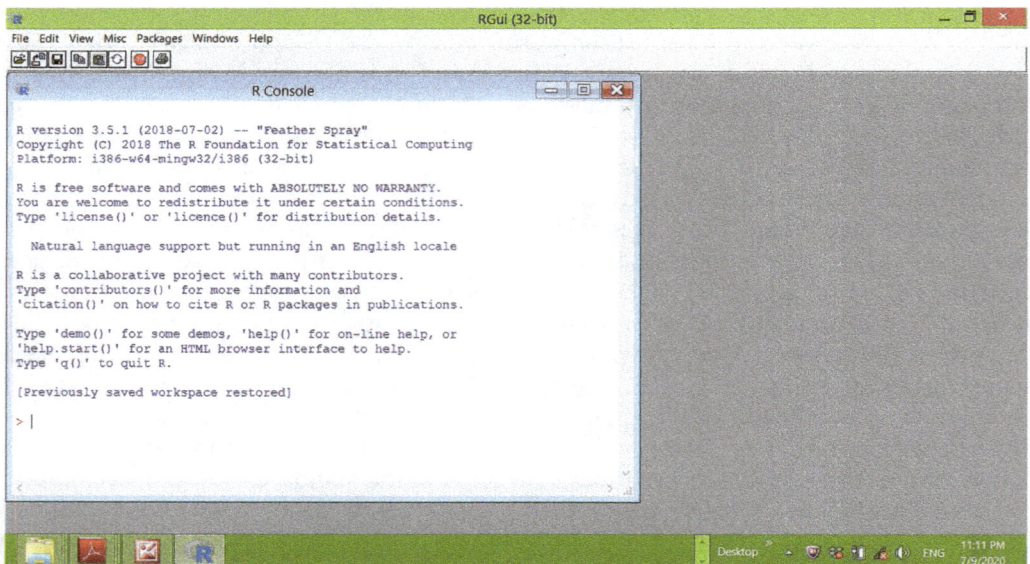

FIGURE 27.2
R 3.5.1 screen.

tydyverse: tidyverse is a collection of essential R packages for data science. The packages under the tidyverse umbrella aid in performing and interacting with the data. There is a whole host of functions available to handle one's data, such as subsetting, transforming, and visualizing. Tidyverse was created by Hadley Wickham and his team with the aim of providing all these utilities to clean and work with data.

gmodels: gmodels includes a variety of R programming tools for model fitting that is vaster than base R. it has the following functions: testing a general linear hypothesis for a regression model, cross-tabulation with a test for factor independence, efficient computation of principal components and singular value decompositions, constructing a user-specified contrast matrix, constructing and estimating linear functions of model coefficients, computing and testing arbitrary contrasts for regression objects, returning model parameters in a data frame, and computing confidence intervals.

kableExtra: The kableExtra package (Zhu, 2019) is designed to extend the basic functionality of tables produced using kable(). The striking feature of kableExtra is that it works perfectly for both HTML and PDF formats.

Amelia: Amelia is R package for handling missing data. It is a tool that multiplies imputes missing data in a single cross-section (such as a survey) from time series (such as variables collected for each year in a country) or from a time-series-cross-section data set (such as data collected by year for each of several countries). It also includes useful diagnostics of the fit of multiple imputation models.

tibble: tibble is a tibble R package developed by Hadley Wickham. The tibble R package provides easy-to-use functions for creating tibble, which is a modern rethinking of data frames. Tibbles retain features that have stood the test of time and dropped the features that used to be convenient but are now frustrating (such as converting character vectors to factors).

panelr: panelr deals with regression models and utilities for repeated measures and panel data. It provides object type and associated tools for storing and wrangling panel data. It implements several methods for creating regression models that take advantage of unique aspects of panel data.

27.7 Sources of Survey Data Sets

In recent times, there are paid data sets that are available, but there are also free survey data sets available for research that are very useful to researchers in developing countries who may not have the funds to purchase such data. The following are sources of survey data sets:

i. National Bureau of Statistics
   The NBS came into being with the merger of the Federal Office of Statistics (FOS) and the National Data Bank (NDB). The creation is part of the implementation of the Statistical Master Plan (SMP), a program document of the Federal

Government of Nigeria (FGN). The document's preparation was funded by the World Bank in 2003. The implementation was designed to span 5 years, 2005–2009. NBS is charged with the following responsibilities: Infrastructure and Equipment, Human Resources Management and Development, Improved Data Production Methodology, Data Management, and Dissemination and Access. NBS has also worked with other agencies in Nigeria such as the Federal Ministry of Health, National Population Commission, and Central Bank of Nigeria (https://www.nigerianstat.gov.ng) and NBS also worked with external agencies such as WHO, United Nations Children's Fund (UNICEF) in survey matters in Nigeria.

In the website of NBS the following data are available for download for statistical analysis: population and vital statistics, including GIS Maps; social statistics, including unemployment data, security survey, crime data, poverty profiling survey, demographic data, and tourism data; and economic statistics, including the production price index, industrial production price index, production indicators, monthly consumer price index, and quarterly national accounts. These data can be used to aid teaching statistics in Nigeria, while students can use these data for their final-year project to aid national development in Nigeria.

ii. Integrated Public Use Microdata Series (IPUMS)

IPUMS is an international project dedicated to collecting and distributing census data from around the world. The project goals are to collect and preserve data and documentation, harmonize data, and disseminate the harmonized data free of charge. IPUMS houses data for 98 countries including Nigeria and 443 censuses and surveys with over 1 billion person records (https://ipums.org).

IPUMS provides census and survey data from around the world integrated across time and space. IPUMS integration and documentation makes it easy to study change, conduct comparative research, merge information across data types, and analyze individuals within family and community context. Data and services are available free of charge.

IPUMS provides excellent survey data sets that can aid teaching statistics in developing countries such as Nigeria; such data are highly convenient, being offered in Excel, SPSS, and STATA formats that can easily be adapted for use in R.

IPUMS provides geographical data and GIS that can be used for spatial analysis. Fertility, mortality, and migration data can be sourced from IPUMS. Research data enclaves and linked censuses are also provided via the IPUMS website free of charge to assist researchers in developing countries such as Nigeria.

iii. The Nigeria Demographic Health Survey (DHS) Program: Country Main

The DHS program is a program supported by USAID to ensure quality information for planning, monitoring, and improving population, health, and nutrition programs by developing nations worldwide (https://dhsprogram.com), and in the collection and use of such data. If these kinds of data are used to teach statistics student in Nigeria, it will go a long way to provide them the necessary skills in statistical computing and analysis using survey data set to proffer solutions to real-life problems in Nigeria.

DHS provides the following data sets: DHSs, AID Indicator Surveys (AIS), Service Provision Assessment (SPA) Surveys, the Malaria Indicator Survey (MIS), Key Indicators Survey (KIS), and other quantitative surveys such as Benchmark

Survey, KAP survey, panel surveys, and other specialized surveys, as well as qualitative research that provides answers to questions that lie outside the purview of standard quantitative approaches (https://dhsprogram.com/data/available-datasets.cfm).

These data are provided in Excel, SPSS, SAS, and STATA formats that can be imported into R for data analysis. Also, GIS data in shapefiles are provided via the DHS website for spatial data analysis.

iv. UNICEF-MICS Dataset

The United Nations Children's Fund (UNICEF), formerly United Nations International Children's Emergency Fund, is a special program of the United Nations (UN) devoted to aiding national efforts to improve the health, nutrition, education, and general welfare of children.

UNICEF operates in more than 190 countries and territories to reach the most disadvantaged children and adolescents and to protect the rights of every child everywhere (https://www.unicef.org).

The UNICEF website has extensive interesting data sets that can aid teaching and learning statistics in Nigeria, including data on sexual violence in childhood, peer violence data, pre-primary data, violence data, birth registration data, and immunization of children surveys (https://mics.unicef.org/surveys), female genital mutilation (FGM) data, child labor data, etc. (data.unicef.org/resources/resource-type/datasets/). Using data from UNICEF website to teach and learn statistics in Nigeria with application of R would help statistics graduates in Nigeria to use data for informed decision-making about children in Nigeria.

27.8 Conclusion and Policy Recommendations

This chapter reviewed statistics education in Nigeria and was found that statistical education is still very poor in Nigeria, thereby affecting the graduate of statistics. This research was undertaken to help change the narrative through Teaching and Learning Statistics with the Aid of Computing and Survey Datasets from International Organizations in order to produce graduates of statistics who can navigate from theories learned in the classroom to solving real-life problem and become employable in the job market and possibly become entrepreneurs. To achieve this, the TEAM framework is a viable tool. The results of a survey conducted in 2018 of the final-year students of statistics at Nasarawa State University revealed that students of statistics in Nigeria required mentorship and preferred computational statistics and applied statistics, which means that Teaching and Learning Statistics with the Aid of Computing and Survey Datasets from International Organizations holds great promise in improving statistical education in Nigeria, while students of statistics would be likely to build careers in statistics after graduation if they are mentor.

Also, since software such as Eviews, STATA, and MINITAB are very expensive, R software is the most viable computing software for students of statistics in developing countries such as Nigeria.

This chapter, therefore, recommends the use of the TEAM framework to help improve Teaching and Learning Statistics with the Aid of Computing and Survey Datasets from International Organizations in Nigeria.

Bibliography

Adelodun, O. A. and Awe, O. O. (2013). Statistics Education in Nigeria: A Recent Survey. *Journal of Education and Practice*, 4(11), 214–220.

Adenomon, M. O. (2018). *Basic Statistical Methods for Physical, Management and Social Sciences with Examples in MINITAB*. Bida: Jube-Evans Books & Publication.

Adenomon, M. O. (2019). *Applied Econometrics with Examples in EViews*. Bida: Jube-Evans Books & Publication.

Adenomon, M. O. (2020). *Basic Statistics and Introduction to Multivariate Analysis using STATA*. Bida: ADECLAR Research Consult and General Contracts Ltd.

Agunloye, O. K. (2014). Bridging the Gap of Manpower Training for Statistics Education in Nigerian Education Colleges of Education: An Empirical Evaluation of Some Selected Colleges in South-Western Nigeria. ICOTS, 9, 1–3.

Awe, O. O., Crandell, I. and Vance, E. A. (2015). Building Statistics Capacity in Nigeria through the LISA 2020 Program. *Proceedings of the International Statistical Institute's 60th World Statistics Congress*, Rio de Janeiro.

Bittmann, F. (2019). *Stata – A Really Short Introduction*. Boston, MA: DeGruyter Oldenboug.

Cochran, J.J. (2006). Workshop on O.R. Education Set for Uruguay. *ORMS Today*.

Cochran, J.J. (2007a). The first ALIO/INFORMS Workshop on OR Education – A Great Success in Montevideo… and Now on to Cape Town! *IFORS Newsletter*, 15(1).

Cochran, J.J. (2007b). Workshop on O.R. Education to be Held in Cape Town. *ORMS Today*, 34(3).

Cochran, J.J. (2008). O.R. in Africa Conference... in D.C. *ORMS Today*, 35(2).

Cochran, J.J. (2009). Progress in the Use of Operations Research in Eastern Africa and the 5th International Operations Research Society of Eastern Africa Conference. *IFORS Newsletter*, 3(4).

Cochran, J.J. (2010a). International Teaching Effectiveness Colloquium Continues to Reap Benefits, *IFORS News*, 4(4).

Cochran, J.J. (2010b). An Update on the INFORMS/IFORS International Teaching Effectiveness Colloquium Series, *IFORS News*, 4(4).

Cochran, J.J. (2013a). How do you find and Prepare for Opportunities to work in Developing Nations? *AMSTAT News*, June.

Cochran, J.J. (2015). An O.R. Perspective on U.S./Cuba Relations. *ORMS Today*, 42(2).

Cochran, J.J. (2018a). Ambassadors and O.R. Education. *ORMS Today*, 45(4).

Cochran, J.J. (2018b). Mongolia Outreach Leads to Nationwide High School Statistics Education. *The International Statistical Literacy Project (ISLP) Newsletter*, September 2019.

Cochran, J.J. (2020). PARIS21 Project Receives Warm Welcome in Caribbean: Buoyed by Success in Grenada, Teaching Team Makes Plans to Take Statistics Literacy Workshops to Rest of Region and Beyond. *ORMS Today*, 47(4), 50–52.

Cochran, J.J. and Cancela, H. (2007). The first ALIO/INFORMS Workshop on OR Education – A Great Success in Montevideo! *ORMS Today*, 34(2).

Eubank, R. L. and Kupresanin, A. (2011). *Statistical Computing in C++ and R*. Boca Raton, FL: CRC Press.

Fienberg, S. E. (1992). A Brief History of Statistics in Three and Half Chapters: A Review Essay. *Statistical Science*, 7(2), 208–225.

Hamilton, L. C. (2013). *Statistics with STATA*. Boston, MA: Cengage.

Ihaka, R. and Gentleman, R. (1996). R: A Language for Data Analysis and Graphics. *Journal of Computational and Graphical Statistics*, 5(3), 299–314.

Khumbah, N.A. and Cochran, J.J. (2013a). IFORS Introductory Operations Research Colloquium held in Cameroon! *IFORS News*, 7(2).

Khumbah, N.A. and Cochran, J.J. (2013b). O.R. Goes to Cameroon. *ORMS Today*, 40(3).

Marques de Sá, J. P. (2007). *Applied Statistics Using SPSS, STATISTICA, MATLAB and R* (2nd ed). New York: Springer.

Ogum, G. E. O. (1998). Statistical Education in Nigeria: Problem and Prospect. Paper presented at the *Fifth International Conference on Teaching Statistics*, Singapore. http://iase-web.org/documents/papers/icots5/Topic5c.pdf.

Oyejola, B. A. and Adebayo, S. B. (2004). *Basic Statistics for Biology and Agriculture Students*. Ilorin: Olad Publishers.

Schwab-McCoy, A. (2019). The State of Statistics Education Research in Client Disciplines: Themes and Trends across the University. *Journal of Statistics Education*, 27(3), 253–264. DOI: 10.1080/10691898.2019.1687369.

Siegmund, D. and Yakir, B. (2007). Background in Statistics. In: *The Statistics of Gene Mapping: Statistics for Biology and Health*. New York: Springer, pp. 3–34.

Sonomtseren, M., Sodnomtseren, A., and Zorigt, K. (2016). O.R. without Borders: Introducing O.R. in Mongolian Schools INFORMS Leaders Join Mongolia's Strategic Efforts to Emphasize Analytics in Its Education System. *ORMS Today*, 43(4).

Stigler, S. M. (1992). The History of Statistics: The Measurement of Uncertainty before 1900. Cambridge, MA: Harvard University Press, p. 410.

Vance, E. A. & Smith, H. S. (2019). The ASCCR Frame for Learning Essential Collaboration Skills. *Journal of Statistics Education*. DOI: 10.1080/10691898.2019.1687370.

Yan, D. and Davis, G. E. (2019). A First Course in Data Science. *Journal of Statistics Education*, 27(2), 99–109. DOI: 10.1080/10691898.2019.1623136.

Part 6

Importance of Statistics in Urban Planning and Development

28

Assessment of Indicators of Urban Housing Quality in Owerri Municipal, Nigeria: A Factor Analysis Approach

Ifeoma Evan Uzoma

University of Uyo

CONTENTS

28.1 Introduction

Housing is a multidimensional concept that has evoked different responses and definitions based on existing environmental and physical conditions. Housing is defined in the Nigerian National Housing Policy (2012) as "the procedure of providing safe, attractive, functional, affordable, comfortable, and identifiable shelter in a proper setting within a vicinity, which is held together by regular and continuous maintenance of the built neighborhood, for the day-to-day functional activities of people and families within the environment while reflecting their social, economic, cultural aspirations and tastes." In another perspective, housing has been defined by the World Health Organization (WHO, 1961) as a "residential vicinity that includes the physical space that man uses for shelter, all necessary equipment, services, facilities, and devices needed for the physical, mental and social well-being of the family and person." These definitions have served as a foundation for the examination of the issue of urban housing quality.

DOI: 10.1201/9781003261148-34

Urban housing is a contentious issue in most developing countries. This is more problematic in areas with rapid population growth like Nigeria and other countries of the African continent. Rapid population growth has been fingered as one of the causes of pressure on urban housing and indirectly on housing quality. In addition to these are urbanization and rural-urban migration (Adeleye and Anofojie, 2011). The Universal Declaration of Human Right (1948) stated that the well-being of humans, which includes quality housing, was a human right, and Sustainable Development Goal 11 posits making human settlements sustainable. This incontestably means that urban housing and housing quality issues are very important issues to be considered.

Onibokun (1999) stated that the major determinants of urban housing quality are the age of the dwelling, the types of building and the materials used in their construction, the varieties and sufficiency of facilities provided in houses, and the ways of handling different aspects of building construction such as land preparation, foundation laying, building of walls, and pattern of roofing. These variables and more were incorporated into the current study to strengthen its findings.

In addition, the application of statistics and statistical techniques to urban housing quality issues is also important as, according to Udofia (2011), there is a need to see patterns, dimensions, and classifications with statistical techniques as a guide. The statistical techniques help the researcher to measure variables with greater precision and accuracy.

28.2 Research Aim and Objectives

The aim of this research was to assess the indicators for housing quality in Owerri Municipal with the following objectives:

i. To investigate the underlying factor structure of the housing quality in the study area.
ii. To classify the houses in the study area based on the existing quality of housing.

28.3 Review of the Literature

Studies abound of housing quality around the world, but are more prevalent in the developing nations of the world, as housing is a major problem in developing countries. Other studies of housing quality served as the foundation of this present study. These are studies carried out on various aspects and indicators of housing quality around the world with the aid of different techniques.

One such study is a nationwide study in Nigeria by Morenikeji et al. (2017). The authors looked at housing quality from a spatial analysis point of view. The study concentrated on the factors responsible for the variation in housing quality in the 36 component units of the federation, including Abuja based on 33 housing quality characteristics. Using the 2006 national census housing quality characteristics, data were collected ng aspects of quality like electricity, water supply, and sanitation and hygiene, and the collected data were analyzed using principal component analysis, yielding three components contributing 38%,

31%, and 7% to the variation in housing quality in Nigeria. The factor loadings were further used as discriminant analysis variables, ing three housing quality regions that corresponded to the North, West, and East flanks of the country. The regions were classified as high (west), low (north), and moderate (east). The study recommended that buildings that do not conform to the quality scale should be checked through enforcement of laws and edicts. The strength of this kind of study is that since it covers the entire country, it gives a fair idea of what is happening in the country in terms of housing quality. The weakness is that it hides existing disparities and inequalities among and within the various regions considered.

In other studies, income was discovered to be a major indicator of the housing conditions of people. Oluwaseyi (2019) examined the housing quality in Osun State, Nigeria, with Oshogbo LGA as the focus. The study was directed at the residential housing quality in the area with a view of finding the sustainability of housing quality. Indicators used in this particular study were the condition of housing, in-house or indoor facilities, and the impact of the socioeconomic level of the people on the housing quality in the area. The study was based on field data and already existing statistics and applied qualitative and quantitative analysis. Multistage sampling was used to distribute 210 questionnaires while using systematic random sampling to stratify the whole study area into 15 geopolitical wards based on records and documents of the planning board. The planning board stratified the area into three density areas: high density (12 wards), medium density (two wards), and low density (one ward). Using these strata, six wards were selected randomly, giving all the buildings equal chances of being selected. The buildings were sampled systematically, such that the tenth house from the first was chosen randomly. The results of analysis indicated that the housing quality in the study area based on the locals' perceptions was low and this level of housing quality was a result of the low income earned by the people in the area. The study finally recommended that micro finance schemes should be set up to assist low-income earners in upgrading their housing quality. The strength of this study lies in the use of both questionnaires (quantitative) and perception (qualitative) instruments of data collection, while its weaknesses lies in the fact that it was limited to descriptive analysis (charts, tables, and percentages), as many inferences cannot be drawn from the study.

Still, on the basis of income and housing quality, studies have been carried out factoring other elements into the income equation. Using a peripheral urban settlement as a case study, Adedire and Adegbile (2018) assessed housing quality in Ibeju-Lekki, a peripheral settlement outside Lagos metropolitan region. The research made use of purposive sampling to select 370 housing units from a sample of a cluster of 16 peri-urban settlements. Direct field data were sourced through structured questionnaires, guided interviews with local planning personnel, and observation schedules administered through a field survey. The information collected concerned building materials, socioeconomic information, neighborhood quality, dwelling characteristics, and locational information. These variables were entered into an SPSS statistical environment and analyzed to generate percentages and frequencies. A statistical transform tool was used to recode the variables and correlation analysis was carried out to demonstrate the factors impacting the quality of housing in the area of study. The results indicated that there was a significant positive correlation between the income of households and the quality of their housing. The quality of housing in Ibeju-Lekki was impacted by building materials, socioeconomic information, neighborhood quality, and locational attributes, but the major determinant was the socioeconomic characteristics of the respondents. Developers were advised to promote other types of building materials to enhance housing affordability by the low-income group.

Building elements have been said to also affect the quality of housing in certain areas. In this light, Awe and Afolabi (2017) went further to perform a critical assessment of housing quality in the urban core of Ado-Ekiti, the capital of Ekiti State. Data for this research were collected from both direct (primary) and indirect (secondary) sources. A total of 300 questionnaires were distributed using random sampling method to ensure effective coverage of the urban core, and 295 questionnaires were retrieved. Descriptive methods of data analysis were used and the findings from the study indicated that building elements like roofing materials, doors, walling, flooring, and ceilings were in pitiable condition, thereby rendering the buildings unfit for human occupation. Therefore, the study advocated building conservation, pointing to the need for governmental strategies for shabby buildings to be resold and for redevelopment of the urban core.

The studies reviewed above gave insights into the ideas used in the current study.

28.4 Conceptual Framework Residential Quality Model

The Residential Quality Model, which served as the model for this study, was developed by Adewoyin (2017) for a housing quality study in Ibadan. The model was constructed using residential housing (condition of buildings, presence of basic infrastructure) and environmental quality (neighborhood characteristics, nearness to the habitats of disease vectors, and social services within the neighborhood) indicators. Major risk factors for unhealthy living conditions were also included in the model construction based on the conditions highlighted by the World Health Organization (WHO) and those of other theories. Disease (malaria) incidence of the people living in the environment was also a major building block of the model. This particular block was brought in to show the influence of housing quality on the quality of life of the people. These indicators were built into a factor analytic model for validation, and the model yielded an 86% validation rate, indicating that the model is useful for measuring the quality of urban residential areas and for identifying health risks in such environments. This model was used in this study, showing that both indoor and outdoor characteristics were important components or indicators of housing quality, each playing distinct roles.

28.5 Methodology Area of Study

The study area, Owerri Municipal, is located in the southeastern part of Nigeria. It is one of the 27 Local Government Areas of Imo State. The headquarters of the municipal are in the city of Owerri. As a former headquarter of the Old Owerri Local Government Area, it became a municipal council on December 15, 1996. The inhabitants are mainly lgbo, but there is the presence of people from other ethnic groups from different parts of Nigeria and beyond. The Municipal lies between latitudes 5°30′N and 5°36′N and between longitudes 7°00′E and 7°12′E. It is bounded on the north by Amakohia, on the northeast by Uratta, on the east by Egbu, on the southeast by Naze, on the south by Nekede, and on the northwest by Irete (Cartography Directorate, Ministry of Land and Housing, Owerri, 2010). Traditionally, the Municipal was an urban setting with one autonomous community made

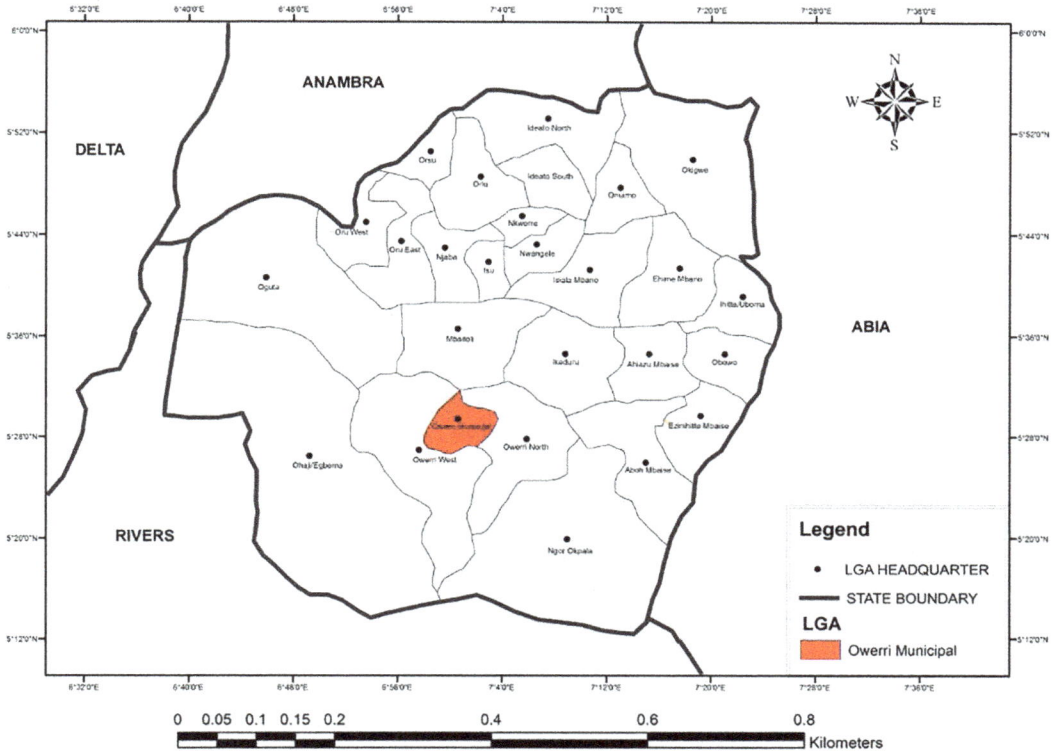

FIGURE 28.1
Imo State showing Owerri Municipal households (NPC, 2006). This suggests that there is rapid population growth without a corresponding increase in physical infrastructures and facilities to keep pace.

up of five indigenous kindred called Owerri Nchi-ise. The five communities are Umuoronjo, Amawon, Umuonyeche, Umuodu, and Umuoyima. The Municipal also covers the following areas: World Bank Area, Ikenegbu, Aladimma, Shell Camp Area, and Works Layout.

According to the 2006 National Population Census, Owerri Municipal has a total population of about 127,213 inhabitants: 62,990 male and 64,223 female with 24,282 (Figure 28.1).

28.6 Research Design

A survey research design was adopted for this study, as this affords researchers the opportunity to collect information on the unique characteristics of the population. Primary data were obtained through direct observation and questionnaire administration, as well as a building demographic and facility survey. The questionnaire constituted the major instrument used in information collection and was designed and administered to elicit information on the socioeconomic characteristics, building conditions, infrastructural facilities, and environmental condition of the study area. Direct observation was also used to validate claims and responses on physical, environmental, and housing conditions of the study area.

TABLE 28.1

Sector Number and Name

Sector No.	Sector Name
1	Are L. Ring Rd–Sam Mbakwe–Onisha Rd–Area M Ring Rd
2	Area M Ring Rd–Area 1 Ring Rd–Sam Mbakwe Avenue
3	Onitsha Rd–Bank Rd–Orlu Rd–New Rd
4	Orlu Rd–Bank Rd/Okigwe Rd–Works Rd
5	Works Layout/Bishop Unegbu Avenue–Okigwe Rd–Works Rd
6	Wetheral Rd–Ikenegbu Layout St–Samek Rd.–Okigwe Rd
7	Ikenna Nzimiro Av–Sam Mbakwe Ave–Porthrcourt Rd
8	Omuma Rd–Port Harcourt Rd–Ikenna Nzimiro Ave–Sam Mbakwe Ave
9	Area N Ring Rd/General Hospital Rd–PH Rd, –Omuma Rd–Sam Mbakwe
10	PH Rd–E.C Iwuanyanwu Ave–Portharcourt Rd–Achike Udenwa Ave
11	E.C Iwuanyanwun Ave–B. Anyanwu Rd–Nekede/Ogor Ave–AchikeUdenwa Ave
12	Ndubisi Kanu Ave–B Anyanwa Rd–Nathan Ogor Ave–Achike Udenwa Ave
13	Ndbuisi kanu Ave–B AchikeUdenwa Rd–Nathan Ogor Ave–PH/Bank Rd
14	Life Chapel Ave/Nathan Ogoh Ave–Nekede Rd–Royce Rd–Bank Rd
15	Royce Rd–Nekede Rd–Douglas Rd
16	Nworie St–Deeper Life Ave–Aba Rd–Nekede Rd
17	Aba Rd–Wetheral Rd–Tetlow Rd–Mbaise Rd
18	Douglas Rd–Tetlow Rd–School Rd
19	Douglas Rd–Bank Rd–School Rd
20	Tetlow Rd–Bank Rd–Wetheral Rd–School Rd
21	Tetlow Rd–Wetheral Rd–School Rd
22	Wetheral Rd–Egbu Rd–ChukwunaNwaoha Rd–MCC/Uratta Rd
23	Chukwuma Nwoha Rd–Egbu Rd–Dick Tiger Ave–MCC Uratta Rd
24	Wetheral Rd–MCC/Uratta Rd–Aladinma Hospital Rd–Ikenegbu Layout St
25	Aladinma Hospital Rd –MCC/Uratta Rd–St. Andrew Rd–Isiukwuato St
26	St. Andrew Rd–Isiukwuato St–Raymond Chikwe Ave–Samek Rd
27	Ikenegbu Lay, Street/Aladinma Hospital Rd–Isiukwuato St–Raymond Chikwe Ave–Samek Rd
28	Okigwe Rd–Ama Wire/Ugwnorji Rd–Samek Rd
29	E.C. Iwuanyanwu Av.–B. Anyanwu Rd–Ndubuisi Kana Ave–Achike Udenwa Ave
30	Omuma Rd–B Anyyanwu Rd–E.C. Iwuayanwu Ave–Achike Udenwa Ave

Source: Researcher's field survey.

According to the 2006 National Population Census, Owerri Municipal had a total population of 127,213 inhabitants with 24,282 residential housing units. The population of this study consisted of all the residential housing units found along the roads and streets within the 30 sector locations of Owerri Municipal Council (Table 28.1).

28.7 Sampling Procedure

From the housing enumeration areas demarcated by the National Population Commission in 2006, Owerri Municipal has 24,282 residential housing units. To get a sample size for the study, Taro Yamane's (1967) formula was used to ensure minimum bias and maximum precision given the resources available.

$$n = \frac{N}{1 + N(e)^2}$$

where
   n is sample size
   N is the total population
   e is the limit of tolerable error (0.05)
   1 is the unity

Applying the Taro Yamane formula,

$$n = \frac{24,282}{1 + 24,282\,(0.05)^2} = 390$$

Hence, using the above formula, the sample size for the study was 390. Since the study area was demarcated into 30 sectors, it therefore implied that,

$$\text{Number of households selected} = \frac{390}{30} = 13$$

A total of 390 respondents (household heads) were randomly sampled using stratified sampling to select 13 respondents from each sector. To select samples in each sector, the detailed map of Owerri Municipal Council was divided into grid cells (quadrates) and a point random approach was applied to sample 13 residential houses from each sector. Because the grid map contains the names of streets and other landmarks, it was possible to identify points sampled.

The study adopted 30 important indictor variables to describe housing quality in Owerri Municipal, including structural elements (walling, roofing, and flooring materials), lighting, age of the building, location of the building, indoor characteristics, outdoor characteristics, safety, security/crime, crowding accessibility, household hygiene/sanitation, waste disposal, and environmental quality.

28.8 Data Analysis

One of the methods used for the data analysis was factor analysis method, which was used to collapse the sets of variables into fewer factors, thereby identifying groups of interrelated variables for housing quality in the study area. In the end, factor analysis reduced the 30 housing indicators for which data were collected into eight factors (Figure 28.2).

Factor analysis is a multivariate dimension reducing statistical technique which primarily enables reduction of a large number of original variables $X_1, X_2, X_3, X_4, \ldots X_n$ into a new set of fewer variables for greater parsimony (Udofia, 2011). The choice of factor analysis method was based on its ability to collapse a large dataset into a smaller number of factors to identify groups of interrelated variables. One of the basic assumptions guiding factor analysis is that the variables should be associated in a linear way, which is usually checked

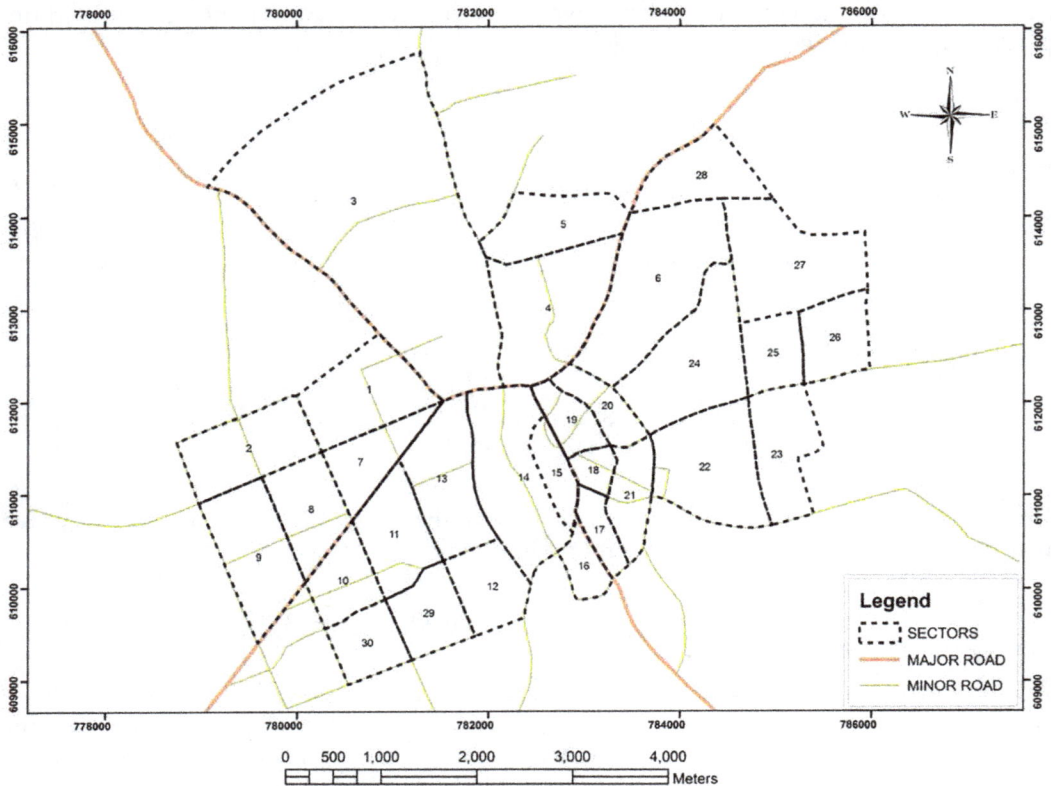

FIGURE 28.2
Housing sectors of Owerri Municipal.

using a scatter plot. The analysis works by linear combination, and the principal components method of extraction used does not extract uncorrelated linear combinations. The factors that were extracted were rotated to make them interpretable and more accurate.

The factors were further subjected to cluster analysis, which yielded three housing quality regions.

Cluster analysis is a multivariate classification technique used to group data into natural associations to ensure coherence and internal consistency (Udofia, 2011). It helps create a structure within a body of data. To derive a structure of housing quality appropriate for the area, cluster analysis was applied as the most suitable method of grouping data.

28.9 Presentation of Results

28.9.1 Summary of the Socioeconomic Data of Respondents

Housing quality has been determined to be a function of the socioeconomic characteristics of the householders. The table below presents a summary of the socioeconomic characteristics of the respondents from whom the data were obtained.

The summary of the socioeconomic characteristics of the respondents reveals that slightly above 9% of the respondents had only primary education, about 41.53% had secondary education, and 48.72% had tertiary education. Housing quality appears to be positively associated with both higher school attainment and some of the behavioral indicators that tend to accompany it. Hence, the level of education of respondents is an indicator of the quality of housing standard in the study area. The occupational pattern indicates that 22.31% of the respondents are civil servant, 36.92 were public servant, 16.15% were craftsmen, 17.95% dealt in trade, 2.05% were farmers, 2.82% were apprentices, and 1.80% were unemployed. The data indicate that most of the respondents were civil and public servants, and this affects their hosing quality because the nature of the occupation determines the income. In the area of monthly income, only 5.64% of the respondents receive below the minimum age of #18,000. 13.33% had an income level of #19,000–#36,000, 18.21% had an income of #37,000–#50,000, 22.31% and 38.97% had an income level of #51,000–#72,000 and above #72,000, respectively. This indicates that most of the respondents were living above the minimum wage which enabled them to be able to afford good housing.

Household size is the number of persons living together, either as a nuclear or extended family unit within the same house and sharing common facilities. This study indicates that 11.03% of the respondents had a household size of 1–3 persons, 47.69% a household size of 4–6, 24.87% a household size of 7–9 persons, and 16.41% a household size of above 9 persons. The United Nations Standard for Nigeria for room occupancy is 2.2; the World Health Organization (WHO, 1990) stipulates between 1.8 and 3.1, while the Nigerian Government's prescribed standard is 2.0 per room (Okoko, 2001). From the result, 70% live in overcrowded accommodations.

Housing quality refers to the character, disposition, and nature of the housing units. It is the attribute, properties, special features, and degree or grade of excellence the housing unit must possess. This section presents data on many attributes of housing quality in the study area.

The structural elements of the buildings were also assessed as shown in Table 28.2. The structural elements affect the comfort, convenience, and security of the occupants of the building. This research indicated that 90% of the study participants had sandcrete walls in their houses, while only 0.51% had mud walls. The wall materials indicated that the houses in the study area had good, strong walls judging from the sandcrete materials involved. The data also showed that 48.97% of the houses had long-span aluminum roofs, while 43.08% had zinc-coated iron sheet roofs. There were no thatched roofs in the area. This result indicates that the houses in the study area had good roofs in line with modern technology and that the people were conscious of their convenience and the esthetics of their buildings.

The data collected also indicated that ceramic tiles (38.46%) and cement screed (35.39%) were the major flooring materials for houses in the study area, followed by marble (14.36%) and terrazzo (11.8%). There were no houses with mud floors. The materials used for flooring in the study area can be attributed to the availability of the materials and the recent trends in flooring of houses. The houses in the study area have quality floors.

In the area of plastering, 95.38% of the houses were plastered inside and outside, 3.6% were plastered inside only, and 0.51% were not plastered at all. The data indicate that based on plastering as a quality indicator, the houses in the study area were of high quality.

Table 28.3 refers to the location of a dwelling unit in relation to noise pollution sources. It was measured by assessing the relative distance of the dwelling unit to the noise pollution sources. The table indicates that more than 90% of the respondents live more than 200 m from sources of noise pollution, showing that the level of noise pollution in the area is low.

TABLE 28.2

Building Materials Indicators

Building Characteristics	Number	(a) Percentage
Wall material mud	2	0.51
Sandcrete block concrete	351	90.0
Bricks	9	2.31
	28	7.18
Roofing material	191	48.97
Long-span aluminum asbestos	31	7.95
Zinc-coated iron sheet thatched	168	43.08
	0	0
Floor material ceramic tiles	150	38.46
Terrazzo	46	11.8
Marble	56	14.36
Cement screed mud	138	35.39
	0	0
Plastering plastered inside plastered outside	14	3.6
Plastered inside and outside not plastered at all	2	0.51
	372	95.39
	2	0.51

Source: Researcher's fieldwork.

TABLE 28.3

Location of Building in Relation to Noise Sources

	Distance to House		
Noise Source	Less than 200 m	More than 200 m	Total
Market	18	372	390
Sound record store	17	373	390
Church	23	367	390
Night club	17	373	390
Major road	76	314	390

The low level of noise is an indicator that the areas used were purely residential areas that are not affected by the noise or influence of other urban activities.

Principal component analysis in varimax with Kaiser normalization rotation was applied, converging in nine iterations to yield eight new factors that, as indicated below, accounted for 79.1% of the total variance in the original dataset. The basis for including these factors was that the magnitudes of their eigenvalues were 1.0 and above. The eight factors were rotated to enable clarification and definition of the factors. Also, the naming of the factors was based on loadings on the initial variables. Udofia (2011) suggested that any component contributing less than 5% to the total variation should be rejected; therefore, factor 8 has the lowest contribution at 5.7% and all factors contributing less than 5% of the variation were excluded. The rule of thumb was to use variables with ±0.5 and above on the factor to name the factor. Due to the factors extracted, all factors with negative values were excluded and the factors were named using the highest loading.

TABLE 28.4

List of 30 Independent Variables

X_1	Wall materials
X_2	Roofing materials
X_3	Floor materials
X_4	Plastering
X_5	Age of building
X_6	Location of building
X_7	Building type
X_8	Indoor temperature
X_9	Accident and safety
X_{10}	Security and crime
X_{11}	Crowding indicator
X_{12}	Size of room
X_{13}	Distance between property
X_{14}	Lighting
X_{15}	Electricity
X_{16}	Sanitary waste bin
X_{17}	Distance between bin and property
X_{18}	Waste evacuation
X_{19}	Hygiene
X_{20}	Toilet
X_{21}	Bathroom
X_{22}	Kitchen
X_{23}	Water supply
X_{24}	Water source
X_{25}	Distance to water vendor
X_{26}	Food store
X_{27}	Cooking fuel
X_{28}	Access road
X_{29}	Compound floor
X_{30}	Stagnant water

28.10 Factor Structure of Housing Quality

The 30 independent variables were entered into the factor analysis model (Tables 28.4 and 28.5).

i. **Accessibility factor**: Factor 1 was named "access road factor," having loaded high and positive on the access road indicator (X_{28}; 2.04). The accessibility factor contributed 26.4% of variance in the dataset with an eigenvalue of 8.97.

ii. **Water supply factor**: Factor 2 was named "water supply factor" having loaded high on water source (X_{24}; 2.39). The water supply factor contributed 9.3% of variance in the dataset with an eigenvalue of 4.2.

iii. **Ventilation factor**: Factor 3 was named "ventilation factor" having loaded high on indoor temperature (X_8; 1.86). The ventilation factor contributed 9.0% of variance in the dataset with an eigenvalue of 2.3.

TABLE 28.5

Rotated Component Matrix

	1	(b) 2	(c) 3	(d) 4	(e) 5	(f) 6	(g) 7	(h) 8	(i) Com
X_1	0.17386	1.27318	−1.86292	−0.50796	−2.08357	0.40252	0.22835	0.03614	
X_2	−0.21051	1.3038	−2.00775	−0.72674	−1.18445	0.22005	1.15382	0.72925	
X_3	−0.17444	0.66098	−0.47044	−0.965	−0.41512	0.05381	−1.31806	−0.25594	
X_4	−0.48919	−0.21294	−0.45518	−1.27314	0.17307	−1.53199	−0.40425	−0.91737	
X_5	−0.82699	0.78309	0.19158	−0.90461	0.42946	−2.14121	0.03096	−0.01679	
X_6	−0.46411	0.39722	1.00653	0.54219	−1.75725	−0.56138	0.88039	−0.33616	
X_7	−0.87156	0.90882	−0.59968	−0.06683	−0.66344	0.4817	0.41439	−0.61735	
X_8	−0.86255	0.23805	1.85857	0.14684	−1.27192	0.34261	0.05849	−0.91681	
X_9	−0.90091	−0.32424	0.57521	−0.62978	−1.07461	0.39764	−2.57851	0.44543	
X_{10}	−0.71125	−0.77383	0.69599	−0.17173	−0.79139	−0.78453	−0.44285	−0.07625	
X_{11}	−0.24671	−0.35914	−0.47368	−0.36173	0.28697	−1.50857	1.63816	1.19797	
X_{12}	−0.81913	−0.48951	−0.11523	−0.25312	0.18714	−0.44058	−0.10222	1.2141	
X_{13}	−0.51822	−1.28415	1.16926	0.5405	−0.74435	−0.11082	1.304	0.50961	
X_{14}	−1.14851	−1.82582	−1.00082	0.98004	−0.29352	0.6504	0.43825	−1.30949	
X_{15}	−0.88176	−0.72594	0.82618	0.10553	0.23132	0.85951	1.13092	−1.22068	
X_{16}	−1.57904	0.7658	0.00888	0.07269	1.95276	0.88633	0.10887	1.32739	
X_{17}	−0.83578	−0.01011	0.61487	−0.38084	0.94369	2.38808	0.22225	0.32426	
X_{18}	−0.57958	0.64364	−0.34227	−0.9237	1.8813	0.45411	0.55663	0.79443	
X_{19}	0.41068	−1.36685	0.57062	−0.8943	1.01207	−1.23417	−0.64965	−1.34082	
X_{20}	−0.54216	−0.291	−0.7061	0.44333	1.25473	0.74522	−1.96224	0.41324	
X_{21}	0.35408	−1.51528	−0.64669	2.16021	−0.15501	−0.0587	−0.53231	−0.09551	
X_{22}	0.97091	−1.09247	−1.79786	1.91334	−0.02209	−0.6243	−0.15186	1.31695	
X_{23}	0.48332	1.2716	0.04608	2.09617	0.22801	−0.25599	−1.37375	0.33917	

(Continued)

TABLE 28.5 *(Continued)*

Rotated Component Matrix

	1	(b) 2	(c) 3	(d) 4	(e) 5	(f) 6	(g) 7	(h) 8	(i) Com
X_{24}	0.61331	2.3864	1.43715	2.15315	0.84933	0.09951	1.00127	-0.81188	
X_{25}	0.97431	0.89387	-0.34676	0.25598	1.42428	-1.92882	0.05473	-1.50703	
X_{26}	1.51621	0.53962	0.62126	-0.28995	-0.6309	0.32487	-1.53534	-0.18	
X_{27}	1.74001	0.10596	0.76938	-0.13917	-0.77549	0.59491	0.25205	1.18296	
X_{28}	2.0404	0.04152	0.97951	-1.39171	0.15957	0.65542	-0.2613	0.40875	
X_{29}	1.71562	-1.30018	0.92257	-0.71168	0.22812	-0.0828	1.0459	1.75357	
X_{30}	1.66965	-0.64212	-1.46825	-0.81799	0.6213	1.70722	0.79291	-2.39113	
EIG	8.79	4.2	2.3	2.1	1.96	1.47	1.32	1.08	
%va r	26.4	9.3	9.0	8.1	7.6	6.6	6.4	5.7	
Cum	29.9	43.9	52.6	59.7	66.2	71.1	75.5	79.1	

iv. **Hygiene factor**: Factor 4 was named "hygiene factor" having loaded high on bathroom (X_{21}; 2.16). The hygiene factor contributed 8.1% of variance in the dataset with an eigenvalue of 2.1.

v. **Sanitation factor**: Factor 5 was named "sanitation factor" having loaded high on the presence of sanitary waste bin (X_{16}; 1.95). The sanitation factor contributed 7.6% of variance in the dataset with an eigenvalue of 1.96.

vi. **Neighborhood factor**: Factor 6 was named "Neighborhood factor" having loaded high on the distance of the sanitary bin from the property (X_{17}; 2.38). The neighborhood factor contributed 6.6% of variance in the dataset with an eigenvalue of 1.5.

vii. **Crowding factor**: Factor 7 was named "crowding factor" having loaded high on the crowding indicator (number of people per room; X_{11}; 1.63). The crowding factor contributed 6.4% of variance in the dataset with an eigenvalue of 1.3.

viii. **Outdoor characteristics factor**: Factor 8 was named "outdoor characteristics factor" having loaded high on compound flooring materials (X_{29}; 1.7). The outdoor characteristics factor contributed 5.7% of variance in the dataset with an eigenvalue of 1.1.

The eight factors cumulatively contribute 79.1% of the variance in the dataset and therefore represent a valid solution for identifying the factor structure of housing in Owerri Municipal. The communalities are highly loaded indicating the suitability of the variables included in the factor solution and the entire factor analysis. The new variables are the general determinants of housing quality in the study area.

28.11 Classification for the Study Area Based on Housing Quality

Hierarchical cluster analysis was conducted to classify the study area based on housing quality. In this study, hierarchical cluster analysis (Ward's method) was employed to aid in classifying the 30 housing sectors into three regions on the basis of the eight housing quality components extracted by factor analysis. This method aided in the determination of housing quality characteristics of sectors, thereby revealing developed and lagging sectors in terms of performances on the housing quality indicators to ensure coherence and internal consistency. The purpose of cluster analysis was to find out whether sectors can be formed into any natural system of group on the basis of regularities and interrelationships among the samples.

According to Udofia (2011), a convenient rule of thumb for selecting the appropriate number of clusters for representing the dataset in the program is to stop where the coefficient between the adjacent cluster suddenly becomes large. In the agglomeration schedule above, such a change occurred at stages 10 and 11 where the number of groups was 3. At that point, the cluster coefficient was 23.132, compared with 19.794 at the next lower stage. It was concluded that the initial 30 sectors could be adequately represented using three housing quality zones from the table of cluster membership.

After obtaining these three different groups, the performance of each group was measured on the eight factors earlier extracted using factor analysis. The groups are considered below.

Group 1: This group consisted of four housing sectors (1, 2, 7, 30) found mainly around the Sam Mbakwe area. The characteristics of these four housing sectors include a strong performance on two major housing qualities, F7 (crowding factor) and F2 (water supply indicator) (Tables 28.6–28.8).

Group 2: This group consists of 12 housing sectors (3, 9, 16, 17, 18, 20, 23, 24, 26, 27, 28, and 29) found around Onitsha-Orlu Road, Aba Road, Iwuanyanwu Avenue, Achike Udenwa Avenue, Wetheral, MCC, and Uratta Road. The characteristics of these 12 housing sectors include a strong performance on four major housing qualities: F2 (water supply indicator), F3 (ventilation characteristics factor), F6 (neighborhood factor), and F8 (outdoor characteristics factor) (Table 28.9).

Group 3: This group consists of 14 housing sectors (4, 5, 6, 8, 10, 11, 12, 13, 14, 15, 19, 21, 22, 25) found around Okigwe Road, Works Layout, Ikenegbu Layout, Port Harcourt Road, Ndubisi Kanu Avenue, and Aladinma Hospital Road. The characteristics of these 14 housing sectors include a strong performance on five major housing qualities: F2 (water supply indicator), F3 (ventilation characteristics factor), F4 (hygiene factor), F5 (sanitation factor), and F8 (outdoor characteristics) (Table 28.10).

A description of the three new groups obtained by cluster analysis is given in Table 28.11.

The cluster analysis indicates that Owerri Municipal is lagging in F1, accessibility. This indicates that accessibility to houses may be a problem in this area. The dendrogram representing a panoramic view of all possible combinations of sectors is displayed here (Figure 28.3).

Housing quality showed a positive association with education level. Hence, the level of education of respondents is an indicator that contributes to the quality of housing standard in the study area. This agrees with the finding of Gambo (2013) on respondents' level of education in Makoko, Lagos State, that slightly above 63% of the respondents had only secondary education, while only about 7% had education beyond secondary school. It was also interesting to see that almost half of the respondents had tertiary education, which is an indication that there is high adult literacy in Owerri. This was to be expected, given that

TABLE 28.6

Agglomeration Schedule

	Cluster Combined		(j)	(k) Stage Cluster First Appears		(l)
Stage	Cluster 1	Cluster 2	Coefficients	Cluster 1	Cluster 2	Next Stage
1	1	2	1.198	0	0	11
2	16	18	2.601	0	0	12
3	4	5	4.150	0	0	15
4	6	8	5.716	0	0	16
5	27	28	7.440	0	0	10
6	10	12	9.247	0	0	14
7	21	22	11.459	0	0	19
8	3	26	13.805	0	0	13
9	13	15	16.639	0	0	16
10	27	29	19.794	5	0	26
11	1	7	23.132	1	0	22
12	16	17	26.890	2	0	23
13	3	9	30.683	8	0	20
14	10	11	34.780	6	0	21
15	4	19	38.946	3	0	18
16	6	13	43.336	4	9	21
17	20	23	47.935	0	0	20
18	4	25	52.736	15	0	27
19	14	21	59.429	0	7	24
20	3	20	66.938	13	17	25
21	6	10	76.078	16	14	24
22	1	30	90.187	11	0	28
23	16	24	104.936	12	0	25
24	6	14	122.578	21	19	27
25	3	16	142.210	20	23	26
26	3	27	162.513	25	10	28
27	4	6	183.754	18	24	29
28	1	3	206.731	22	26	29
29	1	4	232.000	28	27	0

Source: Author's Computation (2018).

TABLE 28.7

Composition of Groups

Group No.	Sector	(m) Number of Sectors
1	1, 2, 7, 30	40
2	3, 9, 16, 17, 18, 20, 23, 24, 26, 27, 28, 29	12
3	4, 5, 6, 8, 10, 11,12, 13, 14, 15, 19, 21, 22, 25	14
		30

TABLE 28.8

Housing Quality Profile for Group 1

Sector	Q_1	(n) Q_2	(o) Q_3	(p) Q_4	(q) Q_5	(r) Q_6	(s) Q_7	(t) Q_8
Score on Major Qualities								
1	1	1	−2	−1	−2	0	0	0
2	−0	1	−2	−1	−1	0	1	1
7	−1	1	−1	0	−1	0	0	−1
30	2	−1	−1	−1	1	2	1	−2

Source: Researcher's computation.

TABLE 28.9

Housing Quality Profile for Group 2

Sector	Q_1	(u) Q_2	(v) Q_3	(w) Q_4	(x) Q_5	(y) Q_6	(z) Q_7	(aa) Q_8
Score on Major Qualities								
3	−0	1	−0	−1	−1	0	−1	−0
9	−1	−0	1	−1	−1	0	−2	0
16	−2	1	0	0	2	1	0	1
17	−1	0	1	0	1	2	0	0
18	−1	1	−0	−1	2	0	1	1
20	−1	−0	−1	0	1	1	−2	0
23	0	1	0	2	0	0	−1	0
24	1	2	1	2	1	0	1	−1
26	2	1	1	−0	−1	0	−2	−0
27	2	0	1	−0	−1	1	0	1
28	2	0	1	−1	0	1	0	0
29	2	1	1	−1	0	0	1	2

Source: Researcher's computation.

TABLE 28.10

Sector	Q_1	(bb) Q_2	(cc) Q_3	(dd) Q_4	(ee) Q_5	(ff) Q_6	(gg) Q_7	(hh) Q_8
Score on Major Qualities								
4	−0	−0	−0	−1	0	−2	−0	−1
5	−1	1	0	−1	0	−2	0	−0
6	−0	0	1	1	−2	−1	1	−0
8	−1	0	2	0	−1	0	0	−1
10	−1	−1	1	−1	−2	0	−3	0
11	−0	−0	−0	−0	0	−2	2	2
12	−1	−0	−0	−0	0	−0	−0	1
13	−1	−1	1	1	−1	−0	1	1
14	−1	−2	−1	1	−2	1	0	−1
15	−1	−1	1	0	0	1	1	−1
19	0	−1	1	−1	1	−1	−1	−1
21	0	−2	−1	2	−0	−0	−1	−0
22	1	−1	−2	2	−0	−1	−0	1
25	1	−1	0	0	1	0	1	−1

Source: Researcher's calculations.

TABLE 28.11

Group No.	Group Name	(ii) Composition	(jj) Number of Sectors
1	Crowded	1, 2, 7, 30	4
2	Highly crowded	3, 9, 16, 17, 18, 20, 23, 24, 26, 27, 28, 29	12
3	Over crowded	4, 5, 6, 8, 10, 11, 12, 13, 14, 15, 19, 21, 22, 25	14
			30

Source: Researcher's calculation (2018).

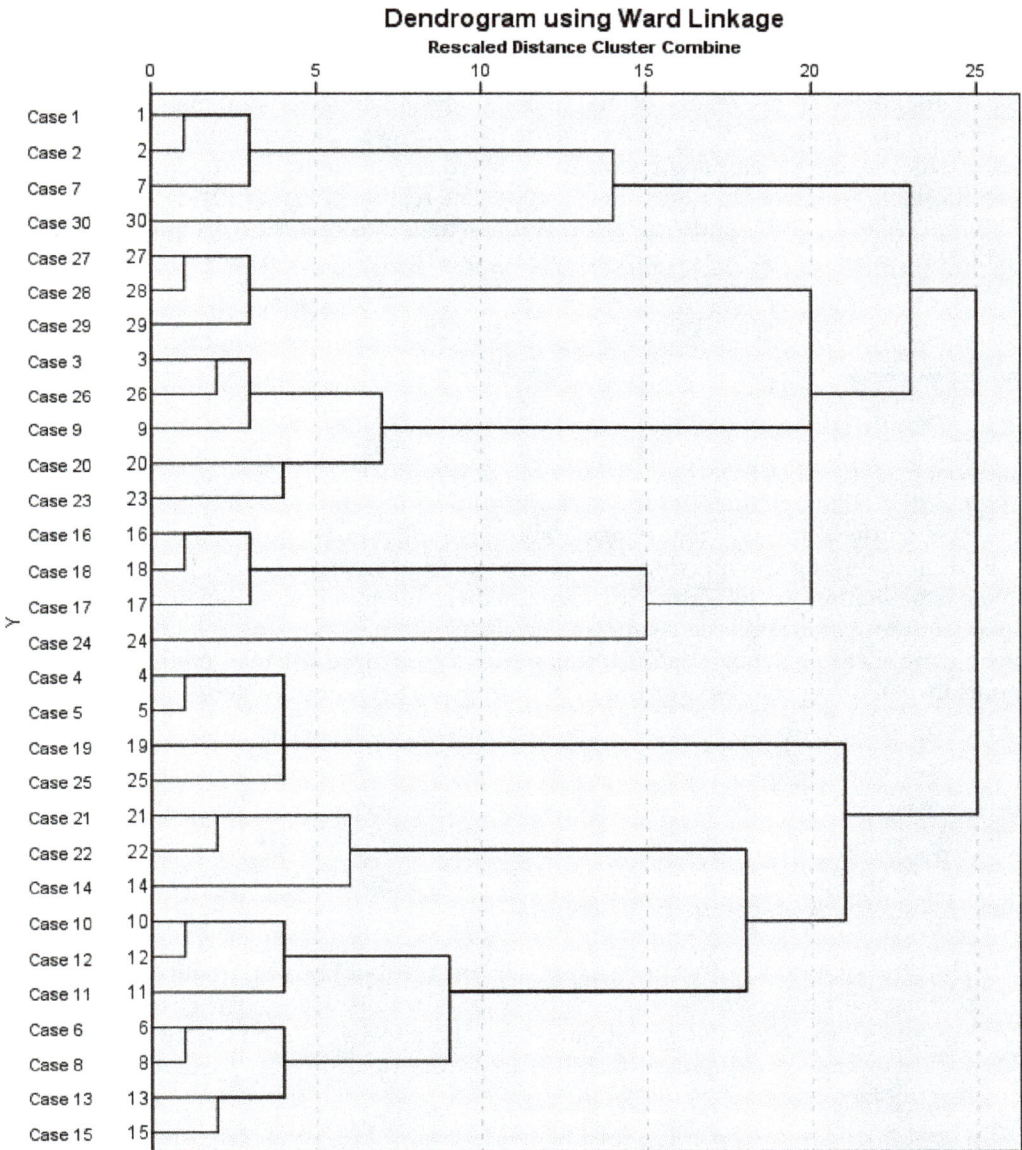

FIGURE 28.3
Cluster classification discussion of findings.

Owerri has a high concentration of higher institutions of learning that provide opportunities for people to acquire higher education.

Moreover, several studies have argued that income plays a great role in the quality of houses people find themselves in (Adewoyin, 2017). The results for monthly income discussed above are contrary to the findings of Omole (2010) on income and occupational distributions in Akure, Nigeria, that about 51.3% receives monthly income below ₦5,000, of whom 19.1% have no stable source of income. About 40.4% receive between ₦5,000 and ₦15,000, while only 8.3% receive above ₦15,000. With this low-income distribution, affording good quality housing, properly maintaining an existing one, and adequate feeding might be very difficult, if not impossible.

On the other hand, the results of this study show that 70% of the study respondents lived in houses with more than 2.0 occupants per room, leading to overcrowding. This is further supported by the results of the cluster analysis that 46% (14 housing sectors) of the 30 housing sectors studied were overcrowded. The crowded areas were spread across the metropolis, meaning that many people in the study area were living in poor housing quality conditions, which could affect their health. The results of this study corroborate those of Morenikeji et al. (2017) and Adewoyin (2018), who found that the differences in housing quality in the country gave rise to a number of health and social issues.

28.12 Conclusions

Analyses performed in this research resulted in the satisfactory demarcation of the study area into three housing quality regions. Perhaps the most important conclusion to emerge from this study is the fact that considerable disparities exist in levels of housing quality indicators among the housing sectors in Owerri Municipal of Imo State, which must devote urgent attention to addressing some of these inequalities, especially in the lagging aspects of crowding and water supply. The distribution of housing quality indicators in the housing sectors are not balanced, as some areas are more advantaged than others. The importance of using statistics for urban planning has been highlighted.

28.13 Recommendations

Thus, this study recommends the following:

1. The study found significant variations in the level of housing quality among the sectors, with regard to the housing quality indicators. Hence, the observed housing quality patterns and performance may be improved upon by channeling the socioeconomic development facilities to appropriate targets, particularly in the lagging factors.
2. The influences of education, income, and household size in the study area were found to be significant. This therefore calls for the consideration of socioeconomic

characteristics of population when implementing housing development projects, rather than working in isolation.

3. The study area was demarcated statistically into three housing quality regions. It is recommended that the low-quality regions be targeted to improve the housing conditions in the area.

References

Adedire, F. and Adegbile, M. (2018). Assessment of Housing Quality in Ibeju-Lekki Peri-Urban Settlement, Lagos State, Nigeria. *Acta Structilia* 25(1), 126–151.

Adeleye, O.A. and Anofojie, A.E. (2011). Housing in Festac Town, Lagos State, Nigeria. *Ife Research Publication in Geography* 10(1), 154–163.

Adewoyin, Y. (2018). A Residential Habitat Quality Model for Population Health Vulnerability Assessment in Urban Nigeria. *International Journal of Scientific Reports* 4(3), 59–67.

Awe, F.C. and Afolabi, F.I.O. (2017). Assessment of Housing Quality in Urban Core of Ado-Ekiti, Nigeria. *Civil and Environmental Research* 9(7), 37–42.

Gambo, Y.L. (2013). Impact of Poor Housing Condition on the Economy of the Urban Poor: Makoko, Lagos State in View. *Journal of Emerging Trends in Economics and Management Sciences (JETEMS)* 3(4), 302–307.

Morenikeji, W., Umaru, E., Pai, H., Jiya, S., Idowu, O. and Adeleye, B.M (2017). Spatial Analysis of Housing Quality in Nigeria. *International Journal of Sustainable Built Environment* 6, 309–316.

National Population Commission (2006). *National Population and Housing Census*. Federal Government of Nigeria, Abuja.

Nigerian National Housing Policy (2012).

Oluwaseyi, O.B. (2019). Assessment of Housing Quality in Osun State, Nigeria. *European International Journal of Science and Technology* 8(5), 69–102.

Onibokun, A.G. (1999). *Housing in Nigeria*. NISER, Ibadan, pp. 58–59.

Udofia, E.P (2011). *Applied Statistics with Multivariate Methods*. Immaculate Publications Ltd., Enugu.

United Nations Declaration (1948). The Universal Declaration of Human Rights (UDHR) Summary.

United Nations Human Settlements Program (UN-Habitat) (2002). Cited in World Economic and Social Survey 2013.

World Health Organization (1961). Housing.

29

Role of Gender Disaggregate Statistics in Urban Planning and Development in Pakistan

Lubna Naz and Muhammad Umair

University of Karachi

CONTENTS

29.1 Introduction

The size of the gender wage gap has always been an important topic in setting development goals because it is a persistent feature of labor markets in developing countries. The Sustainable Development Goals (SDGs) emphasize adopting wage and social protection policies, ensuring equal opportunities, and ending discrimination to achieve greater equality; see SDG 5 and SDG 10.[1] Many studies have compared the gender wage gap based on the socioeconomic and demographic characteristic differentials of workers. Age and education remain the essential determinants of the gender wage gap. Age can be used as a proxy of labor market experience (in the absence of total years of experience in data). Likewise, higher educational attainment improves human capital and thereby increases the productivity of labor (Sabir & Aftab, 2007).

Gender differentials in occupational choice significantly contribute to the gender wage gap. Women remain restricted to particular occupations in the selected sectors of the economy (Horrace & Oaxaca, 2001), while wages of both men and women are higher in urban areas due to a wage premium (Siddiqui & Siddiqui, 1998; Glaeser & Mare, 2001). Some countries have larger gender wage gaps in urban areas (Ahmed & Maitra, 2010), while others have

---

[1] SDG 10: Reduced Inequalities https://www.sdgpakistan.pk/web/goals/goal10.

higher gender wage gaps in rural areas (Ashraf & Ashraf, 1993). In most cases, the discrimination effect may be attributed more to the gender wage gap than the productivity effect.

Moreover, gender wage discrimination is lower in public enterprises than private ones. Public sectors ensure labor rights more than private enterprises (Hyder & Reilly, 2005). Metropolitan status also affects the wage gap. Some research has found that municipal workers' wages are higher than those of their nonmetropolitan counterparts due to differences in the standards of living and experience (Kim, 2004). According to the Labor Force Survey (LFS, 2014–2015), an official data set of labor, the nominal monthly wages of female workers were 24% lower in the urban labor markets of Pakistan: The wages of urban male workers were 17,614 PKR,[2] while the wages of urban female workers were 13,410 PKR.

This study aims to analyze and compare the gender wage differentials in the large and small urban centers of Pakistan. The wages in the large urban centers were 17% higher than in the small urban centers. Further, gender disaggregation shows that this gender wage gap was 31% in the large urban centers, but only 8% in the small urban centers. In the large urban centers, monthly male wages were 18,614 PKR, while female workers earned 12,892 PKR. In comparison, the earnings of male and female workers in the small urban centers were 15,543 and 14,251 PKR, respectively.

The structure of this study is as follows: The introduction in Section 29.1 discusses the context and scope of the study. Section 29.2 presents the literature review. Section 29.3 discusses the methodology. Section 29.4 presents estimates of the main findings, followed by a discussion in Section 29.5. Section 29.6 concludes the study.

29.2 Literature Review

The wage gap across gender is one of the challenging issues for governments and policymakers. Researchers have categorized the gender wage gap into explained and unexplained components. The explained parts are the characteristic differentials based on endowments such as education, experience, training, and other productivity features. These endowments are the dominant determinants of the gender wage gap (Stephan, 2017). In comparison, non-labor-market characteristics lead to unexplained components, such as labor market discrimination. It also includes the effect of unobserved components (Lee, 2012). However, there is no clear distinction between the human capital and discrimination effects of the wage gap (Becker, 1993). Gender differences in time allocation and human capital investment significantly generate the gender wage gap. Despite the same human capital required in the labor market, female working hours are lower than those of their male counterparts due to the former's housekeeping responsibilities. The wage gaps are nonuniform, along with the wage distribution. Workers' occupation and sector of employment also affect wage gaps (Bergmann, 1974; Kee, 2006).

Education increases the productivity of labor. However, the returns to an additional year of schooling may be higher for female workers than their male counterparts. Despite this, female educational opportunities are sparser than their male counterparts (Aslam, 2009a, 2009b). The earnings in large urban centers are higher due to higher human capital and learning by doing work (Roca & Puga, 2017). The higher returns to work experienced in bigger cities largely contribute to the wage gap between large and small urban centers

---

[2] PKR shows Pakistani Rupees.

(Baum-Snow & Pavan, 2012). The sectoral analysis shows that the contribution of primary sectors in small urban centers is higher than its share in the large urban centers (Wang & Cai, 2008).

A few studies have established that the gender wage gap in urban areas is lower than in rural ones. Studies like Bacolod (2017) identified a negative link between the gender wage gap and city size in the United States. The gender wage gap in the large urban centers declined over the period 2000–2010. However, no significant difference was evident in the average skills of workers in large and small urban centers. The reward of agglomeration economies is higher for male than female workers, as male workers are mostly employed in physical-skill-intensive jobs, while females are concentrated in interactive and cognitive skills. The human capital development of female workers could out-earn their male counterparts.

Nisic (2017) used the household panel dataset of Germany to assess the regional characteristics of the male-female earnings differentials over 1992–2012. The gender wage gap is lower in large urban centers. Household responsibilities and partnership ties restrict the spatial mobility of women. Despite smaller human capital differences and no gender discrimination, female workers had to be limited to their local labor markets.

Various studies have estimated the gender wage gap in Pakistan. Sabir and Aftab (2007) examined the gender wage gap in Pakistan over an extended period (1996–2006) using quantile regressions. It used LFS to determine a small change in the gender wage gap (ratio of female to male wages), 0.64 in 1996–1997 to 0.69 in 2005–2006. The gender pay gap was higher in the lower quantiles than the upper ones of the wage distribution. The decomposition analysis also supports the declining wage gap. However, the contribution of the discrimination effect remains higher than the endowment one.

Siddiqui and Siddiqui (1998) analyzed female labor force participation over time. Using the Household Income and Expenditure Survey (HIES), 1993–1994, the study found a 43% gender wage gap in Pakistan. Human capital differences largely contributed to the gender wage gap. Labor market discrimination was higher against female workers, while the returns to education were higher for females. The occupational choice and industrial distribution of male and female workers affected the gender wage gap, and male and female workers earn higher wages in urban areas than rural ones.

Ashraf and Ashraf (1993) analyzed the dynamism in gender earning differentials in Pakistan. The gender wage gap declined from 63% in 1979 to 33% in 1985–1986. The study employed the household dataset (HIES) to conclude that there were higher gender wage gaps in rural areas than urban ones. However, the differences declined at the national and provincial levels, both in urban and rural areas. To the best of the author's knowledge, none of the earlier research studies in Pakistan or elsewhere examined the gender wage gap in large and small urban centers. This study will explore new dimensions of the gender wage gap in the urban labor market.

29.3 Data and Methods

29.3.1 Data Sources

For the descriptive analysis and econometric estimations, the study employed the nationally representative data set for the labor market in Pakistan, the LFS, 2014–2015. LFS

2014–2015 provides data for a large urban center as a separate stratum; the other urban centers (remaining urban areas) were considered small urban centers.[3]

29.3.2 Variables

The study used several socioeconomic and demographic variables, and the selection of the variables was made using the existing empirical evidence on gender inequality. The dependent variable, Wages (W), is the natural log of nominal monthly wages (PKR) of urban workers. The information available on the periodicity of earnings encompasses weekly, daily, and monthly wages, so we converted weekly wages into monthly wages. Age was measured as the age (in years) of an urban worker. Gender shows the gender of an urban worker; it was categorized into male and female. Working hours are the total number of working hours per week in the main occupation. Education was a categorical variable arranged into five groups: no education, primary (schooling up to grade 5), matriculation (schooling up to grade 10), intermediate (schooling up to grade 12), and graduation and above (graduation, masters, and doctorate). The sector of employment of the urban workers was classified into three groups,[4] agriculture, industry, and services. The occupation of an urban worker was categorized into four groups,[5] white-collar high skilled (WHS), white-collar low skilled (WLS), blue-collar high skilled (BHS), and blue-collar low skilled (BLS).

29.4 Blinder–Oaxaca Decomposition

The study followed Jann (2008)[6] in decomposing the wage gap by gender in the large and small urban centers of Pakistan. The gender wage gap is the mean difference in the wages of male and female workers. The mean wages of the two groups are modeled as shown in equations (29.1) and (29.2) as follows:

$$W_{m_i} = X'_m \beta_m + \varepsilon_{m_i} \tag{29.1}$$

$$W_{fi} = X'_f \beta_f + \varepsilon_{fi} \tag{29.2}$$

where W shows the logarithmic monthly wages, X is the vector of independent variables, β shows the parameters, and ε shows the error terms. The mean wage difference between the two groups (ΔW) in equation (29.3) is represented as $\bar{W}$:

$$\Delta \bar{W} = \bar{W}_m - \bar{W}_f \tag{29.3}$$

Equation (29.4) represents the Blinder–Oaxaca decomposition:

$$\bar{W}_m - \bar{W}_f = \left(\bar{X}_m - \bar{X}_f \right) \hat{\beta}_m + \left(\hat{\beta}_m - \hat{\beta}_f \right) \bar{X}_f \tag{29.4}$$

---

[3] See LFS (2014–15) for the list and coding scheme of the large and small urban centres.
[4] For details, see Appendix A1.
[5] For details, see Appendix A2.
[6] For details, see Blinder (1973) and Oaxaca (1973).

where $\left(\bar{X}_m - \bar{X}_f\right)\hat{\beta}_m$ is the differentials in gender characteristics, called the explained wage gap, and $\left(\hat{\beta}_m - \hat{\beta}_f\right)\bar{X}_f$ is the unexplained differentials in gender-related characteristics. The unexplained component is referred to as the discrimination effect, which includes unobserved characteristics as well.

Equation (29.5) specifies the Blinder–Oaxaca decomposition for the large urban centers (*ml* for males and *fl* for females in the large urban centers), while equation (29.6) specifies it for the small urban centers (*ms* for males and *fs* for females in the small urban centers); thus,

$$\bar{W}_{ml} - \bar{W}_{fl} = \left(\bar{X}_{ml} - \bar{X}_{fl}\right)\hat{\beta}_{ml} + \left(\hat{\beta}_{ml} - \hat{\beta}_{fl}\right)\bar{X}_{fl} \tag{29.5}$$

$$\bar{W}_{ms} - \bar{W}_{fs} = \left(\bar{X}_{ms} - \bar{X}_{fs}\right)\hat{\beta}_{ms} + \left(\hat{\beta}_{ms} - \hat{\beta}_{fs}\right)\bar{X}_{fs} \tag{29.6}$$

29.5 Results

The gender wage differentials in the large urban centers from the LFS 2014–2015 are presented in Table 29.1. Females were less educated than their male counterparts. Around one-fifth (20%) of the males and more than one-third (35%) of the female workers had no formal education. Most males had completed their matriculation (36%). The percentage of the graduated females (16%) was greater than their male (22%) counterparts. Most of the males and females were employed in the services sector (50% of the males and 72% of the females). The occupational classification shows that most of the males were BLS workers (28%), while females were WHS workers (44%).

Workers with higher education earned higher wages. The wages of graduated males (33,492 PKR) and females (21,163 PKR) were the highest among all educational levels. However, the mean difference also increased with higher education. Wages in the services sector were the highest for both genders (20,619 PKR for males and 15,959 PKR for females). The gender wage gap was the highest in the industrial sector (10,185 PKR). The occupational classification shows that wages were the highest in the WHS jobs (34,569 PKR for males and 21,189 PKR for females), and the lowest in the BLS jobs (12,490 PKR for males and 5,782 PKR for females). The mean difference was also the highest in the WHS jobs (13,381 PKR). Males (15,280 PKR) in the WLS jobs earned lower than their female (16,069 PKR) counterparts.

Table 29.2 compares the percentages and wages of male and female workers in small urban centers. More than one-quarter of male workers (27%) and one-third of female workers (39%) had no formal education. The percentages of graduated females (20%) was almost double that of their male (10%) counterparts. More than half of the male workers (54%) and a bit less than three-fourths of the female workers (72%) were employed in the services sector. Besides large urban centers, the occupational classification in the small urban centers also shows that most of the males were BLS workers (39%). In comparison, most of the females were WHS workers (53%).

Like large urban centers, both male (30,618 PKR) and female (18,963 PKR) graduate workers earned the highest wages among all educational groups in the small urban centers. The

TABLE 29.1

Gender Wage Differentials in Large Urban Centers

Characteristics	Large Urban Centers (%)		Average Earnings of Male Workers		Average Earnings of Female Workers		Mean Difference	95% Conf. Interval	
	Male	Female	Mean	SD	Mean	SD			
Education									
No education	19.8	35.2	12,177	8,384	5,060	3,191	7,117*	6,018	8,215
Primary	17.7	15.1	11,946	5,830	6,520	4,676	5,426*	4,233	6,619
Matriculation	36.3	17.8	14,634	10,176	9,222	8,397	5,412*	3,593	7,232
Intermediate	10.7	10.4	19,362	11,569	10,809	9,405	8,552*	5,761	11,344
Graduation	15.5	21.5	33,492	24,163	21,163	17,848	12,329*	8,328	16,331
Sector									
Agriculture	0.8	0.5	10,395	3,958	6,776	2,525	3,618**	−44	7,280
Industry	49.5	27.7	16,423	14,306	6,239	5,640	10,185*	8,229	12,140
Services	49.7	71.8	20,619	18,513	15,959	17,514	4,660*	3,055	6,266
Occupation									
WHS	22.5	43.5	34,569	24,914	21,189	19,573	13,381*	10,666	16,096
WLS	24.7	5.2	15,280	9,277	16,096	13,202	−817	−3,762	2,128
BHS	24.8	21.1	14,141	10,727	5,257	3,963	8,884*	7,172	10,597
BLS	28.0	30.2	12,490	7,386	5,782	3,507	6,709*	5,750	7,667

Source: Authors' calculations from LFS (2014–2015).
* Shows $p < 0.01$ for a mean comparison test.
** Shows $p < 0.05$ for a mean comparison test.

gender wage gap was the highest in workers with graduation (11,656 PKR) or workers with no formal education (5,800 PKR). Wages in the services sector were the highest for both genders (19,420 PKR for males and 18,520 PKR for females). The gender wage gap was the lowest in the services (900 PKR). The wages were the highest in the WHS jobs (30,444 PKR for males and 21,715 PKR for females). The mean difference was the highest in the WHS jobs (8,729 PKR), while the lowest in WLS jobs (5,071 PKR). The wages of females in all sectors were fewer than their male counterparts.[7]

A nonparametric approach to estimate the probability density function of a random variable is the kernel density estimate. The kernel density estimates of the gender wage gap for urban workers are exhibited in Figure 29.1. Figure 29.1a shows that the mean wages of urban males tend to be higher than those of their female counterparts, and Figure 29.1b shows that the mean wages of males in large urban centers were much higher than those of females. The kernel density estimate of the mean wages of female workers was bimodal in Figure 29.1c. However, the wages of a smaller number of females were higher than their male counterparts in the urban centers, especially in the small urban centers.

The results of the Blinder–Oaxaca decomposition among the workers of the large and small urban centers are compared in Table 29.3. Female workers earned lower than their male counterparts in the urban centers, a difference of 7,140 PKR in the large urban centers and 4,237 PKR in the small urban centers. Female workers earned 58% lower in the large urban centers, while 30% earned lower amounts in the small urban centers. The mean wages of male and female workers were higher in the large urban centers when

---

[7] Results of OLS estimates are presented in Appendix B.

(a)

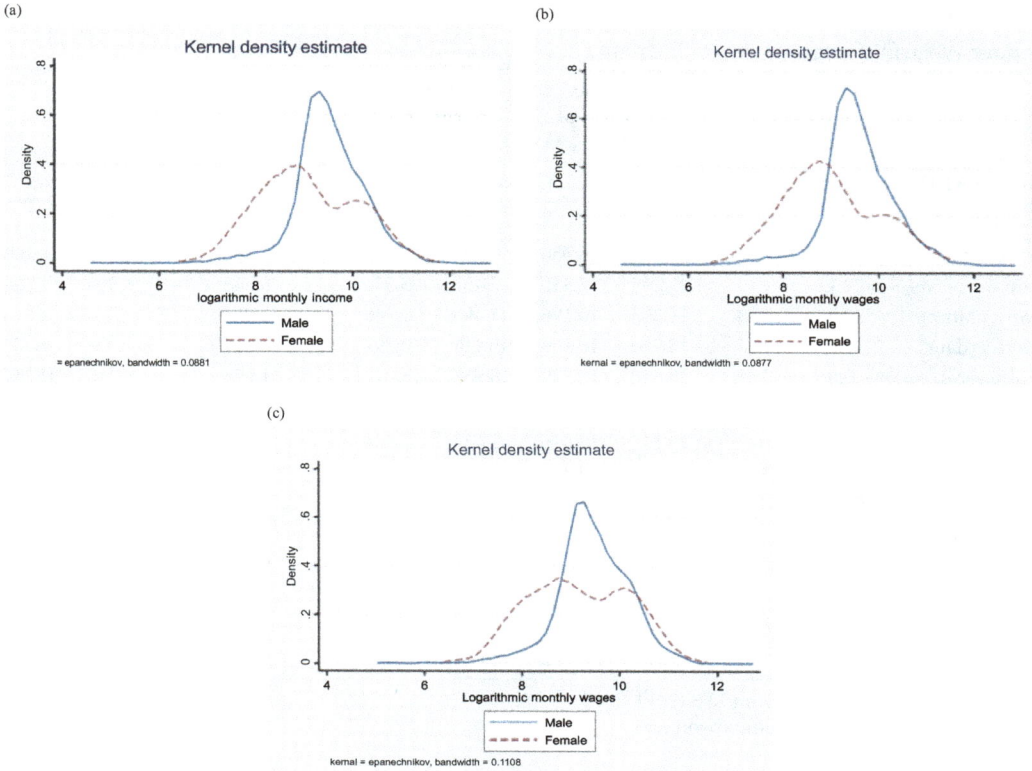

(b)

(c)

FIGURE 29.1
Kernel density estimates of the gender wage gap. (a) Urban centers. (b) Large urban centers. (c) Small urban centers.

compared with their corresponding small urban centers' counterparts. The decomposition of the wage gap in the large urban centers shows that the explained and unexplained components were 19% and 81%, respectively. Age, education, and sector positively contributed to the explained components, while age and education positively contributed to the unexplained components of the wage gap in the large urban centers. The contribution of the explained component was 38% of the gender wage gap in small urban centers. The contribution of the unexplained component was 138%. Age, education, and occupation significantly contributed to the unexplained component.

29.6 Discussion

In the urban labor markets, female workers earned 24% lower than their male counterparts in 2014–2015. The demographic segregation shows that the gender wage gap was higher in the large urban centers (31%), while it was lower in the small urban centers (8%). One of the most important reasons for the gender wage gap is the difference in educational attainment. Education increases human capital and workers' productivity. The workers in the

TABLE 29.2

Gender Wage Differentials in Small Urban Centers

Characteristics	Small Urban Centers (%)		Average Earnings of Male Workers		Average Earnings of Female Workers		Mean Difference	95% Conf. Interval	
	Male	Female	Mean	SD	Mean	SD			
Education									
No education	27.3	38.6	11,494	6,580	5,694	4,406	5,800*	4,728	6,872
Primary	22.1	10.0	11,309	7,880	7,465	5,374	3,843*	1,399	6,288
Matriculation	30.5	19.8	14,395	10,196	11,493	9,945	2,903*	864	4,941
Intermediate	10.3	11.7	18,950	10,859	15,116	10,397	3,834*	993	6,675
Graduation	9.8	20.0	30,618	18,729	18,963	18,214	11,656*	7,716	15,595
Sector									
Agriculture	1.9	6.7	15,511	10,646	5,644	3,521	9,867*	6,255	13,479
Industry	43.8	21.3	13,169	9,372	5,984	4,109	7,185*	5,422	8,949
Services	54.3	72.0	19,420	16,524	18,520	17,460	900	−754	2,553
Occupation									
WHS	18.4	53.0	30,444	21,403	21,715	18,188	8,729*	6,230	11,227
WLS	24.2	3.5	15,312	10,925	10,241	6,967	5,071	479	9,663
BHS	19.0	18.2	13,074	7,722	5,484	3,594	7,590*	6,023	9,158
BLS	38.5	25.4	11,781	6,770	6,066	4,623	5,715*	4,518	6,911

Source: Authors' tabulation from LFS (2014–2015).
* Shows $p < 0.01$ for a mean comparison test.

large urban centers are more educated than those in small urban centers. Therefore, wages in large urban centers are higher than the smaller ones. Males have more years of schooling than their female counterparts in urban centers.

In contrast, educational inequalities are higher among female urban workers than their male urban counterparts. The percentage of female workers with no formal education is greater than that of males with no qualifications in urban centers. All the same, workers with graduate and above qualifications were mostly women, while the difference between the graduated females and males in small urban centers was higher than in the large urban centers. Workers with tertiary education earned the highest among all educational groups. Previous studies have identified a gap in education as one of the main contributing factors to the gender wage gap (Nordman, Robilliard, & Roubaud, 2011).

Most urban workers were employed in the services and industrial sectors. It has been documented that the urban economy more favorably generates jobs for educated and trained workers. The gender wage difference was the lowest among the services sector in large and small urban centers.

Due to a higher number of female graduates in urban areas, more female workers had employment in WHS jobs. However, male managers earned more wages in large urban centers, while female managers earned higher wages in small urban centers.

In contrast, most uneducated females were employed in BLS jobs. Female clerical support workers received higher wages than their male counterparts in large urban centers. In comparison, male clerical support workers in small urban centers earned higher than female ones. Among plant and machine operators, males' earnings were higher than those of females. In short, gender wage differentials were higher in the large urban centers than

TABLE 29.3

Blinder–Oaxaca Inequality Decomposition Results

Differential	Large Urban Centers			Small Urban Centers		
	Coeff.	SE	Z	Coeff.	SE	Z
Male	17,077	218	78*	14,202	199	71*
Female	9,938	488	20*	9,965	526	19*
Gender wage difference	7,140	535	13*	4,237	563	8*
Explained						
Age	2,387	427	6*	−187	423	0
Age squared	−1,339	297	−5*	4	245	0
Working hours	−658	188	−4*	−550	151	−4*
Education	653	214	3*	−224	200	−1
Sector	363	78	5*	53	33	2**
Occupation	−76	133	−1	−695	131	−5*
Total	1,330	422	3*	−1,598	358	−4*
Contribution	19%	–	–	−38%	–	–
Unexplained						
Age	16,185	4,318	4*	5,606	7,892	1*
Age squared	−6,745	2,222	−3*	−3,328	4,326	−1*
Working hours	−15,117	1,960	−8*	−9,849	1,713	−6*
Education	761	1,099	1*	1,858	1,506	1*
Sector	−6,009	1,667	−4*	−933	1,636	−1*
Occupation	−4,129	1,022	−4*	573	1,547	0*
Constant	20,864	3,730	6*	11,909	5,502	2***
Total	5,810	543	11*	5,835	530	11*
Contribution	81%	–	–	138%	–	–

Source: Authors' estimation from LFS (2014–2015).
* Shows $p < 0.01$.
** Shows $p < 0.1$.
*** Shows $p < 0.05$.

the smaller areas. Previous studies have suggested that occupation and industry effects are significant in explaining the gender wage gap (Blau & Kahn, 2017).

The decomposition analysis also supports the argument that males earned higher wages than their female counterparts. The explained component identifies that age and education improve human capital. Age is a proxy variable for the experience or human capital accumulation (Lim et al., 2010; Mitchell et al., 2000). The share of the unexplained component (along with the discrimination effect) was higher in small urban centers than large urban centers.

Besides the characteristic differentials, the gender wage gap is evident in the public and private sectors. The qualification and earnings of public sector workers were higher, while gender discrimination was lower. A large proportion of male and female urban workers were employed in the public sector. The share of public sector workers to the total workers in small urban centers was more than what was found in the large urban centers. The average earnings of female workers in the urban centers were much higher than those of their male counterparts; the gap was more prominent in the small urban centers than the larger ones.

29.7 Conclusion

This study analyzed and compared the gender wage differentials in the large and small urban centers of Pakistan employing LFS 2014–2015. The gender wage gap was 24% in the urban centers, 31% in large urban centers, and 8% in small urban centers. Gender discrimination tends to be more prevalent in the lower tiers of the urban labor market than in the managerial ranks. Workers in large urban centers earn 17% higher wages than workers in small urban centers. The uneducated group is the most vulnerable in the urban labor markets, especially among female workers.

According to the Human Development Report (HDR-2019),[8] Pakistan's Human Development Index (HDI) ranked 152nd in the list of 189 countries. Pakistan's Gender Development Index (GDI) is the lowest (0.75) when compared with Bangladesh (0.90), India (0.83), and the whole South Asian region (0.83). The gender-disaggregated HDI is lower due to considerable differences in a decent standard of living and access to knowledge.

Female workers earn lower wages than their male counterparts in the urban labor markets. However, it is interesting that female workers with higher education make more than their male counterparts. It also implies no signs of a glass ceiling effect against female workers in the urban centers. The reward of agglomerative forces is higher for males than their female counterparts.

The increase in female labor force participation and increased educational opportunities for females will reduce the gender wage gap. A small proportion of females is employed in WLS jobs. SDG 5 is committed to ending all forms of discrimination and ensures active participation in economic and social activities. This is possible with the effective implementation of legislation on gender wage equality in the labor market in Pakistan. Further, with their investment in human capital accumulation, female workers may out-earn their male counterparts in various sectors. Most of the enterprises in Pakistan are equal-opportunity employers, and the gender wage gap is minimal in public sector enterprises (Hyder & Reilly, 2005).

SDG 10 targets the sustained income growth of vulnerable groups by 2030, possibly with increased education and human capital development. Active participation of female workers in the urban labor market will reduce the gender wage gap at the bottom level of the wage distribution. SDG 11 aims to formulate proactive employment policies to tackle the challenges of urbanization. Around one-third of females are unemployed in the urban labor markets, and their employment opportunities need further enhancement. The study suggests an estimation of the gender wage gap in the large and small urban centers of Pakistan at the provincial and district levels for in-depth analysis and policymaking at the grassroots level.

References

Ahmed, S., & Maitra, P. (2010). Gender wage discrimination in rural and urban labormarkets of Bangladesh. *Oxford Development Studies, 38*(1), 83–112. https://doi.org/10.1080/13600810903551611.

---

[8] Pakistan's Human Development Report 2019 http://hdr.undp.org/sites/all/themes/hdr_theme/country-notes/PAK.pdf.

Ashraf, J., & Ashraf, B. (1993). An analysis of the male-female earnings differential in Pakistan. *The Pakistan Development Review, 32*(4), 895–904. https://www.pide.org.pk/pdf/PDR/1993/Volume4/895-904.pdf.

Aslam, M. (2009a). Education gender gaps in Pakistan: Is the labor market to blame? *Economic Development and Cultural Change, 57*(4), 747–784. https://doi.org/10.1086/598767.

Aslam, M. (2009b). The relative effectiveness of government and private schools in Pakistan: Are girls worse off? *Education Economics, 17*(3), 329–354. https://doi.org/10.1080/09645290903142635.

Bacolod, M. (2017). Skills, the gender wage gap, and cities. *Journal of Regional Science, 57*(2), 290–318. https://doi.org/10.1111/jors.12285.

Baum-Snow, N., & Pavan, R. (2012). Understanding the city size wage gap. *The Review of Economic Studies, 79*(1), 88–127. https://doi.org/10.1093/restud/rdr022.

Becker, G. S. (1985). Human capital, effort, and the sexual division of labor. *Journal of Labor Economics, 3*(1), 33–58. https://doi.org/10.1086/298075.

Becker, G. S. (1993). Nobel lecture: The economic way of looking at behavior. *Journal of Political Economy, 101*(3), 385–409. https://doi.org/10.1086/261880.

Bergmann, B. R. (1974). Occupational segregation, wages and profits when employers discriminate by race or sex. *Eastern Economic Journal, 1*(2), 103–110. https://www.jstor.org/stable/40315472.

Blau, F. D., & Kahn, L. M. (2017). The gender wage gap: Extent, trends, and explanations. *Journal of Economic Literature, 55*(3), 789–865. https://doi.org/10.1257/jel.20160995.

Blinder, A. S. (1973). Wage discrimination: Reduced form and structural estimates. *Journal of Human Resources, 8*(4), 436–455. https://doi.org/10.2307/144855.

Dumont, M. (2006). The reliability-or lack thereof-of data on skills. *Economics Letters, 93*(3), 348–353. https://doi.org/10.1016/j.econlet.2006.06.008.

Glaeser, E. L., & Mare, D. C. (2001). Cities and skills. *Journal of Labor Economics, 19*(2), 316–342. https://doi.org/10.1086/319563.

Horrace, W. C., & Oaxaca, R. L. (2001). Inter-industry wage differentials and the gender wage gap: An identification problem. *ILR Review, 54*(3), 611–618. https://doi.org/10.1177%2F001979390105400304.

Hyder, A., & Reilly, B. (2005). The public and private sector pay gap in Pakistan: A quantile regression analysis. *The Pakistan Development Review, 44*(3), 271–306. https://www.pide.org.pk/pdf/PDR/2005/Volume3/271-306.pdf.

Jann, B. (2008). The Blinder–Oaxaca decomposition for linear regression models. *The Stata Journal, 8*(4), 453–479. https://doi.org/10.1177%2F1536867X0800800401.

Kee, H. J. (2006). Glass ceiling or sticky floor? Exploring the Australian gender pay gap. *Economic Record, 82*(259), 408–427. https://doi.org/10.1111/j.1475-4932.2006.00356.x.

Kim, B. (2004). The wage gap between metropolitan and non-metropolitan areas. In *Econometric Society 2004 Australasian Meetings* (No. 189). Econometric Society.

Labor Force Survey (LFS) (2014–2015a). Pakistan Bureau of Statistics, Ministry of Planning, Development and Reform, Islamabad, Pakistan. http://www.pbs.gov.pk/labor-force-publications.

Labor Force Survey (LFS) (2014–2015b). Pakistan Bureau of Statistics (PBS), Government of Pakistan. http://www.pbs.gov.pk/content/lfs-2010-2018-microdata.

Lee, L. (2012). Decomposing wage differentials between migrant workers and urban workers in urban China's labor markets. *China Economic Review, 23*(2), 461–470. https://doi.org/10.1016/j.chieco.2012.03.004.

Lim, D. S., Morse, E. A., Mitchell, R. K., & Seawright, K. K. (2010). Institutional environment and entrepreneurial cognitions: A comparative business systems perspective. *Entrepreneurship Theory and Practice, 34*(3), 491–516. https://doi.org/10.1111%2Fj.1540-6520.2010.00384.x.

Magnani, E., & Zhu, R. (2012). Gender wage differentials among rural-urban migrants in China. *Regional Science and Urban Economics, 42*(5), 779–793. https://doi.org/10.1016/j.regsciurbeco.2011.08.001.

Mitchell, R. K., Smith, B., Seawright, K. W., & Morse, E. A. (2000). Cross-cultural cognitions and the venture creation decision. *Academy of Management Journal, 43*(5), 974–993. https://doi.org/10.5465/1556422.

Nisic, N. (2017). Smaller differences in bigger cities? Assessing the regional dimension of the gender wage gap. *European Sociological Review, 33*(2), 292–3044. https://doi.org/10.1093/esr/jcx037.

Nordman, C. J., Robilliard, A. S., & Roubaud, F. (2011). Gender and ethnic earnings gaps in seven West African cities. *Labor Economics, 18,* 132–145. https://doi.org/10.1016/j.labeco.2011.09.003.

Oaxaca, R. (1973). Male-female wage differentials in urban labor markets. *International Economic Review, 14*(3), 693–709. https://doi.org/10.2307/2525981.

Roca, J. D. L., & Puga, D. (2017). Learning by working in big cities. *The Review of Economic Studies, 84*(1), 106–142. https://doi.org/10.1093/restud/rdw031.

Sabir, M., & Aftab, Z. (2007). Dynamism in the gender wage gap: Evidence from Pakistan. *The Pakistan Development Review, 46*(4), 865–882. https://www.pide.org.pk/pdf/PDR/2007/Volume4/865-882.pdf.

Siddiqui, R., & Siddiqui, R. (1998). A decomposition of male-female earnings differentials. *The Pakistan Development Review, 37*(4), 885–898. https://www.pide.org.pk/pdf/PDR/1998/Volume4/885-898.pdf.

Stephan, M. (2017). *The Australian gender wage gap,* Doctoral dissertation, The Australian National University, Australia. https://openresearch-repository.anu.edu.au/handle/1885/112054.

Wang, M., & Cai, F. (2008). Gender earnings differential in urban China. *Review of Development Economics, 12*(2), 442–454. https://doi.org/10.1111/j.1467-9361.2008.00450.x.

Appendix A

Appendix A1

Sectors and sub-sectors under PSIC (2010)

S. No.	Sector	Section	Divisions	Description
1	Agriculture	A	01–03	Agriculture, forestry, and fishing
2	Industry	B	05–09	Mining and quarrying
		C	10–33	Manufacturing
		D	35	Electricity, gas, steam and air conditioning supply
		E	36–39	Water supply; sewerage, waste management, and remediation activities
		F	41–43	Construction
3	Services	G	45–47	Wholesale and retail trade; repair of motor vehicles and motorcycles
		H	49–53	Transportation and storage
		I	55–56	Accommodation and food service activities
		J	58–63	Information and communication
		K	64–66	Financial and insurance activities
		L	68	Real estate activities
		M	69–75	Professional, scientific, and technical activities
		N	77–82	Administrative and support service activities
		O	84	Public administration and defense; compulsory social security
		P	85	Education
		Q	86–88	Human health and social work activities
		R	90–93	Arts, entertainment, and recreation
		S	94–96	Other service activities
		T	97–98	Activities of households as employers; undifferentiated goods and services-producing activities of households for own use
		U	99	Activities of extraterritorial organizations and bodies

Sectors and sub-sectors under PSIC (2010). Pakistan Standard Industrial Classification (All Economic Activities) PSIC Rev. 4, Government of Pakistan Statistics Division Federal Bureau of Statistics.

http://www.pbs.gov.pk/sites/default/files/other/PSIC_2010.pdf.

Appendix A2

Major groups and sub-major groups under PSCO (2015)

The study followed Dumont (2006) for the classification of occupations

S. No.	Occupation	Major Group	Sub-Major Group	Description
1	White-collar high skilled	Managers	11	Chief executives, senior officials, and legislators
			12	Administrative and commercial managers
			13	Production and specialized service managers
			14	Hospitality, retail, and other service managers
		Professionals	21	Science and engineering professionals
			22	Health professionals
			23	Teaching professionals
			24	Business and administration professionals
			25	Information and communications technology professionals
			26	Legal, social, and cultural professionals
		Technicians and associate professionals	31	Science and engineering associate professionals
			32	Health associate professionals
			33	Business and administration associate professionals
			34	Legal, social, cultural, and related associate professionals
			35	Information and communications technicians
2	White-collar low skilled	Clerical support workers	41	General and keyboard clerks
			42	Customer services clerks
			43	Numerical and material recording clerks
			44	Other clerical support workers
		Service and sales workers	51	Personal service workers
			52	Sales workers
			53	Personal care workers
			54	Protective services workers
3	Blue-collar high skilled	Skilled agricultural, forestry and fishery workers	61	Market-oriented skilled agricultural workers
			62	Market-oriented skilled forestry, fishery, and hunting workers
			63	Subsistence farmers, fishers, hunters, and gatherers
		Craft and related trades workers	71	Building and related trades workers, excluding electricians
			72	Metal, machinery, and related trades workers
			73	Handicraft and printing workers
			74	Electrical and electronic trades workers
			75	Food processing, woodworking, garment, and other craft and related trades workers
4	Blue-collar low skilled	Plant and machine operators, and assemblers	81	Stationary plant and machine operators
			82	Assemblers
			83	Drivers and mobile plant operators
		Elementary occupations	91	Cleaners and helpers
			92	Agricultural, forestry, and fishery laborers
			93	Laborers in mining, construction, manufacturing, and transport
			94	Food preparation assistants
			95	Street and related sales and service workers
			96	Refuse workers and other elementary workers

Source: Major groups and sub-major groups under PSCO (2015). Pakistan Standard Classification of occupations. Pakistan Bureau of Statistics, Government of Pakistan. http://www.pbs.gov.pk/sites/default/files/PSCO_2015.pdf.

Appendix B

Ordinary Least Square (OLS)

The study employed OLS to estimate the factors that can contribute in the mean wages (*WW*) of male and female urban workers in large and small urban centers. There were 6,166 males and 657 females in the sample of large urban centers. The wage functions for male (*ml*) and female (*fl*) workers in the large urban centers are represented in equations (B1) and (B2) as:

$$W_{ml} = \alpha_{ml} + X_{ml}\beta_{ml} + \varepsilon_{ml} \quad \ldots \quad \ldots \quad \ldots \quad \ldots \quad \ldots \quad \ldots \tag{B1}$$

$$W_{fl} = \alpha_{fl} + X_{fl}\beta_{fl} + \varepsilon_{fl} \quad \ldots \quad \ldots \quad \ldots \quad \ldots \quad \ldots \quad \ldots \tag{B2}$$

In comparison, there were 3,878 males and 461 females in the sample of small urban centers. The equations (B3) and (B4) represent the wage functions for male (*ms*) and female (*fs*) workers in the small urban centers as:

$$W_{ms} = \alpha_{ms} + X_{ms}\beta_{ms} + \varepsilon_{ms} \quad \ldots \quad \ldots \quad \ldots \quad \ldots \quad \ldots \quad \ldots \tag{B3}$$

$$W_{fs} = \alpha_{fs} + X_{fs}\beta_{fs} + \varepsilon_{fs} \quad \ldots \quad \ldots \quad \ldots \quad \ldots \quad \ldots \quad \ldots \tag{B4}$$

where *X* is the set of explanatory variables. The slopes and error terms are represented by *β* and *ε*, respectively.

OLS regression estimates for the gender wage gap in large and small urban centers.

| Characteristics | Large Urban Centers | | | | Small Urban Centers | | | |
| | Male | | Female | | Male | | Female | |
	Coeff.	SE	Coeff.	SE	Coeff.	SE	Coeff.	SE
Constant	8.036*	0.104	6.991*	0.379	7.597*	0.108	6.633*	0.388
Age	0.067*	0.003	0.046*	0.010	0.085*	0.004	0.070*	0.017
Age squared	−0.001*	0.000	0.000*	0.000	−0.001*	0.000	−0.001*	0.000
Working hours	0.001	0.001	0.026*	0.003	0.001	0.001	0.022*	0.003
Education								
No education	–	–	–	–	–	–	–	–
Primary	0.056*	0.023	0.268*	0.082	0.046**	0.026	0.444*	0.126
Matriculation	0.142*	0.020	0.385*	0.090	0.231*	0.024	0.240**	0.129
Intermediate	0.308*	0.028	0.452*	0.126	0.372*	0.036	0.506*	0.172
Graduation	0.617*	0.030	0.995*	0.117	0.783*	0.040	0.614*	0.170
Sector								
Agriculture	–	–	–	–	–	–	–	–
Industry	0.144**	0.077	−0.433	0.339	−0.020	0.065	0.136	0.199
Services	0.110	0.077	−0.465	0.330	−0.052	0.066	−0.278**	0.145
Occupation								
WHS	–	–	–	–	–	–	–	–
WLS	−0.352*	0.024	0.009	0.129	−0.211*	0.034	−0.288	0.194
BHS	−0.263*	0.028	−0.294***	0.143	−0.163*	0.040	−0.739*	0.221
BLS	−0.341*	0.027	−0.288*	0.110	−0.178*	0.036	−0.658*	0.162
R^2	0.36		0.44		0.38		0.35	
No. of observations	6,166		657		3,878		461	
F-statistic	$F(12, 6153)=284.60^*$		$F(12, 644)=42.66^*$		$F(12, 3865)=200.92^*$		$F(12, 448)=20.48^*$	

Source: Authors' estimations from LFS (2014–2015).

* Shows $p<0.01$.
** Shows $p<0.05$.
*** Shows $p<0.1$.

30

Youth Empowerment and Sustainable Urban Development

Arun Kumar Sinha

(Formerly at) *Central University of South Bihar*

Markandey Rai

UN-Habitat

CONTENTS

30.1 Introduction

It is usually said that the world belongs to the youth, and this in turn reveals their growing importance in achieving the sustainable developmental goals. At the same time, a person cannot be assumed to remain young forever. Also, as time is a continuous variable, everyone ages continuously. Considering these scenarios, we take into account of only the youth population.

There are many definitions of youth with varying age groups. In fact, "youth" constitutes more a fluid category than a fixed age group. The United Nations (UN) defines "youth" population between the ages of 15 and 24 years. All UN statistics on youth are based on this definition as given by the annual yearbooks of statistics published by the UN system on demography, education, employment, and health.

In India, the youth are defined as those aged 15–29 years as described in the National Youth Policy. Currently, this age group constitutes 27.5% of India's total population. The 2011 Indian Census counted 363 million young people from 10 to 35, according to the 12th Five-Year Plan, volume II (2013). In a report on the youth in India published in 2017 by the Ministry of Statistics and Program Implementation, the Government of India considered 15–34 years as youth.

In India, the share of youth in the total population has been increasing continuously from the level of 19.9% in the year 1971 to 31.2% in the year 2011. As per India's

Census 2011, the youth (15–24 years) in India constitute one-fifth (19.1%) of India's total population.

India has a relative advantage at present over other countries in terms of distributions of youth population. India's advantage in young population is also evident when it is compared with other Asian countries. India is observed to have remained younger longer than China and Indonesia, the two main countries other than India that dominate the demographic features of Asia. However, the proportion of youth among females is generally lower on account of better longevity of females than males.

30.2 Youth Empowerment

Youth empowerment is an attitudinal, structural, and cultural process whereby young people gain ability, authority, and agency to make decisions and implement change in their own lives and the lives of other people, including youth and adults. It is often addressed as a gateway to intergenerational equity, civic engagement, and democracy building. Many local, state, provincial, regional, national, and international government agencies and nonprofit community-based organizations provide programs entered on youth empowerment. The activities involved therein may focus on youth-led media, youth rights, youth councils, youth activism, youth involvement in community, decision-making, and other methods of youth empowerment. In fact, it is also a central theory of the UN Convention on the Rights of the Child, which every country in the world has signed into law.

The United Nations Development Program (UNDP) supports the capacity development of young people and young-led organizations, and the development of youth caucuses in government, parliaments, and other bodies. We engage with relevant stakeholders through outreach, advocacy, global networks, and policy debates, in particular in the context of the post-2015 development agenda, which is called the Sustainable Development Goals (SDGs), to be achieved by 2030. These goals with 169 targets were agreed to by more than 170 heads of state attending the UN General Assembly in September 2015. The main slogan this time is *No One Should Be Left Behind*. We also support the mainstream of youth issues in development planning and interministerial and inter-sectorial coordination.

After serving as cochair of the UN Inter-Agency Network on Youth Development (IANYD) from March 2015 to 2016, it continues to play an active role in IANYD as colead of the UN System-Wide Action Plan on Youth task team, cochair of the UN Working Group on Youth, and the 2030 Agenda and of the Working Group on Youth Political Inclusion and Civic Engagement.

UNDP promotes:

- Inclusive youth participation in effective and democratic governance.
- Economic empowerment of youth.
- Strengthened youth engagement in building resilience in their communities.
- Inclusion of youth in the future development agenda, including through consultations and discussions.

In July 2016, UNDP launched its first **Youth Global Program for Sustainable Development and Peace**, or **Youth GPS (2016–2020)**, a 5-year global program on youth empowerment.

30.3 Strengthening Skill Development

To enhance youth empowerment, we need to focus on health and education. Yoga helps maintain good health while strengthening quality education we need to update the curriculum regularly and include moral and character education. With the growing applications of artificial intelligence (AI), this is an important area that needs to be included in the curriculum. We need to devote greater attention to value-based skill development.

30.4 Indian Population Scenarios

The age structure of the Indian population in 2010 as depicted by Talreja (2014) shows that 61.6% of India's population is in the age group 15–59.

30.5 Urbanization and Urban Development

Urbanization describes the rapid growth of population in the urban areas, and its main reason is the growth in economic activities, which in turn help develop urban areas. For the same reason, migration takes place from rural to urban areas. In a nutshell, the availability of more employment opportunities, as well as health, education, and a higher standard of living in urban areas compared with rural areas spurs urbanization. According to one estimate, half of humanity, i.e., 3.5 billion people, lives in cities today, and 5 billion people are projected to live in cities by 2030.

Though over 50% of the population lives in urban areas today, by 2045 the world's urban population is expected to grow by 1.5 times to 6 billion. In view of this scenario, decision makers need to act accordingly to meet the needs of the growing population. It is said that by 2050, the urban population will become more than double its current size, and 7 of 10 people in the world will live in cities.

TABLE 30.1

Urban Populations of India in Various Census Years

Census Years	Total	Percent Urban
1891	221,172,952	9.2
1901	291,361,036	9.0
1921	318,942,480	10.2
1931	529,837,773	7.4
1941	388,997,955	12.8
1961	439,234,771	18.0
1971	548,159,652	19.9
1981	665,287,849	23.7
1991	846,303,000	25.7
2001	1,028,610,328	27.8
2011	1,210,569,573	31.2

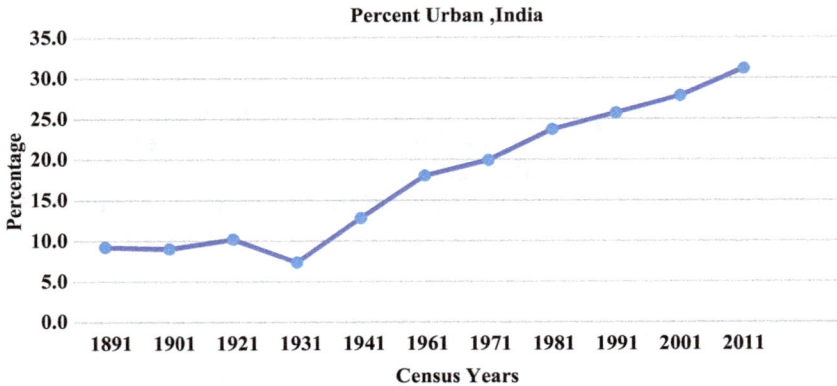

FIGURE 30.1
Percentage urban population, India.

Table 30.1 shows that the level of urbanization has increased from 27.8% in the 2001 Census to 31.2% in the 2011 Census while the proportion of the rural population has declined from 72.19% to 68.84%. Figure 30.1 depicts this more emphatically.

30.6 Sustainable Urban Development

Sustainable development is defined as development that meets the needs of the present without compromising the ability of future generation to meet their own needs.

Sustainable urban development implies a process by which sustainability can be attained emphasizing progress and positive change, incorporating both environmental and social dimensions. This highlights the need to reform the mechanisms to achieve environmental goals and the achievement of a balance with social and economic considerations.

This includes:

- A change in the quality of growth
- The conservation and minimization of the depletion of nonrenewable resources
- A merging of economic decisions with those on the environment
- A strong consideration of the needs of future generations

Goal 11 of the UN SDGs proposes making cities inclusive, safe, resilient, and sustainable. This includes ten targets.

30.7 Monitoring of Youth Empowerment and Sustainable Urban Development

We propose to use ranked set sampling, which is a cost-effective sampling and monitoring method, apart from population projection methods. For more detailed discussions of this sampling method, see Sinha (2016), Sinha (2005), Patil (2002), and Patil et al. (1994). See Ramakumar (2005) and Pathak and Ram (1998) for population projection methods.

Urbanization is the engine that propels the world toward prosperity in the 21st century, and youth are the engineers. Youth are society's most essential dynamic human resources. The youth live by and large in cities and towns, the cities of the developing world accounting for over 90% of the world's urban growth, and youth account for a large percentage of those inhabitants. Moreover, 45% of the world's youth live in the Asia-Pacific region. Growth demographics are seen as assets and no longer liabilities; the global spread of information and communication technology (ICT), including AI, is seen as an avenue to increase youth engagement and the creation of opportunities. The current slogan in the world to encourage youth participation in the development agenda is *Build the Youth: Build the Nation*. The UN Secretaries General have always conveyed the message and given utmost attention to youth and volunteerism.

Improved training and capacity building involve establishing a national vocational educational and training system that takes into account the recent developments in ICT, and further establishing national private sector incentives for hiring apprentices and interns and creating jobs for youth in the ICT sectors. Other measures are to provide support funds specially targeted at youth-led initiatives. To establish youth forums and platforms, a national assembly of youth should be mandatory in all countries. The objective of the youth strategy for enhanced engagement is to present an integrated approach to urban youth development and to provide a road map for the promotion of urban youth empowerment. Young people need acknowledgment, guidance, resources, and training in order to reach their full potential. Youth should be treated as a source of positive energy, and not utilizing it fully in sustainable development is a great loss to any nation in this century that would make it impossible to achieve the SDGs by 2030. The main features of the state of urban youth in 2012–2013 were the importance of job-oriented education to the development of urban youth; and a better match between skills and labor markets through vocational training and with participation of the private sector. The issue of how to make the youth job givers rather than job seekers is of the essence. *Catch them young* is the objective or mantra of the day for any government, including India.

Given their numbers and the impact they can have on international development policies, it is crucial to recognize youth's participation and partnership in the implementation of the SDGs. One important survey in this regard was *My World Survey*, which provided a platform for young people to express their priorities and views for the consideration of world leaders in the process of defining the set of global goals to end poverty and indicate how the global youth population assesses the challenges for our common future. The largest contributors of World Survey were youth worldwide. About 70% of the respondents, both offline and online, were young people below the age of 30 years. This global consultation with youth especially highlighted a number of factors of utmost importance in the implementation of the SDGs. One important finding was that the youth want to contribute to the implementation of the SDG.

Governments thus must make sure that they are meaningfully involved from the beginning, from defining goals to participating in their implementation. India's Niti Aayog, formerly the Planning Commission of India, is responsible for monitoring the progress made in the implementation of SGDs in India, and it has developed its own goals and targets as well as a mechanism to monitor them on the state and district levels. One of their initiatives is SDG Choupal, which is popular and youth-oriented to create awareness among all the people to realize the common slogan of "No one should be left behind."

Youth do not constitute a homogeneous group; their socioeconomic, demographic, and geographical situations vary widely both within and between regions. Notwithstanding these differences, regional-level analysis affords us a general understanding of their development profile. The vast majority of the world's youth, some 87%, live in developing countries and face challenges such as limited access to resources, education, training, employment, and broader economic development opportunities. The *My World Survey 2015* has found that quality education is the greatest priority for more than 75% of the respondents. Furthermore, The *My World Survey* indicates that there is strong willingness for the part of youth to participate in the decision-making process. For this, the youth should be treated as a source of positive energy that would cost a nation a great loss if youth are not utilized properly in its development. The Family and Youth Service Bureau of the US Department of Health and Human Services advocates the theory of Positive Youth Development. This was translated to the Asia-Pacific Ministerial Conference on Housing and Urban Development with the overall theme of Youth and IT in Sustainable Urban Development in December 2012 in Amman, Jordan, under full coordination with UN-Habitat. It was a consensus that nations should address youth issues on a priority basis; otherwise, social pressure like the Arab Spring may be replicated in many more countries. For more detailed discussions, see UN-Habitat (2013a).

The State of Urban Youth 2012–2013 by UN-Habitat, which heads the Youth Program in the UN system, states that the increasing prominence of the youth bulge in most urban countries presents a unique opportunity, as they represent the most dynamic human resources available. Their number today is larger than at any time in human history. Yet this group suffers the most from urban unemployment and often feels that it lacks equal access to opportunities. This is especially acute in developing countries, which have a relatively youthful population that must be mobilized to realize greater economic and social development goals. More than 90 million youth around the world are unemployed, a number that is increasing daily and which makes up 47% of all unemployed, and an additional 300 million belong in the "working prior" category—they are in unskilled, insecure jobs and live in poor conditions. See UN-Habitat (2013b) for more details.

Kalam (1998) also stressed imparting proper and compatible skills to youth based on the demand in the future in *India 2020: A Vision for the New Millennium*. India's human resource base is one of its greatest core competencies. It is India's strength. We can train an unskilled Indian, and if we can impart better skills to a skilled Indian and if we create a more challenging environment for the educated, as well as build avenues for economic activity in agriculture, industry, and the service sector, these Indians will not only meet the targets but exceed them. See Kalam (1998) for more details.

Acknowledgments

We express our gratitude to a reviewer for comments and suggestions that helped improve the content of the chapter considerably. We are obliged and thankful to Dr. Lubna Naz, Chapter Leader, for her patience and interest, and Dr. Mukesh Ranjan, Dept. of Statistics, Pachhunga University College, Mizoram University, Aizawl, Mizoram, India, for his constructive suggestions.

References

Abdul Kalam, A. P. J. (1998). *India 2020: A Vision for the New Millennium*. Penguin Books India, New Delhi.

Pathak, K. B. and Ram, F. (1998). *Techniques of Demographic Analysis*, 2nd Ed. Himalaya Publishing House, Mumbai.

Patil, G. P. (2002). Ranked set sampling. In: *Encyclopedia of Environmetrics*, A. H. El-Sharaawi and W. W. Piegorsch, eds., Vol. 3. John Wiley and Sons Ltd, Chichester, 1684–1690.

Patil, G. P., Sinha, A. K., and Taillie, C. (1994). Ranked set sampling. In *Handbook of Statistics*, G. P. Patil and C. R. Rao, eds., Vol. 12. North-Holland, Elsevier Science B. V., Amsterdam, 167–200.

Ramakumar, R. (2005). *Mathematical Demography*. New Age International (P) Limited Publishers Ltd., New Delhi.

Sinha, A. K. (2016). Ranked set sampling: As a cost-effective and more efficient data collection method. *Statistical Journal of the International Association of Official Statistics (IAOS)*, **32** (4), 607–611.

Sinha, A. K. (2005). Some recent developments in ranked set sampling. *Bulletin of Informatics and Cybernatics*, **37**, 137–160.

Talreja, C. (2014). India's demographic dividend: Realities and opportunities. *Indian Journal of Labor Economics*, **57** (1), 139–155.

UN-Habitat (2013a). *Youth and IT in Sustainable Urban Development*, HS/024/13E, Nairobi, Kenya.

UN-Habitat (2013b). *State of Urban Youth Report*, HS/089/12 E, Nairobi, Kenya.

United Nations (2015). *My World Survey 2015: The United Nations Global Survey for a Better World*.

31

A Statistical Approach to Urbanization: The Case of Turkey

Yakup Ari

Alanya Alaaddin Keykubat University

CONTENTS

31.1 Introduction

Urbanization is a concept used to describe a large movement of the rural population to the urban centers of a country and/or an increase in the growth rate of the urban population relative to the nonurban population or the total population. The urbanization process is generally associated with increasing pressures in urban activities, rapid increases in infrastructure, and demand for goods and services (Aboagye and Nketiah-Amponsah, 2016). Factors affecting the urbanization processes are explained as sociocultural (population and migration), technological, and political factors. The concept of urbanization in a narrow sense explains the increase in the population living in cities. In addition, it is not only a demographic element but also an indicator of economic, social, political, and cultural processes. In addition to urbanization attracting individuals to cities, the new residents must be able to adopt the city's lifestyle. In this context, the urbanization process covers the differences in social, political, and cultural areas as a result of the transformation of urban space and social life (Tunçer, 2015). It is seen that education is open to everyone without discrimination in the cities after industries are established. The status of individuals emerges depending on their rise through education and success (Çan, 2013). Depending on these factors, it is stated that the increase in the urban share of the population accompanies industrialization and rapid increases in income. Economic growth is a concept that is effective in the development of countries, with a continuous increase in per-capita real income. The factors involved in economic growth are technological changes and developments, accumulation of physical capital, trade, financial capital, exchange rates, and inflation. Individuals whose income increased with agricultural production in the rural areas migrated to cities for higher incomes and new job opportunities.

DOI: 10.1201/9781003261148-37

Today, urbanization is one of the strongest indicators of social change, especially in developing and modernizing societies. However, while evaluating the rapid urbanization situation in societies, the speed of urbanization and whether it is progressing parallel to economic development should also be taken into consideration. Unconscious urbanization causes many problems in societies that have not fully reached the stage of being an industrial society and have not completed the modernization process. Therefore, statistical approaches have gained importance in the measurement of the relationship of urbanization with other parameters in urban studies.

The causes of urbanization can generally be listed as economic, political, technological, and socio-political. Economic factors are characterized by the fact that the socio-political and economic advantages offered by cities are greater than in rural areas. With the development of industrial, tourism, and agricultural activities, new attraction centers emerge. In other words, the increasing growth of interregional economic development differences leads the centers that have been receiving immigration to receive still more immigration.

Major changes in social and economic life are expected after the COVID-19 outbreak currently affecting the world. For example, with remote working, the necessity of working in high-rise plaza offices and even the necessities of living in the same city as one's job are being discussed. The COVID-19 outbreak continues to directly affect many sectors in different working branches. Although sectors such as accommodation, retail, tourism, and real estate face global uncertainty, the change in the conditions that make up supply and demand causes a differentiation of the effects of the events on different sectors. The importance of agriculture and animal husbandry can be felt more clearly in this period. Decreasing import–exports and the increasing importance of access to basic foodstuffs positively affect agricultural activities and thus land values. In addition to agriculture, areas such as fattening farms, silos, and fish farming will gain importance in the coming period. It is anticipated that smaller areas will be preferred in office areas with an increase in distance working models. Considering this process, an increase in the unemployment rate is the most important problem. During the epidemic, the fact that people remain closed in the cities has created a need for space, and as a result, the expectation of reverse migration, that is, migration from the city to rural areas, has increased.

Although the effects of the epidemic are beyond the scope of this study, the situation shows that studies of urbanization, unemployment, inflation, and economic growth are more important than ever.

Urbanization accelerated after 1950 in Turkey. However, immigration in the country accelerated urbanization more strongly after 1980. It is known that 75.6% of the population lives in urban areas in Turkey. Although this situation is mostly explained by economic reasons, the effect of political and socio-political reasons cannot be denied. After 1968, as seen in the graph below Figure 31.1, the urbanization rate in Turkey has passed the global average and is now approaching the European Union average. As mentioned earlier, the breakage occurred after 1980. In 1980, the adoption of more liberal policies instead of mixed economic policies and the sale of state-owned economic enterprises through privatization led to the acceleration of this process. In addition, the actions of separatist terrorist groups in the southeast of the country caused internal migration.

It is understood that the acceleration of urbanization rate decreased at the end of the first decade after 1980.

This rapid urbanization brought many problems with it. Another feature that forms the basis of the problems is that it is a social structure figure that is noticeably stratified and integrated with its socioeconomic and sociocultural structure. Due to the fact that the industrialization rate is lower than the urbanization rate, the poverty of the rural

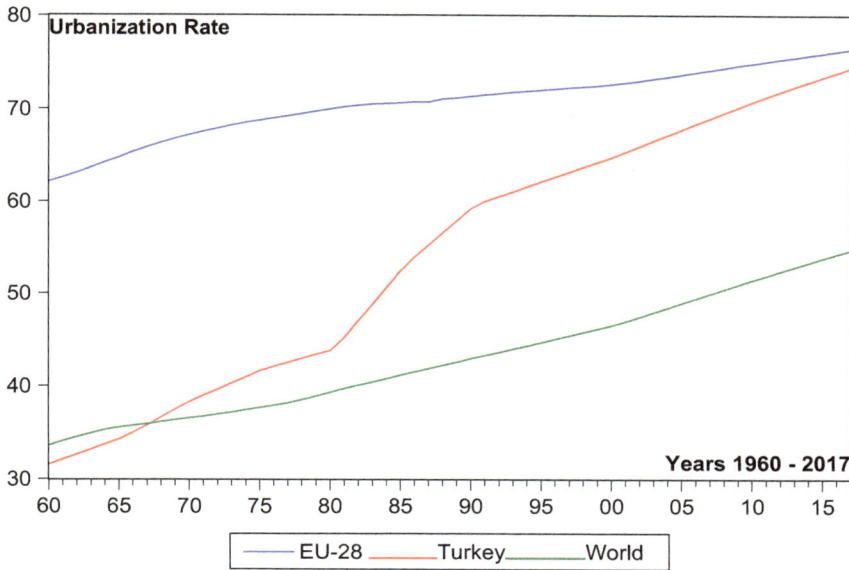

FIGURE 31.1
Plot of the urbanization rate. (https://data.worldbank.org.)

population is carried to the cities and the low industrialization rate does not create sufficient employment opportunities. This incomplete social structure and income injustice
due to unemployment spur the formation of slums surrounding the urbanized city center in large cities and the formation of an imbalanced social structure. Although agriculture still employs an important part of the population today, considering that the share of
agriculture in total employment is around 40%, in addition to the unemployment created
by the new economic structure in cities, the fact that agricultural employment is high
causes unemployment to remain an important problem. The fact that the decrease in the
share of agricultural employment in total employment is generally accompanied by economic development may mean an increase in unemployment especially among the young,
women, and the elderly. While these people were previously under traditional protection
in their family businesses, they may face difficulties in the wage labor market. It is not
easy to find jobs in nonagricultural sectors for these people who are self-employed in the
agricultural field or appear as unpaid family workers. Therefore, with urbanization, new
employment problems arise. However, the variables in the focus of this study will be economic and demographic. These variables will be the ratio of the population living in the
cities to the total population, that is, the urbanization rate, economic growth, the unemployment rate, and inflation.

This chapter aims to reveal the short-run and long-run relationships between the mentioned variables, and also to determine the direction of causality between the variables.
Moreover, it is an example of a statistical approach that can be used in urbanization studies.
For this purpose, the Autoregressive Distributed Lag (ARDL) Bounds Test (Pesaran and
Shin, 1998, 1999; Pesaran et al., 2001) approach is used to reveal the long-run relationship
between variables. After determining the presence of cointegration, an error correction
model (ECM) was used to understand after how long period a deviation from the long-run
equilibrium has been corrected in the short run. Toda and Yamamoto (1995) causality tests
are applied to reveal the direction of causality among the variables.

The rest of the chapter is organized as follows. In the second part, there is a brief literature review that includes the statistical methods used in urban studies and the studies using the method given as an example in this study. Then there will be a short methodology section describing the structure of the model and causality tests used in this study. The fourth section includes the results of the analysis and the interpretation of the results, as well as a short discussion section. In the last section, a general evaluation is made and other statistical approaches that can be used in urbanization studies will be mentioned.

31.2 Literature Review

Some of the studies that will form the basis of this study can be listed as follows:

Faria and Mollick (1996) tried to explain the process of city growth motivated by the tax differentials charged upon the city and the hinterland in a study where they examined urbanization, economic growth, and welfare. They concluded that welfare implications for consumption and urbanization suggest a rather instantaneous migration adjustment. Moomaw and Shatter (1996) investigated urbanization and economic growth in metropolitan cities. As a result, it was determined that urbanization increased with the increase in Gross Domestic Product (GDP) per capita, industrialization, and exports, and urbanization decreased with an increase in the importance of agriculture. Cohen (2006) examined the urbanization of developing countries and revealed the changes from past to present. Pointing out that urbanization has increased in the last 20 years due to production and consumption, Cohen predicted that this trend will continue in the next 30 years.

Henderson (2000) investigated how urban density affects economic growth with the data for 5- year intervals of 80–100 countries in the 1960–1995 period, concluding that the increases in urban concentration are important in places where per-capita income is $5,000 and above. It has been observed that the urban density decreases in places where per-capita income is below $3,000. Tandoğan (2017) investigated the relationship between economic growth in Turkey for the period 1968–2016, urbanization, and the service industry. He used the Toda and Yamamoto causality test and concluded that there is one-way causality between growth and urbanization in Turkey, but a two-way causality between urbanization and the service sector. The study of Öztürk and Çalışkan (2019) aimed to determine the causality relation between the rate of the population living in the urban areas and the economic growth rate of Turkey's GDP in the period 1960–2016. According to the findings, two-way causality is found between the urbanization rate and GDP. Yıldırım (2019) examined the relationship between economic growth and urbanization in Turkey. The existence and direction of the long-run relationship have been tested using the ARDL model. As a result, it has been determined that urbanization is cointegrated with economic growth in the long run. The increase in the rate of urbanization in Turkey is increasing the economic growth rate, but the rate of urbanization is not affected by changes in economic growth. Bayraktutan and Alancioğlu (2019) aimed to determine the relationship between the rate of urbanization and per-capita GDP (economic growth) for BRICS and Turkey considering the interaction between growth dynamics and demographic distribution. As a result of the panel data analysis, a one-way causality relationship from urbanization to growth was found.

Akça and Ela (2012) analyzed the relationship between education in Turkey, fertility, and unemployment, and the results of the study can be stated as follows. Turkey is going through a demographic transition process in recent years. In this process, the fertility rate decreases and the population growth rate is gradually decreasing. The slowdown in the population growth rate affects the labor market and unemployment directly by determining the labor supply. However, the effect of the demographic transformation experienced through the labor market in the short run increases unemployment. Although the population growth rate has slowed down, the working population still continues to increase. In addition, the unemployed workforce living in the country and not working as unpaid family workers migrates to cities, increasing the pressure of unemployment and of population growth on unemployment. Therefore, the continued increase in the population, albeit slowly, and the migration of unemployed youth from the countryside to the city shows that the decrease in population in the short term will not change the pressure on unemployment. The projections made show that the slowdown in the population growth rate will return to stabilization in the future, and the working population will decrease in the longer term.

31.3 Methodology

In this study, the relationship between unemployment, urbanization, inflation, and economic growth for Turkey is investigated using linear time-series analysis methods. When using linear time-series methods, an important issue is to determine the stationarity levels of the variables in order to avoid the spurious regression problem. In this context, the stationarity properties of the variables are determined with traditional unit root methods such as Augmented Dickey–Fuller (ADF) and Phillips–Perron (PP). Another important point to consider when using linear time-series methods is to investigate the short-run and long-run relationships between variables using appropriate linear time-series methods depending on the order of stationarity of the variables. In this context, using traditional methods such as vector autoregressive (VAR) method, Engle and Granger (1987), Johansen (1988, 1991), and Johansen and Juselius (1990) have reached consistent results when the variables are integrated in the same order $I(1)$. If the variables are stationary at different orders, one of the methods that can be used in this case is the ARDL bounds test, since statistically healthy results cannot be obtained with these conventional methods (Pesaran et al., 2001). Compared with traditional linear time-series methods, the ARDL bounds test approach can be used even if the variables are stationary of different orders. In addition, it has important advantages such as achieving statistically significant results in small samples, being suitable for the examination of short-long run relationships together and being practical in reaching an ECM. For this reason, the ARDL-Bounds method has become one of the most preferred econometric techniques in studies in which the relations between variables are examined by using linear time-series techniques in recent years.

After this point, the study of Ari (2020) is followed. In the ARDL models, both AR and DL are included in one regression written

$$\text{ARDL}(p,q): y_t = \beta_0 + \sum_{i=0}^{p} \alpha_i y_{t-i} + \sum_{i=0}^{q} \beta_i x_{t-i} + u_t \tag{31.1}$$

where u_t is an error term with zero mean and x_t is a k-dimensional vector of explanatory variables. Typically, a constant is included as β_0. In the ARDL approach, the dependent and independent variables can be introduced in the model with lags. Thus, AR part of the ARDL approach refers to lags of the dependent variable and the DL part of the ARDL refers to the lags of explanatory variables. Inferentially, this property of the ARDL model states that the effect of a change of the independent variables may or may not be instantaneous. The OLS estimation is applied to determine the parameters of the ARDL model.

The ARDL method is based on the standard least squares regression method, in which the lagged values of both the dependent variable and the explanatory variable(s) are used as the explanatory variable. In the ARDL-Bounds test approach, first, to determine whether there is cointegration between the variables in a model, the model is transformed into an unconstrained ECM based on the ARDL approach and estimated with the least squares (OLS) estimator, and based on this model, the boundary test based on F-statistics is performed. Based on the standard regression model, the ARDL-Bounds test equation was established to determine the cointegration relationship between the variables in the model as follows:

$$\Delta\left(\text{unemp}\right)_t = \alpha_0 + \sum_{i=1}^{p} \alpha_{1i}\Delta\left(\text{unemp}\right)_{t-i} + \sum_{i=0}^{q} \alpha_{2i}\Delta\left(\text{urb}\right)_{t-i}$$

$$+ \sum_{i=0}^{r} \alpha_{3i}\Delta\left(\text{gdp}\right)_{t-i} + \sum_{i=0}^{s} \alpha_{4i}\Delta\left(\text{inf}\right)_{t-i} + \beta_1\left(\text{unemp}\right)_{t-1} + \beta_2\left(\text{urb}\right)_{t-1} \quad (31.2)$$

$$+ \beta_3\left(\text{gdp}\right)_{t-1} + \beta_4\left(\text{inf}\right)_{t-1} + e_{1t}$$

where unemployment, urbanization, economics growth, and inflation are denoted by 'unemp', 'urb', 'gdp', and 'inf', respectively. The alpha coefficients in equation (31.2) represent the short-term and the beta coefficients the long-term dynamics. In order to ensure the stability conditions of the estimation, first the optimal lag lengths (p, q) of the variables in the equation (31.2) are determined with the help of information criteria, and then the bounds test is applied to the model estimated with the appropriate lag. In the bounds test, the null hypothesis, which states that there is no long-term relationship between the variables, is tested with the F-test. The null hypothesis is tested by applying a zero constraint to the coefficients of lagged variables in equation (31.2). Accordingly, in this study, we test the null hypothesis for F-test (H_0: $\beta_1 = \beta_2 = \beta_3 = \beta_4 = 0$) and the alternative hypothesis (H_0: $\beta_1 \neq 0$ for atleast one i; $i = 1, 2, 3, 4$). The F-statistic obtained by the bounds test (Pesaran et al., 2001) according to the structure of the model (restricted, constant, and trend inclusion) and the number of output variables (k) in the model are compared with asymptotic critical values calculated for various confidence levels. The standard F-test used to test the null hypothesis has a nonstandard distribution in some cases.

After determining that there is a long-term relationship between the level values of the variables by the bounds test, the long-term relationship between the variables is examined by the ARDL method. In this study, the ARDL model to be estimated to examine the long-run relationship between variables is as follows:

$$\left(\text{unemp}\right)_t = \alpha_0 + \sum_{i=1}^{p} \alpha_{1i}\Delta\left(\text{unemp}\right)_{t-i} + \sum_{i=0}^{q} \alpha_{2i}\Delta\left(\text{urb}\right)_{t-i} + \sum_{i=0}^{r} \alpha_{3i}\Delta\left(\text{gdp}\right)_{t-i} + \sum_{i=0}^{s} \alpha_{4i}\Delta\left(\text{inf}\right)_{t-i} + e_{2t}$$

$$(31.3)$$

The short-run relationship between the variables is examined with the ECM based on the ARDL method. This model is as follows:

$$\Delta(\text{unemp})_t = \alpha_0 + \sum_{i=1}^{p} \alpha_{1i}\Delta(\text{unemp})_{t-i} + \sum_{i=0}^{q} \alpha_{2i}\Delta(\text{urb})_{t-i} + \sum_{i=0}^{r} \alpha_{3i}\Delta(\text{gdp})_{t-i}$$

$$+ \sum_{i=0}^{s} \alpha_{4i}\Delta(\text{inf})_{t-i} + \delta\text{ECT}_{t-1} + e_{3t} \tag{31.4}$$

where ECT is an error correction term that shows how soon it is possible to correct a short-term equilibrium between dependent and explanatory variables in the model. For the error correction mechanism to work, the coefficient of this variable is expected to be negative and statistically significant.

After revealing the short-run and long-run between the variables, the causality relationship between the variables is examined with the Toda and Yamamoto (1995) causality test, which is one of the econometric analyses used to determine causality between the mentioned variables. The Toda and Yamamoto test has some advantages over other causality tests. First, taking the difference of nonstationary series at the stages of other causality tests, the analysis results in missing information problems in the causality relationship. In addition, if the series used in the study are found to be stationary at the same order, cointegration relations should be examined. In particular, the ECM and vector VECM can be used to analyze cointegration processes. However, ECM and VECM analyzes are sensitive to the lag length used in the studies, and if there is no cointegration between the series, the VECM model is invalid (Rambaldi and Doran, 1996). In studies where the length of the series is low, Monte Carlo simulation was applied to analyze the power of causality tests (Yamada and Toda, 1998). As a result, Toda–Yamamoto causality test was found to give appropriate results.

Toda and Yamamoto (1995) argue that even if the causality test is nonstationary variables, the VAR model is established with the level values of the variables. Therefore, there is no need for unit root and cointegration tests, which are preliminary preparation processes in this test. In addition, the use of the level values of the series in the analysis eliminates the lack of information about the series. The Toda–Yamamoto causality test has several stages. First, the stationarity processes of the variables are examined and the highest stationary level (d_{max}) is determined without taking the difference. In the second stage of the test, the most appropriate lag length (p) of the VAR model is determined in terms of various criteria [primarily Akaike Information Criteria (AICs) and Schwarz Information Criteria (SIC)]. At this stage, the d_{max} level should be less than the lag length (p) as a preflight. After achieving the aforementioned precontrol, the VAR ($d_{max}+p$) model is estimated by summing the maximum stationary level (d_{max}) and the lag length (p). The VAR ($d_{max}+p$) model to be estimated for the case of two variables can be defined with the help of equations below:

$$Y_t = \alpha_{10} + \sum_{i=1}^{p+d_{max}} \alpha_{11}(i+d_{max})Y_{t-(i+d_{max})} + \sum_{i=0}^{q} \alpha_{21}(i+d_{max})X_{t-(i+d_{max})} + e_{4t}$$

$$X_t = \alpha_{20} + \sum_{i=1}^{p+d_{max}} \alpha_{21}(i+d_{max})Y_{t-(i+d_{max})} + \sum_{i=0}^{q} \alpha_{22}(i+d_{max})X_{t-(i+d_{max})} + e_{5t}$$

The H_0 hypotheses are created at the last stage of the causality test. The H_0 null hypothesis is defined as $H_0 = \alpha_{21(i+d_{\max})} = \alpha_{22(i+d_{\max})} = 0$ and the mentioned hypothesis is tested with the help of the modified WALD test. If the H_0 hypothesis is rejected, it is understood that there is a causal relationship between the variables.

31.4 Data and Findings

The data used in the analysis were downloaded from the World Bank website (https:// data.worldbank.org) and consist of annual data for the period 1980–2019. The data set consisting of the unemployment rate, urbanization rate, economic growth rate, and the inflation rate of Turkey is given in Figure 31.2 in the time-series graph.

The various descriptive statistics and the results of the normality tests of the variables are given in Table 31.1. Jarque–Bera statistics show that all the variables are normally distributed. The inflation rate series have the highest standard deviation.

According to the PP and ADF stationarity tests, the results of which are given in Table 31.2, economic growth and urbanization variables seem to be stationary at the level. Since variables have different levels of stationary, it can be analyzed whether there is cointegration between them by using ARDL approach.

The result of the ARDL model, in which unemployment is defined as an independent variable, is presented in Table 31.3. Due to the low number of observations, the maximum independent lag value was chosen as 4 and ARDL (2, 0, 0, 0) model was found to be the most suitable model among 500 models according to AIC. Although the coefficient of urbanization is statistically insignificant, the model's adjusted R^2 or explanatory power

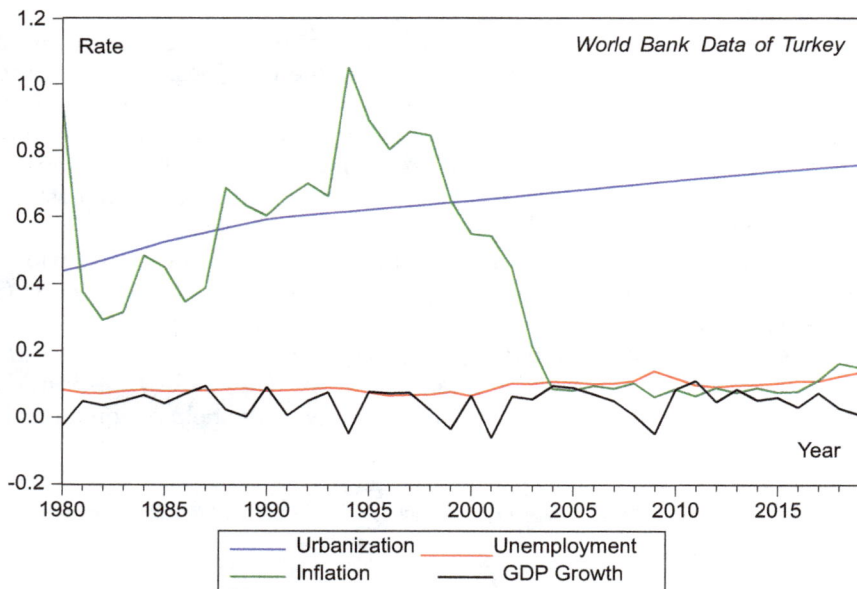

FIGURE 31.2

Time-series plot of the variables. (https://data.worldbank.org.)

TABLE 31.1

Descriptive Statistics and Normality Tests

	URB	UNEMP	INF	GDP
Mean	0.634462	0.092100	0.397440	0.044728
Median	0.644820	0.086000	0.361124	0.051012
Maximum	0.756300	0.140000	1.052150	0.111135
Minimum	0.437800	0.065000	0.062510	−0.059623
Std. Dev.	0.087215	0.018166	0.304699	0.042716
Skewness	−0.605831	0.717244	0.446032	−0.874403
Kurtosis	2.507312	3.070061	1.898023	3.029301
Jarque–Bera	2.851444	3.437774	3.350220	5.098634
Probability	0.240335	0.179266	0.187288	0.078135
Sum	25.37848	3.684000	15.89762	1.789102
Sum Sq. Dev.	0.296651	0.012870	3.620814	0.071162

TABLE 31.2

Unit Root Tests

		Phillips–Perron (PP) and Augmented Dickey–Fuller (ADF) Tests at Level			
Test		**GDP**	**INF**	**URB**	**UNEMP**
With Constant					
PP	*t*-Statistic	−6.8014	−2.0400	−5.4171	−0.7986
	Probability	*0.0000*	*0.2693*	*0.0001*	*0.8084*
ADF	*t*-Statistic	−6.7508	−1.9638	−2.7780	−0.9150
	Probability	*0.0000*	*0.3009*	*0.0734*	*0.7729*
With Constant and Trend					
PP	*t*-Statistic	−6.8007	−2.1860	−4.6314	−2.5235
	Probability	*0.0000*	*0.4837*	*0.0034*	*0.3158*
ADF	*t*-Statistic	−6.6236	−4.0447	−3.6581	−2.7938
	Probability	*0.0000*	*0.0168*	*0.0415*	*0.2082*
Without Constant and Trend					
PP	*t*-Statistic	−3.6467	−1.8996	3.9184	1.0335
	Probability	*0.0006*	*0.0557*	*0.9999*	*0.9180*
ADF	*t*-Statistic	−0.6020	−1.9079	1.4737	0.6614
	Probability	*0.4489*	*0.0547*	*0.9627*	*0.8547*

has a value of 85%. The probability value corresponding to the *F*-statistic value shows that the complete model is statistically significant.

The long-run relationship between the variables, in other words, whether there is cointegration between the variables, is tested by ARDL-Bounds test with restricted constant and no trend. If the *F*-statistic, which is the result of the *F*-Bounds test, is greater than the critical values, the null hypothesis that there is no level relationship is rejected. Accordingly, the *F*-statistic value given in Table 31.4, *F*-stat=8.1646 is greater than any critical value determined for any confidence level from 10% to 1%. Accordingly, the variables mentioned in the study are cointegrated.

The Selected Model ARDL (2, 0, 0, 0) Output

Variable	Coefficient	Standard Error	t-Statistic	Probability*
Dependent Variable: UNEMP				
UNEMP(−1)	0.848989	0.135645	6.258924	0.0000
UNEMP(−2)	−0.291125	0.133334	−2.183431	0.0365
URB	0.032478	0.021488	1.511482	0.1405
GDP	−0.127557	0.028983	−4.401125	0.0001
INF	−0.020223	0.005698	−3.549394	0.0012
C	0.034405	0.013826	2.488413	0.0182
R^2	0.871852	Mean dependent variable		0.092842
Adjusted R^2	0.851829	S.D. dependent variable		0.018307
S.E. of regression	0.007047	Akaike info criterion		−6.928465
Sum squared residuals	0.001589	Schwarz criterion		−6.669899
Log likelihood	137.6408	Hannan–Quinn criterion		−6.836469
F-statistic	43.54236	Durbin–Watson statistic		2.066401
Prob(F-statistic)	0.000000			

TABLE 31.4

ARDL-Bounds Test Output

F-Bounds Test		Null Hypothesis: No Levels Relationship		
Test Statistic	Value	Significance	I(0)	I(1)
		Asymptotic: $n=1{,}000$		
F-statistic	8.164626	10%	2.37	3.2
K	3	5%	2.79	3.67
		2.5%	3.15	4.08
		1%	3.65	4.66
Actual sample size	38	Finite sample: $n = 40$		
		10%	2.592	3.454
		5%	3.1	4.088
		1%	4.31	5.544
		Finite sample: $n = 35$		
		10%	2.618	3.532
		5%	3.164	4.194
		1%	4.428	5.816

When examining the regression equation showing the long-run relationship between the variables, it can be said that urbanization has a positive effect on unemployment, but the coefficient that demonstrates this effect is not statistically significant. The 1% increase in urbanization increases the unemployment rate by 0.73%, while the 1% increase in inflation and economic growth decreases unemployment by 0.45% and 2.88%, respectively.

TABLE 31.5

Long-Run Regression Output

Variable	Coefficient	Std. Error	t-Statistic	Probability
Levels of Equation				
URB	0.073458	0.040944	1.794101	0.0823
INF	−0.045739	0.012286	−3.722785	0.0008
GRGDP	−0.288502	0.114222	−2.525791	0.0167
C	0.077816	0.031474	2.472365	0.0189

It is seen in Table 31.5 that the effects of inflation and economic growth variables are statistically significant in the long run.

Next, the ECM results are used to examine the short-run relationship between variables. The ECM is an approach that reveals how long it takes to return to equilibrium if the equilibrium between the variables is violated. As seen in Table 31.6, the error correction coefficient is between 0 and −1 and statistically significant. According to the error correction coefficient, when the equilibrium between unemployment, urbanization, inflation and economic growth deteriorates, it rebalances after about 2.26 years. In other words, the effect of short-run shocks that disrupt the equilibrium between these variables last a little more than 2 years.

Finally, the results of the hypothesis tests of the model are given in Appendix A. According to Breusch–Pagan–Godfrey and ARCH tests, a problem of heteroscedasticity does not appear in the error terms of the model. According to the Breusch–Godfrey auto-correlation test, there are no autocorrelations among the error terms. In addition, according to the Ramsey Reset Test, there is no error in the specification of the ARDL model. CUSUM and CUSUM-SQ graphs also show the stability of model parameters. Because of the ARDL model structure, there is multicollinearity between the independent variables, and because of that, the VIF values of the parameters are not included.

Before applying the Toda–Yamamoto causality test, the optimum lag length for the VAR model was found to be 2, and causality results were obtained with the VAR (2) model, which also included the 4th lag value of the variables. According to these results, no causality relationship was found between the variables. The results are given in Appendix B.

TABLE 31.6

The Error Correction Model (ECM) Output

Variable	Coefficient	Std. Error	t-Statistic	Probability
ECM Regression				
D(UNEMP(−1))	0.291125	0.113245	2.570749	0.0150
CointEq(−1)*	−0.442136	0.065242	−6.776874	0.0000
R^2	0.554461	Mean dependent variable		0.001632
Adjusted R^2	0.542085	S.D. dependent variable		0.009818
S.E. of regression	0.006644	Akaike info criterion		−7.138991
Sum squared residuals	0.001589	Schwarz criterion		−7.052803
Log likelihood	137.6408	Hannan–Quinn criterion		−7.108326
Durbin–Watson statistics	2.066401			

31.5 Conclusion

Considering that the rate of urbanization in the world is around 55.7% (World Bank, 2020a) and the world population is 7.67 billion (World Bank, 2020b), all socioeconomic and socio-cultural variables become even more important. As the population of the people living in cities increases, problems in education, work, and health are increasing. With liberal market practices, the increase in income injustice is threatening social life. In such an environment, policymakers should put forward projections based on scientific methods. Therefore, data security and transparency of data have become more important than ever. The most painful example of this is seen during the ongoing COVID-19 epidemic. In this context, the importance of the data used in the analysis and the importance of the analysis method is accepted all over the world. With this great epidemic, the contraction in the world economy, or rather, the contraction and unemployment in urban economies, which cannot be self-sufficient, is increasing. The current situation will be more difficult especially for developing countries. With urbanization, it has become more important than ever to provide job opportunities for the masses coming from rural to urban areas.

Turkey as a developing country with 83.4 million population makes up more than 1.09% of the world population (World Bank, 2020c). It has hosted millions of refugees in the last decade, especially due to the civil wars and wars taking place in neighboring countries and its region. Even if Turkey fights well against the epidemic COVID-19, the economy of the country is affected by the recession in the global economy. High short-term foreign debts, increased unemployment, and high inflation cause economic stagnation. For these reasons, every scientific study on the country's economy has become a priority for policymakers.

This study analyzed the relationship between unemployment and urbanization; the results indicate that urbanization increases unemployment. One of the reasons is the slow-down in the growth rate of Turkey's population, which will stabilize in the medium term. One of the important results is that a 1% increase in economic growth reduces unemployment by 2.88%. This case shows that each year Turkey's economy should grow in the 4%–5% range. It is clear that the only way to create employment is an economic model based on production. This study can be developed using different variables. By determining whether these variables have a structural break, alternative models can be used.

References

Aboagye, S., & Nketiah-Amponsah, E. (2016). The Implication of Economic Growth, Industrialization and Urbanization on Energy Intensity in Sub-Saharan Africa. *Journal of Applied Economics & Business Research, 6*(4), 297–301.

Akça, H., & Ela, M. (2012). Türkiye'de Eğitim, Doğurganlık veİşsizlik İlişkisinin Analizi. *Maliye Dergisi, 163*, 223–242.

Ari, Y. (2020). The Impact of USD-TRY Forex Rate Volatility on Imports to Turkey from Central Asia. In Christiansen, B., & Sezerel, H. (Eds.), *Economic, Educational, and Touristic Development in Asia* (pp. 70–89). IGI Global. doi:10.4018/978-1-7998-2239-4.ch004.

Bayraktutan, Y., & Alancioğlu, E. (2019). Kentleşme-Büyüme İlişkisi: BRICS-T İçin Bir Analiz. *Elektronik Sosyal Bilimler Dergisi, 18*(72), 1824–1831. Retrieved from https://dergipark.org.tr/tr/pub/esosder/issue/47063/561654.

Çan, M. F. (2013). Kentleşme, Sanayileşme ve Kalkınma Etkileşimi, 1–11. https://www.fka. gov.tr/sharepoint/userfiles/icerik_dosya_ekleri/firat_akademi/kentle%c5%9e me, %20sanay%c4%b0le%c5%9eme%20ve%20kalkinma%20etk%c4%b0le%c5%9e%c4%b 0m%c4%b0.pdf.

Cohen, B. (2006). Urbanization in Developing Countries: Current Trends, Future Projections, and Key Challenges for Sustainability. *Technology in Society, 28*(1–2), 63–80.

Engle, R. F., & Granger, C. W. J. (1987). Co-integration and Error Correction: Representation, Estimation, and Testing. *Econometrica, 55*(2), 251–276. doi:10.2307/1913236.

Faria, J. R., & Mollick, A. (1996). Urbanization, Economic Growth, and Welfare. *Economics Letters, 52*(1), 109–115.

Granger, C.W.J. (1980). Testing for Causality: A Personal Viewpoint. *Journal of Economic Dynamics and Control, 2*, 329–352. doi:10.1016/0165-1889(80)90069-X.

Hacker, S., & Hatemi, J. A. (2012). A Bootstrap Test for Causality with Endogenous Lag Order: Theory and Application in Finance. *Journal of Economic Studies, 39*(2), 144–160.

Henderson, J. V. (2000). *How Urban Concentration Affects Economic Growth.* World Bank Policy Research Working Paper No. 2326. Available at SSRN: https://ssrn.com/abstract=630698.

Johansen, S. (1988). Statistical Analysis of Cointegration Vectors. *Journal of Economic Dynamics and Control, 12*, 231–254.

Johansen, S. (1991). Estimation and Hypothesis Testing of Cointegration Vectors in Gaussian Vector Autoregressive Models. *Econometrica: Journal of the Econometric Society, 59*, 1551–1580

Johansen, S., & Juselius, K. (1990). Maximum Likelihood Estimation and Inference on Cointegration with Applications to the Demand for Money. *Oxford Bulletin of Economics and Statistics, 52*, 169–210. doi:10.1111/j.1468-0084.1990.mp52002003.x.

Moomaw, R. L., & Shatter, A. M. (1996). Urbanization and Economic Development: A Bias Toward Large Cities? *Journal of Urban Economics, 40*(1), 13–37.

Öztürk, S., & H. Çalışkan. (2019). Kentleşme Gelişiminin Ekonomik Büyüme Üzerine Etkisi: Türkiye Örneği. *Iğdır Üniversitesi Sosyal Bilimler Dergisi, 17*, 673–690.

Pesaran, M. H., & Shin, Y. (1998). An Autoregressive Distributed-Lag Modelling Approach to Cointegration Analysis. *Econometric Society Monographs, 31*, 371–413.

Pesaran, M. H., & Shin, Y. (1999). An Autoregressive Distributed Lag Modelling Approach to Cointegration Analysis. http://www.econ.cam.ac.uk/faculty/pesaran/ardl.pdf.

Pesaran, M. H., Shin, Y., & Smith, R. J. (2001). Bounds Testing Approaches to the Analysis of Level Relationships. *Journal of Applied Econometrics, 16*(3), 289–326.

Pesaran, M. H., & Smith, R. (1998). Structural Analysis of Cointegrating VARs. *Journal of Economic Survey, 12*(5), 471–505.

Rambaldi, A. N., & Doran, H. E. (1996). *Testing For Granger Non-Causality in Cointegrated Systems Made Easy.* Working Papers in Econometrics and Applied Statistics, Department of Econometrics, the University of New England.

Tandoğan, D. (2017). Türkiyede Ekonomik Büyüme, Kentleşme Ve Hizmet Sektörü İlişkisi: 1968–2016 Nedensellik Yaklaşımı. *International Congress on Politic, Economic and Social Studies, 3*, 75–87.

Toda, H. Y., & Yamamoto, T. (1995). Statistical Inference in Vector Autoregressions with Possibly Integrated Processes. *Journal of Econometrics, 66*, 225–250.

Tunçer, P. (2015). Sürdürülebilir Kentleşme Potikaları Ve Türkiye. *Turkish Studies, 37*, 275–290.

World Bank. (2020a). Urban Population (% of Total Population) [Data file]. Retrieved from https://data.worldbank.org/indicator/SP.URB.TOTL.IN.ZS.

World Bank. (2020b). Population, Total [Data file]. Retrieved from https://data.worldbank.org/indicator/SP.POP.TOTL.

World Bank. (2020c). Population, Total – Turkey [Data file]. Retrieved from https://data.worldbank.org/indicator/SP.POP.TOTL?locations=TR.

Yıldırım, S. (2019). Hangisi Lokomotif, Ekonomik Büyüme mi Kentleşme mi? Türkiye Örneği. *Hitit Üniversitesi Sosyal Bilimler Enstitüsü Dergisi, 12*(1), 223–234. doi:10.17218/hititsosbil.462016.

Appendix A

TABLE A1

Heteroscedasticity Test: Breusch–Pagan–Godfrey

F-statistic	1.862386	Prob. $F(5, 32)$	0.1288
Obs*R^2	8.565402	Prob. $\chi^2(5)$	0.1277
Scaled explained SS	6.526433	Prob. $\chi^2(5)$	0.2583

TABLE A2

Breusch–Godfrey Serial Correlation LM Test

F-statistic	0.332015	Prob. $F(4, 28)$	0.8541
Obs*R^2	1.720749	Prob. $\chi^2(4)$	0.7869

TABLE A3

Heteroscedasticity Test: ARCH

F-statistic	0.335717	Prob. $F(4, 29)$	0.8516
Obs*R^2	1.504721	Prob. $\chi^2(4)$	0.8258

TABLE A4

Ramsey RESET Test

	Value	df	Probability
t-Statistic	0.753377	31	0.4569
F-Statistic	0.567577	(1, 31)	0.4569

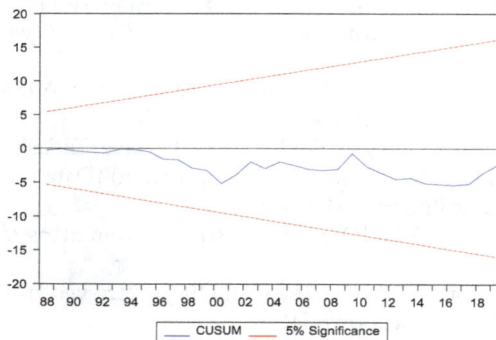

FIGURE A1
The graph of CUSUM test.

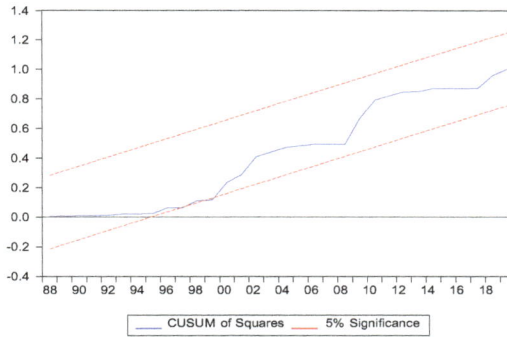

FIGURE A2
The graph of CUSUMSQ test.

Appendix B

TABLE B1

VAR Lag Order Selection Criteria

Lag	LogL	LR	FPE	AIC	SC	HQ
0	227.6292	NA	4.73e−11	−12.42385	−12.24790	−12.36244
1	412.6972	318.7281	3.97e−15	−21.81651	−20.93678*	−21.50946
2	435.8299	34.69912*	2.77e−15*	−22.21277*	−20.62925	−21.66008*
3	448.6312	16.35715	3.63e−15	−22.03507	−19.74776	−21.23673
4	461.5413	13.62732	5.20e−15	−21.86340	−18.87231	−20.81943

TABLE B2

Toda–Yamamoto Causality Tests

Excluded	χ^2	df	Probability
Dependent Variable: UNEMP			
URB	0.319666	2	0.8523
INF	3.137296	2	0.2083
GRGDP	0.008838	2	0.9956
All	7.204609	6	0.3023
Dependent Variable: URB			
UNEMP	1.597933	2	0.4498
INF	7.560771	2	0.0228
GRGDP	0.540712	2	0.7631
All	9.072220	6	0.1696
Dependent Variable: INF			
UNEMP	1.464246	2	0.4809
URB	2.039314	2	0.3607
GRGDP	4.034912	2	0.1330
All	7.113619	6	0.3105
Dependent Variable: GRGDP			
UNEMP	3.477359	2	0.1758
URB	0.118808	2	0.9423
INF	5.568665	2	0.0618
All	7.703733	6	0.2606

Part 7

Statistical Literacy in the Wider Society

32

Statistical Literacy and Domain Experts: Evidence from UI-LISA

Serifat A. Folorunso, Tomiwa T. Oyelakin, Oluwaseun A. Otekunrin, and Olawale B. Akanbi

University of Ibadan Laboratory for Interdisciplinary Statistical Analysis (UI-LISA)

CONTENTS

32.1 Introduction

Interdisciplinary research offers the ability to explore several fields of knowledge from a common perspective so that the phenomenon could be perceived differently (Sajjad et al., 2016). The classification of knowledge into different disciplines is not to distinguish knowledge of one domain from the other as these are the parts of a whole but to make it easy for various disciplines to display their expertise, and to provide space for development and promotion of knowledge. Studies found that interdisciplinary research contributes

DOI: 10.1201/9781003261148-39

tremendously toward enhancing the collaborations and research development of trans-disciplinary scholars in discovering the unseen facts. University researchers' literacy in Information and Communication Technology, statistics, and critical thinking/problem-solving skills seem to be the foundation for successful team productivity (Sajjad et al., 2016). For successful team productivity in interdisciplinary research, it is important for each domain expert to be literate in statistics. There are several definitions given to statistical literacy such as the "ability to understand and critically evaluate statistical results that permeate our daily lives" and the "ability to appreciate the contributions that statistical thinking can make in public and private, professional and personal decisions" (Wallman, 1993). Ben-Zvi and Garfield (2004) differentiate between the three common concepts of statistical literacy, statistical reasoning, and statistical thinking. Statistical literacy is the "ability to organize data, construct and display tables, and work with different representations of data". Statistical reasoning is the ability "to make sense of statistical information" and "fully interpret statistical results". Statistical thinking entails "an understanding of why and how statistical investigations are conducted and the "big ideas" that underlie statistical investigations".

At the University of Ibadan, there are various domain experts across faculties/colleges and UI-LISA is located in the Faculty of Science. There are 96 departments, 13 faculties, and 6 research centers/units/institutes. Most of the domain experts from these departments employ statistics as a tool in their research works without any strong technical knowledge of the application of the subject before the emergence of UI-LISA. Some of them even employ the services of "roadside data analysts", who got some statistical software training without the theoretical knowledge of the nitty-gritty of the discipline. Many domain experts encountered statistical challenges such as abnormalities in research outputs without support from statisticians. For example, in some of the domain experts' research publications, one finds researchers finding the mean of gender (which is a qualitative variable) and this explains the common misapplication of statistics.

The highest level of statistical training attained by most of the domain experts who visited UI-LISA was introductory statistics. Many are not trained to communicate with statistics in solving real-life problems, which is a determinant of statistical literacy level. Since the inauguration of UI-LISA in January 2015, the statistical laboratory has contributed immensely to the growth of statistical literacy in the university community. In April 2015, UI-LISA organized a training workshop on "Basic Statistical Tools for Quality Research in Biological and Physical Sciences" for postgraduate students in the Faculty of Science. Since then, more hands-on trainings have been organized by the laboratory to improve the statistical literacy within the university. Major activities organized by UI-LISA targeted at enhancing statistical literacy can be summarized as (1) seminar on teaching statistics in various disciplines; (2) training on research-based knowledge of statistics for interested researchers; and (3) workshop on the use of statistical software to teach statistics. The activities of UI-LISA have assisted many University of Ibadan researchers to achieve good research objectives.

The goal of this study is therefore to formally assess the degree of statistical literacy among the domain experts in and outside of the University of Ibadan noting their major statistical challenges and the role UI-LISA has been playing in solving these challenges.

32.2 Statistical Literacy

Scholars have given different definitions of statistical literacy in the recent past. According to Wallman (1993),

> Statistical literacy is the ability to understand and critically evaluate statistical results that permeate our daily lives – coupled with the ability to appreciate the contributions that statistical thinking can make in public and private, professional and personal decisions.

In the past three decades, researchers have had various discussions on statistical literacy in an interdisciplinary environment; the discussion is based on the fact that statistical literacy requires many abilities, the most important of which are mathematical and statistical skills, the competency to understand figures correctly, and to distinguish between valid and misrepresented data.

Furthermore, it enables people to assess the information that the figures provide and finally to understand what the actual data reveal about society (United Nations, 2012).

Statisticians only recently started to reflect on the fact that the ability to understand statistics is the prerequisite for successful communication with other users. Therefore, several initiatives are being implemented to improve statistical literacy levels among users of statistics in different strata of our societies. "Statistical literacy", broadly summarized under three dimensions, includes statistical numeracy, communicating statistics, and discovering the use of statistics (United Nations, 2012). The more informed people are about figures and official statistics, the better their assessment of the meaning and quality of the data. Gal (2002) suggests two components to statistical literacy in his research of adults' statistical practices. They are (1) the ability to *interpret* and *evaluate critically* statistical information in a variety of contexts, and (2) when relevant, the ability to *discuss* or *communicate* this understanding in a fashion that can have an impact on decision-making.

Statistical literacy is more than having good numerate skills (United Nations, 2012). Many people have excellent mathematical skills, a skill needed for statistical literacy, but this is not sufficient. They should be able to read and communicate the meaning of data, including words (and pictures) added to data (Murray & Gal, 2002, Gordon & Nicholas, 2006, Bradstreet 1996, Chinnapan et al., 2007). These qualities enhance their decision-making abilities.

Klein et al. (2016) defined the statistical literacy indicator and classified it into three categories: (1) basic consideration involving the adoption of a measurable term to indicate the status and progress toward a particular development outcome, such as the literacy rate, mortality rate, and standard of education; (2) diagnosis which includes the illustration of the state of such a measurable phenomenon and the supposed causes of its variation (e.g., mortality rate of 5 per 1,000 adults, 5% improvement in adult literacy); and (3) statistical analysis, including the use of statistical techniques to represent the status of the measurable phenomenon and the causes of its variability (e.g., 5% increase in statistical literacy is associated with progress in life expectancy, lower interquartile income, and so on). A statistically literate person should be able to demonstrate considerable level of understanding of each of these categories in any measurable phenomenon.

Following this current understanding of statistical literacy, domain experts need to have proper understanding of texts, tables, and graphs included in their research outputs. This would lead to better presentation of various statistical aspects of their research outcomes. The role of UI-LISA, in this regard, cannot be overemphasized. UI-LISA assists domain experts in interpreting tables, charts, and figures, giving their real-life implications (and applications).

32.3 Domain Experts and Statistical Skills

32.3.1 Reports from the University of Ibadan *Compendium of Abstracts* (2005–2008)

One of the core missions of UI-LISA is statistical consulting with domain experts, which is necessarily based on the client's level of statistical literacy. The domain experts have varied levels of statistical literacy because of different levels of exposure to statistics in their various disciplines or their training programs. Nevertheless, some of them have a strong background in statistics both in theory and application. Table 32.1 shows commonly used statistical techniques among PhD students in science-based disciplines of the University of Ibadan (2005–2008). The various disciplines listed are Agriculture, Basic Medical Sciences, Clinical Sciences, Pharmacy, Public Health, Sciences, Technology, and Veterinary Medicine. The *Compendium of Abstracts* contains information about the domain experts and their statistical skills. A review of the abstracts was conducted and some of the inconsistencies exhibited in their research include the use of ANOVA in experimental studies without checking whether assumptions underlying the use of the ANOVA have been violated, use of regression analysis on observational studies that have no continuous variable among others. As previously described, the modern definition of what constitutes statistical literacy is the ability to *interpret* and *evaluate critically* statistical information in a variety of contexts, as well as the ability to *discuss* or *communicate* this understanding in a fashion that can have an impact on decision-making (United Nation, 2012) (Table 32.1).

32.4 UI-LISA Consultancy Services

Since inception of UI-LISA in 2015, the statistical laboratory has been running a vibrant consultancy services unit aimed at assisting researchers with statistical problems on research design, data collection, statistical analysis, and interpretation of results. Services rendered by the laboratory cut across various disciplines including biomedical, biological, social, agricultural, pure and applied sciences, and engineering and technological research areas among others. In this section, an overview of UI-LISA's clients is presented. Also, some common domain experts' statistical challenges are identified using illustrations from specific case studies, and solutions proffered by UI-LISA are highlighted. Furthermore, general guidelines aimed at helping domain experts on the identified statistical challenges are presented.

32.4.1 Overview of UI-LISA Clients

A total of 133 domain experts had consulted the Stat Lab since inception. Of this total, 60.9% (81) are males, while 39.1% (52) are females. The Lab had the highest number (38) of domain experts in the year 2016, while the least (5) was recorded in the year 2020.

TABLE 32.1

Commonly Used Statistical Techniques in Science-Based Disciplines of the University of Ibadan (2005–2008)

Faculty	Department	Commonly Used Statistical Techniques
Agriculture	Agricultural Economics, Agriculture Extension & Rural Development, Agronomy, Animal Sciences, Crop Protection and Environmental Biology, and Wild Life and Fisheries Management	Descriptive statistics, T-test, Z-test, ANOVA, correlation, simple linear regression, chi-square, discriminant analysis, and other multiple analyses techniques
Basic Medical Sciences	Biochemistry, Pharmacology and Therapeutics, and Physiology	T-test, ANOVA, chi-square, and multiple logistic regression models
Clinical Sciences	Community Medicine, Nursing and Physiotherapy	Descriptive statistics, independent T-test, Spearman rank correlation, etc.
Pharmacy	Pharmaceutical Chemistry, Pharmaceutics, and Industrial Pharmacy	T-test, ANOVA, and chi-square test
Public Health	Epidemiology and Medical Statistics, Human Nutrition and Health Promotion, and Education	Descriptive statistics, chi-square, and T-test, correlation and regression
Sciences	Archaeology & Anthropology, Botany & Microbiology, Chemistry, Physics, Zoology, and Computer Science	Descriptive statistics, ANOVA, etc.
Technology	Agriculture & Environmental Engineering, Food Technology, Petroleum Engineering, and Civil Engineering	Survey design, regression analysis, and ANOVA
Veterinary Medicine	Veterinary Microbiology and Parasitology and Veterinary Public Health and Preventive Medicine	Chi-square, T-test, ANOVA, and correlation

Source: Compendium of Abstracts (2005–2008 Ph.D. Theses).

TABLE 32.2

UI-LISA Clients by Affiliations

	Universities					Polytechnics		Industries	Others
Year	PhD	M.Sc.	B.Sc.	PGD	MBA	Lecturer	HND	–	–
2015	5	13	1	11	2	2	1	–	3
2016	13	16	5	2	–	1	–	1	4
2017	15	3	–	–	–	1	–	–	–
2018	4	7	2	–	–	2	–	–	–
2019	10	9	3	–	–	–	–	–	2
2020	1	4	–	–	–	–	–	–	–
Total	48	52	11	13	2	6	1	1	9

Source: UI-LISA Clients' Forms.

The organizational affiliations of these clients are presented in Table 32.2. Majority of the clients are from educational institutions. Only one client from the industry has patronized the laboratory since 2015 (Table 32.2).

32.4.2 Some Common Statistical Challenges

In this section, selected domain experts' works are reviewed to identify common statistical challenges and highlight solutions proffered by UI-LISA. Specific cases categorized according to specific challenges faced by the selected domain experts are discussed below.

TABLE 32.3

Selected Case Study on Research Proposal

Case No.	Affiliation	Status	Domain Specialty	Statistical Needs
1.	University of Ibadan	PhD student	Botany	Writing good research proposal

Source: UI-LISA Clients' Forms.

32.4.2.1 *Writing a Good Research Proposal*

Writing a good research proposal is a common challenge facing domain experts, especially those who are into research at the postgraduate level. An example is presented in Table 32.3.

Case 1:

Research Problem:

The research was on drought tolerance and nutrient use efficiency of white yam (*D. rotundata poir*) in Nigeria. The domain expert had challenges with how to identify the factors/treatments that would be investigated, the choice of experimental design and analytical techniques and how to incorporate these into a standard research proposal.

Solution proffered by UI-LISA:

To get a thorough understanding of the subject area, the client was asked to make presentations on his research during three special seminars organized by UI-LISA. Statistical inputs and professional advice were given by lecturers from the Statistical Design of Investigations Unit of the Department of Statistics, University of Ibadan. Specifically, the client was assisted to select appropriate factors, choice of experimental design, and analytical techniques for the study. The client wrote the proposal after the consultations and it was carefully reviewed by the UI-LISA team before the final version was submitted. The intervention of UI-LISA on this work was highly impactful; the student successfully defended his proposal in his department and other postgraduate students of the department were mandated by their Head of Department to visit UI-LISA for better presentation and research outputs.

32.4.2.2 *Constructing Good Research Design*

This is a statistical challenge common to many domain experts regardless of their affiliations. Selected cases are presented in Table 32.4.

TABLE 32.4

Selected Case Studies on Research Design

Case No.	Affiliation	Status	Domain Specialty	Statistical Needs
1.	University of Ibadan	Undergraduate student	Geography	Determination of sample size
2.	University of Ibadan	Master's student	Civil Engineering	Questionnaire design
3.	University of Ibadan	Master's student	Agricultural Economics	Questionnaire design
4.	Nigerian Institute of Trypanosomiasis and Onchocerciasis Research	Research assistant	Biostatistics	Formulation of research questions that addressed research objectives
5.	University of Ibadan	Undergraduate student	Statistics	Formulation of research topic and objectives

Source: UI-LISA Clients' Forms.

Case 1:
Research Problem:
The domain expert was interested in studying the effect of flooding on agricultural practices in Kebbi State using spatial analysis technique. He selected Argungu River, a major river in Northern Nigeria, as the case study. He was able to identify communities that are worst hit by flood but had challenges with the sample size to be selected in each of the identified communities.
Solution proffered by UI-LISA:
Some members of the UI-LISA had extensive interactions with the client to get a good understanding of the research problem. The client was assisted to come up with the appropriate sampling technique.

Case 2:
Research Problem:
The domain expert wanted to assess rural road network development in Akinyele Local Government Area of Oyo State, Nigeria. He had challenges with how to design a questionnaire that would capture the essence of the research.
Solution proffered by UI-LISA:
A team of UI-LISA members assisted the client to identify the relevant variables and appropriate research questions. He subsequently prepared the questionnaire and this was carefully reviewed by UI-LISA before its final administration.

Case 3:
Research Problem:
The domain expert wanted to study the impact of selected social investment programs of the Federal Government on youths' involvement in agriculture. He had challenges with appropriate research questions and questionnaire.
Solution proffered by UI-LISA:
A team of UI-LISA members assisted the client to identify the relevant variables and appropriate research questions. He subsequently prepared the questionnaire and this was carefully reviewed by UI-LISA before its final administration.

Case 4:
Research Problem:
The domain expert was interested in investigating secondary school students' awareness levels on trypanosomiasis (sleeping sickness) and onchocerciasis (river blindness) in Ibadan, Oyo State, South West, Nigeria. He had challenges with formulation of good research questions.
Solution proffered by UI-LISA:
A team of UI-LISA members had extensive interactions with the client to get a good understanding of the research problem. He was, thereafter, assisted to formulate appropriate research questions and hypotheses.

Case 5:
Research Problem:
The domain expert wanted to obtain estimates of some parameters for the Nigerian population. He had challenges with how to conduct the study, including formulation of the research topic and objectives.
Solution proffered by UI-LISA:
The client was assisted by a team of UI-LISA members to have a good understanding of the subject area. He was, thereafter, assisted to formulate appropriate research topic and objectives.

32.4.2.3 Challenges with Data Analysis

Selected cases on statistical data analysis are presented in Table 32.5.

Case 1:

Research Problem:

The study was aimed at investigating health promotion practices and activities among undergraduate students of the University of Ibadan. The domain expert wanted to establish or refute three hypotheses, one of which was that a student's course of study influences health promotion practices and activities of the student. She had challenges with how to analyse the data obtained for the study.

Solution proffered by UI-LISA

The client was assisted to carry out the statistical analysis.

Case 2:

Research Problem:

The domain expert was interested in evaluating patients' satisfaction with pharmaceutical services in public healthcare facilities in Oyo State, Nigeria. The client had challenges with how to statistically identify attributes of pharmaceutical services that patients considered most important.

Solution proffered by UI-LISA

The client was assisted to carry out the statistical analysis.

Case 3:

Research Problem:

The research problem focused on Islamic economic system, banking, and finance. The client had challenges with analyzing data using the SPSS package.

Solution proffered by UI-LISA

The client was trained on how to use SPSS for statistical data analysis.

Case 4:

Research Problem:

The study was about the mediating influence of strategy execution and the masculinity-feminity dimension on the performance of multinational corporations in Uganda. The domain expert had challenges with the data analysis.

Solution proffered by UI-LISA

The client was assisted to carry out the statistical analysis.

TABLE 32.5

Selected Case Studies on Data Analysis

Case No.	Affiliation	Status	Domain Specialty	Statistical Needs
1.	University of Ibadan	Undergraduate student	Nursing	Data analysis
2.	University of Ibadan	Ph.D. student	Sociology	Data analysis
3.	University of Ibadan	Ph.D. student	Arabic and Islamic studies	Data analysis/ SPSS training
4.	University of Ibadan	Ph.D. student	African studies	Data analysis

Source: UI-LISA Clients' Forms.

32.4.3 General Guidelines for Solving Common Statistical Challenges Writing a Strong Research Proposal

A good research proposal is intended to convince others that you have a worthwhile research project and that you have the competence and the work plan to complete it. Generally, a research proposal should contain all the key elements involved in the research process, which include sufficient information for the readers to evaluate the proposed study. Regardless of your research area and the methodology you choose, all research proposals must address the following: (1) what you plan to accomplish, (2) why you want to do it, and (3) how you are going to do it (Sidik, 2005). Statistical literacy is very important in writing a research proposal so that the major areas where statistics is applied directly and indirectly can be addressed in the research appropriately.

32.4.3.1 Choosing an Appropriate Research Design

Researchers are to identify proper statistical procedures in choosing appropriate research designs before adopting the one that best suits their research work, whether quantitative or qualitative research. Researchers with low levels of statistical literacy could find it challenging to choose the appropriate one. Quantitative researchers tend to seek explanations and predictions that allow for generalization. Therefore, quantitative methods include careful sampling strategies and experimental designs. In quantitative research, researchers' objectivity is of utmost concern because their role is to observe and measure while avoiding personal involvement with the research participants (Fadia, 2017). Some studies may require the use of qualitative research design which involves a proper qualitative literature search to learn about existing research designs and to understand the appropriate theoretical basis on which the research is being founded. Also, researchers should know that some research could involve the use of the triangulation method which is the mixture of both quantitative and qualitative methods to capture the main aim of the research appropriately. Denzin (2006) identified four basic types of triangulation: *Data triangulation* involves time, space, and persons; *Investigator triangulation* involves multiple researchers in an investigation; *Theory triangulation* involves using more than one theoretical scheme in the interpretation of the phenomenon; and *Methodological triangulation* involves using more than one method to gather data, such as interviews, observations, questionnaires, and documents. The choice of research design will determine the type of research instruments that best suit the research work.

32.4.3.2 Statistical Consideration for Choosing Research Topics and Objectives

Researchers often have issues in developing research topics and formulating objectives. Evans (2007) highlighted three simple considerations for evaluating a potential research topic, which is interest and curiosity in the topic, whether it is do-able and worthwhile, and whether there is an adequate time frame to complete the research and gather the necessary information. Statistical considerations like the dependent variable and independent variables required for the topic could suggest the choice of statistical analysis and the information to be collected via the research instruments or secondary data to be used. Irrespective of the discipline, researchers need to consider the statistical possibilities before choosing a research topic and developing objectives.

32.4.3.3 Statistical Standards for Data Analysis

Data analysis involves different techniques which can be qualitative or quantitative, univariate, bivariate, or multivariate summarized into descriptive and inferential statistics and adopted in the research.

Statistical analysis choices could depend on the research objectives, which could determine whether to carry out descriptive univariate (one variable) analysis, or inferential bivariate (two variables), or multivariate (more than two variables) analysis. Univariate analysis is done to describe the variables in the objective, while inferential analysis infers generalizations from statistics (from the sample to the population); for the latter, the probability value (p-value) is often used. The choice of analysis could also depend on the distribution of the data. If the knowledge about the distribution of the data (known or unknown) will determine whether parametric or non-parametric statistics should be used for the analysis. It is appropriate that proper statistical analysis techniques should be used for quality research outputs irrespective of the discipline.

32.5 Conclusion

This study examined the level of statistical literacy among domain experts within and outside the University of Ibadan. Major statistical challenges facing the domain experts were discussed using selected case studies. Also, solutions, provided by UI-LISA, targeted at specific statistical problems were discussed. Furthermore, general guidelines for solving common statistical challenges were provided.

A review of the *Compendium of Abstracts* for the University of Ibadan (2005–2008), before the inception of UI-LISA, showed some inconsistencies at different research stages affecting the overall research output of some domain experts.

This study provided supporting evidence to highlight the important role UI-LISA has been playing in raising statistical literacy levels of domain experts within and outside the university community leading to better research outputs and overall development of our societies.

Furthermore, this study also advocated strict adherence, by domain experts, to guidelines for solving some common statistical challenges.

32.6 Recommendation

Schield (1999) affirmed that it is the responsibility of tertiary institutions to ensure that their graduates are well equipped with the ability to access, use, understand, and appraise statistical information in a flourishing data age. Therefore, relevant statistical courses should be introduced to the different disciplines, and these courses should be taught by qualified statisticians to ensure proper understanding of the theoretical and practical aspects relevant to the specific fields of study. Also, statistical laboratories that reflect the ideals and services of UI-LISA, including walk-in-consulting, mobile statistical clinic, seminars and workshops, and hands-on training, among others, should be established in all tertiary institutions in Nigeria. This will help in raising statistical literacy levels among the populace and increase awareness levels on the importance of statistics for day-to-day living and overall development of the nation.

References

Alhija, F.N.-A. (2017). Selecting an Appropriate Research Design. Retrieved from: https://www.researchgate.net/publication/321491126.

Ben-Zvi, D. & Garfield, J. (2004). Statistical Literacy, Reasoning, and Thinking: Goals, Definitions, and Challenges. Retrieved from: https://www.researchgate.net/publication/226958619.

Bradstreet, T.E. (1996). Teaching introductory statistics courses so that non-statisticians experience statistical reasoning. *The American Statistician*, 50, 69–78.

Chinnapan, M., Dinham, S., Herrington, T. & Scott, D. (2007). Year 12 students and higher mathematics: Emerging issues. In *Paper Presented to Australian Association for Research in Education, Annual Conference*, Fremantle, 25–29 November 2007.

Denzin, N. (2006). *Sociological Methods: A Sourcebook* (5th ed.). New York: Aldine Transaction.

Evans. (2007). Choosing a Topic and the Research Proposal. In (Psychology)-3589-03.qxd (p. 21).

Fadia. (2017). Selecting an Appropriate Research Design. https://www.researchgate.net/publication/321491126.

Gal, I. (2002). Adults' statistical literacy: Meanings, components, responsibilities. *International Statistical Review*, 70(1), 1–25.

Garfield, J. & Ben-Zvi, D. (2007). How students learn statistics revisited: A current review of research on teaching and learning Statistics. *International Statistical Review*, 75(3), 372–396.

Gordon, S. & Nicholas, J. (2006). Teaching with Examples and Statistical Literacy: Views from Teachers in Statistics Service Courses Mathematics Learning Centre, The University of Sydney, Sydney NSW 2006, Australia.

Klein, T. Galdin, A., & Mohamedou, E.L. (2016). An indicator for statistical literacy based on national newspaper archives. In *Proceedings of the Roundtable Conference of the International Association of Statistics Education (IASE)*, Berlin, Germany, July 2016.

Murray, S. & Gal, I. (2002). Preparing for diversity in statistics literacy: Institutional and educational implications. (Keynote talk). In B. Phillips, (Ed). *Proceedings, 6th International Congress on Teaching Statistics (ICOTS-6)*, Cape Town, South Africa, July 7–12, 2002. Voorburg, the Netherlands: International Statistical Institute. (Online: www.stat.auckland.ac.nz/~iase).

Postgraduate School, University of Ibadan. (2010). *Compendium of Abstracts Ph.D Theses 2005–2008*. Ibadan: The Postgraduate School, University of Ibadan.

Sajjad, H., Muhammad, I., Nosheen, F., Nasir, A., & Maksal, M. (2016). The role of multiple literacies in developing interdisciplinary research. *Journal of Applied Environmental and Biological Sciences*. ISSN: 2090-4274. www.textroad.com.

Sidik, S.M. (2005). How to write a research proposal. *The Family Physician*, 13(3).

United Nations (2012). *Making Data Meaningful Part 4: A Guide to Improving Statistical Literacy*. Geneva: UNITED NATIONS.

Wallman, K.K. (1993). Enhancing statistical literacy: Enriching our society. *Journal of the American Statistical Association*, 88(421), 1–8.

33

Media Presentation of Statistical Reports: How Adequate and Accurate?

Serifat A. Folorunso and Saheed A. Afolabi
University of Ibadan Laboratory for Interdisciplinary Statistical Analysis (UI-LISA)

Adewale P. Onatunji
Ladoke Akintola University of Technology

Morufu A. Folorunso
Federal School of Statistics

CONTENTS

33.1 Introduction

People are confronted by statistics in newspapers and magazines, on television, and in general conversations. Nigerians constantly use information or data to make judgments and decisions that affect daily and long-term events. It influences our psychological state of mind and future projections. Therefore, it is better released with fair and adequate consideration at appropriate time, in spite of an existing negative effect by the media. Olubusoye (2014) explained that statistics have become an important part of human life. Nevertheless, if it is not properly delivered, it may give a wrong impression or unexpected interpretation.

DOI: 10.1201/9781003261148-40

Statistical literacy and the media are all encompassing. Media publication of price indexes and other economic variables determines consumers' decision, short- and long-term priorities in demand. The objective of the media is to educate and sensitize the public and government on socioeconomic development and proper placement of priorities based on available statistics, while the government and its agencies need statistical reports for planning and forecasts. Therefore, nearly all sectors look forward to media reports of statistics as a guide.

The adequate and accurate communication of scientific and economic outputs through media depends on appropriate statistical reports as data are a set of facts with the provision of a partial picture of reality (Diong et al., 2018; Giovannini, 2008). Generally, information on data variability and results of statistical analyses are required to make accurate inferences, but the reverse might be the case due to the falsification of reports and some errors through media presentation. According to Steen (2004), the 21st-century world is a world awash in numbers; it is therefore important to understand how to decipher numeric values using statistical reasoning. Media representatives' inability to comprehend and interpret numeric values in comments and to communicate intellectually with tables and graphs indirectly leads to conflict.

The media are expected to follow certain procedures before sensitive information that matters to the public are released. Instances of election reports and consumer price index (CPI) are information that affect all aspects of human life. According to Statistical Act (2007),

> production and dissemination of statistics shall conform to standard, classification and procedures as determined by the Bureau, National Bureau of Statistics (NBS) to enhance comparability of such statistics with other statistics of similar nature, and to minimize unnecessary overlapping or duplication with the collection or publication of statistics by the various agencies (including private organizations) and where the publication of the data thus collected require recognition by the Bureau, the said ministries and other public institution shall be required to provide the Statistician-General with the administrative dataset and copies of report on the compiled data.

Under the Act, a code of conduct was expressly established for official statistics:

> In order to establish public confidence in all official statistics and analyses, the Statistician-General shall issue a code of practice that set out professional standards to be followed by all agencies producing official statistics.

This includes all statistical information such as price index or election reports released in print, electronics, and press releases, describing or announcing numerical or data products. The information must not be released before the scheduled release time.

This study covers several examples of the misreporting of statistical information from news networks that can provide different meanings if not correctly reported.

The present study highlights the significance of the foundation of statistical literacy in media reporting. Reporters who cover statistics-related news such as weather forecasts, election outcome, price indexes, GDP, and other important and sensitive news items that can generate negative reactions or crises need to attend advanced statistical reporting seminars.

Section 33.2 discusses statistical literacy as a tool in media literacy, while Section 33.3 presents the role of media in national statistical system (NSS): Nigeria as a case study. Section 33.4 gives the analysis, and Section 33.5 concludes the chapter.

33.2 Statistical Literacy: A Tool in Media Literacy

Researchers have described statistical literacy in different ways. In today's complexities of our information society, a comprehension of statistical information and methodologies has become crucial for everyday living as well as effective professional involvement, prompting calls for enhanced statistical literacy.

The quality of available statistics can vary considerably so that an understanding of sampling techniques and sources of bias can help to first assess what has been done and second adopt a critical stance on statistics. Because of the "huge amount of uncontrolled, unconstrained information being thrown upon a populace that is generally ill suited to digest the information," raising public awareness about the quality of information consumed via television or newspapers is critical (Makar and Rubin, 2009).

Statistical literacy in media involves the ability to read and interpret statistical data in daily and other media (newspapers, the Internet, radio, television channels, etc.), taking into consideration the relevance of graphs, charts, comments, statistical surveys, and studies (UNECE, 2017). Today, statistical information is perhaps more important than ever for daily life practice.

Some authors confirmed in literature that statistical literacy is one of the major requirements of media practitioners for effective delivery of numerical reports. The role of media in any NSS cannot be underestimated, particularly with respect to sharing statistical information through media coverage. Other authors stated that some national statistical literacy programs for the development of statistics should be developed and implemented at every stage of life (UNECE, 2017; Zwick 2013; Martina et al., 2014; Klein et al., 2016).

Statistical literacy should be a central priority in training students to understand statistical information that is widely disseminated in the media (Merriman, 2006; Mawdsley and Tam, 2013). Recently, statisticians acknowledged the need for statistical literacy to be established so that the public can process the flow of statistical information that affects them. Media reports are a routine method for transmitting statistics to the general public and can also be used as a means of promoting statistical literacy in the wider society. Today's journalists need to improve their statistical literacy in order to deliver information so that consumers of that information can make informed decisions and for rational conclusions.

Media houses in Nigeria should encourage journalists to develop their statistical thinking for carrying out media activities. In general, it is not a common practice for journalists to use statistical investigation for media reporting.

Recent literature frequently uses the term statistical literacy (Watson et al., 1994; Watson, 1997; Gal, 2000; Watson and Callingham, 2003). Most researchers agree that statistical literacy encompasses the competencies, attitudes, and knowledge that allow a person to function in the information age. Statistical literacy generally means the ability to recognize and reason with data-related information or statements (Gal, 2000). Ultimately, statistical literacy can be understood as the ability to interpret and critically assess statistical information and the ability to discuss or communicate responses to it with the perceptions and concerns.

Statistical literacy is analytical thinking about numbers, and figures used as facts in claims, according to Keck (2010); it is also the ability to communicate numbers and view them in reports, surveys, tables, and graphs. Martina et al. (2014) ascertained that statistical literacy is an aspect of media literacy, in particular leading to an increasing proportion of media coverage of statistical knowledge. The present study adopted Gal's (2000) definition, which recommended statistical literacy as a means of integrating skills, knowledge, attitudes, dispositions, and the ability to critically think, discuss, and make judgments.

He emphasized that statistical literacy overlaps with their inherent thinking skills into the wider application areas of statistics and mathematics knowledge bases.

To be statistically literate indicates being able to accurately decipher the statistical information available through the media or through the official statistics services. This is necessary for purposes of personal development and for direct or indirect effects on the development of organizations as well as state governments. Statistical information contains an overview of the situation in various areas so that changes in phenomena can be monitored over time and should form the basis for decision-making in the economic, political, cultural, and personal spheres of life. Statistical literacy has so much more to claim than numeracy. This requires the opportunity to interpret the data and communicate its meaning. This quality makes media reporters literate, rather than merely numerically literate. The lack of quantitative competencies is summarized under the term statistical innumerability (UNECE, 2017).

There is a need for a statistical department in any media outlet that will deal with highly complex information; for accurate and high-quality media information, it is also necessary for all staff in those units to have an adequate basic level of statistical knowledge (Zwick, 2013). Organizing seminars, workshops, and advanced training for staff members has played a prominent role within the statistical offices and should be integrated into any media outlet's statistics department. Statistical research has become increasingly important and has gone beyond National Statistical Offices (Zwick, 2013).

The use of statistical evidence on policy making does not seem to relate to basic economic indicators such as GDP, indicators of governance, and other statistical development indicators such as the Index of Statistical Capability. These indicators are statistically based and should be used for policy decision. However, the methods used for calculating these indicators are not always clear to the media and therefore not clear when presented to the public. Most indicators are sensitive to weights, components, and aggregation methods, and this can alter the outcomes if care is not taken (Nardo et al., 2005).

33.3 The Role of Media in the National Statistical System: Nigeria as a Case Study

In literature, media has a versatile role concerning the NSS. In any country, the media are interested as a public organization in the activities of NSS. They serve as an important statistical data redistributor. In the long run, NSS's major data users are referred to as the media (Helenius, 2010). The media often misinterprets statistical information. Numerous misunderstandings and misinterpretations of NSS data can be seen in media reports such as daily newspaper articles and in both print and online (UNECE, 2017).

In the Master Plan for the Nigeria National Statistical System (2004/5–2008/9), there should be the development of a data dissemination policy providing for advance publication of a release calendar and simultaneous release of data to all stakeholders. It should be the responsibility of the National Bureau of Statistics to disseminate statistical data and information to data users such as media outlets in the form of up-to-date product catalogs, providing telephone, fax, Internet, and electronic mail access to an investigation point. Different techniques should be used to disseminate data, including major bookshops in the country's key zonal offices and universities, as the main outlets for statistical reports, newspapers, electronic media, and press releases whenever new data are formally released.

33.3.1 Presentation of Election Result: How Skillful Is the Media?

The ethical foundation of journalism has always been its commitment to truth, but even with "concrete facts" there can always be different ways to slant them, giving them different inferences by using different news frames, narrative structures, premises of value, vocabulary, and so on (Vobič and Dahlgren, 2013). The principles and methodologies used in the preparation of any text must be understood so that one can interpret them by reading the statistical content. Specialized skills are needed before the meaning of data can be understood in textual form, because communicating with the text will produce the statistical information that the audience must understand. Statistical thinking, which is the ability to "read" the data and give meaning to it, should be improved (Campos, 2008).

Reporting in developing countries is a peculiar job for media before, during, and after an election. It makes the media play a significant role in the usage of graphic representations or figures such as the strength of parties participating in elections, the publicity of candidates, attribution and assessment of issues, and communication of politicians to the general public. Elections and the media have a special relationship; reliable data must be accessible to media for a free and fair election. A notable gap in media presentation and reporting is the efficient use of these data. The data should be accurate, reliable, and accessible to the electorate. Kim et al. (2018) analyzed how the media reflected on the online polls and how people viewed the news. Their study was conducted using analysis of the content, surveys, and experiments.

The credibility of the 2007 election results in Nigeria was put to question by the divergent stories in the newspapers' reports (Oboh, 2016).

The use of statistics as evidence in this argument was abused grossly. There will inevitably be the need to use statistical literacy in media presentation and election reporting, with different statistical methods. The misuse of statistics by the media sector would lead to misinformation about the election in African countries. This may lead to uprising during or after an election.

Olugu et al. (2018) concluded that statistical manipulation has always been harmful and counterproductive to economic and national development. They said that misreporting or misinterpretation of facts in the media and social media is one of many forms of statistical manipulation, which in fact has long-term effects on national development.

Asiru et al. (2018) on the 2011 presidential election in Nigeria focused on the language of the presentation of the election result before the official announcement by the Independent National Electoral Commission (INEC). Consider the following examples:

a. Jonathan won convincingly.

b. Jonathan maintained a comfortable lead.

c. Jonathan still clinched the mandatory 25% (Guardian April 18, 2011).

Example "c" is a quantitative statement that cannot be backed with statistical facts.

33.3.2 Presentation of Consumer Price Index: How Accurate Is the Methodology?

The National Bureau of Statistics Statistician General Dr. Kale said that CPI reports are the main statistical production that is generated on a monthly basis (Kale, 2013). Figure 33.1 shows an average of 306.43 points in the CPI in Nigeria from July 2019 to May 2020, touching an all-time high of 322.20 points in May 2020 and a record low of 289.7 points in June, 2019 (Figure 33.1).

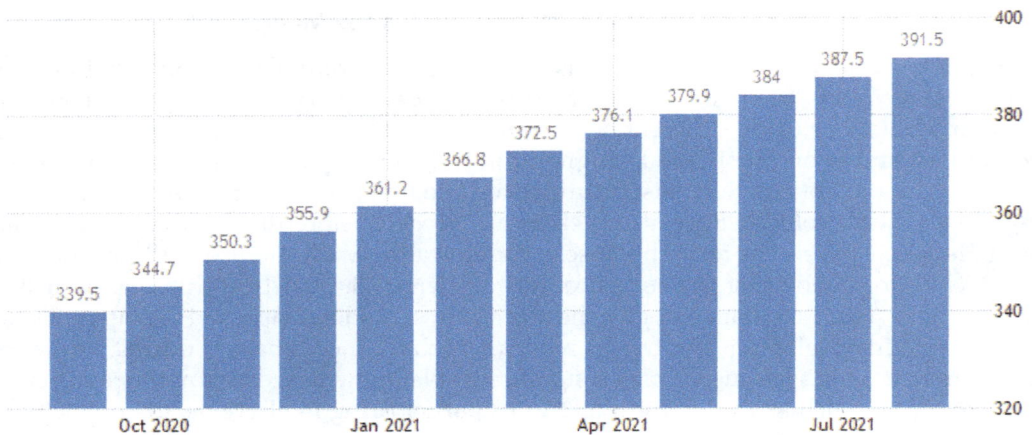

FIGURE 33.1
Nigerian CPI (2019–2020 data).

Greenlees and McClelland (2008) reported while addressing misconceptions about the CPI in the Monthly Labor Review that the CPI is not and can never be a perfect index. This statement means that the concept of CPI is very important and should be handled with care, so media outlet should not simply present it. A thorough understanding of the techniques used in calculating the index should be provided in order to prepare a plan for the dissemination of statistical literacy when presenting the CPI of any nation.

Based on the presentation of the price index to the public, Afolabi (2018) contributed that market price changes are a phenomenon experienced in most parts of the world. Media activities affect not only price but also quantity. If the market price is not determined by ability to pay, then information on social media will go viral about it. This causes incorrect conclusion by the media on this price and thereby leads to the questions on how the index number (both price and quantity) is estimated correctly with the appropriate tools.

33.4 Analyses of Some Review Cases in Nigeria

33.4.1 Analysis of Reports of Media on the Nigeria's Election Results

The analysis in this section comprises some descriptions of the election results reported during the 2019 general elections in Nigeria. Over the years, it has been noticed in Nigeria elections that if any President is to have emerged in the country, he must have won two particular states, which are Kwara and Benue. This was the case in 2007, 2011, and 2015 but in 2019, this did not turn out to be the case. Before the end of the 2019 elections, many predicted the emergence of President Muhammadu Buhari via social media without adequate and accurate details. It was later concluded that the elected President of the 2019 election did not win in Benue state but later won the election by winning 19 out of 36 states, including Federal Capital Territory of Nigeria. Table 33.1 shows the distribution of votes in Kwara and Benue states.

Figure 33.2 shows the graphical representation of the presidential results in three election years in Nigeria.

TABLE 33.1

Frequency Distribution of Popular Winning States by
Nigerian President(s)

Years	Political Party	Popular Winning States	
		Kwara	**Benue**
2019	Winner	308,984	347,668
	Runner-up	138,184	356,817
2015	Winner	302,146	373,961
	Runner-up	132,602	303,737
2011	Winner	268,243	694,776
	Runner-up	83,603	109,680

Source: INEC website.

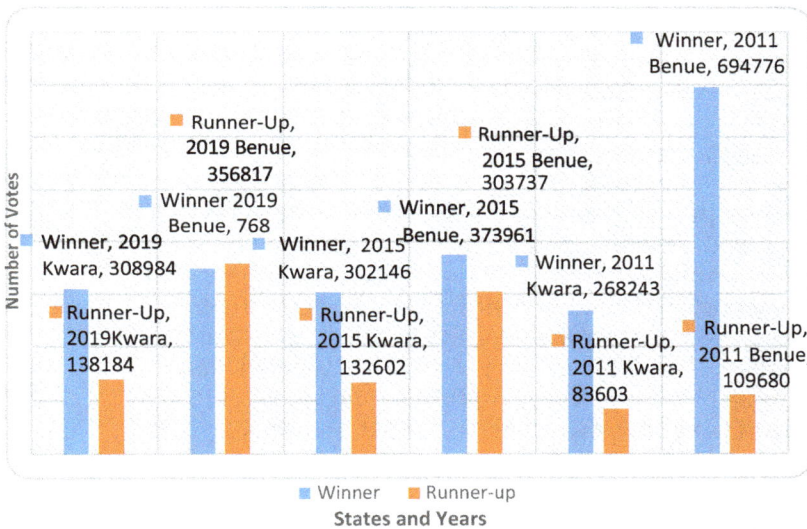

FIGURE 33.2
Distribution of Nigeria elections of 2011–2019.

33.4.2 Analysis of Reports of Media on Index Number

In a report by a journalist at Lagos Guardian, in 2018, he proclaimed that Nigerians would need to brace up for another round of hardship, with the soaring price of food commodities, considered staple foods for households in the country. A bag contains 50 kg of parboiled rice. According to the report, 50 kg of parboiled rice initially sold between #12,000 and #12,500 rose to between #13,000 and #14,500, depending on the area in the last 2 weeks. This report was not reliable because price index was based on the available data in different locations from the questionnaire and that the cost had not really gone up.

It is indicated in Figure 33.3 that the prices of rice rise from short grain to medium grain to long grain via the brand and type at three different areas in Ibadan. However, there is a notable gap in the presentation of the price to the public by the media for the failure of calculating average price of different brands of rice at different locations in the nation.

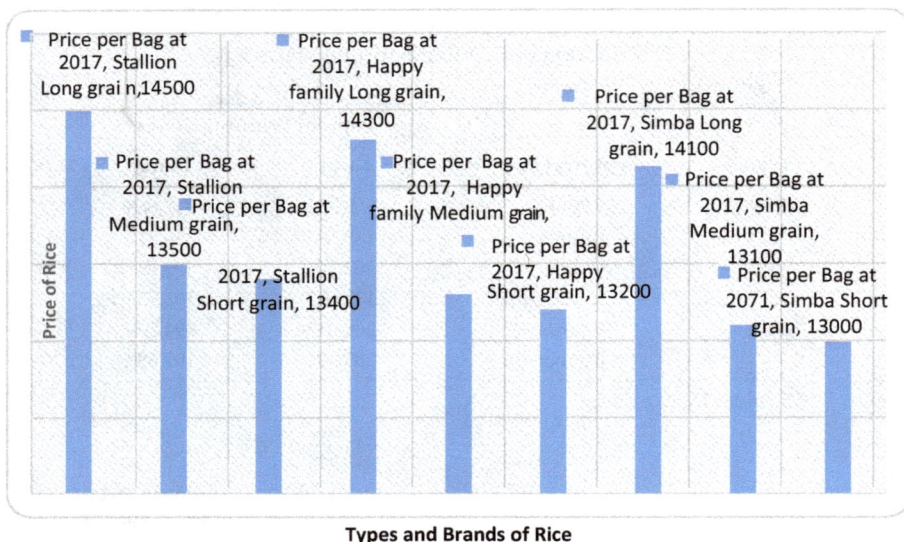

FIGURE 33.3
Analysis of price index of rice.

Table 33.3 gives the estimation of the appropriate statistical tool (index number) to the data in Table 33.2, the estimated values were divided by 1,000 as the majority of the cost of commodities measured in Naira; after that, they were all estimated in percentage (%) for easy interpretation. This is highly relevant because of the appropriate index number tools used (both weighted and un-weighted) to establish the statistical significance in terms of professionalism. From the analysis of Simple Average Relative Index (SPR and SQR), Simple Aggregate Index (SAPI and SAQI), Laspeyres Index (LPI and LQI), Paasche's Index (PPI and PQI), Marshall-Edgeworth Index [ME(P) and ME(Q)], Fisher Ideal Index (FPI and FQI), and Bowley's Index [Bow(P) and Bow(Q)] with the increase in prices of the selected food items estimated to be 74%, 51%, 55%, 65%, 58%, 60%, and 60%, while that of increase in its quantities demanded are 4%, 4%, 3%, 4%, 4%, 3%, and 3%, respectively. This is not a statistical fallacy, but for media people without statistical knowledge of numerals it would not accurately and adequately report this to the public.

33.5 Conclusion

The study identified the role of statistical literacy in media reporting and this was emphasized using Nigerian election results and the need for statistical literacy in the CPI so as to avoid most of the misconceptions in CPI. Media is a very sensitive organization as a major data user of the Nigerian NSS. They must be up to the task to avoid misinterpreting or misinforming statistical facts to the general public. It is now recommended that there is a need for teaching basic statistics to media presenters in order to guarantee and achieve the quality of statistical reporting. Each journalist should at least involve in vocational training in data visualization so that communicating with data will be easy and in turn yield a good statistical output to the populace.

TABLE 33.2

Price and Quantity Index of Some Selected Food Items in 2007 and 2017

S/N	Commodity	Brand	Type	Price (Naira) Per Bag in 2007	No. of Bags Sold per Day (2007)	Price (Naira) per 1.5kg in 2007	Price (Naira) per Bag in 2017	No. of Bags Sold per Day (2017)	Price (Naira) per 1.5kg in 2017
		Stallion	Long grain	8,000	280	265	14,500	78	480
			Medium grain	7,800	89	250	13500	45	450
			Short grain	7,500	310	230	13,400	87	445
1	Rice	Happy family	Long grain	7,800	278	260	14,300	75	450
			Medium grain	7,000	87	230	13,300	43	440
			Short grain	6,500	309	200	13,200	83	400
		Simba	Long grain	7,600	275	200	14,100	65	430
			Medium grain	6,800	92	180	13,100	34	400
			Short grain	6,500	210	150	13,000	23	350
2	Beans	Oloyin		12,000	160	400	19,500	31	650
		Oloyin Pelebe		10,500	40	350	19,500	8	650
		Drum		10,500	80	350	18,000	21	600
		Sokoto		7,500	120	250	15,000	51	500
3	Garri	Egba		5,000	200	50	40,000	61	400
		Oyo		4,000	100	40	35,000	23	350
4	Corn	White		900	260	30	8,100	166	270
		Solo		2,100	230	70	9,000	163	300
			Type per tuber						
5	Yam	Abuja	Per 3-42	6,300	540	150		240	900
6	Yam	Abuja	Per 3-42	6,300	540	150		240	900
			Per 6-42	10,500	300	250		150	12,000
		Isu Oko	Per 3-42	6,300	480	150		120	600
			Per 6-42	8,400	200	200		80	800

Source: Bodija, Oja-Oba, and Oje Markets.

The 10-years interval prices and quantities of five commodities (food items) are given in Table 33.2. In 2007, the prices of the commodities were lower which resulted in higher demand compared to the prices in 2017 which were higher, thereby resulting in lower demand for commodities. Based on the price of the selected products, the media reporting would fail to clearly determine the sample mean of the price at this location, not to talk of whole the country. This consequentially leads to wrong conclusion about the acceptable market price.

TABLE 33.3

Summary of Weighted and Un-weighted Price and Quantity Index

Years	Methods	2007	2017	Increment in % $\left(\dfrac{Est}{1,000} \times 100 \right)$
Price index	SPR	100	736.69	74%
	SAPI	100	513.87	51%
	LPI	100	551.89	55%
	PPI	100	654.899	65%
	ME(P)	100	575.83	58%
	FPI	100	601.19	60%
	Bow(P)	100	603.39	60%
Quantity index	SQR	100	37.865	4%
	SAQI	100	35.5	4%
	LQI	100	30.28	3%
	PQI	100	35.934	4%
	ME(Q)	100	35.067	4%
	FQI	100	32.98	3%
	Bow(Q)	100	33.107	3%

References

Afolabi, S. A. 2018. Changes in the prices of some selected Food Commodities at Bodija Market, Ibadan, Oyo State. *Second International Conference of Professional Statisticians Society of Nigeria (PSSN) [Formerly; Nigerian Statistical Society (NSS)]*. Unpublished.

Asiru, H. T., Ogutu, E. A. and Orwenjo, D. A. 2018. Event and actors representation in selected Nigerian daily newspapers. *Ghana Journal of Linguistics*, 7(1): 84–104.

Campos, P. 2008. Chapter 2: Thinking with data: The role of ALEA in promoting statistical literacy in Portugal. In J. Sanchez (Ed.), *Government Statistical Offices and Statistical literacy*. http://www.stat.ucla.edu/~jsanchez/books/PedroCampos.pdf, downloaded 15th April 2014.

Diong, J., Butler, A. A. Gandevia, S. C. and Héroux, M. E. (2018). Poor statistical reporting, inadequate data presentation and spin persist despite editorial advice. *PLoS One*, 13(8): e0202121. doi:10.1371/journal.pone.0202121.

Gal, I. 2000. Statistical literacy: Conceptual and instructional issues. In D. Coben, J. O'Donoghue, and G. Fitzsimons (Eds.), *Perspectives on Adults Learning Mathematics* (pp. 135–150). Dordrecht, the Netherlands: Kluwer Academic.

Gbenga, A., 2018. Nigeria: Tough time awaits Nigerians, as price of rice soars. The Guardian (Lagos): allAfrica.com. https://allafrica.com.

Greenlees, J. S. and McClelland, R. B. 2008. Addressing misconceptions about the consumer price index. *Monthly Labor Review*, 131: 3.

Helenius, R. 2010. Improving statistical literacy by national and international cooperation. http://iase-web.org/documents/papers/icots8/ICOTS8_7H2.

Kale, Y. 2013. Where are the numbers? National Bureau of Statistics and the reset of the Nigerian National Statistical System, NBS. Retrieved on 20/08/2014 from http://mortenjerven.com/wp-content/uploads/2013/04/SG_NBS-and-the-reset-of-theNigerian-National-Statistical-System-Vancouver-April-2013-FN-small4.pdf.

Keck, W. M. 2010. Statistical Literacy: A Short Introduction.

Kim, S. T., Weaver, D. and Willnat, L. 2018. Media reporting and perceived credibility of online polls. *J &M C Quarterly*, 77(4).

Klein, T., Galdin, A. and Mohamedou, E. L. 2016. An indicator for statistical literacy based on national newspaper archives. In *Proceedings of the Roundtable Conference of the International Association of Statistics Education (IASE)*, July 2016, Berlin, Germany.

Martina, P. S., Nevena, J. M. and Hrvoje, S. 2014. Statistical Literacy as an Aspect of Media Literacy. Medij. istraž. (god. 20, br. 2) (131–153) *PREGLEDNI RAD UDK*: 316.77:311. Zaprimljeno: 30 lipnja.

Mawdsley, F & S. Tam 2013. New ABS strategies to promote statistical education under a new national curriculum for statistics. http://iase-web.org/documents/papers/sat2013/IASE_IAOS_2013_Paper_3.1.3_Mawdsley_Tam.pdf, downloaded 15th April 2014.

Merriman, L. 2006. Using media reports to develop statistical literacy in year 10 students. The University of Auckland, New Zealand merriman@maxnet.co.nz ICOTS-7, 2006: Merriman (Refereed).

Nardo, M., Saisana, M., Saltelli, A., Tarantola, S., Hoffman, A. and Giovannini, E. 2005. *Handbook on Constructing Composite Indicators: Methodology and User Guide*. OECD.

Nardo, M., Saisana, M., Saltelli, A. and Tarantola, S. 2005. *Tools for Composite Indicators Building*. European Commission.

Oboh, G. E. 2016. Reflecting on the Nigerian media, elections, and the African democracy. *SAGE Open*, 6(3): 1–10.

Olubusoye, E. O. 2014. Statistical Literacy and Empirical Modeling for National Transformation, Faculty of Science, University of Ibadan, Nigeria (Faculty Lecture).

Olugu, M. U. Akinyeke, F., Akinwumi, F. F. and Fatoki, F. 2018. Dangers of Statistical Manipulation to National Development. *International Journal of Advanced Research in Science and Engineering*, 7(8).

Statistical Act. 2007. Official Gazette. Federal Republic of Nigeria.

Steen, L. A. 2004. *Achieving Quantitative Literacy*. Washington, DC: The Mathematical Association of America.

UNECE. 2017. Extract of the recommendations on promoting, measuring and communicating the value of official statistics. Note by the Task Force on Value of Official Statistics (Rep. No. ECE/CES/2017/4). https://www.unece.org/fileadmin/DAM/stats/documents/ece/ces/2017/CES_4_E_Value_of_official_stats.pdf.

Vobič, I. and Dahlgren, P. 2013. Reconsidering participatory journalism in the internet age. *Medijskaistraživanja, Zagreb*, 19(2), 9–30.

Watson, J. 1997. Assessing statistical thinking using the media. In I. Gal and J. Garfield (Eds.), *The Assessment Challenge in Statistics Education*, (pp. 107–122). Amsterdam: IOS Press and the International Statistical Institute.

Watson, J. and Callingham, R. 2003. Statistical literacy: A complex hierarchical construct. *Statistics Education Research Journal*, 2(2), 3–46.

Watson, J., Collis, K. and Moritz, J. 1994. *Authentic Assessment in Statistics using the Media*. Report prepared for the National Center for Research in Mathematical Sciences Education – Models of Authentic Assessment Working Group (University of Wisconsin). Hobart, Australia: University of Tasmania, School of Education.

Zwick, M. 2013. EMOS – The European Masters in Official Statistics. http:// www.cros-portal.eu/sites/default/files//NTTS2013fullPaper_231%20zwick.pdf, downloaded 15th April 2014.

Some Online Resources

www.nigerianstat.gov.ng
www.inecnigeria.org
www.m.guardian.ng
www.latestnigeriannews.com

34

The Challenges of Effective Planning in Developing Countries: Appraisal of Statistics Literacy

Morufu A. Folorunso
Federal School of Statistics

Serifat A. Folorunso and Adedayo A. Adepoju
University of Ibadan laboratory for Interdisciplinary Statistical Analysis (UI-LISA)

CONTENTS

34.1 Introduction

National planning is a basic mechanism for socio-economic growth that is efficient and feasible, and statistical findings and research are inevitable resources to design and achieve a credible national plan objective. Proper national planning can be achieved with agencies such as the National Statistical System (hereafter, NSS), international collaboration, institutional or academic statistical training etc. of a country where a signal for high-level statistical literacy is provided from a proper and systematic perspective (Doguwa, 2009).

Planning and development are important parts of the daily activities of every successful government that functions for the benefit of its people. Efficient planning is among the keys to focused, organized and sustainable growth (Khan, 2013). The achievement of any government is assessed by the amount of growth it accomplishes for the country. However, planning and development are no longer limited to government. In reality, making plans to execute organizational goals and work toward achieving specific goals is practiced by all modern companies around the world. This requires the support of strong, specialized statistical departments and agencies across all countries to enhance the level of statistical literacy at all levels (Doguwa, 2009).

DOI: 10.1201/9781003261148-41

The research's main objective is to review the progress on statistical development in the national planning process in Nigeria. This includes assessment and evaluation of some key agencies, such as NSS, international donor agencies collaboration and academia to statistics literacy in Nigeria, as a case study for some developing countries. The study exposes the prevailing circumstances, challenges and prospects for future development.

The remainder of this chapter is organized as follows. In Section 34.2, we present statistical literacy as a device for national development, and in Section 34.3, we describe the Nigeria NSS – recent trends and challenges. Information on the Federal School of Statistics (FSS) Ibadan, Nigeria – a statistical resource center is given in Section 34.4, and Section 34.5 presents donor agencies and statistics collaboration in developing countries. Section 34.6 gives the conclusion and recommendation of the chapter with special attention to assessing statistical literacy.

34.2 Statistical Literacy as a Device for National Development

Statistical literacy is a summary of numerical evidence of any decision taken in any structured society mostly on data sources. Several authors termed it as the ability to understand and critically evaluate statistical outcomes that impact people's daily lives-coupled with the ability to strengthen the involvement that statistical thinking can make in public and private, professional and personal decisions (Wallman, 1993, Gal, 2002, Dodge, 2003, Garfield & Ben-Zvi, 2007). A reliable socio-economic policy can therefore only be achieved through an existing and functioning NSS backed by local and international training and research agencies for effective planning (UN, 2012). In the report of PARIS21 (2015 & 2018), some authors affirmed that there are two main ways, in literature, in which statistics can affect any organization or government's policymaking. The first channel is efficiency, where the accessibility of indicators enables better resource allocation. This allows governments to monitor the use of public resources, for example, the distribution of public investment. The second source is public concern, that is, the degree to which a policy is aimed at fostering community welfare or the degree to which policies benefit the nation. It enables monitoring of the effects of public policies, helping to reinforce those more directly linked to public well-being (Ardanaz et al., 2010).

The existence of good statistics in any governance implies that political-economical factors will further affect statistical capacity development. These include the existence of State departments which need statistics for their functioning, international demands for support and guidance for statistical offices, the role of the executive in promoting statistical development, the effects of both political and economic crises and the presence of NGOs requiring high-quality statistics to hold their governments to account (Ardanaz et al., 2010). These factors relate to the channels through which statistics affect policymaking (e.g., public concern requires an interest in monitoring policies on the part of international organizations, civil society and the private sector).

Several authors agreed that each country needs good statistics for proper national planning that will produce national development in the long run, so the use of statistics was defined as the systematic use of statistical knowledge to inform program design and policy choices, monitor policy implementation and assess policy impacts (Russell and Muñoz-Ayala, 2015, Scott, 2005, Doguwa, 2009). Klein et al. (2016) also categorized Statistical Literacy Indicators into three levels, namely: (1) basic consideration, (2) diagnosis, and (3) statistical analysis.

Badiee et al. (2017) concluded that the provision of appropriate, timely and accessible data is necessary for countries to create goals, make informed choices and adopt better sustainable development policies. They affirmed that the data revolution has placed new techniques and data sources in the hands of statisticians. Data will support both the national statistical offices and the rising number of data users and producers. Building statistical capacity in Nigeria includes the development of statistics at the grassroots level, i.e., involving explicitly the state and local governments (Hamadu et al., 2012). With the emergence of the State Bureau of Statistics, a draft on how this could be achieved has been issued. Consequently, if created, state and local statistical systems should be integrated with the National Bureau of Statistics (NBS) as the integrating agency into the Federal statistical system.

34.3 Nigeria National Statistical System (NSS) – Recent Trends and Challenges: The Evolution and Recent Growth of Nigeria National Statistical System

At the African Economy Development Summit in 2013: Measuring progress and failure in Canada, the Former Nigerian Statistics Bureau's Statistician-General, Dr. Yemi Kale gave a detailed historical evolution of the Nigerian National Statistical System in his paper presentation titled: Where are the numbers? NBS, a Reset of the Nigerian National Statistical System. Kale (2013) presented the full description of the NSS historical evolution from the colonial era until when the Statistical Act was passed in 2007.

The ultimate objective of any national statistical framework is to raise awareness and importance of statistics on the government's development agenda and evaluate the citizens' standard of living (Olubusoye et al., 2015). The objectives of Statistical Act of 2007 are to understand the practice, prospects and gaps in Nigeria. Implementation of the Statistical Master Plan and National Strategy for Statistical Development is a means of enhancing statistical capacity and use.

The Statistics Act of 2007 aims to raise awareness of the importance and role of statistical information to society; collecting, processing, analyzing and disseminating statistically relevant information; promoting the use of best practices and international standards in statistical production; management; and distribution. Initially, the National Statistical Master Plan (NSMP) was scheduled to last 5 years, 2005–2009. Nevertheless, in 2007, the legislative structure was passed which was expected to provide the basics for the development of the proposed autonomy for the sensitive agency to be called the NBS.

The act recognized NBS as Nigeria's principal national agency responsible for the development and management of official statistics. The 2007 Statistical Act aimed to control the development and implementation of NSMP, the National Strategy for Statistical Development (NSSD) and the recent trend (Doguwa, 2009, Kale, 2013, Olubusoye et al., 2015).

34.3.1 The Prospect of Nigerian National Statistical System

This subsection systematically explains the growth of Nigerian NSS from the middle stage to date. It showcases the development of statistics as a necessary tool of government for socio-economic development and measuring the direction of growth of the economy.

In view of the enactment of the Statistical Act and the adoption of the statistical master plan, the role of statistics in national social-economic growth has been further appreciated. The NSS added value to the importance of statistics by providing it with good leverage for proper and systematic development. It also helped to strengthen the infrastructure that would enable academics and professionals to play their roles within the system framework.

The partnership between statistical agencies such as NBS, Central Bank of Nigeria, National Population Commission, academics, etc., where appropriate, has continued to offer the anticipated value to research and statistics. The partnership makes it possible to compare and harmonize research results and to set targets for socio-economic development. Table 34.1 shows the improvement in statistical activities on the internet through various performance indicators, a benefit of Nigerian NSS on the national statistical literacy level. Kale (2013) identified this growth in NBS.

The legal provision of the Statistical Act has gone a long way to encouraging professionalism. The establishment of NBS was backed by the Statistical Act of 2007. The NBS is to serve as the coordinating agency for the National Statistical System. According to law, NSS comprises four elements: (1) The Producer of Statistics, i.e., NBS in line with ministries, etc.; (2) Data Users; (3) Data Suppliers; and (4) Research and Training Institutions. By extension, the statistical master plan was developed for 2004/2005 to 2008/2009, as approved by Federal executive's councils in April 2004 to build a huge capacity.

Increased demand for statistics by the Federal Government of Nigeria, as well as for data for performance and outcome measurement and strategic planning, has added value to the feeling for numbers. The search for growth, new opportunities and demand from international brands entering the Nigerian market for a standard measurement of economic prospects and challenges also boosts the profile of the usefulness of statistics.

The development of the Nigerian Statistical System translates into an increase in study and statistical research at different academic levels. Currently, the demand for statistics as a study course in the different universities and polytechnics has increased tremendously. Technology has reduced the time-consuming, expensive and inefficient aspects of the process of data processing and analysis. Data collected can be sent online to NBS headquarters or zonal headquarters for analysis within a short period of time.

Recently, particularly in the advent of the democratic regime to date, price index reports and publications are available to allow an understanding of the economic structure, the growth indices and sectors where investment and resources should be channeled (Ajakaiye, 2012). This means that demand and supply can be expressed in terms of goods and services' prices. The international community is now measuring the progress of democracy and good governance in each country based on the results of a socio-economic

TABLE 34.1

Demand for Nigerian Statistics 2005 and 2012

	Performance Indicators	2005	2012
1	Reports downloaded	48,479	1,015,6454
2	Request for data onsite	23	334
3	Request for data email	106	4,882
4	Visits to website/no. of hits	36,280	4,486,112
5	Number of times NBS mentioned in the media	73	3,365

Source: Kale (2013), Where are the numbers? National Bureau of Statistics and a reset of Nigerian National Statistical System, NBS.

development statistics report. A call for data revolution was initiated in 2013 by a high level of eminent individuals gathered by the United Nations Secretary-General on the rising need for statistics and access to information.

As a pioneering professional body, the Nigerian Statistical Association (NSA) has played a vital role in promoting statistics. One of the NSA's objectives is to improve NSS capacity by enhancing academic and professional statistical advancement in Nigeria. Acknowledging the importance of the NSA as a stakeholder and professional organization, the 2007 Statistical Act made it mandatory for NSA president or its representative to be a member of the governing board of the NBS. The NSA has also aligned itself with the government of Nigeria and other statistical professions to ensure that the right and competent individual is named as head or board member of statistical agencies. The Chartered Statistician Institute of Nigeria bill, however, is currently top on the NSA's agenda. It is a way of ensuring that reliable and timely statistics are accessible to producers and suppliers of statistics.

34.3.2 The Challenges of Nigerian National Statistical System

The post-Statistical Act enactment and challenges facing the Nigerian NSS cannot be underestimated. In fairness, statistics professionals are all over the place; unfortunately, employment and capacity to employ professionals and related manpower to carry out the NSS task are lacking both in quality and quantity.

The Nigerian National Statistical System is centralized by design; however, the state and local governments do not have a statistical structure in the mold of the NBS at the national level. This is contrary to the objective of state and local government generating useful information for decision-making. The structure and standards at the national, state and local government levels are not equal. This makes grassroots statistics assessment difficult or even impossible. Therefore, it is very difficult to integrate information from the state and local government levels into the national database (Hamadu et al., 2012, Olubusoye et al., 2015).

Olubusoye et al. (2015) explained that the Nigerian government also funds the collection of ad-hoc data when specific data is required to achieve a policy objective. This act is regarded as a "quick fix approach" to statistical research. Officially, local government statistics departments exist in all the 774 local governments, but unfortunately, they are not performing the role of primary sources complimentary to NSS. Meanwhile, these are the supposed institutions that are presented to donor agencies for the successful implementation of statistical findings. Data entry and processing are often run on alternative power supplies that are expensive and functionally unreliable.

Poor regional planning poses challenges to timely and effective data collection, as coordination of information at the community level is difficult because high density settlements are poorly laid out. Therefore, those involved in urban planning need to be sensitized to look beyond structures in order to capture the population. In the view of Jacobs (1961), having more high-quality data about the activities of citizens, households and businesses in an urban environment and having access to this data in a timely and open fashion can undoubtedly be of great help in designing service delivery.

Lobo (2015) affirmed that a community process with trustworthy and verifiable data on the physical, social and economic characteristics of the neighborhoods and their needs is the simplest and most effective means of achieving laudable objectives. In terms of training and development, poor or lack of incentive to data collectors remains a challenge.

Data or census on population and household composition which constitute the sampling frame for the investigation of socio-economic dimension of the people are generally not frequent and often behind schedule in proportion to the international standard. Censuses are done every ten years; this can be explained by high cost, lack of will and poor consideration for population and household census.

In Nigeria, several statistical operations are driven by donor agencies. Projects and initiatives run by donor agencies are typically short term and often take priority over long-term planning, which may distort national statistical development priorities. Devarajan (2013) argued that African statistical institutions lack capacity, lack secure state funding and a disruptive impact of donor agency funding. He emphasized the diminishing allocation of human and financial capital for statistical development by the states. Political influence and other negative factors are clogged in the process of recruiting the right professionals. Therefore, the methods of recruiting manpower are often not transparent and unethical. The entry point qualification of some personnel is irrelevant. It eventually results in ineptitude and poor service delivery.

34.4 Federal School of Statistics (FSS) Ibadan, Nigeria – A Statistical Resource Center

The role of the FSS in the training and development of potential and professional statisticians cannot be underestimated. The school was established in 1947 and later accredited by the National Board for Technical Education under the Statistical Act, 2007. According to the 2007 Statistical Act, the school is expected to enroll students from public and private sectors. It is to award professional diplomas in statistics and serve as a research center for statistics and related fields, such as computer science and geo-informatics. Presently the school is awarding Ordinary National Diploma and Higher National Diploma.

By extension, the demand for professionals in statistics in the public and private sectors has enhanced the growth of the school. The FSS is located along with the University of Ibadan-Ajibode Sasa Road, Ibadan, Oyo State, south-western Nigeria. It is a school of high-level academic standards, which has enabled her to produce seasoned scholars and professionals. In the past and until recent times, graduates of the school seek advanced studies in various Nigerian universities and abroad. In the past, it was a center for training and examination for the Royal Statistics Society of the United Kingdom.

The students are better grounded in statistics because of the practical skills and knowledge acquired in the course of training. The school is presently under the management of the NBS, Nigeria. The school has two other campuses which are located at Kaduna and Enugu. However, the FSS Ibadan is the pioneer and center of attraction because of its contribution to the promotion of statistical studies in Nigeria. The school offers various statistical major courses at intermediate and higher levels.

Table 34.2 depicts that FSS is a source of improving statistical literacy in Nigeria. FSS is becoming an emerging significant middle level institution that trains NBS staff and private students to become a resourceful statistician.

TABLE 34.2

2019/2020 Federal School of Statistics Admitted Students

Department	Level	Male	Female	Total
Statistics	Ordinary National Diploma	32	29	61
	Higher National Diploma	18	13	31
Total		50	41	91

Source: Federal School of Statistics (FSS) 2019/2020 Handbook of Matriculation Ceremony.

34.4.1 The Role of International Organizations

The role of international organizations for the development of statistics must be emphasized. The collaboration provides an opportunity for support and assessment on the progress and the adequacies of statistical information being provided by the NSS in developing countries. It provides the opportunity to identify the capacity gaps that have to be bridged. Most developing countries look forward to and enjoy technical, financial, material and institutional supports from international organizations and advanced countries to chart a path toward statistical development.

According to the Main Statistical report of the Collaboration between the National Bureau of Statistics/Central Bank (2010), the major sources of funding for the research institutes are from the Federal and State Governments, World Bank, Department for International Development, European Union, United Nations Children's Fund and other international development partners.

International donors or agencies usually channel support through government specialized agencies by way of surveys or projects; therefore, they play a vital role in research, policy studies and development (Muluh et al., 2019). Olubusoye et al. (2015) confirmed that Nigeria is the greatest recipient of international financial aid as established by PRESS 2009. This argument is showcased in Tables 34.3 and 34.4.

TABLE 34.3

Countries Receiving Most Aid (Commitments)

Recipient Country	Amount (US$M)	Project/ Program Period
Nigeria	58.8	2004–2013
Mozambique	49.7	2002–2014
Kenya	25.6	2004–2010
Sudan	23.8	2004–2013
Tanzania	23.8	2000–2013
Ethiopia	21.0	2004–2011
Burkina Faso	21.0	2004–2013
Rwanda	18.2	2003–2012
Mali	18.2	2004–2012
Malawi	16.3	2000–2011

Source: Partner report on Support to Statistics (PRESS), pg. 12 PRESS 2009.

TABLE 34.4

Countries Receiving Most Aid (Disbursement)

Recipient Country	Amount (US$M)	Project/ Program Period
Nigeria	97.5	2004–2013
Kenya	32.8	2004–2009
Mozambique	32.3	2002–2014
Sudan	19.1	2004–2013
Ethiopia	17.7	2004–2011
Malawi	12.9	2000–2011
Tanzania	11.6	2000–2013
Mali	10.8	2004–2012
Rwanda	10.7	2003–2012
Burkina Faso	10.6	2004–2013

Source: Partner report on Support to Statistics (PRESS), pg. 12 PRESS 2009.

34.5 The Donor Agencies and Nigeria Development

One of the key donor agencies to Nigeria is United States Agency for International Development (USAID), and in its 2015–2017 Integrated Country Strategy, the US Government Mission in Nigeria identified four priority goals – improving governance, furthering economic development, enhancing stability and expanding opportunity. In support of that vision, the USAID Mission has established its Country Development Cooperation Strategy goal as Reduced Extreme Poverty in a More Stable, Democratic Nigeria. The USAID goal will be pursued through three Development Objectives (DOs):

- **DO 1**: Broadened and inclusive growth
- **DO 2**: A healthier, more educated population in targeted states
- **DO 3**: Strengthened good governance

The second objective is focused on literacy, and therefore, most donor agencies always put developing countries' literacy in mind.

Muluh et al. (2019) identify the following as some key objectives for donor agencies.

- Improvement of statistical research through better harmonization and to avoid duplication of tasks.
- Use of modern and professional implementation mechanisms before, during and after the lifespan of a statistical survey.
- Statistical support through manpower training and input supply.
- Provide an opportunity for policy engagement with relevant government ministries and agencies, and the statistical department as the steering agency.

This subsection examines the role and nature of foreign statistical support and development in Nigeria as a case study of some developing countries; this includes the influence

TABLE 34.5

Donors and Usage of NBS Data

Donor	Data Set
World Bank	Education data, health data, agricultural data, poverty data
UNIDO	Industrial account for publication in the statistical year book
EU	Change management, information technology (IT) and statistical methods
USAID	CPI-inflation, exchange data, national accounts, poverty data
UNHABITAT	Household survey, economic survey data, demographic and environmental data
UNIFEM	Poverty profile
ILO	Statistics on employment, unemployment, underemployment, statistics on child labor, labor market
UNDP	Poverty profile, socio-economic data on Nigeria, national living standard surveys
FAO	Data on agriculture and rural development

Source: National Bureau of Statistics.

and importance of the database for project development and effective national planning, and the post statistical support and numerous barriers faced by the statistical agencies and the governments of the various countries. It will evaluate the significance and impact of foreign statistical support. The efforts will be directed at understanding the planning process, implementation and sustainability of the statistics system of Nigeria as a case study.

Statistical support comes in various forms and is meant to help establish a scientific foundation and assessment of DOs for the socio-economic condition of developing countries. These are given in the form of financial aid or grants, not payable, technical support in training or research partnership, and input for specialized surveys. The type of specialized funding for the improvement of statistical literacy is controlled by The International Statistical Literacy Project (ISLP), which is the only international program that aims to promote statistical literacy worldwide. The International Statistical Institute (ISI) created it in 1991 with the name "World Numeracy Project". Today, the ISLP works under the guidance of the International Association for Statistics Education, which is one of the sections of ISI (Klein et al., 2016, UN, 2012). There is a specific program that is funding the Laboratory for Interdisciplinary Statistical Analysis (LISA) 2020 project and that is the "Accelerating Local Potential" program within the USAID Development Lab. Many LISA laboratories benefitted from this funding. Table 34.5 shows various sources of support to enhance statistical research using various data sets.

34.5.1 The Challenges of Donor Agencies in Developing Countries

There are significant constraints to the capacity and will of various governments to develop and sustain statistics development. Switching and overloading of expenditure on a research project by the various government agencies is a common phenomenon. This may be attributed to poor material or financial support from the government. The implication is poor execution of surveys and unrealistic outcomes which may require a high level of unnecessary adaptation.

Lyson et al. (2001) emphasized that lack of sustainability remains a barrier. Sustainability connotes the magnitude of inheritances after the donor support, the ability of the government to take over supported projects after evaluation and continuation of survey and research upon phasing out of the donor support.

Olubusoye et al. (2015) observed that Nigerian statistical development is donor-driven and their programs are usually short term; these activities take precedence over long-term national priorities for statistical research.

Niyankuru (2016) explained how donors prefer to release their funds to address specific problems that are not host government priorities, and they further use their expertise to implement their objectives without consideration for the peculiarities of the host environment. Data collection and statistical reports require a systemic and consistent process over a period of time. Oftentimes, government financial and material supports are either inadequate or untimely for the timely outcome of the research.

34.6 Conclusion and Recommendation

The chapter appreciates the positive development in statistical literacy of Nigeria, as a case study of some developing countries, particularly the recent development, i.e., post-2007 Statistical Act. This includes the improvement and definition of the role of NSS, International Donor Agencies support and collaboration and the role of statistical institutions, such as FSS, Ibadan, Nigeria.

The school can be upgraded with international support to a specialized professional statistical institution, with applications from computer science, geo-informatics and academic research: a model for Africa, international collaboration and a regional center for statistics training and development. This recommendation will further boost the literacy and awareness of statistics in Nigeria and beyond. The school can be improved to a specialized intermediate and advanced statistical institute for professionals, academics and research, a permanent regional center for statistics training and development for Africa.

The improvement in statistical literacy has translated to a quantum-leap request for statistical works and services by data producers, users and academic research. However, perennial issues, such as "politicization" of statistics, poor findings and the cold attitude of the average citizen toward statistics need to be addressed by sensitization and implementation of research outcomes. A recommendation should be made by NBS to appropriate authorities on how regional planning, private sector support etc. affect the efficiency and reliability of data collection, as well as security monitoring. Further studies should be conducted on how far the legal framework of NSS had been implemented by the government, data producers, users and academics. The sustainability roles and recommendations of NSS are germane to effective national planning.

References

Ajakaiye, O. 2012. Enhancing data generation for national development in Nigeria: Institutional and Structural Issues. *CBN Journal of Applied Statistics*, 3(1), 139.

Ardanaz, M., Scartascini, C. and Tommasi, M. 2010, "Political institutions, policymaking, and economic policy in Latin America", Inter-American Development Bank Working paper No. IDB-WP-158, http://www20.iadb.org/intal/catalogo/PE/2010/04914.pdf.

Badiee, S, Jütting, J., Appel, D., Klein, T. and Swanson, E. 2017. The role of national statistical systems in the data revolution. *Development Co-operation Report* 2017 Data for Development © OECD 2017.

Bédécarrats, F., Cling J. and Roubaud F. 2016. The data revolution and statistical challenges in Africa: Introduction to the special report. *African Contemporary*, 258(2), 9–23. ISSN 0002-0478 ISBN 9782807390072.

Devarajan, S. 2013. Africa's statistical tragedy. *Review of Income and Wealth*, 59, S9–S15.

Dodge, Y. 2003. *The Oxford Dictionary of Statistical Terms*. OUP. ISBN 0-19-920613-9.

Doguwa, S.I. 2009. Statistics for national development. *Journal of Applied Statistics*, 1(1).

Europe, U. N. 2012. *Making Data Meaningful Part 4: A Guide to Improving Statistical Literacy*. Geneva: United Nations.

Gal, I. 2002. Adults' statistical literacy: Meanings, components, responsibilities. *International Statistical Review*, 70(1), 1–25.

Garfield, J. and Ben-Zvi, D. 2007. How students learn statistics revisited: A current review of research on teaching and learning statistics. *International Statistical Review*, 75(3), 372–396.

Hamadu, D., Okafor, R. and Oghojafor, B. 2012. Building the Nigerian statistical system capacity for poverty reduction and sustainable development in the new millennium. *Journal of Sociological Research*, 3(2). doi:10.5296/jsr.v3i2.2567; ISSN 1948-5468.

Jacobs, J. 1961. *The Death and Life of Great American Cities*. New York: Random House.

Kale, Y. 2013. Where are the Numbers? National Bureau of Statistics and the Reset of the Nigerian National Statistical System, NBS. Retrieved on 20/08 /2014 from http://mortenjerven.com/wp-content/uploads/2013/04/SG_NBS-and-the-reset-of-theNigerian-National-Statistical-System-Vancouver-April-2013-FN-small4.pdf.

Khan, S. 2013. Statistics in planning and development. *Pakistan Journal of Statistics*. https://www.researchgate.net/publication/292732616.

Klein, T., Galdin, A. and Mohamedou, E.L. 2016. "An indicator for statistical literacy based on national newspaper archives". In *Proceedings of the Roundtable Conference of the International Association of Statistics Education (IASE)*, July 2016, Berlin, Germany.

Lobo, J. 2015. Urban planning and community data collection efforts in the developing world: Data as a facilitator. *Social Sciences Research Council. Items Insights*.

Lyson, M., Smut, C. and Stephens, A. 2001. Participation, empowerment and sustainability: How do the link work? *Urban Studies*, 38, 1233–1251.

Muluh, G.N., Kimengsi, J.N. and Azibo, N.K. 2019. Challenges and prospects of sustaining donor funded projects in rural Cameroon. *Sustainability* 11, 6990. doi:10.3390/su11246990; www.mdpi.com/journal/sustainability.

Nigeria Statistical Act, 2007. Promulgated SG 57/25.06.1999, amended and supplemented SG 42/27.04.2001, amended SG 45/30.04.2002, amended SG 74/30.07.2002, amended SG 37/4.05.2004, effective 4.08.2004, SG No. 39/10.05.2005, effective 11.08.2005, amended and supplemented, SG No. 81/11.10.2005, supplemented, SG No. 88/4.11.2005, SG No. 100/30.11.2007, effective 20.12.2007, amended and supplemented, SG No. 98/14.11.2008, supplemented, SG No. 42/5.06.2009, amended, SG No. 95/1.12.2009, effective 1.01.2010, amended, SG No. 97/10.12.2010, effective 10.12.2010.

Niyankuru, F. 2016. Failure of foreign aid in developing countries- a quest for alternative. *Business and Economics Journal*, 7(3), 1–9.

Olubusoye, O.E., Korter, G.O. and Kesinro, O.A. 2015. Nigeria Statistical System- the evolution, progress and challenges. doi:10.13140/RG.2.1.3136.4569; https://www.researchgate.net/publication/283715250.

PARIS21, 2015. "A scoring system to measure the use of statistics in the policy-making process", PARIS21, Paris, www.paris21.org/sites/default/files/Scoring_System_Use_Of_Data_2015_DFID.doc.

PARIS21, 2018. Proposing a Use of Statistics indicator for National Development Plans

Partner Report on Support to Statistics (PRESS), PARIS21 – PRESS, 2009. Nigerian Statistical Master Plan 2004/5–2008/9. Retrieved on 29/08/2014 from https://unstats.un.org/unsd/dnss/docViewer.aspx%3FdocID%3D2287.

Russell, M. and Muñoz-Ayala, J. 2015. Un estudio exploratorio para medir el uso de estadísticas en el diseño de política pública (Working Paper No. IDB-BP-374) Washington, D.C.: BID.

Scott, C. 2005. "Measuring up to the measurement problem: The role of statistics in evidence based policy-making", PARIS21, Paris. www.paris21.org/node/672.

USAID/Nigeria. 2013. Mid-term Performance Evaluation of the Leadership, Empowerment, Advocacy and Development (LEAD) Project, December 2013.

USAID/NIGERIA. Country Development Cooperation Strategy 2015–2020.

Wallman, K.K. 1993. Enhancing statistical literacy: Enriching our society. *Journal of the American Statistical Association*, 88(421), 1–8.

35

Role of Statistics in Policymaking for National Development

Oluwadare O. Ojo
Federal University of Technology

Tolulope T. Osinubi
Obafemi Awolowo University

Serifat A. Folorunso
University of Ibadan Laboratory for Interdisciplinary Statistical Analysis (UI-LISA)

CONTENTS

35.1 Introduction

Statistics, which entails collection, organization, presentation, analysis, and interpretation of data, has a critical role to play in making any economic decision. Put differently, most economic problems can be solved through the help of statistics as it involves the provision of empirical data used in economic research, whether descriptive or econometric. In addition, statistics aids the formation of theories and models by providing evidence in economics. It can be summarized that economic facts can be presented in a precise and definite form through statistics. However, if care is not taken, lack of statistical literacy, that is, inability to understand and reason with statistics and data, can adversely affect the conclusions and policy recommendations to be drawn from the data. Consequent to this, the decisions based on these conclusions and policy recommendations would be incorrect and might not reflect the true picture of any economy relying on these data.

According to Shangodoyin and Lasisi (2011), "statistics is an indispensable tool for national development, growth, and planning". In examining the role of statistics in economic development, studies (such as Shangodoyin & Lasisi, 2011) show that economic

DOI: 10.1201/9781003261148-42

development relies on adequate statistical information for implementing a formidable evidence-based policy. This implies that formulation of policies, either by the government or private individuals, in ensuring economic development depends largely not only on data but also on the ability to understand and reason with the data. This is because measures such as income per capita (gross domestic product), literacy rate (human capital), and life expectancy at birth can be used to verify if a country is developing or not. These measures are therefore obtained through the help of statistics.

Furthermore, there is a need to understand the workings of an economy through data in achieving the macroeconomic objectives of high and sustainable economic growth; low rate of inflation and unemployment; equilibrium balance of payment; low government borrowing; stable exchange rate; low level of inequality; and environmental protection. Nevertheless, the lack of statistical literacy on most macroeconomic variables might affect the achievement of these objectives in most developing countries, including Nigeria. This hinges on the fact that statistical literacy helps to make sense of statistics, that is, to reason and understand statistics. To support this fact, most economic models are built based on the available statistical information, and these models help in monitoring the performance of the economy and the social well-being of the people (Shangodoyin & Lasisi, 2011). This, then, means that the misuse of statistics will negatively affect the performance of the economy through policy recommendations emanated from the wrong use of data.

The preceding description reveals that the role of statistics cannot be underestimated in making policies for national development. For instance, it aids the development of National Statistical Systems (NSS) to efficiently produce good statistics. Good statistics is important to assess and identify issues, support the choice of interventions, forecast the future, monitor progress, and evaluate the results and impacts of policies and programmes (Sanga, 2014).

The present work contributes to the existing knowledge by examining the roles of statistics in measuring socioeconomic indicators, especially in the area of policymaking at both national and international levels. To achieve this objective, the benefits of statistical literacy will be looked into through some concepts in different human endeavours like finance, education, and business.

The study is organized as follows. Following this section is Section 35.2 which presents the literature review. Section 35.3 provides the benefits of statistical literacy with some facts. Section 35.4 concludes the chapter.

35.2 Literature Review Conceptual Review

There is a need to look at the concept of statistical literacy together with its importance under this section. This is because a lack of statistical literacy, according to Wallman (1993), can bring about "misunderstandings, misperceptions, mistrust, and misgivings about the values of statistics for guidance in public and private choices". In addition, Shaughnessy and Pfannkuch (2004), Shaughnessy (2007), and Makar and Rubin (2009) establish that statistical literacy with respect to understanding statistical information and techniques is very crucial for day-to-day activities and active involvement in the workplace.

The international community is not left out in the call for statistical literacy. For instance, the United Nation's Secretary-General, in the Post-2015 Agenda in 2014, comments that their analysis should be based on credible data and evidence, and they should enhance data capacity, availability, disaggregation, literacy, and sharing (United Nations, 2014).

However, different authors define statistical literacy differently. This means that there is no consensus regarding its definition (Batanero, 2002). To start with, statistical literacy, in simple terms, is the ability to understand and reason with data. Looking at a broader definition of statistical literacy, Wallman (1993) conceptualizes it as "the ability to understand and critically evaluate statistical results that permeate our daily lives—coupled with the ability to appreciate the contributions that statistical thinking can make in public and private, professional and personal decisions" (see also Trewin, 2005). From this definition, the need for statistical literacy in people's day-to-day activities and the society at large cannot be downplayed. Also, statistical literacy is not just about numbers and formulas, but it typically deals with words and evidence (Milo, 2004; Watson & Kelly, 2003).

There are five knowledge elements of statistical literacy (Gal, 2002). These are "literacy skills", "statistical knowledge", "mathematical knowledge", "context/world knowledge", and "critical skills". Literacy skills comprise the ability to understand statistics in terms of written and oral text. In other words, having literacy skills means the ability to comprehend and derive meaning from statistics. Statistical knowledge is knowledge about "basic statistical and probabilistic concepts and procedures, and related mathematical concepts and issues". Having mathematical knowledge means knowing some of the "mathematical procedures underlying the production of common statistical indicators, such as percentage or mean". In terms of context/world knowledge, anyone is said to be statistically literate if he or she can interpret the results obtained using statistics. Lastly, critical skills represent the way people make use of the conclusion drawn from the statistical results. In order words, these skills represent the policy recommendations given to the policymakers. In sum, these knowledge elements of statistical literacy are overlapping, implying that they cannot work separately.

35.2.1 Empirical Review

Even though statistics is applicable to all disciplines such as economics, medicine, psychology, law, sociology, and political science, there is a dearth of studies on the role played by statistics in national development. There is a claim (Kelegama, 2016; Khan, 2013; Shangodoyin & Lasisi, 2011; Kegame, 2007) that statistics is now a means of achieving national and international development. To achieve an enabling environment, there is a need not just for statistics but also for statistical literacy. In other words, with the aid of statistics, the government would be able to develop appropriate policies that can lead to economic development. Also, with statistics, the government would be able to monitor the policies' progress and make evidence-based decisions about the allocation and management of scarce resources. Apart from doing this, the performance of the government can be measured and monitored using statistics.

In furtherance, previous studies have shown that statistics have indeed helped in making appropriate policies for the development of a nation. Kelegama (2016) states that "World Bank sees statistics as the evidence on which policies are built and without statistics the development progress is blind". Also, Kagame in his speech in 2007 acknowledges the need for statistics as a basis for effective policymaking in Africa. He asserts that "evidence-based policymaking as a means of policies and programmes intend to improve lives based on clearly defined, time-bound, and measurable milestone". This can be achieved through the help of statistics according to him. Khan (2013), in his study "statistics in planning and development", argues that statistics is relevant in various sectors of the economy by setting developmental goals, assessing the progress of the goals, monitoring programmes, and undertaking follow-up initiatives. However, statistics could result in wrong decisions if the users abuse it or are ignorant of its importance.

Using Botswana and Nigeria Statistical Systems, Shangodoyin and Lasisi (2011) reveal that there is a need for the empowerment of the NSS before any substantial national development can occur. From the authors' point of view, statistics is very crucial for the development process because it helps in the implementation, execution, and monitoring process of the plans towards achieving national development. In China, Bourguignon (2005) observes that the statistical system in the country spurs a scientific approach to development work by providing information to policymakers, academics, policy analysts, the general public, and the growing outside world. In India, Mahalanobis (1965) sees statistics or operational research with the help of statistics as a key element for data collection over a specific time horizon on the required plans needed to achieve economic development. Apart from gathering information on these plans, statistics also helps in assessing the plans' progress and monitoring the plans' implementation through execution for the continuation of the plans or introduction of drastic changes in the plan. Looking at the role of statistics on education, which is one of the indices of national development, Makwati et al. (2003) claim that statistics improves the quality of education in sub-Saharan Africa. According to the authors, statistics aids quality assurance in education with respect to "educational planning, policy formulation, management, monitoring, and evaluation of the education systems". Similarly, statistics negatively influences poverty as one of the measures of development. The Organisation for Economic Co-operation and Development (OECD, 2007) examines "the role of statistics in world development by counting down poverty" and finds out that statistics indeed reduces poverty and helps in world development. This is premised on the fact that statistics provide information on the state of people's daily well-being. For instance, it gives information on the location of the poor, the reason for being poor, and their present situation. According to OECD, this information would help in developing and monitoring effective development plans.

Health is another aspect of national development, and Rice (1977) studies the role of statistics in the development of healthcare policy. According to the author, the expectation is that health statisticians and health data are to enhance rational decision-making. Thus, Rice (1977) concludes that statistics and statisticians play a continuous role in developing healthcare policies that would help in improving national development.

Conclusively, the role of statistics in policymaking for national development is actually very important as argued by extant studies. Specifically, studies like Makwati et al. (2003), OECD (2007), and Rice (1977) use education, poverty, and health as measures of development and they find that statistics leads to policies that would help in improving education and healthcare and alleviating poverty. This shows national development is tied to statistics. However, these extant studies fail to account for the importance of statistical literacy in examining the role of statistics in national development. This study intends to fill the gap by taking into consideration the position of statistical literacy in statistics to make policies that would call for national development.

35.3 Benefits of Statistical Literacy in Society

In this section, the contributions of statistical literacy using three concepts will be examined in some human endeavours which include finance, education, and businesses. This is because statistics is crucial in all fields. Those three concepts are

- Statistics and planning,
- Statistics and process monitoring, and
- Statistics and decision-making.

35.3.1 Finance

Statistics and planning: The role of planning in the area of finance is a key aspect of any successful nation. There must be a good development scheme for planning. No modern government can deliver its services without a developmental agenda for the planning of finance. Statistical methods are capable of determining indicators that are necessary to plan financial commitments. For instance, Figure 35.1 gives a pie chart on how the United States wants to spend on the Washington State Department of Health 2017–2019 biennial operating budgets. These funds include the general fund state, dedicated funds, free accounts, and federal. This is just a demonstration of how the use of statistics simplifies communication of budget commitments.

Statistics and process monitoring: Statistics helps to monitor the state of the economy of a country from time to time. Process monitoring facilitates the evaluation of ongoing economic reforms via statistical data. There is a need to know if there is a recession or if the economy is booming. Recently, the Nigerian government had to review its plans on budget by cutting down 20% of the capital expenditure due to the Covid-19 pandemic. This was made possible as a result of reliable statistical data through process monitoring.

Statistics and decision-making: Decision-making is critical for any government of a nation in terms of finance. This involves how to distribute the wealth of a nation especially in terms of budget. Statistics helps the government to make critical decisions on how the budget should be designed. With good statistics, the government of any nation will be able to learn from its mistakes. Figure 35.2 gives a pie chart showing ten ministries with the highest budget allocation in Nigeria for the year 2017.

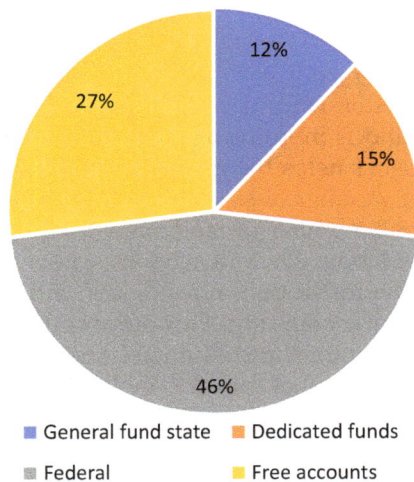

FIGURE 35.1

Department of Health 2017–2019 Biennial operating budget by fund $1.2 billion. (Washington State Department of Health website (www.doh.wa.gov).)

10 MINISTRIES WITH HIGHEST ALLOCATION IN 2017 BUDGET

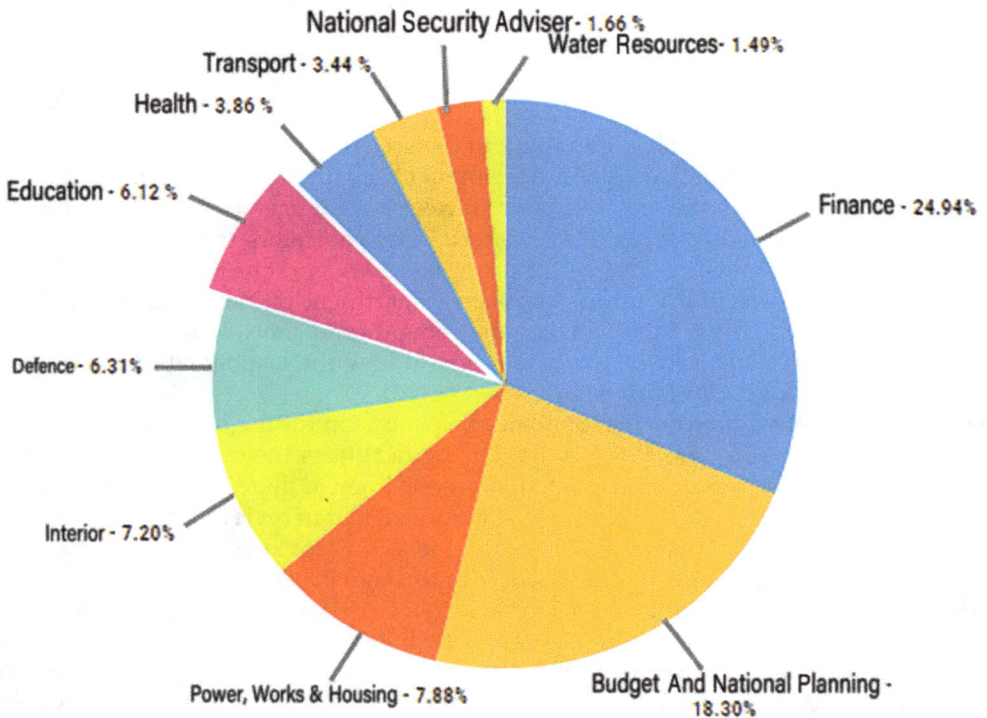

National Security Adviser - 1.66 %

Water Resources - 1.49%

Transport - 3.44 %

Health - 3.86 %

Education - 6.12 %

Finance - 24.94%

Defence - 6.31%

Interior - 7.20%

Power, Works & Housing - 7.88%

Budget And National Planning - 18.30%

FIGURE 35.2

A pie chart showing budget allocation to some ministries in Nigeria for 2017. (National Bureau of Statistics [NBS].)

35.3.2 Education

Statistics and planning: Education is fundamental in any society (Gunar, 1965). The education system is improving and is more recognized than in the past for both social and economic development. Statistics helps the government to know the number of pupils in elementary/primary schools, number of students in high/secondary schools, number of students in higher institutions, number of students who drop out of school, number of graduates in universities, the labour force in schools, school feeding programme, school building equipment, etc. Higher institutions nowadays in different countries have an office exclusively dedicated to academic planning. This office collects and analyses information to generate reports like annual budget and estimates and also sends periodic statistics to different authorities of education in that country. Tables 35.1 and 35.2 show the budgetary allocation to education in Nigeria and how beginners can plan and manage a school budget in the United States. This shows the use of statistics to simplifying communication of planning in education.

Statistics and process monitoring: Statistics is a veritable tool for process monitoring for qualitative education in any nation. If there are timely, accurate, and reliable statistics on education, it will help the government to sustain the already planned goals at each level.

TABLE 35.1

FG Budgetary Allocation to Education (2009–2018)

Year	Budget (#Trillion)	Education Allocation (#Billion)	% of Budget
2009	3.049	221.19	7.25
2010	5.160	249.09	4.83
2011	4.972	306.3	6.16
2012	4.877	400.15	8.20
2013	4.987	426.53	8.55
2014	4.962	493	9.94
2015	5.068	392.2	7.74
2016	6.061	369.6	6.10
2017	7.444	550	7.38
2018	8.612	605.8	7.03

Source: National Bureau of Statistics website.

TABLE 35.2

Statistics for Beginners on How to Plan and Manage a School Budget in the United States

Budget Heading	Total Budget	Projected to Date	Projected to Date (%)	Actual to Date	Actual to Date (%)	Variance
Teaching staff	1,795,086	1,795,086	100	1,795,086	100	0
Supply teachers	50,350	16,783	33.3	24,360	48.4	7,576.67
Education support staff	800,00	800,000	100	800,000	100	0
Technicians	352,640	352,640	100	352,640	100	0
Admin and clerical staff	626,430	626,430	100	626,430	100	0
Caretakers	51,066	51,066	100	51,066	100	0
Building refurbishment	320,919	256,735	80	230,220	71.7	26,515.20
Maintenance	104,640	34,880	33.3	40,000	38.2	5,120
Energy	87,609	24,000	27.4	26,600	30.4	2,600
Catering	69,482	69,482	100	69,482	100	0
Classroom supplies	94,000	37,600	40	37,600	40	0
Other resources	220,000	110,000	50	110,000	50	0
Total	**4,609,444**	**4,211,924.53**		**4,200,706**		**11,218.53**

Source: www.theguardian.com/teachers-network.

Statistics and decision-making: Statistics of teachers and pupils with educational testing data helps to measure the efficiency of the school itself. For example, teachers can know the number of children who should be promoted to different classes, especially in primary/ elementary and secondary/high schools, with the use of averages and ranking. Statistical information on education guides the government to decide on the number of new schools needed across the country, and the number of buildings to be constructed, among other things.

35.3.3 Business

Statistics and planning: Operating a business is a complex thing. Any business requires long-term planning through the use of statistical data to know which products or services to venture into. Statistics provides managers with the confidence to deal with uncertainties

FIGURE 35.3
A pie chart showing accounts receivable turnover for a company. (www.storytelling.com.)

(Williams, 2019). A businessman has to draft a business plan and financial documents; this requires data and statistics to form and back the financial statements. If guesswork and incorrect estimates are utilized, by bringing out fictitious figures, the business plan will be subjective and weak.

Statistics and process monitoring: According to Suez (2018), "in any modern business, almost everything is measurable. Marketing is measurable; it can be moved to the internet. Sales and internal processes in the company are also measurable". With the use of big data analytics, large and varied data sets, or big data, it is possible to uncover and monitor information such as hidden patterns, unknown correlations, market trends, and customer preferences. Therefore, new technology should be introduced for timely and accurate data. Figure 35.3 gives a pie chart showing account receivable of a company for process monitoring.

Statistics and decision-making: Statistics is essential to decision-making in any business. Management of companies has to make day-to-day decisions to know if the goals and objectives of the business are met. Statistics helps to measure effectiveness, performance, and customer satisfaction in order to make necessary decisions; these can be monitored through the net promoter score sheet, average product star ratings, or percent of sales that are repeated. Another important aspect of decision-making in businesses is that statistics helps to measure and control production processes in order to minimize variations which may lead to error or waste, and ensures consistency throughout the process.

35.4 Conclusion

This article examined the roles of statistics in measuring socioeconomic indicators in the area of policymaking, at both national and international levels. The benefits of statistical literacy have also been examined through some concepts in different human endeavours in order to appreciate the roles of statistics in policymaking for national development. These concepts are statistics and planning, statistics and process monitoring, and statistics and decision-making, while the human endeavours considered were finance, education, and businesses.

Planning is important for the development of any nation. This is made possible by accurate and reliable statistics. Statistics is widely used for the design, implementation, and monitoring of already designed plans and targets by individuals and governments for national development. It was also observed that the application of various techniques in statistics makes the raw data become meaningful and also helps in decision-making.

Statistics will help policymakers to learn from their mistakes, especially in the preparation of budgets in the area of finance. That is why there is a need to have legislative laws that will back up the integrity of data in different countries.

Businesses are developing through the application of statistical methods to raw data sets in order to understand business performance. Statistical methods have also been discovered to be of help in the analysis of big data with the use of big data analytics to monitor information. It is important that new technology should be developed for timely and accurate data.

Finally, statistics is now recognized all over the world as a necessary tool for development. That is why there is a need to improve statistical literacy for national development. There must be a regulation on the use of statistics by the government, and development of more statistical agencies at various levels of the government and creation of statistics laboratories at universities should be encouraged.

References

Batanero, C. (2002). Discussion: The role of models in understanding and improving statistical literacy. *International Statistical Review, 70*, 37–40.

Bourguignon, F. (2005). The role of statistics in the scientific approach to development. Being a Presentation at the National Bureau of Statistics of China on 18th May, 2005. Retrieved from www.stats.gov.cn on 24th May, 2020.

Gal, I. (2002). Adults' statistical literacy: meanings, components, responsibilities. *International Statistical Review, 70*(1), 1–51.

Gunar, B. K. (1965). Statistics needed for educational planning. Statistical commission and economic commission for Europe conference of European Statisticians. UNESCO SS/6/72/WP 3. Conf. Eur. Stats/WG.23/7.

Kelegama, S. (2016). Role of statistics for the economic and social development of a country. A Keynote Address (edited version) delivered at the Academic Sessions and the *5th Annual General Meeting of the Institute of Applied Statistics Sri Lanka* at the OPA Auditorium, December 20, 2016. Retrieved from http://www.ft.lk/article/587967/Role-of-statistics-for-the-economic-and-social-development-of-a-country on May 24, 2020.

Kegame, P. (2007). The importance of statistics as a basis for effective policy-making in Africa. *The African Statistical Journal, 4*, 167–171.

Khan, S. (2013). Statistics in planning and development. *Pakistan Journal of Statistics, 29*(4), 513–524.

Mahalanobis, P. C. (1965). Statistics for economic development. *The Indian Journal of Statistics, 27*(1/2), 179–188.

Makar, K., & Rubin, A. (2009). A framework for thinking about informal statistical inference. *Statistics Education Research Journal, 8*(1), 82–105.

Makwati, G., Audinos, B., & Lairez, T. (2003). The role of statistics in improving the quality of basic education in Sub-Sahara Africa. Working Document, Doc.2.C, from the Association for the Development of Education in Africa, ADEA, Biennial Meeting held in Grand Baie, Mauritius on December 3–6, 2003.

Milo, S. (2004). Information literacy, statistical literacy and data literacy. *International Association for Social Science Information Service and Technology*, 28(2), 7–14.

Nwagbara, C. (2020). FG to reduce N1.5 trillion from 2020 budget due to coronavirus. Retrieved from www.nairametrics.com/2020/03/19/fg-to-reduce-n1-5-trillion-from-2020-budget-due-to-coronavirus/ on June 13, 2020.

OECD (2007). Counting down poverty. The role of statistics in world development. PARIS21-OECD/DCD, 2 rue André Pascal-75775 Paris Cedex 16.

Rice, D. P. (1977). The role of statistics in the development of health care policy. *The American Statistics*, 31(3), 101–106.

Sanga, D. (2014). Role of statistics: Developing country perspective. In: Lovric M. (eds) International Encyclopedia of statistical science. Berlin, Heidelberg: Springer.

Shangodoyin, D. K. & Lasisi, T. A. (2011). The role of statistics in national development with reference to Botswana and Nigeria Statistical Systems. *Journal of Sustainable Development*, 4(3), 131–135.

Shaughnessy, J. M. (2007). *Research on statistics learning and reasoning*. In F. K. Lester Jr. (Ed.), *Second Handbook of Research on Mathematics Teaching and Learning* (pp. 957–1009). Greenwich, CT: Information Age and NCTM.

Shaughnessy, J. M. & Pfannkuch, M. (2004). How faithful is old faithful? Statistical thinking: A story of variation and prediction. *The Mathematics Teacher*, 95(4), 252–259.

Trewin, D. (2005). Improving statistical literacy: The respective roles of schools and the National Statistical Offices. In M. Coupland, J. Anderson & T. Spencer (Eds.), *Twentieth Biennial Conference of the Australian Association of Mathematics Teachers* (pp. 11–19). Adelaide: AAMT.

United Nations (2014). The road to dignity by 2030: ending poverty, transforming all lives and protecting the planet. A Synthesis Report of the Secretary-General on the Post-2015 Agenda, New York.

Wallman, K. K. (1993). Enhancing statistical literacy: enriching our society. *Journal of the American Statistical Association*, 88(421), 1–8.

Watson, J. & Kelly, B. (2003). Inference from a pictograph: statistical literacy in action. In: Bragg, Leicha et al. (Eds.), MERGA 2003: mathematics education research: innovation, networking, opportunity. *Proceedings of the Annual Conference of the Mathematics Education Research Group of Australasia*, 26, Geelong, 6th–10th July 2003. Geelong: Deakin University, 2003. pp. 720–727.

Williams, J. T. 2019. The importance of Statistics in management decision making. Retrieved from www.smallbusiness.chron.com on June 12, 2020.

Some Online Resources

www.doh.wa.gov
www.nigerianstat.gov.ng
www.openresource.sue
www.storytelling.com
www.theguardian.com/teachers-network
www.nairametrics.com/ 2020/03/19/fg-to-reduce-n1-5-trillion -from-2020-budget

36

Building Biostatistics Capacity in Developing Countries

O. Olawale Awe
University of Campinas
Global Humanistic University

Adeniyi Francis Fagbamigbe
University of Ibadan

Egwim Evans
Federal University of Technology

Abdulhakeem Abayomi Olorukooba
Ahmadu Bello University

Oluwafunmilola Deborah Awe
University of Nairobi

Mumini Idowu Adarabioyo
Afe Babalola University

Olumide Charles Ayeni
Anchor University

Bayowa Teniola Babalola
Kampala International University

CONTENTS

DOI: 10.1201/9781003261148-43

An essential component of any development planning is data. Without data, a country's efforts to plan for future growth and welfare of its people cannot be grounded in reality and therefore may be severely flawed.

—*Hon. Prof. Peter Anyang' Nyong'O.*
(Minister for Planning and National Development, Kenya)

36.1 Introduction

Biostatistics, otherwise called medical statistics in certain climes, is a branch of statistics that focuses mainly on the development of statistical procedures and methodologies and their application in medicine, health sciences, epidemiology, public health, forensic medicine, and clinical research (Henley et al. 2019). Its application has been extended to other fields such as agriculture and biomedical engineering (Albert 2004). Biostatistics is a field of science concerned with the planning, collection, organization, summarization, and analysis of data and with drawing inferences about a body of data when only a part of the data is observed in health-related disciplines (Elston and Johnson 2008).

Biostatistics has become a cornerstone of medical-related and health sciences, as it has remained a critical decision-making tool. It has contributed substantially to the management of medical uncertainties (Ahmad et al. 2019). Statistical reasoning provides the theoretical basis for extracting knowledge from data in the presence of variability and uncertainty. It encompasses the design of biological experiments and the interpretation of the results (Carlin et al. 2020). Biostatistics plays a major role in shaping and providing scientific information for vital decision-making in health planning, clinical and public health research, and beyond. Modern decision-making, be it for an individual patient, hospitals, clusters of people or facilities, countries, and regions, is mostly based on Biostatistical methods to improve the timeliness, accuracy, quality, and generalizability of information and decisions (Begg and Vaughan 2011).

Biostatisticians help to transform the complex mathematical findings of clinical trials and research- related data into valuable information. This information is used to draw data-driven conclusions, make public health decisions, and deliver effective evidence-based healthcare. Biostatistics is used mostly in the life sciences (Li et al. 2014, Keiding 2005).

The explosion of scientific research over the past decades, however, has led to the misuse and abuse of statistical methods. Biostatistics, a blend of mathematics, biology, logic, and judgment, has significantly contributed to the development of the biomedical sciences by providing the quantitative and accurate dimension that the development of biomedical sciences requires (Brimacombe 2014).

36.2 Why Is the Discipline of Biostatistics Important?

Biostatistics is a subdiscipline of statistics that has steadily grown to be a scientific discipline of its own and not merely a tool for biological or clinical research (Zaugg 2003). The role of biostatistics is better understood through the scope of the discipline. Biostatistics traverses study design, sample size determination, sampling, data mining, data simulation, data collection, data abstraction, data curation, data management, data analysis, data interpretation, reporting, and dissemination (Hian and Tan 2005). The discipline is central to the conduct of health sciences research which is rampant in every clime. Biostatistical reasoning is required to provide the theoretical basis for extracting knowledge from any available data in the presence of such variability and uncertainty. Biostatistics is a critical element of most empirical research in public health and clinical medicine, with the integration of biostatistical inputs on every aspect of health-related studies from design to data management as well as dissemination of findings. It has also been found useful in the field of Economics (Awe et al. 2021). The use of biostatistical methods reinforces vital public health research disciplines, such as epidemiology and health services research, a role that reflects the core nature of the discipline of biostatistics. Similarly, bioinformatics and computational biology are important new areas in data-intensive biomedical research that are fortified by statistical concepts and methods, along with aspects that are deeply informed by other core disciplines such as computer science and mathematics (Abdurakhmonov 2016).

All this stresses the importance of biostatistics as a discipline. Sound biostatistical work requires not only an understanding of mathematics, probability, and sources of bias, which underpins statistical theory and methods, but also (and increasingly) extensive technical skills, including computing skills (Pollard et al. 2019, Rencher and Christensen 2012).

Despite the central role of biostatistics in the planning, implementation, and validation of studies in the health sciences, it was regarded as an ancillary service rather than as an academic discipline for many years (Zaph et al. 2020). However, the tides changed a few decades ago in Europe and the United States, where many universities have developed curriculums and established master's degrees in medical statistics or biostatistics (Glantz 1989, DeMets et al. 2006). Some universities in Africa have also been running postgraduate training in biostatistics, while nearly a score of universities in the United Kingdom run master's degrees in biostatistics/medical statistics, fewer African universities run such a program, and only one university in Nigeria does (Brimacome 2014, Altman 1994, Altman et al. 2002).

36.3 Biostatistics Training

Virtually all university programs in biostatistics are taught at postgraduate levels (Kilic and Celik 2013). Biostatistics curriculums are designed to build on knowledge gained during undergraduate training. Only a few introductory biostatistics courses are taught to life science students at the undergraduate levels to introduce the basic biostatistical concepts to the students (Gezmu et al. 2011). Depending on the courses available in different universities, biostatistics postgraduate programs are usually domiciled in the Schools of Public Health, and at times in veterinary sciences, pharmaceutical sciences, forestry, agriculture, and others (DeMets et al. 2006).

However, some universities have dedicated biostatistics departments (like many universities in the United States), while many others just integrate biostatistics faculty into statistics or other departments, such as epidemiology. In the latter situation, most departments bearing the name "biostatistics" usually exist under many different entities. For instance, relatively new biostatistics departments have been founded with a focus on bioinformatics and computational biology, whereas older departments, typically affiliated with schools of public health, will have more traditional lines of research involving epidemiological studies and clinical trials as well as bioinformatics. Some large universities around the world have separate statistics and biostatistics departments, with little or no collaboration between these departments, whereas in other universities, there is a very high degree of partnership. Generally, statistics programs and biostatistics programs differ in the sense that statistics departments usually host theoretical/methodological research that is not the focus of biostatistics programs, and besides biomedical applications, statistics departments have lines of research in industry, production engineering, business, economics, and biological areas other than medicine. Biostatistical training has advanced in recent years with masters and doctoral degrees awarded in biostatistics. It is worth mentioning that students with an undergraduate background in numerate disciplines may find postgraduate training in biostatistical easier than other students (Macnab 2003).

36.4 Biostatistical Methods Used in Life Sciences

The subject and application of biostatistics started in the mid-17th century with the analysis of vital statistics. After the early developments in vital statistics, the field of genetics also benefitted most from the new statistical ideas emerging from the works of Charles Darwin (1809–1882), Francis Galton (1822–1910), Karl Pearson (1857–1936), and Ronald A. Fisher (1890–1962). Currently, biostatistics has several fields of application and areas of concern, including bioassay, demography, epidemiology, clinical trials, human population surveys, community diagnosis, and biomathematical modeling, among many others. Results and conclusions of good research must be well presented, and a good presentation is as much a part of research as the painstaking collection and analysis of data (Awe 2012).

The quality of research significantly depends, among many other aspects, on valid statistical planning of the study, analysis of the data, and reporting of the results, which is usually guaranteed by a biostatistician (Lash 2017). Essentially, any research study in the life sciences, medical sciences, and public health uses biostatistics from beginning to end, because it provides empirical evidence and a language for communicating the result of scientific research (Keiding 2005).

Statistical reasoning provides the theoretical basis for extracting knowledge from data in the presence of variability and uncertainty. It is a critical element of most empirical research in public health and clinical medicine, with the best studies incorporating biostatistical inputs on aspects from study design to data analysis to reporting. Biostatistical methods underpin key public health research disciplines, such as epidemiology and health services research, a role that reflects the core nature of the discipline of biostatistics (Begg and Vaughan 2011).

The principles of statistics that are used in biostatistics and research methodology serve to provide orderly and objective approaches to collecting and interpreting research data. In nearly all areas of research, the proper use of statistics is used to reach valid conclusions.

Biostatistics focuses on the development and use of statistical methods to solve problems and to answer questions that arise in human biology and medicine. Thus, it expands statistical theory and adapts it to bring specific methods to bear on questions of importance to the community of scientists, practitioners, and policymakers who have an interest in health and all health aspects of the human community (Begg and Vaughan 2011).

As described in the preceding sections, the use of biostatistics starts from study design to dissemination but in practice does not end, as biostatistical skills are needed to interpret the findings, which could be used as a source of data for another study. Therefore, biostatistics and biostatisticians must be consulted and engaged as soon as a study is conceived and the goals established.

Biostatistics methods are of two major categories: descriptive and inferential statistics.

Descriptive statistics are used to collect, organize, summarize, and graphically present the sample for better understanding. Descriptive statistics is subdivided into measures of central tendency and measures of variability.

Inferential statistics are used to draw valid conclusions about a population on the basic information obtained from a representative sample of the population. Most of the tools used in inferential statistics hail from a general family of statistical models known as the general linear model. These include the *t*-test, regression analysis, analysis of variance, analysis of covariance, and many of the multivariate methods like factor analysis, cluster analysis, and discriminant function analysis.

Besides these two major arms of biostatistics, there are other methods embedded in the discipline.

Study design: The definition of a research problem provides the headway to the type of investigation to be carried out. A well-defined research problem points to the method to be adopted for the investigation. There are different study techniques, known as study designs, that could be used for different studies. The choice of a particular study design is dependent on the study objectives and the amount of information known. Study designs range from simple, fast, and inexpensive to complex, expensive, and time-consuming.

However, the selection of a technique involves a series of trade-offs between cost, time, resources, and the quality of information obtained. These factors must be fully explored and considered when choosing a study design.

The commonest study designs are surveys, observation, experiment, and quasi-experiment. Each method has its advantages and disadvantages. However, health studies can be described under different concepts: the purpose of the study (exploratory/formulative, descriptive, or explanatory studies), uses of the study (basic and applied study), time dimension (cross-sectional and longitudinal studies), data collection techniques (quantitative and qualitative studies), and data collection procedures (routine or ad-hoc). Details of the types of study designs in biostatistics are shown in Figure 36.1.

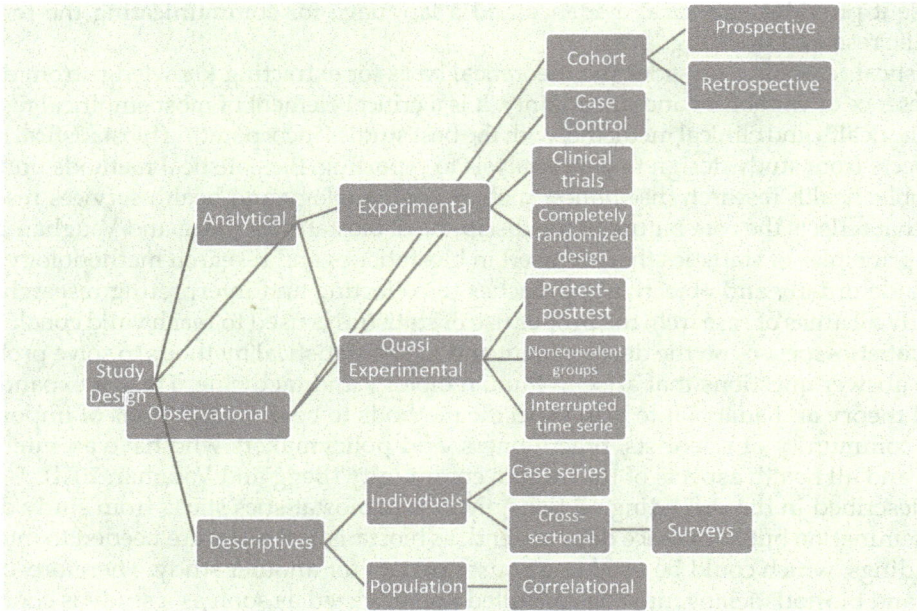

FIGURE 36.1
Types of study designs in biostatistics.

36.5 The Concept of Population and Sample

A population includes all people, subjects, or items with the characteristic of interest in a certain place and specific time. However, as it is often impracticable due to the constraints of time, money, personnel, and availability of subjects to gather information from everyone or everything in a population, the natural option is to find and use a representative sample of that population. Hence, a sample is a subset or a fraction of the population. A sample is a smaller representative collection of units from a population used to determine truths about that population. The population may be finite or infinite. A population is finite if the exact numbers of what constitute the population are known and it is infinite if a population consists of an endless number of individuals (Elston and Johnson 2008) (Figure 36.2).

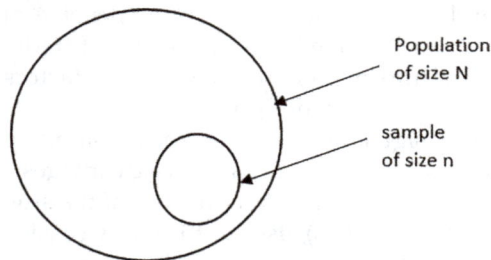

FIGURE 36.2
Population and sample.

Sampling is a statistical procedure concerned with the selection of a subset of individuals within a population of interest to estimate the characteristics of the entire population. Sampling can be further categorized into two types: probability and non-probability sampling. Probability sampling is a sampling technique in which every unit in the population has a chance (greater than zero) of being selected in the sample, and this probability can be accurately determined. A researcher chooses samples from a larger population using a method based on random selection. In this, every unit of the population has an equal chance of being included in the sample. Examples of probability sampling are simple random sampling, systematic random sampling, stratified random sampling, multistage random sampling, and cluster sampling. Non-probability sampling is a sampling technique in which the researcher selects samples based on his or her subjective judgment rather than random selection. As a result, not every sample has an equal chance of selection. Examples of non- probability sampling are judgmental, quota, convenience, snowball, and extensive sampling.

36.6 Sampling Process

The sampling process comprises several stages, and each stage must be addressed adequately:

- Define the population of concern, such as under-five children in Lagos, Nigeria
- Specify the sampling frame, list of all under-five children in Lagos, Nigeria
- Determine the sampling units: individuals, households, streets, etc.
- Specify a sampling method for selecting items or events from the frame
- Determine the sample size
- Implement the sampling plan
- Obtain the sample and collect the data
- Review the sampling process

Sample size determination is the process where the minimum sample size for carrying out a research is determined. For the findings of a research to be reliable, the minimum sample size must be met for the research to produce reliable results. Studies should be designed to include a sufficient number of participants to adequately address the research question. Researches that have either an insufficient number of participants or an excessively large number are both extravagant in terms of participant and researcher time, funds, and other resources to conduct the assessments, analytic efforts, and so on. These conditions can also be viewed as unethical, as participants may have been put at risk as part of research that was unable to answer a vital question. Studies that are considerably larger than necessary to answer research questions are also extravagant.

Sample size determination depends on many factors. The most important factor is the study design. The study design stipulates whether the study is cross-sectional, intervention, or otherwise. The sample size for any health sciences research also depends on the primary outcome of the research, the number of sample groups to be used for the research, the type of hypothesis, etc. Sample size determination is a crucial aspect of any study, and it relies heavily on biostatistical formulas to be able to deliver accurate results. The statistical formulas used, the desired level of significance and precision, the desired power for the test, the population size, whether the population is finite or not, the proportion with the event of interests in the population, and the mean estimate and its standard error are all major issues in this. For instance, a smaller standard deviation requires a larger sample size and vice versa (Glantz 1989).

36.7 Data Management and Biostatistics

Data management consists of all activities involving data. One of the crucial activities in biostatistics is data management.

Key data management consist of

- Data ownership
  - Defines and provides information about the rightful owner of data assets and acquisition
  - Defines the data owner's ability to assign, share, or surrender all of these privileges
  - Implemented in medium to large enterprises with huge repositories
- Data collection
  - Gathering of raw data from the primary sampling units in a survey or use of different measurement devices to extract information from subjects after interventions in experimental studies
- Data storage
  - Stored data safeguards research and research investment
  - Allows future access and use
- Data protection
  - Limit access to data
  - Proper keeping of questionnaires
  - Assure privacy and anonymity
  - Protect systems: up-to-date antivirus
- Data retention
  - How long should data be kept?
  - Continued storage
  - Destroying data
- Data analysis
  - All activities to generate the needed information from the raw data
- Data sharing/reporting
  - Before or after publication?
  - Researchers are not obligated to share findings while research is ongoing

Data management encompasses data entry, data validation, data storage, data retrieval, merging of data from different study centers and data clerks, data cleaning (handling missing data), data analysis, and data presentation. Data cleaning is as important as data analysis.

36.8 Data Analysis

Data analysis is the most visible role of biostatistics. It involves the use of several statistical techniques with consideration for the study designs to extract information from the raw data.
Considerations for data analysis

1. **Study design**: clinical trial, survey, etc.
2. **Study objective**: descriptives/analytical/hypothesis
3. **Number of comparative groups**: $k=1$, $k=2$ or $k>2$
4. **Type of outcome/variable**: qualitative/quantitative, categorical/non-categorical, dichotomous/polytomous, ordered or unordered, dates/time, etc.
5. **Level of measurements**: nominal, ordinal, interval, ratio

36.8.1 Types of Data Analysis

As described in the previous section, biostatistics comprises mostly descriptive statistics and inferential statistics, which have been employed over the years in biostatistics and have remained relevant until today. In addition to these methods of analysis is the rising use of spatial analysis, in which case the geographical location is used in the analysis. It involves the use of coordinates (latitude and longitude) to provide a succinct description of the distribution of any outcome of interests. Also, advances have been made in statistical genomics and health data science. In the next section, some of the data analysis techniques are described briefly.

Descriptive statistics: This includes the use of pictograms, charts, mean, median, mode, percentages, proportions, tables, standard error, standard deviation, variance, and probability theories, to describe data.

Inferential statistics: It is involved in drawing conclusions about a population using information collected from a representative sample of the population. It is usually engaged to test hypotheses. Inferential statistics include the use of confidence intervals, hypothesis testing, survival analysis, experimental designs, decomposition analysis, longitudinal analysis, time series, Bayesian analysis, statistical genomics, and spatial analysis.

36.9 Biostatistics and Health Data Science

The last few years have witnessed the emergence and sharp rise in the influence of data science across the globe. Data science is an interdisciplinary field that engages and uses scientific and statistical methods, apparatus, techniques, processes, algorithms, and systems to extract information, knowledge, and insights from large amounts of structured and unstructured data. Data science encompasses data mining, machine learning (ML), and big data handling. Health data science also became famous simultaneously. Just as biostatistics is related to statistics, health data science is a subdiscipline of data science. Health data science is an emergent discipline from the intersection of biostatistics, computer science, and the health sciences. Health data science can be described as both the art and science that provide tested data-driven solutions through comprehension of complex real-world health problems, employing critical thinking and analytics to derive knowledge from both structured and unstructured big data. Indeed, biostatistics is the backbone of health data science (Elston and Johnson 2008). Health data scientists strive to improve the efficiency and usability of biostatistics methodologies using more intense computer-intensive methods to analyze big and complex data.

36.10 Contributions of Biostatistics to Other Disciplines

Biostatistics concepts and methods assist researchers in these fields to (1) understand the nature of variability and (2) derive general laws from samples.

Statistical modeling methods are popularly used in clinical science, epidemiology, and health services research to analyze and interpret data obtained from clinical trials as well as observational studies of existing health records (MvCullough et al. 2018). Diagnostic and prognostic inferences from statistical models are very important for researchers to advance science, for clinical practitioners to make informed care decisions, and for administrators and policymakers to positively impact the health-care system with improved quality, enhanced access, and reduced costs (Lash 2017).

Scientific research generates huge amounts of data that have to be analyzed and interpreted to provide insights that will shape intelligent decisions. Biostatistics plays a vital role in this process. A few applications are summarized below:

1. **In pharmaceuticals**: Biostatisticians help to design clinical trials and to measure the efficacy of treatments and drugs. Biostatisticians are not only engaged in clinical trials but also help in observational and complex interventional studies using statistical models for prediction and extrapolation.

2. **In public health**: Health-care professionals deal with numerous biological systems that contain inherent variability, and biostatisticians develop statistical tools and methods necessary to draw reasonably accurate inferences, despite the uncertainty inherent in the biological systems (Matthews and Farewell 1996). These inferences are used to improve the quality of public health and to advance evidence-based healthcare.

3. **In epidemiology**: Various factors influence the cause, outbreak, and distribution of a disease. Epidemiological studies aim to identify these factors by collecting various data related to the disease and deriving the link between the cause and the effect. Biostatistics is vital in understanding certain epidemiological principles such as incidence, prevalence, odds, risk ratio, and number needed to treat (Luscombe et al. 2001). The Center for Disease Control reports that "Epidemiology is the basic science of public health and therefore uses statistics to reach conclusions about diseases within certain population groups" (Hosmer et al. 2013).

4. **In medical sciences**: Biostatistics is the science that helps to manage medical uncertainties. A sound knowledge of biostatistics is important for medical and nursing students, so that they can design epidemiological studies accurately and draw meaningful conclusions. An important aspect of a doctor's job is to understand and condense the provided data, analyze findings, and engage in research activities. An important tool for all these tasks is biostatistics! Biostatistics is a branch of applied statistics, and it must be taught with a focus on its various applications in biomedical research (Indrayan and Malhotra 2017).

5. **In nutrition**: Statistics is employed to handle and quantify the variations and uncertainties that stem from the different responses to dietary patterns or nutrition interventions administered to patients. Additionally, statistics help to provide empirical evidence and a language for communicating scientific outcomes (Saracino et al. 2013).

6. **In oncology**: Biostatisticians help to plan cancer clinical trials and also analyze the data collected. The collected data will help to detect the causes and characteristics of cancer, which will be used to recommend treatments for cancer patients. They seek to identify the roles that several factors like drug interaction, diet, and nutrition play in cancer prevention and treatment.

7. **In genome sequencing**: Genome sequencing generates huge amounts of data that help scientists to understand complex traits. Data from the variant genome is compared to another. Biostatistics helps to study the variant frequencies and draw conclusions on genetic make-ups.

8. **In genetics**: Geneticists utilize statistical theories and methods in the genetic study to investigate and validate whether one or more genes might predispose people to an increased risk of developing a specific disease. It is also used to understand the progression and pathways of diseases (Zhang and Liu 2013).

36.11 The Biostatistician and Biostatistical Consulting

Who is a biostatistician? What role does he or she play? These are questions that need answers in today's world of advances in research in life sciences. In contrast to the clearly defined educational and professional career steps of a physician or other profession in life sciences, there is no unique way of becoming a biostatistician. Only very few universities indeed offer studies in biometry, which is why most people working as biostatisticians studied something related, subjects such as mathematics or statistics, or applied subjects such as medicine, psychology, or biology in their undergraduate studies. Most often than not, then, a biostatistician cannot be defined by his or her education but must be defined by his or her expertise and competencies. In the current state of research and understanding of life sciences, biostatisticians play a vital role. Given the growing intricacy and extent of health-related data, the importance of fast-tracking clinical and translational science, and the necessity of conducting reproducible research, the requirement for the considerate development of biostatistics resources is increasing rapidly. A biostatistician is a specialist who has the expertise and competency to provide statistical advice in quantitative research involving life sciences. On top of this role, a biostatistician is supposed to be able to have skills that allow him or her to provide professional assistance to individuals, students, and researchers in life sciences or other related disciplines who have difficulty in applying statistical methods to their work or research (Russell 2001). In other words, biostatisticians should be able to provide consultancy services for individuals in need of help.

The concept of statistical consulting is as old as the discipline of statistics itself (Awe 2012). Biostatistical consulting is a specialized branch of statistical consulting that deals with problems related to research data in biology, medicine, and public health. Not all life science researchers are grounded in the skills to apply advanced statistical techniques that could help make use of the important data collected in their research (Perrie and Sabin 2000). Thus, in many instances, only a few meaningful deductions come out of data that has been collected in life science research. In most cases, researchers will have to consult biostatisticians to be able to help properly analyze their data. These statisticians would also need to have some specific sets of skills to be able in turn to assist these researchers. This type of consultation is both an art and a science because it involves both statistical and nonstatistical skills. There is a desperate need for a crop of biostatisticians who will fill this void.

However, this kind of skill is not easy to come by and is most times not available at least formally. Universities are the best places where these skills can be acquired. Most often, there are no specific courses that specifically deal with statistical consulting. There should be courses specifically teaching students how to provide consultation services to researchers who need help with the statistical aspect of their study. All universities and polytechnics should at least have a firm or lab where researchers can take their research data to access the correct help. Most researchers in developing countries rely on contacting their friends (who are "knowledgeable") to help them out with their analysis.

In recent years, emphasis has been placed on the training of statistical consultants in developing countries. However, this training should also be done specifically for biostatisticians right from the undergraduate period by teaching the principles of statistical consulting. Without it being introduced at an early stage, it will be difficult for individuals to develop this skill at the graduate level.

36.12 Overview of the Differences between Biostatistics and Medical Statistics

Usually, biostatistics is often confused with medical statistics in developing countries. Biostatistics is the application of statistical methods in biological, medical, and public health domain, while medical statistics is the application of statistics to medicine and other health sciences such as public health, epidemiology, forensic medicine, and clinical research (Brimacombe 2014).

Commonly biostatistics refers to application of statistics to biology, while medical statistics plays a key role in medical investigations. It not only provides a way of organizing information on a wider and more formal basis than relying on the exchange of anecdotes and personal experience, but also takes into account the intrinsic variation inherent in most biological processes (Indrayan and Malhotra 2017).

Medical statisticians are often referred to as biostatisticians, and they work in various medical and public health fields. They often conduct statistical research to increase medical knowledge, track and mitigate disease transmission, and ensure improved management of patients by utilization of better adjuvant investigations and therapy (Macnab 2003).

Here are five major but succinct differences between medical statistics and biostatistics:

1. Medical statistics deals with the applications of statistics to medicine and health sciences, which comprise epidemiology, public health, forensic medicine, and clinical research, while biostatistics deals with the use of statistics to biological sciences such as botany, conservation, ecology, evolution, genetics, marine biology, medicine, microbiology, molecular biology, physiology, and zoology.
2. Medical statistics is more commonly used and well recognized in the United Kingdom and Australia for more than 40 years, while biostatistics, on the other hand, is more commonly used in North America.
3. Medical statistics is a subdiscipline of the field of statistics, while biostatistics comprises all applications of statistics to biology.
4. Medical statistics is the science of collecting, summarizing, presenting, analyzing, and interpreting data in medical practice, and using them to estimate the magnitude of associations and testing hypotheses, while biostatistics is the science of collecting, summarizing, presenting, analyzing, and interpreting data in biological practice, and using them to estimate correlation analysis and testing hypotheses.
5. Medical statistics has a common role in medical investigations. It does not only provide a way of arranging information on a broader and more formal basis than relying on the exchange of anecdotes and personal experience but also can take into account the intrinsic variation inherent in most biological processes, while biostatistics may be used to help learn the possible causes of a cancer or how often a cancer occurs in a certain group of people. In some climes, it is also called biometrics or biometry.

36.13 Career Opportunities around Biostatistics

Biostatistics: Importantly, biostatistics as a subdiscipline of statistics is an established scientific discipline of its own. This requires not only an understanding of mathematics, probability, and sources of bias that underpins statistical theory and methods but also (and increasingly) extensive technical skills, including computing. In-depth training is needed to develop these skills along with the understanding required to conceptualize problems and navigate the tricky waters between real-world health questions and complex techniques. Such training would be very difficult to achieve for most clinicians (Elston and Johnson 2008). A superficial understanding of statistics can easily lead to unscientific practice (recently characterized as "cargo-cult statistics") and may be seen as responsible in large part for the current "crisis of reproducibility" in research (Stark and Saltelli 2018).

Health data science: Unarguably, data science has become the jewel of the 21st century. It depends heavily on biostatistical techniques in addition to some computational skills with little or no knowledge of health sciences. Biostatistics has made major contributions to the field of data science in a substantial way. This application of biostatistics in health data science is merging and analysis of complex electronic health data.

Statistical genetics: Statistical methods and mathematical models play a large role in genetic epidemiology research. This will undoubtedly continue. However, consistent with the notion that future genetic epidemiology will embrace evolutionary and population concepts to a greater degree than before, there will probably be greater use of statistical and mathematical constructs oriented around evolutionary and population genetics phenomena.

Machine learning (ML): This is a multidisciplinary subject involving many disciplines, including probability theory, statistics, approximation theory, convex analysis, and algorithm complexity theory (Zou et al. 2018). With recent development in artificial intelligence and ML, this has brought about unprecedented access to research in life and health sciences (Dukhi et al. 2021). For instance, ML is used to automatically detect cancers on scans (Rajkomar et al. 2019).

Bioinformatics: This is a new discipline that applies computational tools to organize, systematize, analyze, understand, visualize, and store information associated with biological macromolecules (Abdurakhmonov 2016). This field seeks the development and usage of computational algorithms to understand and interpret biological processes based on genome-derived molecular sequences and their interactions.

Computational biology: Computational biology deals primarily with the modeling of biological systems by developing software for biological uses as well as conducting the analysis and interpretation of biological data using the software and particular algorithms. The study and knowledge of this course can open a world of possibilities for research and employment. Without them, the scope and ambitions of biological research are unnecessarily constrained.

Biological computation: This is a field that specializes in developing biological computers using the advances of bioengineering, cybernetics, robotics, and molecular cell biology. The area of biological computation involves the development of new, biological techniques for solving difficult computational problems.

Data mining: The sole purpose of data mining is to extract useful information from large databases or data warehouses. Currently, different data mining algorithms applied in the health-care sector play a significant role in the prediction and diagnosis of different diseases. Data mining is becoming gradually integrated into life science processes. There is an opportunity for data mining researchers with a focus on biological processes to contribute to the development of the life and clinical sciences by creating novel computational techniques for discovering useful knowledge from large-scale real-world biomedical data.

36.14 Biostatistical Resources

Below is a list of some current journals that publish up-to-date biostatistical educational materials and research:

- *Biostatistics*
- *International Journal of Biostatistics*
- *Journal of Epidemiology and Biostatistics*
- *Biostatistics and Public Health*
- *Statistical Applications in Genetics and Molecular Biology*

- *Statistical Methods in Medical Research*
- *Pharmaceutical Statistics*
- *Statistics in Medicine*

36.15 Strategies for Developing Biostatisticians in Developing Countries

Biostatistics training is abysmally low in developing countries. Biostatistics is applied to the understanding of health sciences and biology because it provides basic and powerful tools for the development of important health- and biology-related research questions and research design, defining the procedure for implementation, analyzing data, and interpreting research findings. In achieving meaningful results in biostatistics research, the following strategies must be identified and developed to achieve the desired results in developing countries.

36.15.1 Training of Experts in Biostatistics

No progress can be achieved in biostatistics research without the involvement of qualified biostatisticians with the requisite ability to quantify uncertainty in the dataset and to generate sound statistical inferences for management decisions and policy implementation. Therefore, because of the wider scope and increasing complexity in handling big data in health- and biology-related research, the demand for biostatistical experts is expanding day by day.

There is a need to develop a biostatistics curriculum that will train biostatisticians at the undergraduate level to provide experts that will be fully involved in driving health- and biology-related research instead of regarding them as an auxiliary service provider for the health- and biology-related disciplines. As it is today, we need statisticians with strong communication skills who can communicate in a statistical language and interpret their findings for nonstatisticians in industry and in health and related disciplines.

In the present day evolution of biostatistical practice, a would-be biostatistician must be well trained and equipped in the following areas:

- Survival analysis and history of events
- Multilevel and latent variable modeling
- Advanced epidemiological techniques
- Linear regression and robustness
- Statistical theory
- Multivariate and cluster analysis
- Generalized linear models
- Time series and spectral analysis
- Bayesian statistics and causality
- Statistics genomics

36.15.2 Creating a Conducive and Attractive Environment for Biostatisticians

Biostatisticians are statistical analysts whom a conducive and attractive working environment will aid, facilitating their responses to the statistical needs of their clients. There is a need for inclusiveness of biostatisticians in policy decisions and implementation concerning appointment, promotion, career advancement, training and retraining, and other such opportunities that may be available within the working environment. The situation where a non-biostatistician takes a leading role in matters relating to statistical analysis should be discouraged. Therefore, biostatisticians should not be confined to and perceived as a service role. Such mismatches in expectations will negatively affect hiring and, ultimately, retention. Medical research should include biostatisticians, including those just building their careers, who should be recognized and not isolated from other biostatisticians, who may act as mentors, as doing otherwise may limit their opportunities for advancement.

Furthermore, the advancement of new technologies in areas as diverse as imaging, nucleic acid sequencing, and electronic health records is outpacing the development and dissemination of applied statistical methods. Thus, the need for the efficient use of biostatistics resources is increasing.

36.16 Challenges of Biostatistical Capacity Building in Developing Countries

Lack of a career guide for biostatisticians: Unlike other professions, there is no clearly defined educational and professional career guide for becoming a biostatistician. Only a handful of universities indeed offer biostatistics as a course in developing countries, which is why most people working as biostatisticians studied something related, such as mathematics or statistics, or application subjects such as medicine, psychology, or biology. The teaching of biostatistics in the university is mainly done by academics who have little or no training/certification in biostatistics. This is a drawback for the field of sciences. If a professionally trained statistician develops and applies statistical methods in a field where he has little or no understanding of its subject matter expertise (i.e., the life sciences), those methods should be questioned. Conversely, if a professionally trained life scientist with little or no knowledge in statistics develops and applies statistical methods, those methods should be questioned.

Inadequate knowledge of statistics: In developing countries, most biologists only have a rudimentary understanding of statistics and insufficient understanding of how to accurately use commercial software for the analysis of their data. They would rather do it their way or, at best, collaborate with a statistician. This is not good enough for research. Researchers are frequently faced with data on which they must base their contextual judgments for appropriate conclusions. Reviews of the medical literature have revealed invalid conclusions because incorrect statistical arguments were used. Most analytical studies published in well-respected medical journals were termed "unacceptable," in that the conclusions drawn were not valid in terms of the design of experiment used, the type of analysis performed, or the applicability of the statistical tests used. After some years, it was again found that about half of the articles published in medical journals used statistics incorrectly. More recently, this has been extended to investigated articles published in three high-impact clinical journals that had been cited in over 1,000 other published peer-reviewed journals.

Lack of data miners and machine learners for life science: Due to the explosion of data in our world today, computers are used to find possible patterns in the huge amounts of data generated by biological experiments and research. The techniques used in finding such patterns and classifications are popularly regarded as "data mining." Most statisticians find it difficult to develop relevant algorithms for the field of life sciences. To develop such algorithms, they throw away useful and valuable components because of insufficient knowledge of life sciences, which consequently compromises the aim of the research.

The outdated curriculum in most universities: The curriculum (undergraduate and postgraduate) currently in use has not been changed for decades. Newer aspects like data science, ML, and artificial intelligence that rely heavily on statistical methods have not been incorporated into many university curricula.

Poor understanding of the relevance of biostatistics among life science students: Students do not understand the importance of biostatistics and are thus focused more on the core aspects of their courses, relegating biostatistics to the background. This is a major problem with most students, especially in health science. This is also worsened by the generally poor attitude of students toward any mathematics-related subject.

Lack of formal avenues and resources for problem-solving: Many researchers in the life sciences involved in quantitative research lack formal avenues where they can solve their data problems when they arise. Most also fail to consult with biostatisticians at the inception of their research until they run into problems, and when problems arise, it is often too late to be able to make appropriate corrections to the research work, which affects the quality of the research output. There are no certified/registered statistical consulting firms available in or around universities. Moreover, statistical consulting is not taught as a course to biostatisticians or even life science students. This is also primarily due to poor or even nonexistent collaboration between the statistics and life science departments.

36.17 Recommendations

We hereby recommend the following in order to boost the capacity of biostatistics in developing countries:

36.17.1 Establishing Departments of Biostatistics in Developing Countries

The above challenges are interrelated and indicate a great need for higher education institutions to have separate departments of biostatistics, where biostatistics and research orientations are taught in detail and in-depth. Biostatistics should no longer be just a course under the auspices of different departments but rather a discipline. It should be handled by experienced lecturers with strong statistical and biological backgrounds, which may help in bridging the gap between life science and statistics.

36.17.2 Developing and Instituting Guidelines for Biostatisticians

When the above is instituted, the understanding of such a field and the scientific context of the research problem becomes essential for biostatisticians. Furthermore, specific professional expertise is inevitable and also proficiency in the use of current statistical software is very important. Qualification, that is biostatistical expertise, covers methodological background (mathematics, statistics, and biostatistics), biostatistical application, medical documentation, and statistical programming. The experience relates to consulting, planning, conducting, and analyzing medical studies. The above is the guideline for biostatisticians in the medical sciences, but it can also be implemented by researchers in the life sciences. A biostatistician should have a thorough knowledge of statistics and a fair knowledge of the biological sciences. Biostatisticians are integral members of research teams who are needed for their skills in writing competitive grant proposals, executing statistical procedures, conducting advanced data analysis, publishing in high-profile journals, and teaching biostatistics at the undergraduate and postgraduate level.

This will greatly enhance the efficiency of the study and the scientific credibility of results from the field of life sciences. There is no amount of statistical expertise that can salvage the results of a poorly planned or executed study.

36.17.3 Improving the Teaching of Biostatistics in Tertiary Institutions

Biostatistics as it is taught in the various schools should be improved upon by updating the curriculum, investing in biostatistician education training, and employing the latest techniques of teaching. This can be enhanced by improving the collaboration between the core statistics departments and the faculty of health sciences, thereby ensuring a pool of biostatisticians who can be used to teach and practice biostatistics. Biostatistical teaching should be based on problem-based learning. There is also an urgent need to update the existing curriculum to include newer aspects of data science.

36.17.4 Encouraging Biostatistical Consulting in the Universities

One important measure that can aid biostatistics competency and skill set is the establishment of centers where biostatisticians and students can consult a specialist to help solve their statistical problems related to their education or research. This can be done by establishing statistical labs (like that of LISA 2020) where researchers can meet biostatistical consultants (specialists) who can profer advice and solutions. These centers can provide regular standard training and collaborations.

36.17.5 Developing and Implementing Strategies to Improve Biostatistical Resources

There is a need to raise awareness of the need for the thoughtful management of biostatistics resources in tertiary institutions in developing countries. There is an urgent need to take a focused and systematic approach to developing these resources. This can be done by recruiting and retaining experienced biostatisticians and creating a mentorship program to help mentor young and vibrant faculty members to take on biostatistician roles as well as ensure the efficient usage of available biostatistics resources. This is relevant to all tertiary institutions, from those with few or dispersed resources to those that already support a centralized biostatistics unit in developing countries.

36.18 Conclusion

This chapter has highlighted the nature and importance of biostatistics and the need for its development in developing countries. It has been shown that biostatistics plays a critical role in biology- and health-sciences-related research by providing scientific information for vital decision-making in health planning, clinical and public health research, and beyond. It is unequivocally indisputable that biostatisticians are lacking in number in the developing countries, and it has posed a major challenge to modern research and decision-making in this region. Therefore, urgent intervention is needed in other to bridge the knowledge gap in biostatistics. One of the most urgent interventions is to provide scholarship for qualified candidates in developing countries to pursue degrees in biostatistics. A long-time intervention would be the promotion among pupils in both elementary and higher schools in developing countries the learning of mathematics and statistics for mental and emotional development and preparedness to make them fit to pursue future careers in biostatistics and data science at both undergraduate and postgraduate levels.

Finally, biostatisticians should be better remunerated to make them compete favorably with their peers in other disciplines.

References

Abdurakhmonov, IY. 2016. *A Textbook on Bioinformatics: Basics, Development, and Future.*

Ahmad DA, Ahmad KN, Mohib-ul-Haq M, Nayak BG, Maqbool LM. 2019. Some applications of biostatistics to medical research. *International Journal of Advanced Research, 7*(2), 28–31.

Albert A. 2004. Contribution de la biostatistique au développement des Sciences biomédicales [Biostatistics contribution to the development of biomedical sciences]. *Bulletin et Mémoires de l'Académie Royale de Medecine de Belgique1, 59*(5–6), 317–326.

Altman DG, Goodman SN, Schroter S. 2002. How statistical expertise is used in medical research. *Journal of the American Medical Association, 287,* 2817–2820.

Altman DG. 1994. The scandal of poor medical research. *British Medical Journal, 308,* 283.

Awe, OO. 2012. Fostering the practice and teaching of statistical consulting among young statisticians in Africa. *Journal of Education and Practice, 3*(3), 54–59.

Awe OO, Ayeni OC, Sanusi GP, Oderinde LO. 2021. A comparative time series analysis of crude mortality rate in the BRICS countries. *BRICS Journal of Economics, 2*(2), 17–32.

Begg MD, Vaughan RD. 2011. Are biostatistics students prepared to succeed in the era of interdisciplinary science? (And how will we know?) *The American Statistician, 65,* 71–79.

Brimacombe MB. 2014. Biostatistical and medical statistics graduate education. *BMC Medical Education, 14*(1), 1–5.

Carlin J, Kasza J, Moreno-Betancur M, Simpson J, Bartlett J, Metcalfe C, Lee K. 2020. A potpourri of biostatistical research: Special Issue for ISCB ASC 2018. *Biometrical Journal, 62*(2), 267–269.

Center for Disease Control and Prevention. 2011. An Introduction to Epidemiology. http://www.cdc.gov/excite/classroom/intro_epi.htm/.

DeMets DL, Stormo G, Boehnke M, Louis TA, Taylor J, Dixon D. 2006. Training of the next generation of biostatisticians: A call to action in the U.S. *Statistical Medicine, 25,* 3415–3429.

Dukhi N, Sewpaul R, Sekgala MD, Awe OO. 2021. Artificial intelligence approach for analyzing anemia prevalence in children and adolescents in BRICS countries: A review. *Current Research in Nutrition and Food Science, 9*(1), 1–10.

Elston R, Johnson W. 2008. *A Textbook on Basic Biostatistics for Geneticists and Epidemiologists: A Practical Approach.*

Gezmu M, DeGruttola V, Dixon D, Essex M, Halloran E, Hogan J, ... Neaton, J. D. (2011). Strengthening biostatistics resources in sub-Saharan Africa: Research collaborations through US partnerships. *Statistics in Medicine, 30*(7), 695–708.

Glantz SA. 1989. Biostatistics: How to detect, correct and prevent errors in the medical literature. *Circulation, 61,* 1–7.

Henley SS, Golden RM, Kashner TM. 2019. Statistical modeling methods: Challenges and strategies. *Biostatistics & Epidemiology, 4*(1), 105–139.

Hian CK, Tan G. 2005. Data mining application in healthcare. *Journal of Healthcare Information Management, 19*(2)

Hosmer DW, Lemeshow S, Sturdivant RX. 2013. *Applied Logistic Regression.* 3rd ed. Wiley-Interscience.

Indrayan A, Malhotra RK. 2017. *Medical Biostatistics.* Chapman and Hall/CRC.

Keiding N. 2005. Roles of statistics in the life sciences. *International Statistical Review, 73*(2), 255–258.

Kilic I, Celik B. 2013. The views of academic staff on biostatitistics education in health sciences. *International Journal of the Health Sciences (Qassim), 7*(2), 142–149.

Lash TL. 2017. The harm done to reproducibility by the culture of null hypothesis significance testing. *American Journal of Epidemiology, 186,* 627–635.

Li X, Ng S, Wang JTL. 2014. *A Textbook of Biological Data Mining and its Applicaion in Healthcare.*

Luscombe NM, Greenbaum D, Gerstein M. 2001. What is bioinformatics? A proposed definition and overview of the field. *Methods of Information in Medicine, 40,* 346–358.

Macnab JJ. 2003. *Graduate Program in Epidemiology and Biostatistics* (Doctoral dissertation, The University of Western Ontario London).

Matthews DE, Farewell VT. 1996. *Using and Understanding Medical Statistics.* 3rd rev. ed. Basel, Switzerland: Karger.

MvCullough JPA, Lipman J, Presneill JJ. 2018. The statistical curriculum within randomized controlled trials in critical illness. *Critical Care Medicine, 46,* 1985–1990.

Perrie A, Sabin C (eds). 2000. *Describing Data. Medical Statistics at a Glance.* Blackwell Science Ltd.

Pollard DA, Pollard TD, Pollard KS. 2019. Empowering statistical methods for cellular and molecular biologists. *Molecular Biology of the Cell 30,* 1359–1368.

Rajkomar A, Dean J, Kohane I. 2019. Machine learning in medicine. *New England Journal of Medicine, 380,* 1347–1358.

Rencher AC, Christensen WF. 2012. *Methods of Multivariate Analysis.* 3rd ed. Hoboken, NJ: Wiley.

Russell KG. 2001. The teaching of statistical consulting. *Journal of Applied Probability, 38,* 20–26.

Saracino G, Jennings LW, Hasse JW. 2013. Basic statistical concepts in nutrition research. *Nutrition in Clinical Practice, 28*(2), 182–193.

Stark PB, Saltelli A. 2018. Cargo-cult statistics and scientific crisis. *Significance, 15*(4), 40–43.

Zaph A, Rauch G, Kieser M. 2020. Why do you need a biostatistician? *BMC Medical Research Methodology, 20,* 23.

Zaugg CE. 2003. Common biostatistical errors in clinical studies. *Schweizerische Rundschau für Medizin Praxis, 92*(6), 218–224.

Zhang SY, Liu SL. 2013. *Brenner's Encyclopedia of Genetics.* 2nd ed.

Zou Q, Qu KY, Luo YM, Yin DH, Ju Y, Tang H. 2018. Predicting diabetes-mellitus with machine learning techniques. *Frontiers in Genetics, 9,* 10.

37

Technology and Multimedia in Statistical Education and Collaboration

Ezra Gayawan, Olabimpe B. Aladeniyi, Seyifunmi Michael Owoeye,
Oluwatobi Michael Aduloju, Adetola Adedamola Adediran,
and Olamide Seyi Orunmoluyi
Federal University of Technology, Akure (FUTA)

Omodolapo Somo-Aina
University of North Carolina

CONTENTS

DOI: 10.1201/9781003261148-44

37.1 Introduction

The world has, over the years, experienced great changes and developments that impacted the educational delivery system. Generally, technological advancement affects several aspects of society and has also improved human life in many ways especially in the last six decades. Inventions such as televisions and computers have recently had a huge impact on the teaching and practice of statistics beyond the previously known conventional teaching methods. The invention of multimedia has also continued to revolutionize teaching, learning, and pursuing collaborations at various levels of the educational system.

Multimedia technology is fast becoming a popular and effective means of teaching, learning, and collaboration. Unlike the conventional method of teaching, the multimedia approach is "student-centered". Learning no longer depends on teachers' mood or students' perception of the teacher and those who do not learn as fast as others can work at their own pace and time. These methods also give room for more detailed and tailored instructions, encourage students by allowing them to participate actively, and even give feedback that improves collaboration between teachers and students. Learning through visual imagery can play a powerful role in accelerating learning because complicated concepts and processes can be conveyed in simpler forms using visual formats, which, in turn, motivates interest, enhances improved understanding and expression of ideas. Statistics is considered an important subject that is required at the introductory level by many disciplines. Consequently, in most institutions of higher learning, statistics introductory classes are usually large making the conversational teaching and learning of the subject difficult. Under such circumstances, statistical models, methods, and assumptions become difficult for most learners, who would later need them for research or other engagements. Introducing multimedia and technology in the teaching and learning of statistics right from the introductory classes would stimulate interest and enhance understanding of the subject and make the class flashier, for instance, in the aspect of introducing graphs and charts to the students. Students will most likely be interested in understanding how to create such colorful charts unlike when the teacher simply sketches them on a whiteboard.

Multimedia is a system that combines all or several texts, images, audio, motion graphics, animation, and video for expressing ideas or facilitating a technique. The user can control the speed of the presentation and decide the location and time of usage. These digital tools comprise not only resources to share and build knowledge, but also instruments that can shape our way of thinking and learning. Thus, using images, videos, and animations alongside text inspire the brain. Students' devotion and retention increase under these circumstances. In a multimedia learning environment, students can discover and solve problems more effortlessly. These cultural and symbolic artifacts enable users to go beyond themselves, get in contact with other minds, and move forward together in the creation of ideas that turn out to be jointly satisfying. Adaptation of multimedia in teaching, learning, and collaboration implies the exhibition of a series of high complexity mental processes on the part of the teacher and students, and this is even more so successful when information and communication technology are used collaboratively. The involvement of textual and pictorial information demands that the students assume a high level of active participation and take on accountability for their learning. Collaborative learning constitutes a socially mediated activity in which knowledge is built in a collective form of learning that allows for a dynamic interaction established between the learners, the environment in which the activity is developed, and the instruments used for such activity.

Multimedia is a rich medium that accommodates numerous instructional strategies. It addresses many of the challenges of training in both academic and corporate environments. It is available over distance and time and provides a medium for consistent delivery. In education, multimedia can be used as a basis for information. Teachers can use multimedia presentations to make lessons more remarkable by using animations to highlight key points. Multimedia provides students with an alternate way of acquiring knowledge designed to improve teaching and learning through various mediums and platforms. The technology allows students to learn at their own pace and gives teachers the ability to observe the individual needs of each student. The use of multimedia in multidisciplinary settings is anchored on the idea of creating a hands-on learning environment through the use of innovative technology. Lessons can be adapted to the subject matter as well as be personalized to the students' varying levels of knowledge on the topic. It allows for the management of learning content through activities that exploit and take advantage of multiple multimedia platforms. This kind of learning encourages interactive contact between students and teachers and opens feedback channels, introducing an active learning process particularly with the frequency of new media and social media. Technology has impacted multimedia as it is largely linked with the use of computers or other electronic devices and digital media due to its capabilities concerning research, communication, problem-solving through simulations, and response opportunities.

Collaborative processes are often required to facilitate the materialization of productive interactions that are multimedia based. When multimedia is used in collaborative learning contexts, the digital texts can promote the launch of highly elaborated exchanges about the contents being considered and the explicitness of abilities.

37.1.1 Collaboration and Teamwork

Teamwork is the working together of two or more people with the sole purpose of achieving a common goal. It involves people of similar interests and skills in which individuals have well-defined roles, working together toward an already set goal. Collaboration involves working together with people with different skills or areas of specialization in achieving a preset goal (Jeong & Hmelo-Silver, 2016). It is the process of problem-solving by a group of people with different skill sets. Collaboration becomes necessary because it is quite impossible to have a single person with the skills, knowledge, and resources to solve today's real and complex problems. However, successful collaboration can only be achieved when collaborators develop a cooperative spirit and mutual respect, and team members function effectively as part of the team and are willing to balance personal achievement with the group's goals. When properly done, it has several benefits, including that it saves money, time, and leads to efficient and robust solutions to the problem being solved.

As different researchers and institutions have varied research practices, collaboration brings them to common practice and enables them to learn from and leverage on one another. Collaborators usually bring their different skills and talents and thus, combining their different perspectives can better solve the common problem. Collaborations can be done by researchers within or between departments of an institution, between institutions, and with public or private organizations. Scientists initiating into collaboration have multiple expectations to support their strategic plans, to facilitate the management of the scientific process, to have a positive or neutral impact on scientific outcomes, to provide advantages for scientific task execution, and to provide personal conveniences especially when collaborating across distances (Sonnenwald, 2003). Solving today's problems in virtually all walks of life involves some multidisciplinary approaches where people of different backgrounds and perspectives contribute to their expertise. In multidisciplinary research, for instance, researchers from different disciplines within or between institutions come together to address a research problem, thereby providing a robust solution to research findings (Cummings & Kiesler, 2005).

Unlike collaboration, teamwork brings together members of similar skills, usually people from the same discipline. When members of a department within an institution form a group with another set of people from the same department and work together, it is called teamwork as the people coming together are all in the same field of study with similar skill sets. However, when members of a department from an institution form a group with another department entirely, whether from the same institution or not it leads to collaboration. A food scientist can work with a physician to produce a particular product that can only be taken by hypertensive patients; statisticians can form a collaboration with marketers of a company to examine if a new market policy would have a negative or positive effect on the company; pharmaceutical companies, epidemiologist, statisticians, and other research experts can form a collaborative group to find a vaccine for the COVID-19 pandemic; statisticians and economists can collaborate to quantify the effects of COVID-19 on the economy. All these are examples of collaborations that involve different approaches to solving a common problem based on the fields of study involved.

Most successful endeavors are a product of teamwork or collaboration. To have a successful collaboration, the following needs to be put in place:

1. There should be a well-written agreement on the terms of the collaboration.
2. Each member of the team must have a clear definition and agreement on his/her role.

3. Each member must have good communication skills without hoarding information from others.

4. Methods to be used for accomplishing tasks must be agreed on.

5. There must be mutual respect among members.

6. Cooperation among members is a must.

7. Holding a grudge against any member of the team destroys collaboration.

It is worthy to note that globally, there has been a rapid increase in the number of research activities focusing on technology-supported collaborative learning. This is not a shock since technology and multimedia have ultimately made collaborative learning easier especially because it allows for overcoming geographical barriers in research collaboration. A little look into literature shows that the term computer-supported collaborative learning (CSCL) was first used by O'Malley and Scanlon in 1989 and was later recognized as an important area of research (Hakkarainen et al., 2004). CSCL is an interdisciplinary field of research that focuses on how technology can make the teaching, learning, and creation of knowledge easier through interactive and collaborative learning processes (Resta & Laferrière, 2007). Primarily, CSCL aims to provide an environment that facilitates collaboration between students to greatly improve their learning processes, collective learning, or group cognition (Kreijns et al., 2003; Stahl, 2006).

Researchers have identified greater benefits of CSCL when compared with individualistic learning. These benefits include higher academic achievements (Johnson et al., 1989; Johnson et al., 1990), higher levels of learning and thinking (Benbunan-Fich et al., 2003; Wegerif, 2006), greater satisfaction and motivation (Springer et al., 1999), and more detailed and complete report of the task carried out (Benbunan-Fich et al., 2003). In a nutshell, technology and multimedia have made collaboration easier and faster as different researchers can connect and work efficiently from their location through different platforms, unlike in the 1990s where fax machines were the major technologies available and used to send data and documents between researchers.

37.2 Types of Teamwork

Teamwork brings together members of similar skills to work together in the same field of study. In this section, we discussed cooperative learning and cooperative faculty groups as types of teamwork that statistics departments in developing nations of the world can adopt to enhance statistics education.

37.2.1 Cooperative Learning

Cooperative learning is a teamwork that brings together students to learn similar academic disciplines or subjects. In order to have a good understanding of what cooperative learning entails, Statistics teachers must first understand that all groups are not cooperative groups. Groups may be structured in a way that learning is either hindered or promoted, and in some other ways, groups may initiate disharmony among students, and dissatisfaction about the quality of their classroom. Cooperative learning leverages on the epistemic and motivational process between students rather than within individual students. Alternatively, cooperative learning approaches are focused on the way each students' unique goals are connected to other students' goals, instead of relying on individual inquisitiveness, or the stimulating nature of the school curriculum (Baldwin & Chang, 2007; Fiechtner & Davis, 1992).

The general layout of a cooperative learning environment requires that teachers take the following steps in succession:

Step 1 – Specify the academic and social objectives of the lesson: Target social skills sets must be capable of building and sustaining positive relationships among students. Teachers should make conscious efforts to tackle social norms that make students lose interest in academics, and enhance inclusive learning environments for students to equally participate in statistics classrooms.

Step 2 – Explain task and cooperative framework to students: Here, the teacher will assign students to small groups consisting of not more than five people to enhance equal participation; instructors must be clear about the criteria by which student performance will be evaluated; and make cooperative goals explicit, by stating clearly individual and group responsibilities.

Step 3 – Monitor each learning group: Once an activity begins, instructors must monitor and, when necessary, intervene in students' group work. At this point in cooperative learning, teachers should observe interaction among group members, and assess both academic progress and the appropriate use of interpersonal skills.

Step 4 – Evaluate students achievement: Students should be allowed to reflect and evaluate what they found beneficial and worthless in completing the task. For instance, the teacher could ask about the strength and weaknesses of each group, and how the group could be improved for better performance in the future.

Evaluations like this will help to determine whether the objectives of the lesson were attained, and strengthen classroom ethics, and expectations.

37.2.1.1 Cooperative Learning in Practice

Cooperative learning in practice requires that instructors of statistics execute the following actions to achieve target lesson objectives:

1. **Assign students to different groups**: For short tasks, students can work with their friends. But for extended activities, it is advisable that teachers assign students to random groups. These random groups can be formulated using class roster or having the students count numbers in turn.

2. **Get groups started to work on tasks**: Instructors should ensure that groups do not suffer uncomfortable silence in starting assigned tasks. Rather than having social roles take over the class activities – i.e., the verbose students always talking and the reserved students keeping quiet. Instructors should subdue social rules to promote interdependent roles and purpose structures.

3. **Assign roles to members of each group**: Teachers should ensure that each student has a specific role in the group task. Allocating roles to students can help to discourage conventional social roles and promote critical reasoning skills. For instance, students can be assigned roles like summarizing lessons learnt in their

task, checking through what others did, looking for elaboration by connecting lessons with past lessons or experience, and other contextual roles.

4. **Help students build teamwork skills**: When students display good teamwork skills, the quality of learning is improved. In practice, teamwork skills are specifications about behaviors that are and are not suitable for individuals and learning groups.

5. **Help a group that is not united**: Students can vary in their preparedness, motivation, and priorities for learning. Instructors should attend to groups that are not working well, to figure out the cause of the problem and proffer solutions.

6. **Task**: Here, the students will first meet in their assigned groups to discuss the study questions and reach consensus about the relevant answers. The collective goal is to ensure that all members in the group understand how to tackle the study questions appropriately. Next, students will do the individual portion of the task and converge later to complete the group portion of the task. In situations where group members cannot come to an agreement on an answer, teachers need to intervene or urge them to make comparisons with a neighboring group.

7. **Evaluation**: This provides an opportunity to assess the impact of cooperative activities on learning. The teacher may ask students to specify a minimum of two behaviors that were beneficial to the group and a minimum of two behaviors that could help the group perform more effectively in subsequent time. Instructors could say, "signify by raising your hands if all your group members participated in performing the task".

37.2.2 Cooperative Faculty Group

Research has shown that cooperative faculty groups at the department level impact students' learning in a positive way, just like the use of cooperative learning groups positively influences learning among students (Johnson, Johnson, & Smith, 1991). Aside from the fact that statistics education must reflect statistical practice, which is by nature a collective undertaking, the use of cooperative faculty groups will also help to develop quick and responsive changes to teaching and curriculum. A cooperative teaching group constitutes various instructors in the same field, coming together to share and discuss teaching experiences, provide support for one another, and work together to achieve a collective goal of quality education (Baldwin & Chang, 2007; Rumsey, 1998). The faculty cooperative group is beneficial because it provides opportunities to:

1. Accomplish more at an advanced level than working as an individual. For instance, better instructional materials and teaching approaches are produced through the contribution of various instructors in the group with diverse backgrounds and experiences.

2. Reflect on personal teaching styles, by discussing and justifying personal beliefs and practices, which will also lead to questioning those beliefs and practices.

3. Get motivative and supportive environments that encourage individuals to make changes that may be frienthening to try alone. Members can reflect on these changes for confirmation and forge ahead, rather than abandon efforts because they are not successful instantly.

4. Build and maintain consistency within the same course of learning. In the course of developing educational materials as a team, there will be conversations about what is essential for students to be learning, consequently ensuring a tight connection between learning objectives, syllabus, and assessment.

5. Build a community that can improve the work environment and job satisfaction. The team members will develop positive feelings toward the group and commitment toward working together.

6. Give support and guidance to new instructors. New teachers could benefit from the support in the group and have a more positive early career as statistics teachers.

37.2.2.1 Cooperative Faculty Group in Practice

Cooperative faculty groups can be in the form of periodical meetings or professional discussions about students learning among faculty members and graduate teaching assistants (Baldwin & Chang, 2007; Rumsey, 1998). Practical tips on how to establish and sustain a cooperative faculty group include:

1. **State clearly and agree on the structure of the group and expectations of members**: The group pioneers should create an agreed-upon framework, specifying clearly feasible individual and group expectations.

2. **A faculty member may start a group by requiring that graduate teaching assistants participate**: The group can start with faculty members and students who show interest to join.

3. **Appoint someone who is willing to schedule and run the meetings and coordinate the group**: It is better to appoint individuals who are willing to schedule and manage the meetings and also help to coordinate the group.

4. **Encourage members of the group to observe each other's classes**: Group members should be encouraged to observe each other when teaching, to witness how the same lesson can be taught in different ways.

5. **Then compare the different outcomes of the same lesson taught by different teachers**: It is quite illuminating to compare diverse outcomes of the same lesson that different instructors taught in different settings.

6. **Implement actions based on lessons learnt**: Group members should have time to reflect on the lessons learnt in their meetings and observations and apply what is beneficial in improving statistics learning for students.

37.3 Types of Collaborations

To justify the need for collaboration in statistics education and get the desired results, a right approach is needed. In this session, we extensively discussed five types of collaborations in statistics education, collaboration within an academic institution, collaboration between academic institutions, collaboration between academic institutions and government, collaboration between an academic institution and private industry, as well as international collaboration.

37.3.1 Collaboration within Academic Institutions

In an era of fast and dynamic nature of the world system, professors in colleges and universities need continuous training to engage a diversified population of students in learning, through a wide range of innovative and technology-enhanced teaching strategies. However, the majority of higher institutions of learning around the world have limited resources at the departmental level to train students, especially with the emergence of various educational curriculum, courses of study, technology, multimedia, and innovations. In situations like this, collaboration becomes a powerful tool to advance departmental learning, promote professional training, and effectively maximize the effect of resources institutions invest in faculty (Kezar, 2006). Departmental collaborations can enhance and sustain an active institutional atmosphere that promotes an appropriate learning climate for students.

Research collaborations within academic institutions can include faculty members, administrative staff, and students in different departments or across disciplines. Departments working on research may decide to involve colleagues with common interests, related expertise, and high profile to participate in new or ongoing projects. It is assumed that individuals or parties involved in conducting the research have an understanding of the basic concept of the field. Research work is mostly focused on multifaceted themes, necessitating multidisciplinary efforts to develop innovative strategies to resolve difficult tasks. Collaborative research is initiated between department and students or among students in the following ways:

1. **Department and students' collaborative research**: This can occur in various ways like; a task for research technique courses, instruction on fundamentals of action research, mentoring, membership in collaborative initiatives with educational institutions, government agencies/departments, and private organizations. A study conducted by Felecia et al. (2018) on the process involved in collaboration at a historically Black university in the United States confirmed that collaborative effort initiatives in under-funded institutions between faculty and students' affairs will promote understanding among the departmental staff and provide a better learning environment for students. Students gain practical research skills, learn from successful or failed investigations in a genuine world context, and get stimulated to pursue collaborative opportunities in the future.

2. **Students collaborative research**: The collaboration takes place among students of similar research experience or between a student with higher expertise assisting another student with lower research experience in choosing a suitable research design, supervising implementation, performing analysis and assessment, and writing reports.

37.3.1.1 Opportunities for Collaboration within Academic Institutions

Collaboration is used as a reliable tool to enhance learning and advance professional development within academic institutions. In their study, Baldwin and Chang (2007) identified various opportunities for collaborative activities within academic institutions, namely:

1. **Co-mentoring**: This involves co-mentoring projects specifically designed to give support to subordinate and experienced faculty members to work in partnership on research, teaching, and service-learning of common interest. These

partnerships are based on the notion that purposeful mentoring can help provide learning opportunities to everyone, regardless of diverse levels of expertise.

2. **Student/research assistant**: Some career advancement funds allow professors to employ graduate students to support their research or projects. These funds provide students with the opportunities to gain from apprenticeships and their department will be able to promote vital projects.

3. **Intellectual community advancement**: Some universities use a portion of their educational development funds to provide resources within their institutions for intellectual society. This requires that various departments formed discussion or reading forums centered on multidisciplinary themes.

4. **Leadership enhancement**: Academic institutions organize workshops for heads of departments to learn and share ideas on how to manage their units effectively.

5. **Research collaboration**: Research groups from different faculties partner on a research project. Usually, work on the project is equally shared among the parties involved and there may be no external financial support to execute the project. Members meet regularly to review progress and strategize on how to publish their research findings.

37.3.1.2 Motivations for Collaboration within Academic Institutions

One of the prominent reasons for collaboration within academic institutions is to actualize goals that cannot be attained all alone. However, beneath this overarching reason, the rationale for collaboration can be categorized into four overarching points; broadening influence, cost-effectiveness and resource networking, and advancing learning.

1. **Broadening influence**: From an ambitious point of view, individuals collaborate with others to boost their individual or institutional prestige, unite against shared opponents or fight for a common interest. Collaboration can be initiated to form a league against a common opponent.

2. **Cost-effectiveness and resource networking**: The limited fund available to higher institutions of learning necessitates the need to adopt collaboration as a tool to optimize inadequate resources while sustaining high-quality performance.

3. **Advancing learning**: Collaborative initiatives connect individuals and organizations to share knowledge rather than spending more resources to access the desired knowledge as individuals. Innovative learning styles or perceptions that readily fit and respond to the dynamic world system are often produced through collaboration.

37.3.1.3 Challenges of Collaboration within Academic Institutions

Anytime individuals or parties work together, disputes are not inevitable. Some of the challenges attributed to collaborations within academic institutions are cultural variation, inability to attain similar goals and interests, geographical limitations, time constraints, and power differentials.

1. **Cultural variation**: It is a major barrier to communication in collaboration, especially in situations of wider cultural variability.

2. **Inability to attain shared goals and interests**: It is seen as one of the factors that make collaboration among partners to be challenging.

3. **Time constraints**: These are one of the major challenges in collaboration within academic institutions, as more time is needed to prepare proposals, resolve disputes, and complete the collaborative initiative.

4. **Geographical disparities**: These poise a lot of threats to communication among participants, as these relate to time and cultural differences.

5. **Power differentials**: These occur among partners when a certain party has superior authority in decision-making and relevant processes over others, there is likely a negative impact on the whole relationship in the team.

37.3.1.4 Elements of Successful Collaboration within Academic Institutions

Collaboration within academic institutions can be implemented or strengthened if it features trust, effective communication, a perception of common goals and interests, and clearly defined roles and expectations (Baldwin & Chang, 2007).

1. **Trust**: The level of trust individuals have among themselves is one of the determinant factors of how collaboration will succeed, because persons who perceived that others in collaboration are excessively cunning or overly opportunistic or high-conflict will be hesitant to participate fully in the collaboration for fear of being abused.

2. **Effective communication**: This is pivotal to promoting and sustaining trust between partners, as well as aligning individual interests and aims.

3. **A perception of common goals and interests**: The collaboration leadership is saddled with the duty of ensuring collective responsibility in attaining target goals, through common vision well-defined goals collectively endorsed task and manner of approach, joint decision-making procedures, and capacity to compromise.

4. **Clearly defined roles and expectations**: This will help to create boundaries as to what individuals are expected to do within the confinement of the purpose of collaboration.

37.3.1.5 Deliverables in Collaboration within Academic Institutions

Common deliverables associated with the interpersonal collaborations within higher institutions of learning are either tangible or abstract.

1. **Tangible deliverables**: This includes research journal articles, enhanced curriculum, as well as new and updated courses. Modern and updated courses derived from the collaborative effort of professors within academic institutions enhance the schools, syllabus, and students' opportunities for learning.

2. **Abstract deliverables**: This entails outcomes like community enhancement, enhanced productivity, cooperative interaction or teamwork, and learning. Studies have shown that activities such as reading groups, and other joint learning methods, creativity, and innovations as individuals meet, share ideas, and collectively learn.

37.3.1.6 *Findings on Collaboration within Academic Institutions in Practice*

Despite the usefulness of collaboration within academic institutions to advance learning and professional training, it can be time-intensive, expensive, and disappointing if not correctly executed and managed. A study on a Mellon Foundation collaborative departmental career advancement program in about 23 universities highlighted that choice of partners, socialization, monitoring and assessment, and flexibility are the components of successful collaborations (Baldwin & Chang, 2007). When partners are carefully selected based on shared interests and goals, the likelihood of conflict is minimized.

Socialization promotes communication among partners, especially in the presence of refreshments, and other opportunities, as excellent ideas are bound to emerge amidst dialogue. The dynamic nature of collaboration requires that participants regularly take a critical assessment of how activities evolve, recognize difficulties, and resolve crises. A failure to critically appraise collaborations can hinder the initial purpose for which it was established. The impact of flexibility cannot be underestimated in collaborative learning within institutions, as it cushions the effect of any eventualities.

37.3.2 Collaboration between Academic Institutions

When people from different institutions work together, different research practices are brought together leading to more insight on the topic at hand, access to different software, database, and equipment as resources lacking in one of the institutions may be seen in other institutions; if, for instance, one of the institutions is well known for research excellence, it helps the growth of the other collaborating institutions and also helps institutions or individuals/researchers to expand their research network.

Considering the unrestrictive application of collaboration, which appears to cover a broad range of combined activities, many sources of literature have focused on identifying the characteristics of collaborations to explain or classify different approaches endorsed by institutions to work collectively. For example, Robinson et al. (2000) identified cooperation, competition, and coordination as types of collaborative engagement. Cooperation entails combining various and distinctive institutions to work together as unique organizations to achieve a higher result that could not be achieved independently. Competition is rivalry among institutions for economic rationality and provides learners with choices. And coordination occurs among institutions working to make the best use of available resources, to maximize the commitment of all participating institutions to the common interest. Thomas and Woodrow (2002) also highlighted four collaborative models:

1. **Bridge**: It is characterized by institutions with similar systems, quality, and scale. The collaborative work involved in the model is substantial and long-lasting. Relatively, it usually takes a long time to gather momentum and advance, because there is no single leading institution, which makes it difficult to develop a common vision and reach agreement on metrics for success.

2. **Pyramid**: It depicts a leading academic institution that controls how other participating institutions relate to each other. In this situation, collaborative approaches are enforced; there is a considerable level of coordination and clarified directives from the early stages of development. These highlighted features of the pyramid model relatively give room for easy progress assessment and monitoring.

3. **Marriage**: It features collaboration work that is usually prompted by either or both central and political obligations, with the anticipation for economic ratification.

The collaborative activities involved in the marriage model are an amalgamation and maybe profitable or otherwise.

4. **Spider**: It is characterized by diverse organizations collectively working from various parts within their sectors and sometimes outside the sector. Participating organizations can be nonidentical in magnitude and quality. The model is integrated on a large scale but challenging, based on the intricacy triggered by the numerous and varying programs of the organizations. The collaborative work involved in this model exists because of the presence of interpersonal relationships rather than institutional connection; procedures for operation rely on individuals involved and not on institutional partnership.

37.3.2.1 Opportunities for Collaboration between Academic Institutions

Opportunities provided for interinstitutional collaboration among educational institutions are:

1. **Intellectual consultation grants**: The professors are provided with funds to travel to consult with a fellow somewhere else who has a common research area or they invite the fellow to the grant holder's home institution.

2. **Aid for collaborative studies**: Various institutions develop strategies to enhance research collaborations across institutions. This form of aid for cross-institutional project collaboration is specifically beneficial for departments who could be the only ones at their institution who majors in a particular field.

3. **Field-based workshops and seminars**: Departments from different universities come together to design workshops and conferences on common areas of expertise or interdisciplinary interests. These jointly scheduled meetings will be attended by various institutions to share knowledge, promote research skills, and exchange creative educational strategies.

4. **Aid for the intellectual community**: Many universities set aside some resources to provide time and facilities for the intellectual community on their campuses. This is usually done to ensure continuous discussion between scholars and the community, by creating reading groups centered on broad themes and conducting seminars on multidisciplinary themes on different campuses.

5. **Collaborative leadership development**: The similarities among many institutions give room to utilize limited career development resources as they exchange knowledge and resourceful ideas that could boost their labor productivity through collaborative leadership development programs.

37.3.2.2 The Motivation for Collaboration between Academic Institutions

Factors motivating collaboration among academic institutions range from responding to external factors to drawing on business opportunities and resource networking to develop work efficiency and opportunities for learners. In their work on the consequences of cross-institutional relationships, Robinson et al. (2000) classified motivations for interinstitutional collaboration into four, pragmatism, evangelism, synergy, and market imperative. Other motivational factors of interinstitutional collaboration highlighted in literature are shared understanding of common needs by Silverman (1995) in his study about determining factors of institutionalized collaboration and financial strain by (Martin & Samels, 1993).

1. **Pragmatism**: This is based on the expectation that collaboration renders great commercial sense. Institutions work together to prevent duplication of resources like technological and structural resources.

2. **Evangelism**: Institutions collaborate with the notion that good results will emerge from collective work.

3. **Synergy**: The strength and weaknesses of parties in collaboration are harmonized to give all the vital factors to make collaborative activities successful. In this category, all participating institutions have the required features to collectively unite for the success of the collaborative initiatives.

4. **Market imperative**: Collaboration is initiated to capitalize on available opportunities. For instance, institutions working together develop a new curriculum ahead of rival institutions in reaction to the legislative transformation that is required for a career to attain new proficiency standards.

5. **A shared understanding of common needs**: What is essential for collaboration is understanding among member institutions on how their collective effort can resolve their common needs. Usually, parties are united in needs and goals.

6. **Financial strain**: Institutions collaborate to manage inadequate resources rather than restraining an educational opportunity.

37.3.2.3 Challenges of Collaboration between Academic Institutions

Common challenges of collaboration between academic institutions are problems in meeting workforce evolving needs, difficulties in the institutionalization of a new order, inequality in educational systems, and challenges in communication and budgetary allocation (GudeButucha et al., 2014).

1. **Problems in meeting workforce evolving needs**: Each country of the world lays down its job qualification requirements, which are not exactly the same across the globe. Therefore, it would be necessary that both job qualification requirements and workforce operations were standardized, which may not be practicable (Hugonnier, 2007).

2. **Difficulties in the institutionalization of a new order**: Academic institutions are usually faced with difficulties in institutionalizing a new order of opportunities that gives room for greater flexibility and even collaborative initiatives. Therefore, it would require more time to build infrastructure, train staff, and bridge cultural segmentation (Wallace and Pocklington, 2002).

3. **Inequality in education systems**: All over the world, the educational systems are not equitable, considering the three levels of equitability, which are accessibility of education to everyone, equal access to quality education, and equality of educational outcomes. Hence, there are barriers to collaborative ventures between students across institutions, because the socioeconomic status of many of them has a major influence on their ability to carry out certain work (Hurgonnier, 2007).

4. **Challenges in communication and budgetary allocation**: Academic institutions in collaboration find it difficult to sustain the momentum partnership

because there were no clear rules of communication and no adequate funds allocated by each party to support the collaborative ventures (GudeButucha et al., 2014).

37.3.2.4 Elements of Successful Collaboration between Academic Institutions

Elements of successful cross-institutional collaboration proposed by Thomas and Woodrow (2002) include common vision, endorsed strategy and harmonized incentives, good choice of partner, senior managers' commitment, prioritizing, progress metrics, and adopting practical techniques in situations where there are short-duration opportunities.

1. **Common vision**: Institutions collaborating have common goals.
2. **Endorsed strategy and harmonized incentives**: Institutions involved in collaboration have a thorough understanding of the responsibilities of each party and the techniques to execute the initiative.
3. **Effective communication**: Good communication among parties enhances collaboration activities.
4. **Good choice of partner**: Choice of institution to partner is well considered rather than being in partnership with any available institution.
5. **Senior managers' commitment**: Actively committed senior managers rather than just being eloquent.
6. **Prioritizing**: Emphasize activities that are crucial to the successful outcome of the initiative, above other activities involved in the entire collaborative process.
7. **Outcomes and progress indicators**: Clearly defined outcomes and progress indicators for collaborative initiatives to assess benefit rather than mere description.
8. Adopting practical techniques in situations where there are short-duration opportunities. This entails taking a straightforward collaboration technique for the self-supporting design-type initiative.

37.3.2.5 Deliverables of Successful Collaboration between Academic Institutions

The deliverables in interinstitutional collaboration initiatives cut across various research areas, depending on the type of educational programs offered in the partner institutions (Fraser et al., 2015). These deliverables are either tangible or abstract.

1. **Tangible deliverables**: This includes research results and publications, as well as new resources that would enhance practice.
2. **Abstract deliverables**: These are new or expanded professional networks, inspiring ventures that stimulate individual daily roles, and skill development for career advancement.

37.3.2.6 Findings on Collaboration between Academic Institutions in Practice

A consortium of three higher educational institutions was formed to serve the population of the city of Detroit, Michigan in the United States, with a special interest in welfare, health, and economic opportunities. The consortium, ReBUILDetroit was designed to change the culture of higher institutions of learning in Detroit, by training students from various and socio-economically underprivileged backgrounds to turn out to be the future generation of researchers in the biomedical field.

Despite the economic hardship, involving students in high schools, with innumerable strengths and weaknesses, coupled with a well-matured college-going job market, the collaborative initiative served as a model for academic institutions in major cities to broaden their workforce and offer more opportunities for career growth for its divergent populations (Andreoli et al., 2017).

37.3.3 Collaboration between Academic Institutions and Government Agencies

In addressing health, socio-economic, educational-, environmental-, and developmental-related issues, government agencies need to collaborate with academic institutions to proffer solutions to current challenges. Academic institutions need to be approached for the technical solution of the problem while the government funds the project, government agencies can create a competitive fund for a particular problem to be solved by researchers, and institutions can also approach the government for collaboration on a research project. Formal agreements on the collaboration between academic institutions and public agencies are initiated through contracts and memoranda of understanding (Livingood et al., 2007).

Collaboration between academic institutions and government agencies/departments has received more attention in recent years. To make sure that innovations resulting from this kind of collaboration advances into employment creation and benefit the entire populace, government agencies and academia must be in partnership during the process of innovation. Innovations are beneficial if corresponding outcomes like inventions, technology, and processes to address needs can be reproduced at a profitable cost. However, in persisting financial constraints, declining resources, environmental issues, and the rapid pace of science and cognitive activities, matching innovation advancement demands effective collaborations between public or government agencies and academia (Saguy, 2011).

37.3.3.1 Opportunities for Collaboration between Academic Institutions and Government Agencies/Departments

Government agencies sponsor research that promotes collaboration between academia, industry, and federal laboratories. These partnerships aim to connect expertise and resources from various origins to resolve pressing challenges in society. However, each agency or institution has a wide range of programs that enhance collaboration and serve their own needs. Some of the ways through which government agencies support, facilitate, or funds collaboration are:

1. **Collaborative research agreements**: Are collaborative efforts on research and development (R&D) projects between two or several institutions. Primarily for exploration or to achieve a task or create commercial goods (Williamson et al., 2016).

2. **Utilization of resources agreement**: Promote partnership efforts through resource-sharing. These partnerships are usually for a shorter period during which parties involved can borrow or utilize available resources rather than purchase or build by themselves, which can be expensive and take a lot of time (Livingood et al., 2007).

3. **Educational agreements**: Entail personnel development programs that fit the mission of all participants. Instances of this are internships, fellowships, and professional developments designed to enhance parties involved to gain critical experiences that they would not have the chance to acquire (Livingood et al., 2007).

4. **Consortia**: Assemble institutions to combine expertise and research to address issues that are beyond the reach of anyone party (De Francesco et al., 2002).

37.3.3.2 The Motivation for Collaboration between Academic Institutions and Government Agencies/Departments

The collaboration between academic institutions and government agencies/departments developed based on addressing issues related to insufficient training of the public workforce, inadequate field experience for students, insufficient resources, and low productivity (Livingood et al., 2007):

1. **Workforce development**: The collaboration is initiated to provide training opportunities to employees in the government agencies, especially those with little or no formal training in specific professional practices.

2. **Internship**: Government agencies provide students with opportunities to acquire field experience in their respective degree programs.

3. **Resources mobilization**: As a result of the increasing global risk of contagious diseases and diminishing resources, a collaborative initiative can enhance government agencies and academic institutions to leverage financial, human, and intellectual resources.

4. **Increase productivity**: Capacities to serve can be expanded through collaboration, as the workforce acquire more educational skills, students get access to various degree programs and hands-on experience, while the outcomes of research can be used for solving societal problems.

37.3.3.3 Challenges of Collaboration between Academic Institutions and Government Agencies/Departments

The challenges common to a collaborative initiative between the academic institutions and government agencies/departments are unequal partners, administrative burden, unclear vision, and resistance to change (Livingood et al., 2007).

1. **Unequal partner**: Parties involved in collaboration may not have similar goals and commensurable benefits from the partnership.

2. **Administrative burden**: There are difficulties in budgeting, funding arrangement, and securing the liability of parties involved in the collaboration.

3. **Unclear vision**: Occasionally, public agencies are unclear as to what they are expected to accomplish in collaborative initiatives

4. **Resistance to change**: Public agency employees with inadequate training are usually resistant to more impactful collaborations with academia and are opposed to changes resulting from an academic-health departmental model of running the agency.

37.3.3.4 Elements of Successful Collaboration between Academic Institutions and Government Agencies/Departments

The elements of successful collaboration between academic institutions and government agencies/departments include common vision, mutual benefit, recognition process, agreement protocol, and model law (Livingood et al., 2007).

1. **Common vision**: The two parties in collaboration must have common goals.
2. **Mutual benefit**: Partners have to benefit from the collaboration commensurably.
3. **Recognition process**: This is a review and accreditation process, through which the public agencies create awareness and provide support for the collaboration with academia.
4. **Agreement protocol**: These are guidelines and recommendations for contracts and memorandum of understanding between academic institutions and the public agencies.
5. **Model law**: This will enhance the government to expand and sustain the collaboration at various levels.

37.3.3.5 Deliverables in Collaboration between Academic Institutions and Government Agencies/Departments

The expected outcomes of the collaboration between academic institutions and government agencies/departments are either academic-based or public-agency-based (Livingood et al., 2007).

1. **Academia-based outcomes**: Academic institutions offer research services, evaluation activities, and other assessment programs to the government agencies
2. **Public agency-based outcomes**: Most public agencies provide educational options to the academia through; sites for field-work, adjunct professors for teaching in specific programs of study, advisory support for curriculum development, and health-care services in cases where the health agency/department was involved in the collaboration.

37.3.3.6 Findings on Collaboration between Academic Institutions and Government Agencies/Departments in Practice

The collaboration between local government health agencies and academia in Florida, a state in the southeastern region of the United States of America, involved county health departments (CHDs) and a significant proportion of academic institutions in the state. The CHDs provided the academic institutions with supports like location for fieldwork, adjunct professor to teach in the program of study, advisory assistance for curricula development, and health-care services. On the other hand, the academic institutions only provided research services to the CHDs.

Although the CHDs provided more educational support to academia than they got in return, the collaboration between the dual was able to improve the state health-care system and community-based public health-care service capacity. Also, the services offered to the academic institutions were central to the functions of the CHDs (Livingood et al., 2007).

37.3.4 Collaboration between Academic Institutions and Private Industry

Collaborations between academic institutions and private industry occur in many diverse ways, an industry can fund research with an academic institution in solving a target problem, industries can consult institutions on a solution to their problems, industries help the institution with resources while the institutions form alliances for research, assist in publications, guest lecturing, staff training among others. Some industries collaborate with some academic faculties for staff training whenever they want to acquire knowledge in a new area; new staff members in some industries are also trained by academicians.

Collaboration between academic institutions and private industry entails interaction between higher institutions of learning and private industry and the main motive is to promote knowledge and technology and this collaboration has, in recent times, gained ground in different disciplines. In industries, the surge for this partnership is attributable to pressures of rapid changes in technology, the shorter life cycle of products, and competitive global systems, while in educational institutions, partnerships are initiated in response to the pressures of the rapid increase in modern knowledge and inadequate funding (Giuliani & Arza, 2009). The forms of academic institutions and industries partnership commonly used in practice are network, consortium, alliance, and joint venture (Barringer & Harrison, 2000).

1. **Network**: This depicts a hub and steering wheel structure, where a local institution at the hub organizes the interrelatedness of a complex assembly of other institutions.
2. **Consortium**: This specific cooperative venture involves diverse structures. A consortium is a group of institutions or industries geared toward resolving problems and technological processes.
3. **Alliance**: This is an agreement between two or several institutions that established an exchange union but has no co-ownership.
4. **Joint venture**: This is constituted when two or several institutions merge parts of their resources to form a distinct but collectively owned firm.

37.3.4.1 Opportunities for Collaboration between Academic Institutions and Private Industry

Several companies had the opportunities to partner with academic institutions through combined efforts on R&D, technology transfer, intellectual property utilization, and assistance innovative projects (Perkmann & Salter, 2012).

1. **Combined efforts on R&D**: Industry-academic institutions collaborations are initiated for both parties to access a diverse wealth of knowledge, and reduce R&D expenses.
2. **Technology transfer**: Technologies realized through the complete cycle of innovation at various institutions of learning are disseminated for industry use.
3. **Intellectual property utilization**: Academic institutions-generated intellectual properties usually stimulate more innovations in industries when put to use.
4. **Assistance on innovative projects**: Academic institutions help companies during the entire process of their innovative projects.

37.3.4.2 The Motivation for Collaboration between Academic Institutions and Private Industry

The motivations for academic institutions, especially the universities to collaborate with industries can be classified into six overarching themes, compulsion, reciprocation, productivity, invariability, legitimation, and asymmetrical (Ankraha & AL-Tabbaa, 2015).

1. **Compulsion**: Governments are compelled by the rapidly evolving world system to implement actions in support of research collaborations with academic institutions with the expectation that such partnership will aid in economic growth, as they diffuse expertise and knowledge across industries joined in collaboration.

2. **Reciprocation**: Academic institutions provide comprehensive accessibility to various research structures and expertise, while industries provide comprehensive accessibility to various product design/marketing, market awareness, and job opportunities for college graduates.

3. **Productivity**: Inadequate funding from public funding sources has prompted academic institutions to search for alternative sources of financing for core research and facilities, by commercializing departmental research or project, patent licensing, exploiting academic property rights. Also, educational institutions are more motivated to partner with industry than government agencies, because it usually requires less bureaucracy.

4. **Invariability**: The paradigm shifts to the era of knowledge-focused economy have motivated industries to partner with colleges. Academic institutions enhance the ability of industries to tackle complex issues, create, and stimulate technology firms for business improvement.

5. **Legitimation**: Industries boost their prestige by entering into a partnership with notable institutions. It is believed that association with prestigious institutions will promote authenticity in the presence of other strong parties involved.

6. **Asymmetrical**: Industries are motivated to collaborate with academic institutions to market institution-based technologies for economic benefit. Hence, industries want exclusive ownership of technologies generated and control the path of higher institutions of research efforts.

37.3.4.3 Challenges of Collaboration between Academic Institutions and Private Industry

Studies have shown that the benefits of the partnership between academic institutions and private industry outweigh any possible shortcomings. However, it is paramount to highlight the likely shortcomings so that guiding principles can be developed to avoid failure and achieve the success of this type of collaboration in the future. Consequently, the shortcomings are grouped into four, a diversion from goal or objectives, quality-related issues, dispute/disagreement, and risk (Harman & Sherwell, 2002).

1. **Diversion from goal or objectives**: This occurs through jeopardized research autonomy or loyalty to market benefits that can have adverse effects on academic institution commission and open science. Relatively, delayed educational bureaucracy can suppress the commercialization of technology, lower industry performance, and slow the achievement of industries' goals.

2. **Quality-related issues**: Divided attention and energy of the research team on both academic activities and research work could impact the quality of fundamental research. Likewise, the low mental quality of some contractual work produces impractical resolutions considering that academic institutions are too hypothetical and not pragmatic while the industry is focused on crucial circumstances demanding a quick response.

3. **Dispute/disagreement**: Dispute could emerge between academia and industry over disseminating unfavorable results, conflict might be in the form of flawed relationship among researchers during project execution and disagreement over copyright and patent ownership.

4. **Risk**: The risk of whether to publish results for short-lived income or restrained to license with the dilemma of result becoming outdated, risk of financial support from industry, leaked copyright information, incomplete hand-over, or failure of research outcomes/results.

37.3.4.4 Elements of Successful Collaboration between Academic Institutions and Private Industry

Although collaborations between academic institutions and industries usually produce fascinating results, it has been observed that those results had little or no positive influence on companies' efficiencies. Hence, to get the maximum benefit from university and industry collaborations, measures to be implemented for a positive effect on organizational commodity and procedures include the effective selection process, investment in long-term partnership, effective communication, creating awareness about the project in the industry, and providing internal support.

1. **Effective selection process**: Entails a clear definition of industries' research strategic framework, specific partnership deliverables that can add value to the organization, and recognize the internal beneficiaries of the deliverables at the operational level.

2. **Nominate versatile managers**: Research managers equipped with extensive knowledge of the technology required in the project domain, the capability to connect to operational and institutional boundaries, and inclination to link research and application.

3. **Share a common vision**: Academic institutions are well informed on how the research will help to improve the industry, researchers who can comprehend the industry practices and project goals should be engaged and university partners must appreciate the research critical framework.

4. **Investment in long-term partnership**: Collaboration between industries and academia should be on a multiyear plan, build purposeful relationships with researchers in target academic institutions, even though the industry is not providing direct support to their research work.

5. **Effective communication**: Cultivate and maintain strong communication networks with academic institutions and organize regular in-person or virtual meetings with the research team. Promote extensive staff exchange or a visitor from industry to university and vice versa.

6. **Create awareness about the project in the industry**: Foster interaction between the research team and various operational areas of the industry, enhance effective

feedback mechanism to aid the research team in aligning projects with industry needs.

7. **Provide internal support**: Industry should give suitable assistance for technical and administrative supervision all through the contract period, offer internal support until the research output can be utilized, and incorporate responsibilities for industries utilization of research outcomes as components of the project supervisory duty.

37.3.4.5 Deliverables in Collaboration between Academic Institutions and Private Industry

The deliverables in the partnership between educational institutions and private industries can be classified into commercial and inter-organization (Harman & Sherwell, 2002).

1. **Commercial**: Products of collaboration between academic institutions and private industry include copyright revenue, monetary benefits to researchers, job opportunities, modern or improved commodity/method, wealth creation, and access to the public subsidy.

2. **Inter-organization**: The collaborative initiatives between academia and industry provide the following: practical experience and exposures to professors and students about real-life problems that impact educational curriculum positively, test bed for research evaluation and enhancement, motivation for research in thematic areas, collectively published materials, accelerated innovations, and enhanced prestige for industries.

37.3.4.6 Findings on Collaboration between Academic Institutions and Private Industry

In the year 2017, John DesJardins developed a program for bioengineering students at Clemson University to assess the medical equipment needs in an African country as one of the academic institution's developing countries biomedical equipment innovation cooperative program. The initiative was structured to partner bioengineering students from Clemson University with biomedical companies based in the United States, to improve on existing medical equipment and develop the latest innovations that respond better to the needs of medical practitioners and patients around the globe.

The collaborative initiative served both parties involved. Considering that industries had no time or resources needed to develop relationships with hospitals in those developing nations of the world, while students spent a considerable amount of time and effort in the field, with a focus on the relationship between actual medical equipment needs and what engineers thought was needed. Through the collaboration, students' research findings helped the industries to know the imperfections in their products and ways to improve on them. Likewise, students were able to boost their research expertise through real-life experience (Perkmann & Salter, 2012).

37.3.5 International Research Collaboration

International research collaboration is important to solve a global problem. This kind of collaboration could be between individuals, groups, departments, institutions, companies, regions, or countries. International collaboration helps in establishing and sustaining relationships, getting things done, and sharing different views on certain issues

(Shore & Groen, 2009). For example, the need to research ways to reduce global warming; it's a global challenge that could facilitate the need for international research collaborations. This kind of collaboration encourages access to large scientific laboratories, funding, specific research skills or expertise, data, and software that is not available in a company, region, or country.

An effective way to execute international research collaboration is through strategic alliances. A strategic alliance is an agreement between two or several independent organizations to undertake a project that is mutually advantageous to each of them. The alliance enables each party to pool their resources together while focusing on individual competitive benefits and at the same time expanding their businesses. These resources may include goods, manufacturing capabilities, delivery channels, intellectual abilities, and funds (Martyak, 2014).

37.3.5.1 Opportunities for Domestic and International Research Collaboration

In many cases, researchers can collaborate at an international level through incentives from overseas-based programs and enormous domestic research ability (Adams & Gurney, 2016; UK National Academies Report-Opinion Leader Survey, 2017).

1. **Incentives from overseas-based programs**: Incentives offered by several framework programs in countries like the United Kingdom increase their international research collaboration with countries like the United States of America, France, Germany, and Mexico.

2. **Enormous domestic research ability**: The United States of America is mostly the regular international collaborator for other nations of the world, due to its large domestic research ability, which creates a wide range of research opportunities.

37.3.5.2 The Motivation for International Research Collaboration

International research collaborations are motivated through a personal relationship with foreign researchers, accessibility to resources outside the country of origin, strategic collaboration, a quest for prestige, increased productivity and updates on new developments (García de Fanelli, 2016; Kwiek, 2020).

1. **Personal relationship with foreign researchers**: Interpersonal connections between researchers working in different countries of the world spur collaboration on an international scale.

2. **Accessibility to resources outside the country of origin**: Gaining access to resources that are not available within the home country motivates collaboration with partners abroad.

3. **Strategic collaboration**: In collaborative initiatives, academic institutions in the European countries and the United States of America provide funds for infrastructure and in some cases traveling expenses.

4. **Quest for prestige**: The desire for the prestige that international contacts add to research institutions inspires collaboration abroad, as it helps to strengthen domestic and external legitimization in academia.

5. **Increased productivity**: most researchers tend to believe that across border associations enhance their productivity since the impact of their research would go beyond the domestic academic community.

6. **Update on new developments**: In certain areas, the only way to keep updated on recent developments is to publish in international journals and take part in key conferences.

37.3.5.3 Challenges of International Research Collaboration

The significant impact of international research collaboration in enhancing knowledge and providing business opportunities globally is increasing, owing to Internet connectivity, worldwide development, and higher ideas, human and knowledge transfer. Although international research partnership is growing, some shortcomings that can present obstacles to global collaborations are the absence of national lead, prestige and project management, difficulty in ensuring long-term commitment, difficulty in dividing costs and benefits proportionately, technology transfer, sociocultural diversity, and management complexity (Congress, 1995).

1. **Absence of national lead, prestige, and project management**: The ownership of large amenities depicts power. Relatively, various countries and researchers will only unite to develop these amenities, if there are no other options. This is based on the fact that huge scientific projects are strongly correlated to national authority and reputation.

2. **Difficulty in ensuring long-term commitment**: The most common obstacle to sustaining international collaboration has been the challenge of ensuring long period commitment from all partners involved in the project. Countries are not always willing to collaborate on costly, long-term projects except they are convinced about the commitments of their potential partners. Also, the lack of trust in the dependability of partners makes it hard to develop the mutual confidence required to carry out the best research.

3. **Difficulty in dividing costs and benefits proportionately**: Allocation of funds to contribute and contracts may as well hinder collaboration. All partners in international or domestic collaboration need to be convinced that project funding is well structured. To successfully collaborate, parties involved must be content about the policies on how funds are disbursed.

4. **Technology transfer**: Regarding science and business, countries and industries are not always willing to take part in projects that could lead to the transfer of technologies to potential opponents in which they had a technical or commercial advantage.

5. **Sociocultural diversity**: Differences in language pose a major threat to clear communication that should serve as a vital factor to hold day-to-day institutional discourse and develop trust-based relationships and collegiality that can enhance collective creativity. Other differences that could negatively impact collaboration are a way of living, work practices, and adaptability to a foreign lifestyle.

6. **Management complexity**: The management of international collaborative ventures is more complicated than the management of domestic type. These complexities are expressed through high operational cost, high intricacy of the international decision-making process at both scientific and administrative levels, and in a few cases, there is less financial control and accountability.

37.3.5.4 Elements of Successful International Research Collaboration

In many cases, research collaboration across international boundaries can be most grati-
fying, both vocationally and from an individual perspective. However, international collaborative initiatives are executed with unique challenges and obstacles that parties need
to be cognizant about a priori. Hence, building critical awareness of these pitfalls will
provide profitable returns to modern researchers who are interested in international collaboration as it continues to gain ground across the globe. In this study, measures attributable to a successful outcome in research collaborations among scientists based in different
nations of the world are clarification on the need for international collaboration, reflection
on the qualities of potential collaborator, utilization of pragmatic procedures in building
a relationship, clarification on the level of resources to share in collaboration, specification about the desired type of collaboration, clarification on established goals and end
products, keeping abreast on the barrier to building collaboration or networking, early
agreement on intellectual property right and dissemination rules, preventing a conflict of
interest, and awareness about available funding opportunities (de Grijs, 2015).

1. **Clarification on the need for international collaboration**: International research
   teams may consequently provide access to additional prospects, experience, and
   knowledge. On the other hand, domestic research focus may also result in individual countries' variations in terms of accessibility to resources or facilities beyond
   what can be accessed domestically.

2. **Reflection on the qualities of a potential collaborator**: This is crucial in cases
   where there are language barriers or cultural diversities, which might make
   communication difficult. In the modern world, situations like this encourage the
   increasing number of international collaboration networks along with the shared
   language. However, cultural diversities can enrich collaborative initiatives if parties involved are transparent, flexible, and possess substantial knowledge about
   cultural issues.

3. **Utilization of pragmatic procedures in building relationships**: Opportunities
   for interaction in domestic or international collaboration can occur at meetings or
   conferences. However, more effective ways of establishing effective networks are
   through visiting programs, usually run by some research institutions employing
   workshop series.

4. **Clarification on the level of resources to share in collaboration**: High-ranking or
   seasoned researchers are quite busy people, who get a lot of requests to undertake
   collaborative research yearly. And one of the factors that can gain the readiness
   and engagement of such types of collaborators is contributing to innovative ideas,
   funds, personnel, proprietary information, and accessibility to special equipment.

5. **Specification about the desired type of collaboration**: There are multiple choices
   of collaboration depending on the type of research question to be addressed, the
   purpose of the study, type of expertise required, administrative measures of party
   institutions, priorities of the funding organization, and past experiences with the
   potential collaborator.

6. **Clarification on established goals and end products**: During the developmental
   stages of collaborative research, target goals and outcomes are clearly defined to
   specify the requirement for success and to evaluate every facet of the collaboration
   within the context of the international research partnership.

7. **Keeping abreast of barriers to building collaboration or networking**: In creating domestic research collaborations, there are quite several challenges for various reasons, these shortcomings may be higher in international research collaboration. Researchers planning to initiate international collaboration need to be aware of these factors, even if parties to be involved in the partnership are presumably comparable.

8. **Early agreement on intellectual property rights and dissemination rules**: The likelihood of being constrained by the rules of institutions of each party involved in collaboration necessitates the importance of knowing the boundary conditions that apply to intellectual property rights and marketing of research products at the early stage. Developing active dissemination rules demands attaining agreement on authorized parties to speak on account of collaboration, core audience, and content of information to share, both within domestic and international contexts.

9. **Preventing conflict of interest**: To protect research integrity and prestige, complete transparency is recommended in declaring the conflict of interests at an early stage of development to funding institutions, organizations, collaborators, and journal article editors.

10. **Awareness about available funding opportunities**: Research funds come with different requirements, pressures, amounts, and types depending on the nature of collaboration. In many cases, internal funding institutions are the first point of contact. Nevertheless, typical research funds often cover benefits for international expenses and exchanges. A number of these funding institutions run locally and internationally and usually focus on countries or territories in the third world. Preferably, research institutions or world organizations are the most effective starting point to initiate these potential opportunities.

37.3.5.5 Deliverables in International Research Collaboration

The various types of values developed in domestic and international research collaborations are in the form of products, processes, and services (Klerkx & Guimón, 2017).

1. **Products**: These include scientific outputs like patents and publications.
2. **Processes**: It entails innovation outputs, such as patent licensing, contracts with industry, and spinoff effects.
3. **Services**: These are consulting as well as organizational capabilities like internal collaboration, and management structure.

37.3.5.6 Findings from International Research Collaboration in Practice

ePals is an educational network connecting students and teachers in about 200 nations and territories. Teachers looking for collaborative projects can obtain direct access to community platforms and can connect to impactful discussions while interacting with several like-minded participants virtually.

One of the largest ePals collaborative efforts was Spark! Lab Invent It Challenge, a yearly collaborative event that allows K-12 students all over the world to identify and proffer solutions to real-life problems. In partnership with a study center for invention and innovation, participants in the events were made to follow some specific measures in their invention processes, as well as problem detection, research, model building and examination placing products on the market and display, among others. Although competitive in nature, the event had distinguished students in several ways, and several technologies were invented (Wolf, 2014).

37.4 Technology as an Important Tool for Communication in the 21st century

The crux of development in the 21st century is based on collaboration and communication; these two enable the sharing of thoughts, ideas, and knowledge across platforms and disciplines. It is becoming increasingly necessary for people from different geographical settings and backgrounds to unite together in solving emerging problems and challenges facing mankind. Thus, the need to connect with people of long-distance communication with little or no barrier necessitates the use of technology in communication. Information and telecommunication technologies (ICTs) generally condense temporal and spatial barriers, allowing fast communication and exchange of ideas within the shortest time possible. According to Nickerson (1995), notable events in the recent history of long-distance communication have been the building of computer-based communication networks and the development of technologies that have made possible the implementation and exploitation of these networks.

Technology in itself has enhanced communication, especially long-distance communication, since the early 1990s. Before this period, long-distance communication was delayed for weeks or months as the major means of communication was by posting letters through the traditional post office resulting in long delays before feedback was obtained. However, the globalization of the world through the Internet and ICT has helped to streamline the time taken to share ideas, knowledge, and information between people.

37.4.1 What is Communication?

The word "communication" stems from the Latin word communicate, which means to share or to make common. Thus, communication is defined as the process of sharing and understanding meaning. Effective communication leads to an understanding of the concept being communicated. Several components are involved in effective communication, including:

- **Source**: This describes the root of the communication to be passed across
- **Message**: The main crus of communication, what is to be passed on to other people
- **Channel**: The means through which the message is to be communicated
- **Receiver**: The person or persons to whom the message is meant or communicated
- **Feedback**: The response of the person to whom the message was communicated
- **Environment**: This entails the physical location, condition and time the message was communicated
- **Context**: This involves the circumstance, scene, and desires of people engaged in communication
- **Interference**: The noise or disturbance that hinders or thwarts the effective interpretation of the message communicated.

Communication can be described as the sharing of ideas, knowledge, or message from a source to the receiver through a channel between environments without any interference, and that warrants feedback.

37.4.2 The Importance of Communication in Learning and Collaboration

The central feature of learning is embedded in communicating ideas between people. Effective learning involves the sharing of ideas, knowledge, and information for human development. All the components of communication need to be maximized during learning and collaboration. If one element is left out, the effectiveness of the communication process will be hampered, and the results cannot be achieved. How the information is conveyed helps to complete the communication process.

In learning and collaboration, communication helps the sender and receiver or the people who desire to work together, to establish good self-esteem which will enhance the process. The feedback from the receiver in the form of questions or contributions would boost the self-esteem of the sender, increasing his confidence to be able to share more of what he has with the receiver or other people requiring his ideas with the knowledge that what is being communicated is well received. The feedback given by the sender to the receiver in the form of response to questions or contributions and ideas gives the receiver the energy and will to desire more communication with the sender.

Communication also enhances the relationship between the sender and the receiver as it enhances mutual trust and enables both parties to freely share ideas and information that could be of help in their respective fields.

In collaboration, effective communication between all the members of the group helps to develop interpersonal skills within the team that makes every member function effectively; without it, the aim of collaboration would be defeated. One of the rules for having an effective collaboration is to have accurate and unhindered information flow among the team members. Once a member of the team does not effectively benefit from the communication that takes place within the team members, he would be left in the dark, things would gradually fall apart in the team.

37.4.3 Communication for Learning and Collaboration Communication for Learning

Once technology is involved, the use of multimedia cannot be left apart, this is because multimedia plays a huge part in conveying the clear message of the sender to the receiver. The following are ways technology has enhanced communication in learning.

37.4.3.1 Long Distance Learning (Online Courses and Degrees)

Distance learning programs are increasingly becoming useful to individuals who intend to acquire additional learning or certificates but, because of other engagements, cannot attend the regular educational programs. Technology has boosted the availability of degrees and professional courses through distance learning programs, removing barriers arising from the conventional classroom settings that entail everyone converging for learning to take place. Educationists can, through suitable technology and multimedia, make their curriculum content available to distant learners without any hindrance in communication.

The learners, on the other hand, can access information and knowledge within their comfort zone, without being under any intense pressure. They can have the full benefits of a student; in terms of the lectures, answers to questions, and the ability to relate with other learners. Further, both the learners and their teachers can explore and share ideas as though they are in the same place, questions and answers are not delayed till the next meeting, and learners can have their questions answered by multiple people at the same time without the restriction of time during the study. Materials, assignments, and other information are accessible to the students at a convenient time. The medium allows students to be introduced to a community comprising old and new students which enhances learning and connecting people of like minds.

37.4.3.2 Interactive Classes

Technology has made it possible for classes to be interactive, whether it is physical or virtual. Teachers can easily get instant feedback from students and vice versa without having to wait until the next lesson or the end of the semester. It encourages participants to engage in practical and collaborative activities through various multimedia tools. When some of the tools are incorporated into online courses, it increases students' interaction and engagement with learning. For instance, the use of audiovisuals enhances students' participation in in-class activities and understanding.

37.4.3.3 Fast and Time Saving

Time is essential in the 21st century; all the advancements in technology have helped in saving time, and there is no exception when it comes to learning. The use of technology in learning has helped in saving the time of the teacher and the student simultaneously.

The use of technology in communication during learning has helped to save time. The teacher is allowed to enhance teaching with aids that will make the message conveyed faster and will provoke an adequate response on time. The response is not going to be delayed but can be placed anytime whether in the class or outside the class, as long as the message gets to the receiver on time.

37.4.3.4 Expertise at Your Fingertips

Right from anywhere in the world, anything can be learned from the beginning to the point of mastery. There is ease in accessing materials and information as a result of the use of technology in learning. The student has access to everything on a subject matter from the beginning to the very recent idea about any field of endeavor. There is no longer a need to search far and wide or be physically present to learn any skill in any field. Another aspect of this is that you can access mentorship at your fingertips. You can connect with mentors to learn from without being physically present with them.

37.4.3.5 Encourages Instant Practice and Creativity

One of the things that come with technology-aided learning is the willingness to quickly practice all that was learned. Due to the ease in understanding produced by technology-aided learning, the student is ready to practice the things taught, the added materials to enhance the learning process by the teacher make the student eager and ready to apply the lessons to real-life situations.

It also stirs up the creativity in the student to match the knowledge gained to fit the applicable real-life scenarios.

37.4.3.6 *Lowers the Cost Involved in Learning and Providing Knowledge*

Due to the availability of content easily to the student, the cost of learning is reduced on the part of the student and the university. The university does not have to spend so much on making classrooms, libraries, and other facilities available to the learner. There is a reduced cost for the learner also as the student can access learning content at their convenience.

37.4.4 Importance of Technology and Multimedia in Communication for Collaboration

Communication in collaboration is very essential as it forms the core part of teamwork. Due to the fact, no team can work effectively without having to communicate, the use of technology to communicate ideas, knowledge, and information helps the team to function better. Some of the ways technology and multimedia have helped communication during collaboration are described in the following sections.

37.4.4.1 *Ease of Communication of Ideas*

Ideas from either the team lead or any of the team members are communicated easily to the team due to the aid of technology and multimedia. The sender can express his ideas easily due to the relationship created by the presence of good communication in the team and also use multimedia like audiovisuals, images, and videos to perfectly illustrate the ideas that will provoke feedback from the receiver (team members).

37.4.4.2 *Time Constraints and Difference Removed*

Time is one of the major constraints in learning and collaboration; technological advancements over the years have helped to ease these constraints over time by ensuring that tasks can be done on the go with minimum time either physically or virtually.

The beauty of collaboration is when teams of different disciplines, and in different countries, can work together to achieve a common aim. With this kind of collaboration, the constraint of time difference comes to play and can be overcome easily with the use of technology-aided meetings.

37.4.4.3 *Improves Productivity*

Communication improves the morale of every team member and can easily improve the productivity of the team. Members feel among and they are motivated to work fast to achieve the aim of the project. Because communication of ideas and information in the team is done on time and fast, the efficiency of the team is increased. The team can work effectively and get quick feedback for their tasks because of technology-aided collaboration.

37.4.4.4 *Encourages the Involvement of More People*

With technology-aided collaboration, the size of the team is not an issue, this is since information can flow seamlessly to everyone without any interference. Everybody can easily share their ideas and knowledge with the other members of the team irrespective of the team size.

37.4.4.5 *Easy Access to Learning Materials and Research Data*

Technology has made learning in research collaboration easier and more enjoyable with researchers having access to a wide variety of primary and secondary resources that will enable the team to perform optimally which will, in turn, lead to the achievement of their objectives are easily accessible. These materials could be in the form of a book, journals, research data, etc. By using a repository or an online storage facility, members of the team can get access to research materials. Also, if there is a structure of reporting for each team member regarding their tasks, the team leader does not need to experience time lag in accessing the reports of the team members but would just visit the repository to get the report. Some examples of online data repositories are Havard dataverse, Figshare, Mendeley Data, Dryad digital repository, Open Science Framework, Data World, Awesome Public Data Set, Google Dataset Search, Kaggle; Zenodo.

Technology also provides researchers with access to classes, online degrees or diplomas within or outside their geographical location. Instructional technologies can also serve the instructional needs of researchers in statistics collaboration. Virtual colleges, online college credit courses, all make courses available to researchers and students through the Internet. Through an online program, researchers can obtain their diplomas, learn new concepts without attending a particular school. Some examples of online statistics and data science online courses or diplomas are:

1. World Quant University Data Science
2. Data Science Specialization – John Hopkins University @ Coursera
3. Statistics and Data Science MicroMasters – MIT @ edX
4. Applied Data Science with Python Specialization – UMich @ Coursera
5. CS109 Data Science – Havard
6. Data Science MicroMasters – UC San Diego @ edX
7. Machine Learning Certication – Stanford University @ Coursera
8. Berkeley Program on Data Science and Analytics UC Berkeley Executive Education
9. Alison, Dataquest, Codewar, BitDegree, Code Academy.

37.4.4.6 *Reduces the Cost of Hosting Team Meetings*

With technology-aided collaborations, the cost of hosting physical meetings is reduced, especially when the team members are not in the same region. With technology, teams can meet regularly and effectively.

37.5 Collaboration in Statistics Using Technology and Multimedia

37.5.1 Collaboration Technologies

Collaboration technologies are software, tools, and systems designed to make teamwork more effective both in-office and remotely. They are tools that help in communication and file sharing.

This software is beneficial to teamwork, especially when teamwork is done remotely (i.e. when team members are far from each other but need to connect).

Collaboration technologies can be described as groupware that is designed to make group work optimal, from assigning tasks and responsibilities, to sharing of documents, to checking and approving projects. This groupware makes the overall team workflow easy, allowing the coordination of the achievement of the team's goals. CTI (2019) and Olson and Olson (2013) categorized collaboration software into four which are:

- Communication technology
- Conferencing technology
- Coordination technology
- Computational infrastructure

37.5.1.1 Communication Technology

Communication technology is a software collaboration technology aimed at sharing messages, group chats, and discussions in the text between individuals and groups irrespective of the location of the individuals. In the context of learning and collaboration, communication technology could be divided into two major segments, communication for learning and communication for collaboration.

Communication technology could be instant messaging applications or emailing applications, instant messaging applications, depending on the preference of the class or team. With emailing applications, classes and teams can create individual and group discussions in historical or non-real-time (asynchronous) to enhance productivity. Examples of emailing apps are Gmail, Yahoo Mail, YandexMail, Microsoft Outlook, AOL mail, etc. With instant messaging apps, classes and teams can create individual and group discussions in real time (synchronous) to enhance productivity. Examples of instant messaging apps are Slack, DingTalk, Skype, HipChat, Google Hangout chat, Basecamp 3, Microsoft Teams, and others. WhatsApp and WeChat could also be used in this manner. With these technologies, classes and team members can share information in text and any file needed for the productivity of their members and the security information shared across these platforms. It should be noted that the security of the information shared across these platforms is strong; therefore, no third party can gain access to the information unless granted (Figure 37.1).

37.5.1.2 Conferencing Technology

Conferencing technology enables real-time voice and video communication among a large number of people. This technology enables individuals or collaborating teams to conduct meetings and training remotely in real time. These technologies have screen-sharing devices that enable the host or any other participating member to share ideas with the team in real-time. Some of the text communication technologies also fall in this category, they can as well accommodate voice calls and video calls with a large number of people without restrictions. Conferencing technologies include Zoom, Google Meet, DingTalk, Microsoft Meet, Slack, Uber Conferencing, Cisco Webex Meetings, and Skype.

37.5.1.3 Coordination Technology

Coordination technologies are designed to enable and integrate teamwork and taskwork functions. They allow for interpersonal connections between individuals and group

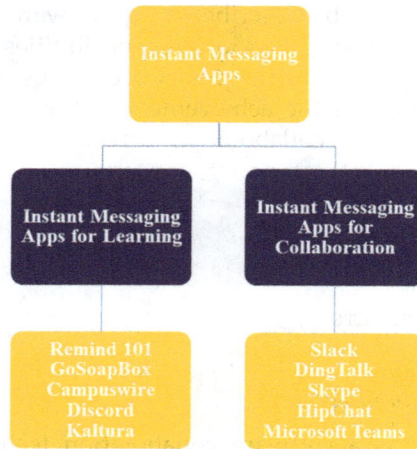

FIGURE 37.1
Instant messaging apps for learning and collaboration.

members and coordinated activities, making end-to-end-work more efficient. Coordination tools help with the assigning of tasks and approval of work, scheduling meetings, and carrying out tasks in real time.

Coordination technologies help the team leader to seamlessly undertake regular coordination of daily team activities, prevent several copies of the project, and enable the ability to track changes made to the project by each team member. Some examples of coordination technologies are Trello, Google Calendar, GitHub, DingTalk, Hive app, and Ryver. These technologies are mainly Internet based and allow team members to work seamlessly while being in different locations.

37.5.1.4 Computational Infrastructure

With the recent increase in the need for large-scale computations in most research and works due to the availability of large data, there is the need for people of computations and development is done in large teams comprising people with different disciplines located at different places, large computation is quite limiting when it is done by everybody and at different places. Not only does it create unnecessary duplicates of the project, but it also drags the project as other team members have to wait for one person to finish before they continue.

Cloud computing has made the job easy, with large computational infrastructures able to store terabytes of data easily and also able to process computations needing days in minutes. These computational technologies are accessible to every member of the team, as their project and research are protected by layers of encryption and checks. Every member of the team has access to these platforms and can work on their part of the project at convenient times. Examples of these computational technologies are Amazon Web Services, Google Cloud Platform, Microsoft Azure Cloud, Alibaba Cloud, Qwiklabs Platforms, Cloudera, and other cloud computational technologies.

37.5.2 Collaboration in Statistics: Possible Options Available Using Multimedia and Technology

The continuous advances in technology have redefined the way researchers obtain information and conduct research; they can now collaborate more widely and efficiently.

The flexibility in the use of multimedia and technology has opened new channels and opportunities that allow statisticians to enhance research collaboration with colleagues from a variety of disciplines. Today, statistical collaboration has become more productive because new enabling collaborative technologies have made it possible for researchers to exchange ideas, data, reports, news, and other resources more rapidly. Multimedia and technology have done much in removing the constraints of computational speed, cost, and distance from research.

Today, collaboration is more of a prerequisite than a preference in research, because it allows a given problem to be solved from several dimensions, and the use of multimedia and technology is increasingly appealing to collaborative research. New collaborative technologies appear every day; however, most researchers do not take full advantage of the various options which can support their research collaborative efforts. Consequently, we present several possible technologies and multimedia collaborative tools open to aid statistical collaboration. These tools are collectively referred to as social network sites (SNS).

SNS are web-based sites that connect a community of people who hold common interests and activities across the economic, political, and geographical borders. SNS have three basic functionalities that turn the sites into useful tools for collaborative research as they allow for identity and network management, information management, and communication.

Recently, specific SNS that intentionally cater to the needs of researchers have emerged. These SNS are called *social research* network sites (SRNS). Bullinger et al. (2010) defined SRNS as SNS that allow researchers to construct public or semipublic profiles within a bounded system, identify other researchers with whom they share a connection, and communicate, share information with other researchers, and collaborate within the system. Several SRNS, including ResearchGate, Academia.edu, Publon, Mendeley, and Zotero, have gained popularity over the past decade. A detailed description of these SRNS and how they effectively aid statistical research collaboration is given below.

37.5.2.1 Academia

Academia is a research directory site that focuses on identifying researchers according to criteria such as research areas and competencies in theory and method. With Academia, statisticians could share their research, follow other researchers with similar research interests, and be able to view other various ranges of topics. Through Academia, researchers can publish their activities on other SNS like LinkedIn. This feature could be used by researchers to expand their research network without having to physically do much.

37.5.2.2 ResearchGate

ResearchGate is a research awareness site aimed to help researchers, including statisticians, stay informed within their network or field of research. This platform allows the researcher to build a profile by making available detailed information about their research works (current or past) and interests and also connect with other researchers within or outside their scope of research interest. It provides statisticians with options to connect and collaborate with research scholars in and outside of statistics, share, and access research outputs and publish data via messages. ResearchGate aids collaboration by suggesting researchers with similar interests or within your research area and also supports social bookmarking. The bookmarking feature allows users to keep track of the activities of interested researchers.

37.5.2.3 Mendeley

Mendeley is a research management and awareness site and desktop application that helps researchers gather, manage, and share references. It helps researchers, including statisticians, organize and manage their research activities, and simultaneously make connections with other researchers by combining its functionalities of the reference manager tool and those of the social research networks. Mendeley creates a social network between its members and its references database by combining its reference library with a social network of researchers. This unique feature makes it possible to share references and make connections with other researchers in the process. On Mendeley, statisticians can create a shared collection where only a selected group of researchers have access to and can contribute in the form of updating details of articles, annotating files, and adding new projects or papers.

37.5.2.4 Zotero

Zotero is an SRNS that allows researchers to collect, organize, cite, and share their work. Zotero allows researchers to organize their research collection through a drag and drop interface to their style, create instant references, and bibliographies through word processor plugins that support more than 9,000 citation styles. With Zotero, researchers can create and join research groups, cowrite an article with other researchers, distribute research materials with collaborators, or build a collaborative bibliography. Zotero sets itself apart from other SNS by being the only SRNS that automatically senses research articles on the web.

37.5.2.5 MethodSpace

MethodSpace is a multidimensional research network site for researchers to network and share research, resources, and debate topics. Similar to other SRNS, MethodSpace allows researchers to create a profile, share their research, follow, and connect with researchers with similar research interests. This site allows researchers to create an interest group for research or field and invite other researchers to join the group. Researchers registered on MethodSpace are granted free access to relevant journal articles, book chapters, and emerging topics in a field.

37.5.2.6 Quora

Quora is an American question-and-answer website where questions are asked, answered, followed, and edited by Internet users, either factually or in the form of opinions. Quora allows users to ask questions, invite users who are experts in that field to answer such questions and vote for answers that are helpful to their questions. Quora also allows users to open, post, and also follow topics and rooms that are inclined toward their interest.

37.5.2.7 Stack Overflow

Stack Overflow is a question-and-answer site for professional and enthusiast programmers. It allows members to post questions relating to their programming challenge and get them answered by both professionals and members who have had a similar problem. These answers are voted by both members of the site based on relevance to the question asked.

37.5.3 Guidelines for Effective Statistical Collaboration through Technology and Multimedia

Research collaboration is an integral component of knowledge development in statistics education and statistical research. Several schools of thought have proposed different models for research collaboration and identified challenges and processes in building a research collaboration team. There are several issues that researchers need to consider when establishing, planning, and conducting collaborative research.

37.5.3.1 Establish Partnership

The following could guide researchers wanting to take full advantage of available technologies to maximize collaborative research efforts:

1. Explicitly state the goals, benefits, expected contribution, and expected results of the collaboration, taking into consideration the view of each researcher.
2. Assess the communication tool available to research partners and the best time for discussion about the research if researchers are not within your region. Also, identify what type of communication system to utilize.
3. Identify and access the resources needed to completely carry out the research. These resources include personnel resources (e.g. data collectors if needed), financial resources needed, and available funding.
4. Identify and clearly state the role and responsibility of each collaborator at the start of establishing a partnership.

37.5.3.2 Execute the Project

After establishing an effective collaborative partnership, it is required to implement the project. In implementing the project, researchers must be able to carefully plan, conduct, and manage the project. In the process of implementing the project, the following guidelines could be adopted:

1. Draw out an effective design for the project, setting out a time frame for the completion of each of its aspects.
2. Establish mechanisms to deal with delays. If a setback occurs, other partners could assist members in missing deadlines. Also, revisit deadlines that seem unattainable.
3. During the implementation of the project, set meeting schedules to discuss the progress of the project.

37.5.3.3 Evaluate the Project

Once project implementation is complete, it is important to evaluate the outcome of the project. The findings from the research must be carefully evaluated. The significance and application of findings to knowledge development and scientific practices should also be evaluated. The project should be carefully evaluated for possible extension or additional areas of research.

37.6 Conclusion

In the overview, the use of technology and multimedia in statistics education and collaboration can have both positive and negative effects. Some of the benefits derived include improved understanding among students on topics related to random processes and seemingly complex statistical concepts through the application of digital simulations and models, easy accessibility to information and skills, deeper learning through multimedia-enhanced classroom engagement and knowledge retention, better communication among teachers and students, flexible approach to learning, less time and effort demanding educational assessment, increased opportunities for professional development among teachers, increased connections among scholars and researchers regardless of geographical and cultural disparities, as well as reduced plagiarism.

On the other hand, the negative impacts of using technology and multimedia-enhanced collaboration in statistics education entail the expensive cost of acquiring and maintaining technology and multimedia tools, limited understanding among students, due to the supply of computations with little or no insight into the underlying statistical method, reduced access to teacher's guidance and support, when outside the classroom, greater tendency to be addicted to unprofitable activities or distracted from educational work, disconnections from face-to-face relationships when classes are often taken online, encourage laziness among students who surf the Internet for solutions to assignments, promote idleness among students and teachers that rely on duplicating statistical information online instead of painstakingly learning the techniques of some methods, and can replace the lecturer/tutor's presence when many students rely on it.

References

Adams, J. (2012). Collaborations: The rise of research networks. *Nature*, 490, 335–336.

Adams, J., & Gurney, K.A. (2016). The Implications of International Research Collaboration for UK Universities. *Digital Science: Digital Research Reports*.

Akman, O., & Powell, M. (2018). A model for cross-institutional collaboration: How the intercollegiate biomathematics alliance is pioneering a new paradigm in response to diminishing resources in academia. *Letters in Biomathematics*, 5, 91–97.

Andreoli, J.M., Feig, A., Chang, S., Mathur, A., & Kuleck, G. (2017). A research-based inter-institutional collaboration to diversify the biomedical workforce: ReBUILDetroit. *BMC Proceedings*, 11, 23. doi:10.1186/s12919-017-0093-6.

Ankraha, S., & AL-Tabbaa, O. (2015). Universities–industry collaboration: A systematic review. *Scandinavian Journal of Management*, 31, 387–408.

Baldwin, R.G., & Chang, D.A. (2007). Collaborating to Learn, Learning to Collaborate. *PeerReview*, 9, 26–30.

Barringer, B., & Harrison, J. (2000). Walking a tightrope: Creating value through inter-organizational relationships. *Journal of Management*, 26, 367–403.

Bartczak, L. (2015). *Building Collaboration from the Inside Out*. Washington, DC: Grantmakers for Effective Organizations.

Benbunan-Fich, R., Hiltz, S.R., & Turoff, M. (2003). A comparative content analysis of face-to-face vs. asynchronous group decision making. *Decision Support Systems*, 34, 457–469.

Bullinger, A.C., Hallerstede, S.H., Renken, U., Soeldner, J.-H., & Moeslein, K.M. (2010). Towards research collaboration – A taxonomy of social research network sites. In *Association for Information Systems (AMCIS) 2010 Proceedings*. (Vol. 92).

Congress, U.S. (1995). *Office of Technology Assessment, International Partnerships in Large Science Projects.* Washington, DC: Government Printing Office.

CTI. (2019). *Type of Collaboration Technology.* Consolidated Technologies, Inc.

Cummings, J.N., & Kiesler, S. (2005). Collaborative research across disciplinary and organizational boundaries. *Social Studies of Science*, 35, 703–722.

De Francesco, S., Bowie, J.V., Frattaroli, S., Bone, L.R., Walker, P., & Farfel, M.R. (2002). The community research, education, and practice consortium: Building institutional capacity for community-based public health. *Public Health Reports*, 117, 414–420.

de Grijs, R. (2015). Ten simple rules for establishing international research collaborations. *PLoS Computational Biology*, 10, e1004311.

Díaz-Gibson, J., Civís-Zaragoza, M., & Guàrdia-Olmos, J. (2014). Strengthening education through collaborative networks: Leading the cultural change. *School Leadership & Management*, 34, 179–200.

European Commission. (2018). Networks for learning and development across school education: Guiding principles for policy development on the use of networks in school education systems. ET2020 Working Group Schools.

Felecia, C., Marybeth, G., Clifton, C., & Thai-Huy, N. (2018). A case study of collaboration between student affairs and faculty at Norfolk State University. *Frontiers in Education*, 3, 1–39.

Fiechtner, S.B., & Davis, E.A. (1992). Why some groups fail: A survey of students' experiences with learning groups. In A. Goodsell, M. Maher, V. Tinto, and Associates (Eds.), *Collaborative Learning: A Sourcebook for Higher Education*. University Park: National Center on Postsecondary Teaching, Learning, and Assessment, Pennsylvania State University.

Fraser, C., Honeyfield, J., Breen, F., Protheroe, M., & Fester, V. (2015). *Getting on: A Guide to Good Practice in Inter-Institutional Collaborative Projects*. Regional Hub Project Fund (RHPF), Ako Aoteraoa.

Frykedal, K.F., & Chiriac, E.H. (2018). Student collaboration in group work: Inclusion as participation. *International Journal of Disability, Development and Education*, 65, 183–198.

Gair, C., Grantmakers for Effective Organizations, & Hope, I.i.E.a. (2012). Strategic Co-Funding: An Approach for Expanded Impact. Scaling What Works.

García de Fanelli, A. (2016). International Research Collaboration: Motivations, facilitators and limitations. *The world view: A blog from the center for international higher education*.

Giuliani, E., & Arza, V. (2009). What drives the formation of 'valuable' university—industry linkages? Insights from the wine industry. *Research Policy*, 38, 906–921.

GudeButucha, K., Balyage, Y., & Hotamo, F. (2014). *Benefits and Challenges of Collaboration in the Institutions of Higher Learning in the East-Central Africa Division*. Baraton: School of Education University of Eastern Africa.

Hakkarainen, K., Paavola, S., & Lipponen, L. (2004). From communities of practice to innovative knowledge communities. *Lifelong Learning in Europe*, 2, 75–83.

Hanleybrown, F., Kania, J., & Kramer, M. (2012). Channeling Change: Making Collective Impact Work. Stanford Social Innovation Review.

Harding, B. (2009). From Bridges to Coalitions: Collaboration between academic advising units and offices that support students of color. NACADA Clearinghouse of Academic Advising.

Harman, G., & Sherwell, V. (2002). Risks in university-industry research links and the implications for university management. *Journal of Higher Education Policy and Management*, 24, 37–51.

Hugonnier, B. (2007). Globalization and education. In Suarez-Orozco, M. M. (Ed.), *Learning in the Era of Global Era*. Berkeley, CA: University of California Press.

Jeong, H., & Hmelo-Silver, C.E. (2016). Seven affordances of computer-supported collaborative learning: How to support collaborative learning? How can technologies help? *Educational Psychologist*, 51, 247–265.

Johnson, D.W., Johnson, R.T., & Smith, K.A. (1991). Cooperative learning: Increasing college faculty instructional productivity. *ASHEERIC Reports on Higher Education*.

Johnson, D.W., Johnson, R.T., & Stanne, M.B. (1989). Impact of Goal and Resource Interdependence on Problem-Solving Success. *The Journal of Social Psychology*, 129, 621–629.

Johnson, D.W., Johnson, R.T., Stanne, M.B., & Garibaldi, A. (1990). Impact of group processing on achievement in cooperative groups. *The Journal of Social Psychology*, 130, 507–516.

Kalia, V. (2019). Strategic Alliance: What is it, Types, Benefits & Why You Need it. WorkSpan.

Kania, J., & Kramer, M. (2011). Collective Impact. Stanford Social Innovation Review.

Kenton, W. (2019). Public-Private Partnerships. Investopedia.

Kezar, A. (2006). Redesigning for collaboration in learning initiatives: An examination of four highly collaborative campuses. *The Journal of Higher Education*, 77, 804–838.

Klerkx, L. & Guimón, J. (2017). Attracting foreign R&D through international centres of excellence: Early experiences from Chile, *Science and Public Policy*, 44(6), 763–774.

Kreijns, K., Kirschner, P.A., & Jochems, W. (2003). Identifying the pitfalls for social interaction in computer-supported collaborative learning environments: A review of the research. *Computers in Human Behavior*, 19, 335–353.

Kwiek, M. (2020). Internationalists and locals: International research collaboration in a resource-poor system. *Scientometrics*, 124, 57–105. doi:10.1007/s11192-020-03460-2.

Livingood, W.C., Goldhagen, J., Little, W.L., Gornto, J., & Hou, T. (2007). Assessing the status of partnerships between academic institutions and public health agencies. *American Journal of Public Health*, 97, 659–666.

Martin, J., & Samels, J. (1993). The new kind of college mergers. *Planning for Higher Education*, 22, 31–34.

Martyak, M. (2014). Strategic Alliances: How They Can Benefit Your Business. POWERLINX.

Nickerson, R.S. (1995). *Emerging Needs and Opportunities for Human Factors Research*. Washington, DC: The National Academies Press.

Olson, J., & Olson, G. (2013). *Working Together Apart: Collaboration over the Internet*. Pennsylvania: Morgan & Claypool Publishers.

Perkmann, M., & Salter, A. (2012). How to create productive partnerships with universities. *MIT Sloan, Management Review*, 53(4), 79.

Resta, P., & Laferrière, T. (2007). Technology in support of collaborative learning. *Educational Psychology Review*, 19, 65–83.

Robinson, D., Hewitt, T., & Harriss, J. (2000). Why inter-organizational relationships matter. In D. Robinson, T. Hewitt, & J. Harriss (Eds.), *Managing Development: Understanding Interorganizational Relationships*. London: Saga Publication Ltd.

Rosenthal, B., & Mizrahi, T. (1994). Strategic Partnerships: How to create and maintain interorganizational collaboration and coalitions. Education Center for Community Organizing at Hunter College School of Social Work.

Rumsey, D.J. (1998). A cooperative teaching approach to introductory statistics. *Journal of Statistics Education*, 6(1)

Saguy, S.S. (2011). Academia-industry innovation interaction: Paradigm shifts and avenues for the future. *Procedia Food Science*, 1, 1875–1882.

Shore, S., & Groen, J. (2009). After the ink dries: Doing collaborative international work in higher education. *Studies in Higher Education*, 34, 533–546.

Sonnenwald, D.H. (2003). Expectations for a scientific collaboratory: A case study. In *Proceeding of the 2003 International ACM SIGGROUP Conference on Supporting Group Work*, pp. 68–74.

Springer, L., Stanne, M.E., & Donovan, S.S. (1999). Effects of small-group learning on undergraduates in science, mathematics, engineering, and technology: A meta-analysis. *Review of Educational Research*, 69, 21–51.

Stahl, G. (2006). *Design of Computer Support for Collaboration*. MIT Press.

State of Victoria. (2017). *High Impact Teaching Strategies: Excellence in Teaching and Learning*. Melbourne: Department of Education and Training.

Thomas, L., & Woodrow, M. (2002). Pyramids or spiders? Cross-sector collaborations to widen participation: Learning from experiences. *Collaboration to Widen Participation in Higher Education*: Cromwell Press.

UK National Academies Report-Opinion Leader Survey (2017). The role of international collaboration and mobility in research: Findings from a qualitative and quantitative study with Fellows and grant recipients of the Royal Society, British Academy, Royal Academy of Engineering and the Academy of Medical Sciences. *The Royal Society.*

Wallace, M., & Pocklington, K. (2002). *Managing Complex Educational Change: Large-Scale Reorganization of schools.* London: Routledge Falmer.

Wegerif, R. (2006). A dialogic understanding of the relationship between CSCL and teaching thinking skills. *International Journal of Computer-Supported Collaborative Learning,* 1(1), 143–157.

Wienkoop, N. (2020). Cross-movement alliances against authoritarian rule: Insights from term amendment struggles in West Africa. *Social Movement Studies.* doi:10.1080/14742837.2020.1770068.

Williamson, H.J., Young, B.R., Murray, N., Burton, D.L., Levin, B.L., Massey, O.T., et al. (2016). Community-university partnerships for research and practice: Application of an interactive and contextual model of collaboration. *Journal of Higher Education Outreach and Engagement,* 20, 55–84.

Wolf, K. (2014). 4 Examples of Global Collaboration: Pure Imagination. *Redbooth.* https://redbooth.com/blog/4-examples-of-global-collaboration-pure-imagination.

38

Statistical Approaches to Infectious Diseases Modelling in Developing Countries: A Case of COVID-19

Ezra Gayawan
Federal University of Technology, Akure (FUTA)

Adeshina I. Adekunle, Anton Pak, and Oyelola A. Adegboye
James Cook University

CONTENTS

38.1 Introduction

Emerging infectious diseases remain public health challenges with serious socioeconomic and political consequences [1,2]. COVID-19, a highly contagious viral infection caused by severe acute respiratory syndrome coronavirus 2 (SARS-CoV-2) was first detected in the city of Wuhan, China, in December 2019. The disease is mainly transmitted between people through direct or indirect (through contaminated objects or surfaces), or close contact with infected people via mouth and nose secretions [3–5]. Although many countries closed their borders to international travellers at the onset of the pandemic, this was not before the virus was transmitted outside of China [6,7]. The majority of the initially reported cases in most countries were linked to China, prompting countries to increase detection, surveillance and evaluate the risk of importations. Similar to the 2009 pandemic of avian influenza H1N1 (swine flu), COVID-19 was declared a pandemic by the World Health Organization (WHO) on March 11th 2020 [8]. As of September 13th 2020, COVID-19 pandemic has spread to all regions of the world, infecting more than 28.6 million people and causing 917,417 deaths (case fatality ratio, CFR=3.2%) across 217 countries [9].

The scope of this chapter focuses on approaches to COVID-19 pandemic, especially in low- and lower middle-income countries (L-LMICs). We first explore the early transmission of the disease outside the initial Wuhan epicentre. Secondly, we take a closer look at various responses to COVID-19 in L-LMICs and give specific examples from India, the Republic of Mauritius and Nigeria. Thirdly, we introduce different mathematical models used in most of these countries. Fourthly, we synthesize various economic responses considered in some L-LMICs. Lastly, we conclude with a discussion of how several countries and regional areas are joining resources to fight the disease, how researchers are collaborating to provide a scientific explanation to the spread of the disease and assess the effectiveness of the mitigation strategies used in these countries.

38.1.1 Early Transmission of COVID-19 in the Low and Lower Middle-Income Countries (L-LMICs)

The literature on early estimates of the transnational spread of COVID-19 is extensive and mostly based on air travel data. Air-travel volumes were keys to evaluate the risk of importation to other regions. It is unclear why the arrival of COVID-19 to many L-LMICs was delayed. L-LMICs are not new to disease outbreaks; the majority of the outbreaks in the last decade occurred in the L-LMICs (Table 38.1). Some studies have linked the slow arrival of COVID-19 in these countries to low detection, preparedness, resources, capacity, travel volume, younger population and perhaps an unidentified genetic factor [6,10–15].

Using early-confirmed cases reported in Wuhan, a study suggested that once there are at least four independently introduced cases, the likelihood that an outbreak will be established in the population is almost certain [16]. Thailand, Cambodia, Malaysia, Canada, the United States and Australia and some pan-European countries were found to have a high risk of importation from China, while Africa and South America had very low risk [10]. Similarly, whereas Asia-Pacific is more at risk (high risk to extreme) from the China epicentre, South America and Africa are more at risk from the Italian epicentre [6]. Preparedness and vulnerability of African countries were linked to the risk of importation from China; African countries with the highest importation risk have moderate to high capacity to respond to such outbreaks [12].

Although COVID-19 pandemic has caused disruptions to health services in most countries, L-LMICs have experienced the greatest difficulties [9]. Prior to the COVID-19 pandemic, L-LMICs have already been dealing with concurrent outbreaks of infectious diseases and were more vulnerable to continuous political instabilities [17–21]. Additionally, routine immunization in L-LMICs is the lowest globally [22,23]. Together, these factors

TABLE 38.1

Major Recent or Ongoing Outbreaks (or Unusual Large Cases of Infection), 2010–2020[a]

Disease	Pathogens	Transmission	Year[b]	Country
Avian influenza	H1N1[c] H7N9	Direct contact with infected animals or contaminated environments	2009/2010 2013–2017	Pandemic China
Chikungunya	*Chikungunya virus*	Bite of an infected female mosquito (*Aedes* spp.)	2013	St. Martin
			2015	Senegal
			2016	Argentina
			2017	France, Italy
			2018	Kenya, Sudan
			2019	Congo
Cholera	*Vibrio cholerae*	Ingestion of contaminated food/water	2010	Central Africa, Haiti, Pakistan
			2011	Congo and DRC
			2012	Sierra Leone and DRC
			2013	Mexico
			2014	South Sudan
			2015	DRC, Iraq, Tanzania
			2017	Kenya and Zambia
			2018	Zimbabwe, Nigeria, Algeria, Cameroon, Somalia, DRC, Mozambique, Tanzania
COVID-19	SARS-CoV-2	Direct, indirect (through contaminated objects or surfaces), or close contact with infected people via mouth and nose secretions[d]	2019/2020	Pandemic
Dengue	Dengue virus	Bite of an infected female mosquito (*Aedes* spp.)	2012	Portugal
			2015	Egypt
			2016	Burkina Faso, Uruguay
			2017	Burkina Faso, Cote d'Ivoire, Sri Lanka
			2018	Reunion
			2019	Sudan, Pakistan, France, Jamaica
			2020	Mayotte, French Guiana, Guadeloupe, Martinique, and Saint-Martin

(Continued)

TABLE 38.1 (*Continued*)

Major Recent or Ongoing Outbreaks (or Unusual Large Cases of Infection), 2010–2020[a]

Disease	Pathogens	Transmission	Year[b]	Country
Ebola	Ebola virus	Direct contact with infected animals, body fluids of an infected individual (dead or alive) or contaminated environments. Sexual transmission from semen of men who have recovered from the disease	2011/2012	Uganda
			2012	Democratic Republic of Congo (DRC)
			2014	West Africa, DRC
			2017	DRC
			2018	DRC
			2019	DRC, Uganda
			2020	DRC
Lassa fever	Lassa virus	Direct contact with body fluids of an infected individual, urine or faeces of *Mastomys* natalensis rats or contaminated environment	2012–2020	Nigeria
Measles	Measles virus	Direct, indirect (through contaminated objects or surfaces), or close contact with infected people via mouth and nose secretions	2011	Regions of the Americas, Europe and Africa
			2015	The Americas, Europe
			2018	Brazil and Japan
			2019	Global situation
			2020	Palestinian territory, Central African Republic, Mexico, Burundi
Meningococcal	*Neisseria meningitidis*	Direct or close contact with infected people via the mouth and respiratory secretions	2010/2011	Chad
			2012	African Meningitis Belt
			2013	African Meningitis Belt
			2015	Niger, Nigeria
			2017	Liberia, Nigeria, Togo
Middle East respiratory syndrome	MERS-CoV	Direct, indirect (through contaminated objects or surfaces), or close contact with infected people via mouth and nose secretions[d]	2012	Saudi Arabia
			2013-2020	Middle East
			2015	South Korea
Plague	*Yersinia pestis*	Bite of an infected flea, contact with contaminated fluid or tissue, infectious droplets	2010	Peru
			2014/2015/2017	Madagascar
			2020	DRC
West Nile virus infection	West Nile Virus	Bites of an infected mosquito (*Culex* spp.), direct contact with body fluids/tissues of infected animals or *In-utero* transmission	2011	Europe

(Continued)

TABLE 38.1 (*Continued*)

Major Recent or Ongoing Outbreaks (or Unusual Large Cases of Infection), 2010–2020[a]

Disease	Pathogens	Transmission	Year[b]	Country
Poliomyelitis	Wild poliovirus	Faecal-oral route or ingestion of contaminated water or food	2011	Pakistan
			2010	Angola
			2011	China
			2011	Chad
			2010	Tajikistan
			2010	DRC
			2010	The Republic of Congo
			2010	Central Asia and the Russian Federation
			2013	Syrian
			2013	Niger
Yellow fever	Yellow fever virus	Bite of an infected female mosquito (*Aedes* spp., *Haemogogus* spp.)	2011	Senegal
			2010	The Democratic Republic of the Congo
			2011	Sierra Leone
			2010/2011	Côte d'Ivoire
			2010/2011/2012	Uganda
			2010	Guinea
			2010/2012	Cameroon
			2012	Ghana
			2010	Senegal
			2012	Sudan
			2012	The Republic of Congo
Zika	Zika virus	Bite of an infected female mosquito (*Aedes* spp.), in-utero transmission, or contact with genital fluids of an infected individual	2015	Americas

[a] *Source*: WHO disease outbreaks (https://www.who.int/csr/don/archive/year/en/) Assessed 16 September 2020.
[b] Confirmed date and not necessarily start/end of the outbreak.
[c] Occasional outbreaks around the world.
[d] Ongoing.

have further weakened the already fragile healthcare system, making it very difficult for these countries to provide adequate and timely measures to mitigate the current pandemic. Moreover, unlike most L-LMICs, Africa has been spared the scale of devastation experienced in some L-LMICs, particularly in Asia and South America, perhaps because Africans are not new to confronting epidemics [24] (Figure 38.1). Thus far, India is the second most severely affected by COVID-19 globally and the worst affected in L-LMICs (accounting for more than 68% of the total COVID-19 cases in L-LMICs). The trajectory of COVID-19 pandemic in L-LMICs has been described as a "slow burn" [25].

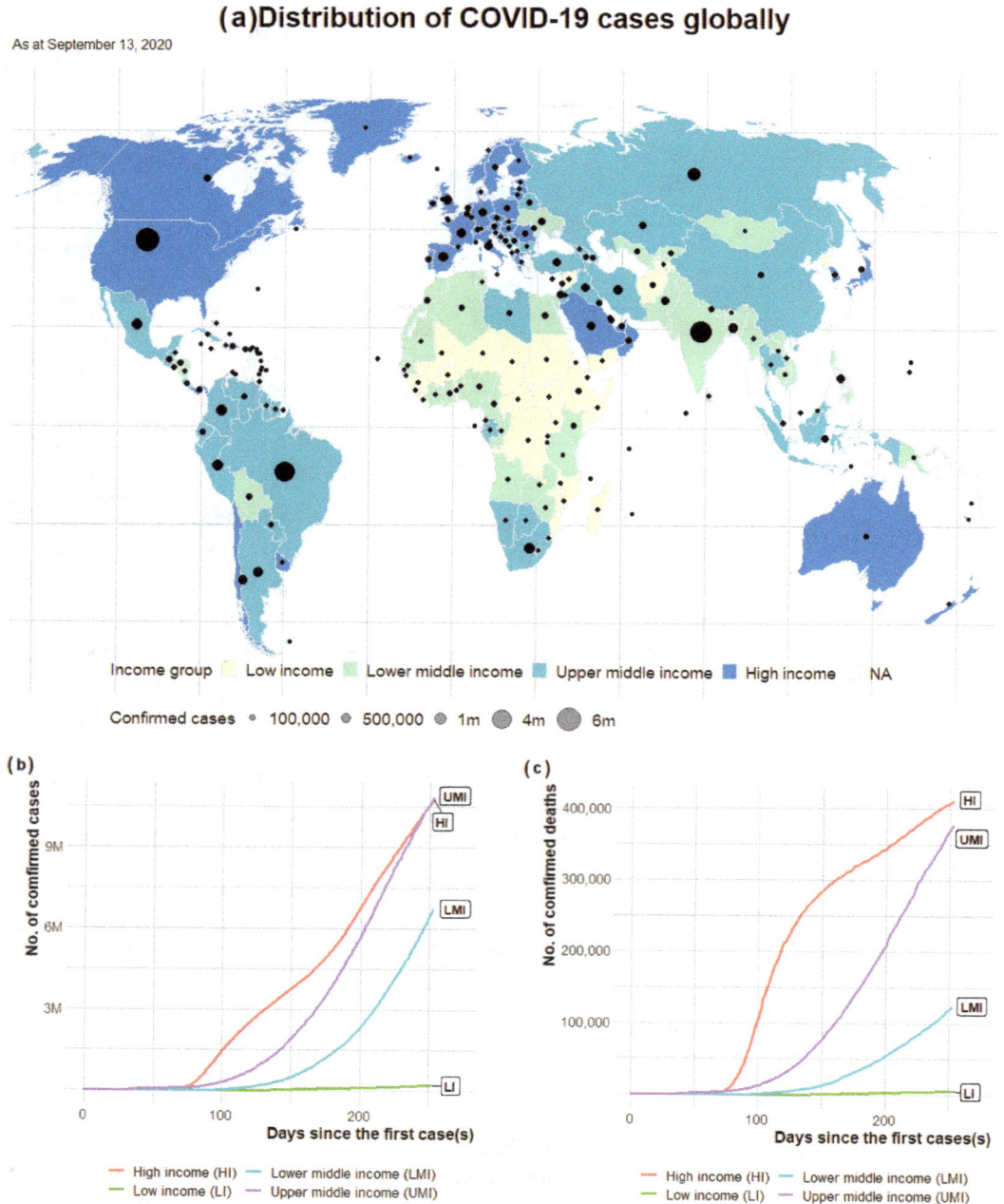

FIGURE 38.1

(a) World map showing the geographical distribution of COVID-19 cases globally, (b) response system of low-income countries could be [26]. Time-series plot of COVID-19 cases from days since the first cases, (c) time series plot of COVID-19 deaths from days since the first cases by World Bank income group.

38.1.2 Responses To and Challenges of COVID-19 in L-LMICs

The COVID-19 pandemic has demonstrated how people from different parts of the world are connected. The global nature of the impact of the pandemic has attracted a global response from public health professionals and experts across disciplines to ensure the disease was brought under control. However, a tremendous increase in cases was still recorded, with most countries experiencing an exponential rise in cases and deaths within a short period. In the absence of an effective vaccine, compliance with WHO recommended public health guidelines for prevention, case identification, quarantine, and treatment remains the effective way of containing the virus.

Before the advent of COVID-19, most L-LMICs have already been facing public health challenges, of which the most devastating ones include Ebola, cholera, malaria, measles, haemorrhagic fever, and meningitis (Table 38.1). The outbreak of Ebola virus diseases and the case fatality rate recorded in some West African countries demonstrate how weak the infectious disease surveillance and response system of low-income countries could be [26].

Collaborative efforts and commitments of national and international governments and agencies were consequently deployed to prevent the global calamity that could have been caused by the virus [27]. However, the continued remnant of the virus in places such as the Democratic Republic of Congo demonstrates that the considerable experience of local health services to identify and deal with emerging pathogens can be hampered by geographical and sociopolitical instability [28].

Prior to the spread of COVID-19 to L-LMICs, the health systems of most of the countries were operating at maximum capacities with the huge workload on their hospital, clinics and health workers [29]. In Table 38.2, we present some healthcare capacity indicators for

TABLE 38.2

Comparison of Some Health Capacities of Selected L-L-LMICs

Country	Density of Medical Physicians[a, b]	Density of Nursing and Midwifery Personnel[a, b]	Density of Pharmacists[a, b]	Hospital Beds[b, c]
Afghanistan	2.8	3.2	0.5	3.9 (2017)
Bangladesh	5.3	3.1	1.6	7.9 (2016)
Benin	1.6	6.1	<0.1	5.0 (2010)
Brazil[d]	21.5	97.1	6.8	20.9 (2017)
Cameroon	0.9	9.3	0.1	13.0 (2010)
Egypt	7.9	14.0	4.3	14.3 (2017)
Ethiopia	1.0	8.4	0.1	3.3 (2016)
Ghana	1.8	12.0	3.6	9.0 (2011)
Haiti	2.3	6.8	0.3	7.1 (2013)
India	7.8	21.1	6.8	5.3 (2017)
Kenya	2.0	15.4	0.5	14.0 (2010)
Liberia	0.4	1.0	0.1	8.0 (2010)
Mauritius	20.2	33.8	3.9	34.0 (2011)
Nigeria	3.8	14.5	0.9	5.0 (2004)
Philippines	12.8	33.3	6.2	9.9 (2014)
South Africa	9.1	35.2	1.5	23.0 (2010)
Zimbabwe	0.8	1.2	0.3	17.0 (2011)

[a] *Source*: WHO World Health Statistics 2019, Annex 2, Part 4.
[b] 10,000 Population, 2009–2018.
[c] *Source*: WHO Global Health Observatory data repository: Hospital bed density.
[d] The recent World Bank income group definition classified Brazil as a middle high-income country.

selected L-LMIC countries. The 2005 International Health Regulation (IHR) indicated a list of actions to be carried out by international organizations and individual countries and regions in order to be prepared for any unfolding public health threat [30]. Several countries were considered to have limited IHR capacities, and a universal improvement was required across the majority of the sub-Saharan African countries [31]. The identified areas of improvement by the report, coupled with the lessons learned from previous outbreaks such as Ebola virus disease and massive investments in surveillance and preparedness, could have positioned most of the countries in a better states to deal with the outbreak of COVID-19.

In responding to the threat posed by the COVID-19 pandemic, most countries of the world adopted the measures implemented in Wuhan, China, where the pathogen was first reported. This includes strict lockdown, social distancing and control of mobility, economic and social activities, all of which were in line with the recommendations of the WHO. However, considering the wide differences in preparedness to public health issues, specific responses by countries are bound to differ. In what follows, we review the responses to COVID-19 pandemic in selected L-L-LMIC countries, namely India, Nigeria and The Republic of Mauritius. India was selected because of its large population, high burden of COVID-19 and proximity to China. Similarly, we considered Nigeria because of its population size in Africa and large traffic between China and sub-Saharan Africa, while the Republic of Mauritius was considered to showcase how a small and Island country responds to the pandemic.

38.1.2.1 India

Countries in the Southeast Asia region have been prone to climate change and emerging infectious diseases, partly because the region is characterized and shaped by differing environmental, ecological and economic factors [32,33]. In the recent past, the region was considered a hotspot for emerging infectious diseases, including those with the potential to transform into a pandemic [33,34]. They are also more susceptible to COVID-19 because of their proximity to China, where the pandemic was first reported. Consequently, to enhance their surveillance, response and other contingency plans against public health challenges, member countries of the Southeast Asia region conduct regular simulation exercises and annual self-assessment with external partners. In September 2019, the countries adopted the "Delhi Declaration" to strengthen their preparedness to respond to emerging public health challenges [35]. The key components of the declaration evolve around collaboration between and among individuals and sectors when responding to public health issues. The four key components include (1) identify risk by mapping and assessing vulnerability for evidence-based planning; (2) invest in people and systems for risk management; (3) implement plans; and (4) interlink sectors and networks to engage and involve all, beyond the sector, who can and have a role in responding to public health emergencies.

The first case of COVID-19 in India was reported on January 30th 2020, the same day that WHO declared the virus a public health emergency of international concern. Earlier, India implemented surveillance measures on January 17th 2020, even before the first case was reported, followed by a series of travel advisories and restrictions [36]. However, it was not until March 25 that the first major control measure that ensured a complete lockdown was put in place. By this time, there were 320 cases and 10 deaths from COVID-19, mostly confined to a few regions [37]. The lockdown was imposed by the Central Government, invoking the Disaster Management Act 2005, which empowers the government to adopt rapid policy decisions and impose restrictions on people to manage any disaster. India also put in place a "five P" response measure for the pandemic. These include (1) proof of concept with a social environment: this ensures complete lockdown with proper experimentation and communication; (2) a proactive approach to ban international flights and screening all international passengers arriving in the country; (3) people management: a "#9PMfor9Minutes" challenge was declared on April 5th 2020 by the Indian Prime Minister to turn off lights and lighting of diyas "oil lamps" and candles for 9 minutes beginning 9 p.m. to demonstrate resolve and resilience for every Indian in the collective fight against COVID-19, and to indicate the country's gratefulness to the frontline workers battling the virus; a cause that was widely supported; (4) partnerships: regular interaction between stakeholders, state governments and the G20 groups that ensured exchanges of ideas and knowledge that further empowered local authorities to confront the pandemic; (5) preparation and collaboration: the country devised and implemented several strategies to contain further spread and death. The health facilities were equipped with personal protective equipment (PPE), ventilators and establishment of isolation wards and COVID-19 care centres. Further, the country, through the National Informatics Centre, came up with an open-source COVID-19 app, "Aarogya Setu" deplored for contact tracing, syndromic mapping and self-assessment. The app tracks the movements of infected persons, alerting a nearby person of their presence, and provides updated travel advisory and containment measures [32,38]

While Indian's proactive efforts at containing COVID-19 might not be adjudged as completely successful going by the astronomic rise in numbers of cases and death around July 2020, the efforts helped to get time for preparation to handle possible widespread situations and were applauded by the WHO [38,39].

38.1.2.2 *The Republic of Mauritius*

The outcome of the responses to COVID-19 in Mauritius demonstrates that a well-implemented and early "hard lockdown" backed by a strong political will could be effective in managing a highly contagious infectious disease outbreak [40]. Mauritius is an African island nation in the Indian Ocean, with about 1.3 million people, and is considered among the most densely populated countries in the world [41]. Long before any case was reported in the island country or even in Africa, the government of Mauritius began the screening of people on arrival at its international airport on January 22nd 2020, introducing fever measurements and separation of people considered to have travelled from countries considered to be at-risk, especially China [42]. The government moved further on February 28th 2020, to quarantine visitors from countries with a high number of cases, even though there was yet any reported case locally. At the same time, the authorities increased accommodation capacity to quarantine suspected COVID-19 cases [43].

The first three cases were detected on March 18th 2020, and the following day, the Mauritius instituted some stringent measures that ensured the closure of all schools and all its borders to travellers, including its citizens [43]. Beginning March 20th 2020, a national lockdown was enforced by the authorities and this was transformed into a curfew that lasted until May 30th 2020. Following WHO recommendations, all confirmed cases were moved to isolation facilities and were closely monitored. The health authorities prioritized contact tracing of people who had been in physical contact with infected patients to identify and test [42]. The country set up a National Communication Committee (NCC) led by the Prime Minister with stakeholders mainly from the Ministry of Health, Ministry of Commerce and the Police Force, who worked collaboratively, addressing the nation regularly, providing COVID-19 statistics and informed the population about the measures put in place and the need for everyone to play their role in the fight against COVID-19. The Ministry of Information and Communication also developed a mobile app called "beSafeMoris" that provided updates and useful tips on the virus and was available for download and used free of charge. The daily data communication by the NCC was used to quantify the effectiveness of measures put in place to contain COVID-19 outbreak in the country. By May 11th 2020 (Day 55 from the date the first cases were reported), there was no active case of COVID-19 in Mauritius [43]. Thus, the Mauritian government was able to achieve the objectives of containing a surge in positive cases of COVID-19 in the country to save the lives of the citizens and prevent the available healthcare facilities from being overwhelmed. The country's prompt interventions and the evident success recorded were highly applauded by WHO and other international organizations [44].

38.1.2.3 Nigeria

Nigeria was one of the West African countries affected by the outbreak of Ebola virus disease in 2014, but the country was notable for its swift action, which demonstrated the importance of adequate preparation and coordinated response in containing infectious disease outbreaks [45]. The successes drawn from the containment of polio and Ebola strengthened the country's healthcare capacity in deploying high-quality surveillance and temperature screening machines at the major ports of entry using equipment acquired during the Ebola outbreak [45]. The human resources, technical expertise, disease surveillance, community networks and logistical capacity used in curtailing polio in the country were equally available for deployment to curtail COVID-19 [46].

Nigeria was prompt in recognizing the risks posed by COVID-19 a few days after the first case was reported in Wuhan, China. The country is seen as a major destination of movements between China and sub-Saharan Africa, as the number of such movements rapidly increased over the past decade [6]. Consequently, a multi-sectoral National Coronavirus Preparedness Group was set up by the Nigeria Centre for Disease Control (NCDC) on January 7th 2020, a week after China's case was reported [47]. The Group met daily to assess the risk COVID-19 posed to the country and review its response to it. The diagnostic capacity for COVID-19 was promptly established in three laboratories. The country's Ministry of Health equally activated Emergency Operation Centres (EOCs) to coordinate the outbreak response activities. The EOCs were organized under six functional units, following the patterns that were adopted for the Polio EOC, namely: Management and coordination, epidemiology, and surveillance, case management, laboratory services, risk communication and point of entry. The EOCs' priorities were to develop the capacities of clinicians, port health officials, point of entry crews and other relevant groups on infection prevention and control, decontamination and contact tracing [46]. Further, the Polio programme has an SMS-based application called "AVADAR" short for "auto-visual AFP detection and reporting" which was used by a network of health workers and community volunteers in hard to reach areas to support disease surveillance through filling in a simple form after receiving a notification [46,47]. This device was deployed by adding relevant disease surveillance questions on COVID-19 to the mobile app.

The index case of COVID-19 was reported in Nigeria on February 27th 2020, and within 2 days, a laboratory diagnostic test for SARS-CoV-2 was set up [48]. Rapid response teams were deployed within the states to lead contact tracing and response activities, while the index case was evacuated to a health facility for treatment and close monitoring [48]. At the onset of the response efforts, the testing capacity of the country was low and limited to symptomatic cases but the number of testing laboratories was increased from 5 to 13 across the six geopolitical zones of the country, leading to more decentralized testing [49]. The testing capacity was also enhanced by donations of testing kits by an individual and a private Biotech company, 54Gene [50,51]. All these efforts notwithstanding, the number of reported cases in the country surged within a few days.

Nigeria initially placed a travel ban on 13 countries considered to be a high risk of COVID-19 and subsequently suspended all international flights in and out of the county on March 23rd 2020. Through a presidential proclamation, two states, Lagos and Ogun, and the Federal Capital Territory, Abuja, considered hotspots during the early outbreak, were placed on total lockdown, and governors of other states implemented similar measures to curb the spread [52]. Restrictions were placed on inter-and intrastate movements, and social and religious gatherings of all forms were banned. All schools and universities were also hurriedly closed. A multi-sectoral and intergovernmental Presidential Task Force on COVID-19 was also put in place. The Task Force ensures the enforcement of all presidential directives related to COVID-19 and equally issues guidelines and regulations weekly. Contact tracing and other public health measures were intensified while the NCDC further deployed an open-source mobile web application for disease outbreak detection, notification, management and response called Surveillance Outbreak Response Management and Analysis System [53]. Daily data collection from all the states, analysis and reporting were equally prioritized by the NCDC to inform the citizens of events around their neighbourhood and to adjudge the progress being made. However, the impact of the lockdown on the economy and, in particular, on the majority of the citizens, especially the most vulnerable became harder by the day as the majority of Nigerians are artisans who depend on daily wages and the daily survival sustenance provided by the government, was grossly inadequate. The lockdowns were consequently relaxed in phases, and citizens were instructed to use face masks and maintain social distance in public places. Despite increasing the number of testing laboratories within the country, low levels of testing to identify infected cases remained an issue in most Nigerian states. Thus, the reported cases, considered fewer when compared with what was reported for some developing countries, may not be true reflections of the number of cases.

38.1.3 Different Mathematical Models Used in L-LMICs

Implementation of public health interventions to mitigate the impact of COVID-19 in developing and developed countries has largely relied on disease spread and its mathematical projections. Mathematical models are essential tools for assessing the underlying mechanisms that govern the spread of infectious diseases, in particular COVID-19 [54]. They were extensively used during Ebola outbreaks from 2013 to 2016 and were instrumental to the development of control policies that helped in curtailing the outbreaks [55]. The Ebola outbreak was mainly in West-African countries, which are all low-income countries. Despite the challenges that face the applicability of mathematical modelling in this region, adjusting for such challenges in modelling exercises still allows mathematical modelling to be readily adapted for changing policies. In the case of COVID-19, there have been limited mathematical models developed to understand the underlying dynamics of the pandemic in L-LMICs [56]. However, those models developed still cover the main modelling types that can be used to determine the epidemic spread and estimate key parameters that aid intervention implementation.

There are two popular modelling approaches that have been adopted for modelling COVID-19. They are deterministic and statistical modelling methods. Deterministic models are mostly compartmental models, while statistical models require the formulation of probability distribution to capture the disease dynamics. The models used in L-LMIC range from simple linear regression to metapopulation models. Also, a single application may require combining more than one modelling approach. We discussed the two approaches below and how they have been applied to COVID-19 in L-LMICs and the model complexities and validity.

38.1.3.1 Compartmental Models

Compartmental models are the most widely used approaches for COVID-19 epidemic modelling [57–59]. In this approach, the population is divided into infection status classes and the disease spread through the population via infection, contact and recovery processes. One commonly used compartmental model is the Kermack and McKendrick SIR (S – susceptible, I – Infectious, and R – recovery) model [60]. This model applies to infectious diseases where individuals become immune after treatment or natural recovery, for example, each strain of flu, measles and COVID-19. Early in the COVID-19 outbreak in February 2020, the decisions of most governments were based on the risk of importation and changing in epicentres [6,12]. Adegboye et al. [6] used a compartmental model to determine where the next epicentre will be and the risk of importation of COVID-19 to Africa. The compartmental model used was a SE1E2I1I2R (S – susceptible, E1 – early exposed, non-infectious, E2-late exposed stage and infectious, I1-early infectious stage, I2-late infectious stage, R – recovered individuals) model. The model used 2014 air travel data between countries to determine the rate of movement of people from one country to another. The dynamical equations for movement between two countries developed were:

$$\frac{dS^1}{dt} = \frac{-\left(\beta_1^1 E_2^1 + \beta_2^1 I_1^1 + \beta_3^1 I_2^1\right)S^1}{N^1} + c_d^1 \gamma_2^1 I_2^1 - \frac{m_{12}S^1}{N^1} + \frac{m_{21}S^2}{N^2} \tag{10.1}$$

$$\frac{dE_1^1}{dt} = \frac{\left(\beta_1^1 E_2^1 + \beta_2^1 I_1^1 + \beta_3^1 I_2^1\right)S^1}{N^1} - \sigma_1 E_1^1 - \frac{m_{12}E_1^1}{N^1} + \frac{m_{21}E_1^2}{N^2}$$

$$\frac{dE_2^1}{dt} = \sigma_1 E_1^1 - \sigma_2 E_2^1 - \frac{m_{12}E_2^1}{N^1} + \frac{m_{21}E_2^2}{N^2}$$

$$\frac{dI_1^1}{dt} = \sigma_2 E_2^1 - \gamma_1^1 I_1^1 - \frac{m_{12}I_1^1}{N^1} + \frac{m_{21}I_1^1}{N^2}$$

$$\frac{dI_2^1}{dt} = \gamma_1^1 I_1^1 - \gamma_2^1 I_2^1$$

$$\frac{dR}{dt} = \left(1 - c_d^1\right)\gamma_2^1 I_2^1 - \frac{m_{12}R}{N^1} + \frac{m_{21}R}{N^2}$$

The parameters in the model are described in Table 38.3. With this modelling effort, the risks of importation of COVID-19 from China and Italy – the then-new epicentre – were determined. The analysis showed that the risk of travellers infected with COVID-19 coming to the L-LMICs was higher from Italy than from China. These prompted many L-LMICs to consider closing their borders and cancelled international flights [52,61].

TABLE 38.3

Description of the Parameters Used in the Model

Parameter	Description
$\beta_1, \beta_2, \beta_3$	Transmission rates for E_2, I_1, I_2 classes
σ_1	First stage incubation rate
σ_2	Second stage incubation rate
γ_1	First stage of recovery
γ_2	Second stage of recovery
c_d	COVID-19 case fatality

The compartmental modelling approach was also used in evaluating non-pharmaceutical intervention controls for COVID-19. COVID-19 is unique in its transmission dynamics, and there were no vaccines for preventing infection and disease. Thus, non-pharmaceutical control measures such as social distancing, contact tracing and lockdowns were introduced [62]. For example, Iboi et al. [63] used this model to investigate the effects of various social distancing measures on reducing COVID-19 cases in Nigeria, where the authors found a moderate level of social distancing and face mask could help control COVID-19 in Nigeria.

38.1.3.2 Statistical Modelling

Another common COVID modelling approach is the statistical models. Early dynamics of COVID-19 in many L-LMICs were evaluated to check the effectiveness of the non-pharmaceutical control measures implemented. One of the parameters for examining this is the effective reproduction number (R_t). An $R_t < 1$ implies that the control measures are effective and ineffective if $R_t > 1$ [64]. This parameter has been estimated for many developed and L-LMICs [62,64–69]. There are many approaches to estimating R_t. Some researchers used compartmental models [68], while others used a probabilistic modelling approach [64]. For example, Adekunle et al. [65] used a Bayesian approach to estimate the R_t for Nigeria between February 27th and May 7th 2020. The model assumed a Poisson distribution for the number of locally observed cases adjusting for imported COVID-19 cases. The estimate showed that as of May 7th, 2020, Nigeria needs to re-evaluate their control measures as R_t was above one. The Bangladesh estimate also shows a similar pattern [70]. For this approach, we let $L(t)$ be locally acquired infections and $\lambda(t)$ be the rate of infection. Thus,

$$\lambda(t) = \left(R_t \sum_{s=1}^{t} \varphi(s) C(t-s) \right) \qquad (10.2)$$

$C(t)$ is the total daily cases, which includes the imported cases, and $\varphi(s)$ is the serial interval distribution. Using equation (10.2), the conditional likelihood via Poisson distribution can be formulated for estimating R_t.

38.1.3.3 Model Complexity and Parameter Sensitivity

Apart from the modelling choice, level of complexities can be added to get robust estimates of disease parameters. The WHO multi-model comparison guide identified the required criteria in adopting a particular model choice to solve real-life problems [71]. However, most of these models are mechanistic with age distribution in transmission but not accounting for sub-populations and comorbidities [72]. The models vary in how COVID-19 transmission, contact patterns and interventions are incorporated. These models are either deterministic or stochastic [72]. For example, Prem et al. [73] used synthetic contact data in a deterministic compartmental model to characterize COVID-19 epidemics in 177 countries, including many L-LMICs. They evaluated three intervention scenarios: 20% physical distancing, 50% physical distancing and shielding, and compared with the unmitigated epidemic [73].

Similarly, the modelling work of Walker et al focused L-LMICs [74]. The model is also a dynamic compartmental model that incorporates age classes and contacts, health system challenges and comorbidities. In most L-LMICs, prior to COVID-19 pandemic, there were other diseases such as malaria, tuberculosis, yellow fever that were weakening the health systems. These existing challenges play a critical role in whether mitigation measures will be effective or not. Walker et al. [74] have to a greater extent, incorporated such factors in their modelling work.

Parameter sensitivity is another aspect that could influence modelling results. Estimated model parameters will vary from setting to setting. For instance, age-dependent contact rates, health capacities and infection fatality rates will differ from country to country or geographically within a country. There are many mathematical methods for conducting parameter sensitivity analysis [75]. This aspect has not been extensively explored for COVID-19 by many modelling groups as the disease is evolving, and detailed information about the dynamics of the virus is still unknown. Conducting country-dependent robustness analysis of the uncertainty surrounding COVID-19 key parameters will be important for future usage of the developed mathematical models for policy change.

38.1.3.4 Disease Mapping

According to Tobler's [76] first law of geography, "everything is related to everything else, but near things are more related than distant things". Geographical locations are important components in decision-making when it comes to prioritizing support systems for emergency health services or situating emergency health facilities, particularly in the face of meagre resources available to most developing countries. Consequently, detecting spatial clusters in disease is an essential component that allows for evidence-based response in containing infectious diseases [77,78]. The goal of disease mapping (also referred to as spatial epidemiology) is to produce spatially smoothed maps that display the variation in disease risk [79]. Apart from helping to determine locations of potential disease clusters, the approach helps to ascertain if the clusters at the different locations are due to the magnitude of the outbreak or resulting from random fluctuations in case count. Approaches to disease mapping can range from a simple mapping of observed disease rate (number of observed cases per unit population) from the different geographical entity to the use of a variety of Bayesian spatial models or other spatial smoothing techniques that estimate risks at different geographical settings.

A number of disease mapping techniques have been deployed to understand the spread of COVID-19 in L-LMICs, ranging from a simple method that maps the incidence and fatality rates to more complex approaches that smooth the spatial variations. In an extensive review of disease mapping approaches to COVID-19, Franch-Pardo and colleagues [80] categorized the approaches into the following groups: spatiotemporal analysis, health and social geography, environmental variables, data mining and web-based mapping. For instance, Arab-Mazar et al. [81] map the incidence and fatality rates for the first 20 days of COVID-19 in Iran using the provinces of the country as the spatial units. On their part, Gayawan et al. [82] used a two-parameter hurdle Poisson model to investigate the spatio-temporal pattern during the first 62 days of COVID-19 in Africa. The Bayesian approach adopts a Markov random field prior for the spatial components that ensure spatial continuity among the spatial units and split the spatial variations into structured and unstructured random effects.

With respect to collaboration in infectious disease modelling, models used for COVID-19 are very diverse. The two models mentioned in this chapter are typical examples of such modelling efforts. Each of these modelling approaches present has its advantages and disadvantages. For instance, compartmental models will give an average representation of disease dynamics in a population, but stochastic models will show the posterior distribution of the parameters. Seeking the average representation of disease dynamics is not as computationally intensive as seeking the posterior distribution. As there are many epidemiological mathematical models employed by many countries to inform policy and control of COVID-19, there are tendencies for such a wide range of modelling efforts to lead to conflicting results and generate alternative opinions. Hence, depending on the researchers' interest and computational power, any of these modelling choices will serve well in helping to understand the dynamics of infectious diseases, as seen in the case of COVID-19.

This implication has led to lead the WHO, the World Bank, the Bill and Melinda Gates Foundation and, the international Decision Support Initiative to form the COVID-19 multi-model comparison collaboration (CMCC) [83]. This initiative was hosted by the Centre for Global Development, the Royal Thai Government and other partners, including the Department for International Development and the following modelling teams: the University of Basel, Institute of Metrics and Evaluation, Imperial College London, Institute for Disease Modelling, London School of Hygiene and Tropical Medicine and the University of Oxford Modelling Consortium came together to form the COVID-19 Multi-model CMCC [83]. They aimed to identify key policy decisions that can benefit from epidemic modelling, resource requirement under different modelling scenarios, data requirement and key assumptions-driven potential differences in predictions, and selecting meaningful models.

38.1.4 Economic Responses to COVID-19 in L-LMICs

Statistical modelling and mathematical projections of the COVID-19 spread have been useful in designing and implementing public health responses by the governments aiming at limiting the spread of the virus and saving lives. As earlier mentioned, common measures deployed at the onset of the pandemic included travel restrictions, bans on mass gatherings, social distancing requirements, school closings and a temporary shutdown of businesses [1,84–86]. Consequently, these containment measures have led to significant economic costs arising from the reduction in economic activities, increase in unemployment, supply chain disruptions and a high level of uncertainty [87,88].

While the COVID-19 pandemic has created economic challenges worldwide, L-L-LMICs might be particularly vulnerable due to their limited health systems capacity, lack of resources and low economic resilience. In this section, we examine economic challenges and policy responses in developing countries, discuss emerging empirical evidence and stress the importance of cooperation and support from international organizations.

38.1.4.1 Challenges to Economic Policies and Public Finances in L-LMIC

COVID-19 pandemic has negatively affected the demand and supply for goods and services. Companies and workers, especially in the industries in which social distancing requirements are difficult to maintain, such as travel and entertainment, have reduced production and work activities. The health crisis has also raised economic uncertainties for economic agents and reduced household income, thus lowering consumer spending and capital investments [89].

To prevent a spiralling economic crisis into a deep economic recession, most governments quickly adopted fiscal and monetary measures. In many developed countries, central banks quickly provided ample liquidity to financial organizations, and some reduced their interest rates easing borrowing constraints [90]. Targeted discretionary fiscal measures were also essential. In March- April 2020, G20 governments committed to provide fiscal support and financial assistance of over US$7.3 trillion to people and firms most affected by the COVID-19 crisis [91]. Expansion of healthcare provisions, unemployment benefits and social assistance payments were among the most common measures to support citizens. Governments of most L-L-LMICs also supported businesses from affected sectors, such as hospitality and transportation, by providing temporal wage subsidies and tax deferrals, as well as supporting businesses through loan relief programmes and credit holidays [92]. The summary of main economic responses by country is described by the International Monetary Fund (IMF) Policy Tracker [93].

These aggressive fiscal and monetary measures were targeted to avoid the economy spiralling by keeping workers in employment and preventing the destruction of valuable economic linkages between businesses. However, the ability to deploy fiscal stimulus measures and deliver broad economic support programmes in L-L-LMICs has been primarily limited by their insufficient fiscal space and capacity [94]. Thus, the economic government assistance in response to the pandemic in developing countries have translated to around 4% of GDP on average, and countries with better fiscal position have averaged 2% of GDP higher in stimulus spending [92].

To finance these new expenditures, governments need to borrow money. However, L-LMICs might experience severe borrowing constraints. Developing countries borrow at higher interest rates due to lower credit ratings and have more significant exposure to exchange rate and maturity risks. Borrowing costs are also increasing in developing economies as public and private debt reached record numbers and now corresponds to 170% of GDP [92]. Less developed financial markets also limit domestic sovereign borrowing. Funding from foreign investors may also be less available for developing countries, as investors tend to move the capital to safer investments during times of high uncertainty and recession. This further raises the cost of borrowing for developing countries. Constraint on public finances makes it critical to use a targeted approach in providing financial support. However, with many developing countries having large informal sectors, less efficient tax administration and lacking social protection systems, the implementation of policy measures aiming to support the most affected firms and people is much more complex to implement than in developed economies.

38.1.4.2 Labour Market and Household Income

Labour markets in developed countries have been quickly responsive to public health measures. For example, in the United States, the weakness of the labour market was evident by the large spikes in unemployment insurance claims across states and a steep drop in job postings [95]. Fiscal stimulus measures for both firms and workers have provided financial support to survive the pandemic and prevent the country from falling into a deep recession. Wage subsidies offered by many governments were targeted to keep workers employed even during the lockdowns when firms were not able to operate. This measure prevents the destruction of many valuable employer-employee connections and enables firms to bounce back quickly during the recovery stage.

However, the situation in the developing countries is more challenging due to their limited fiscal space, the structure of their economies and labour market arrangements. In many L-LMICs, the informal sector accounts for a large part of the economy. The lockdown measures, travel restrictions and reduction in aggregate demand have created a negative shock on jobs and income, with more than 1.6 billion informal sector workers being affected [96]. To mitigate the burden of income loss, most governments in developing countries implemented or widened their unemployment and social security initiatives. The timing of these payments is of high importance as many low-income households might not have sufficient savings and have limited access to borrowing. Borrowing from social networks (e.g. relatives, friends, colleagues) is also likely to be difficult when the pandemic has created widespread economic disruptions and uncertainties. However, the effectiveness of government assistance programmes would be reduced if the rapid roll-out of payments were not implemented to the eligible recipients. To improve the delivery of social benefits, developing countries have started to simplify administrative procedures and increase technological capacity to collect information online and process payments electronically [96].

Job losses were particularly severe in many developing countries whose economies are less diversified and who have been relying on export-oriented industries, such as hospitality, tourism and light manufacturing. The majority of these jobs attract low wages, and workers in these occupations and industries are highly vulnerable to losing their jobs or being suspended without pay as they are least able to transition to "work from home" arrangements. For example, in Bangladesh, the domestic labour market suffered from severe contraction caused by public health restrictions and a sharp decline in exports, particularly from the garment industry. About 46% of Bangladesh producers reported a large number of their orders had been cancelled during the pandemic, and as a result, more than 1 million garment workers lost their jobs or were temporarily suspended without pay [97].

The situation is also exacerbated as these low wage and informal workers could not rely on unemployment benefits as unemployment protection mechanisms are either non-existent or very limited in the majority of L-LMICs (see Figure 38.2). In the early stages of

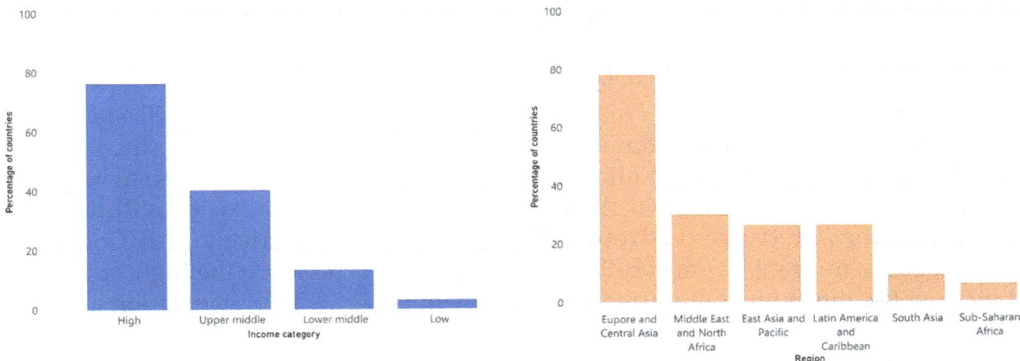

FIGURE 38.2 Availability of unemployment protection by income and region. (World Bank. https://blogs.worldbank.org/developmenttalk/its-time-expand-unemployment-protections.)

the pandemic, it was estimated that the informal workers experienced a drop of 60% in their income [98].

In addition to the heterogeneity in unemployment levels in industries, the impacts of the pandemic on income levels have also been different in rural and urban areas. This is largely attributed to higher human mobility, interconnectedness and population density in urban areas; hence more restrictive containment measures were put in place to mitigate the spread of COVID-19. As a result, for example, in Bangladesh, more than 72% of urban households reported that they lost their main source of income compared to 54% of rural households [99].

38.1.4.3 Migration Flows and Remittances

The economic crisis induced by the COVID-19 pandemic had a significant impact on migration and remittance flows. Lockdowns, travel restrictions and social distancing requirements significantly reduced economic activities in many developed and upper-middle-income countries, which are target destinations for low-waged labour migrants. In addition to the reduction of job opportunities, migrants also face elevated health risks of getting COVID-19 as they often live in suboptimal and overcrowded conditions with no or limited health care in their host countries [100]. Despite a strong public health rationale to extend COVID-19 testing and treatment initiatives to all residents, including migrants, many governments have preferred to prioritize their citizens' given constraints on health systems [101]. This situation was exemplified in the Gulf States with many low-waged migrants restricted to move out of their living compounds and thousands had already been dismissed or furloughed [49]. These migrant workers faced severe financial challenges, as they often do not have regular employment contracts nor can rely on unemployment insurance or social assistance payments.

This is especially challenging for L-LMICs as migrant remittances often account for a large part of their economy and provide an economic lifeline for poor and vulnerable households [102]. Many countries in which remittances account for a large part of foreign exchange revenues might also experience negative macroeconomic implications with respect to inflation, trade and investments. For example, Lesotho and Liberia have very high exposure to foreign exchange, with 86% and 82% of their foreign exchange revenues from international remittances [103]. Furthermore, remittances are important as they can result in poverty reduction in the countries of origin [104,105] and act as an insurance mechanism and a source of investments [106].

Due to the global nature of the COVID-19, with both migrant target destinations and countries of origin being affected, remittances have followed procyclical (rather than countercyclical) behaviour together with the economic downturn in the countries of origin. It is expected that remittances to L-LMICs will drop by $109 billion (around 20% from the 2019 level) in 2020 [102]. This significant reduction in income is reflected in the fall in migrant wages and the drop in the number of jobs available due to the subdued demand for goods and services. While the decline in remittances is expected to be severe in all regions, Europe and Central Asia region and sub-Saharan Africa will be worst affected, with the remittances to drop by 27.5% and 23.1%, respectively [102]. Compared to other regions, migrants from sub-Saharan Africa also experience the highest transfer cost (8.9% of the remitting amount) to send money home [102]. Implementation and promotion of digital and electronic payment systems to drive down the transfer costs have the potential to save significant amounts and increase the disposable income of vulnerable families.

38.1.5 International Collaboration on COVID-19

Many have advocated for extensive collaborative responses to the ongoing pandemic in an attempt to contain its impact and limit further transmission [1,24,56]. The cooperation between international development organizations and national governments has played an important role in addressing health and economic threats by sharing knowledge, developing medical treatments, providing technical expertise and supporting vulnerable countries financially.

Given the limited financial resources of L-LMICs, the provision of official development assistance and emergency financing from international development organizations has played an important countercyclical role aiming to mitigate the health and economic impact of the pandemic. The IMF has made $250 billion available to member countries through multiple lending facilities and debt relief programmes allowing L-LMICs to access concessional borrowing [107]. Similarly, the World Bank Group has planned to provide up to $160 billion in new fast-tracked funding targeting health, economic and social shocks, including $50 billion in grants and low-cost loan facilities for the world's poorest countries [108]. The G20 countries also committed to temporarily suspend debt servicing for the poorest countries [109]. Many other international organizations, including regional development banks, have initiated similar funding and technical assistance programmes to L-L-LMICs. Furthermore, the international development community has advocated for appropriate trade responses in developed and developing countries, which will facilitate the cross-border flow of sensitive goods, such as medical and food supplies [110].

In addition to economic development cooperation, the inter-agency and international collaborations among the scientific community and policy experts have also been extensive during the pandemic. Holistic cooperation within the scientific/academic world and inter-agency collaboration bridged the gap between desks/laboratories and industries (pharmaceuticals), desks/laboratories and politicians, industries (pharmaceuticals) and politicians cannot be overemphasized during this period.

38.2 Conclusion

In this chapter, we have reviewed the literature on approaches to COVID-19 pandemic in L-LMICs. Within the limit of their available resources, L-LMICs took steps to mitigate the challenges posed by the COVID-19 pandemic. It is evident that although the strict restrictions, containment and mitigation efforts put in place to reduce COVID-19 transmission and save lives may have associated economic consequences, especially in L-LMICs [1,111,112]. Prior disease outbreaks in these countries enabled them to prepare infectious disease surveillance protocols that were readily available for deployment during the COVID-19 pandemic. There are similarities and differences in the manner in which the countries responded, leading to the effectiveness or otherwise of similar measures in different countries. For instance, while the lockdown imposed in Mauritius was properly managed and, in combination with other measures, assisted in containing the pandemic in less than 2 months from the time the first cases were declared in the country, such similar measure was not as effective in the other countries leading to a surge in the number of confirmed cases.

The adverse effect of lockdown, such as the accompanying economic and psychosocial issues that challenge the daily livelihood and survival of the majority of the populace, could be a major factor in considering how effective lockdown could be in mitigating infectious disease. International and regional collaboration is essential to mitigate the consequences of the pandemic on a global scale and bounce back quicker on the road to recovery. This is clear from both economic and public health perspectives, as control of COVID-19 is a global public good. In addition to financial and technical support provided to L-LMICs by international development agencies and developed countries, cross-border collaborations and trade cooperation have been crucial in facilitating the flow of essential medical supplies and avoiding food shortages in many L-LMICs. Also worthy of note is that most L-LMICs, as evident from the three countries whose approaches were reviewed in this chapter, adopted multi-sectoral collaborative measures that ensure exchanges of ideas in confronting the virus. Appropriate measures might focus on trade-offs between public health and the economy [25].

References

1. Pak, A., et al., Economic consequences of the COVID-19 outbreak: the need for epidemic preparedness. *Frontiers in Public Health*, 2020. **8**(241): pp. 1–4.
2. Nii-Trebi, N.I., Emerging and neglected infectious diseases: insights, advances, and challenges. *BioMed Research International*, 2017. **2017**: p. 5245021.
3. World Health Organization, *Transmission of SARS-CoV-2: Implications for Infection Prevention Precautions: Scientific Brief*, 09 July 2020. 2020, World Health Organization, Geneva.
4. World Health Organization, *Modes of Transmission of Virus Causing COVID-19: Implications for IPC Precaution Recommendations: Scientific Brief*, 27 March 2020. 2020, World Health Organization, Geneva.
5. Ezechukwu, H.C., et al., Lung microbiota dysbiosis and the implications of SARS-CoV-2 infection in pregnancy. *Therapeutic Advances in Infectious Disease*, 2021. **8**: p. 20499361211032453.
6. Adegboye, O., et al., Change in outbreak epicenter and its impact on the importation risks of COVID-19 progression: a modelling study. *Travel Medicine and Infectious Disease*, 2020. **40**, p. 101988.
7. Adegboye, O.A., A.I. Adekunle, and E. Gayawan, Early transmission dynamics of novel coronavirus (COVID-19) in Nigeria. *International Journal of Environmental Research and Public Health*, 2020. **17**(9): p. 3054.
8. World Health Organization, *WHO Director-General's Opening Remarks at the Media Briefing on COVID-19*, 11 March 2020. 2020, World Health Organization, Geneva.
9. World Health Organization, *Coronavirus disease (COVID-19): weekly epidemiological*, 4 September 2020 [cited 2020 September 2020]. Available from: https://www.who.int/emergencies/diseases/novel-coronavirus-2019/situation-reports.
10. Haider, N., et al., Passengers' destinations from China: low risk of novel coronavirus (2019-nCoV) transmission into Africa and South America. *Epidemiology & Infection*, 2020. **148**: p. e41.
11. Wu, J.T., K. Leung, and G.M. Leung, Nowcasting and forecasting the potential domestic and international spread of the 2019-nCoV outbreak originating in Wuhan, China: a modelling study. *The Lancet*, 2020. **395**(10225): pp. 689–697.
12. Gilbert, M., et al., Preparedness and vulnerability of African countries against importations of COVID-19: a modelling study. *The Lancet*, 2020. **395**: pp. 871–877.
13. De Salazar, P.M., et al., Identifying locations with possible undetected imported severe acute respiratory syndrome coronavirus 2 cases by using importation predictions. *Emerging Infectious Diseases*, 2020. **26**(7): p. 1465.

14. Davies, N.G., et al., Age-dependent effects in the transmission and control of COVID-19 epidemics. *Nature Medicine*, 2020. **26**(8): pp. 1205–1211.
15. Ogunleye, O.O., et al., Response to the novel corona virus (COVID-19) pandemic across Africa: successes, challenges and implications for the future. *Frontiers in Pharmacology*, 2020. **11**: p. 1205.
16. Kucharski, A.J., et al., Early dynamics of transmission and control of COVID-19: a mathematical modelling study. *The Lancet Infectious Diseases*, 2020. **20**: pp. 553–558.
17. Murshed, M. and A. Ahmed, An assessment of the marginalizing impact of poor governance on the efficacy of public health expenditure in LMICS. *World Review of Business Research*, 2018. **8**(1): pp. 147–160.
18. Ndokang, L.E. and A.D. Tsambou, Political instability in Central African Republic (CAR) and health state of the Cameroon population. *Journal of Life Economics*, 2015. **2**(2): pp. 113–129.
19. Fosu, A.K., Instabilities and development in Africa, in *Towards Africa's Renewal*, J. Senghour and N. Poku, Editors. 2020, Ashgate Publishing Limited, Farnham. pp. 209–224.
20. Buchwald, A.G., et al., Aedes-borne disease outbreaks in West Africa: a call for enhanced surveillance. *Acta Tropica*, 2020. **209**: p. 105468.
21. Kapiriri, L. and A. Ross, The politics of disease epidemics: a comparative analysis of the SARS, zika, and Ebola outbreaks. *Global Social Welfare*, 2020. **7**(1): pp. 33–45.
22. Utazi, C.E., et al., Mapping vaccination coverage to explore the effects of delivery mechanisms and inform vaccination strategies. *Nature Communications*, 2019. **10**(1): pp. 1–10.
23. Peck, M., et al., Global routine vaccination coverage, 2018. *Morbidity and Mortality Weekly Report*, 2019. **68**(42): p. 937.
24. El-Sadr, W.M. and J. Justman, Africa in the path of Covid-19. *New England Journal of Medicine*, 2020. **383**(3): p. e11.
25. McBryde, E.S., et al., Role of modelling in COVID-19 policy development. *Paediatric Respiratory Reviews*, 2020. **35**: pp. 57–60.
26. Tomori, O., *Ebola in an unprepared Africa. BMJ*, 2014. **349**: p. g5597.
27. Elmahdawy, M., et al., Ebola virus epidemic in West Africa: global health economic challenges, lessons learned, and policy recommendations. *Value in Health Regional Issues*, 2017. **13**: pp. 67–70.
28. Aruna, A., et al., Ebola virus disease outbreak – Democratic Republic of the Congo, August 2018–November 2019. *Morbidity and Mortality Weekly Report*, 2019. **68**(50): pp. 1162–1165.
29. Kapata, N., et al., Is Africa prepared for tackling the COVID-19 (SARS-CoV-2) epidemic. Lessons from past outbreaks, ongoing pan-African public health efforts, and implications for the future. *International Journal of Infectious Diseases*, 2020. **93**: pp. 233–236.
30. WHO Regional Committee for Africa, *Recurring Epidemics in the WHO African Region: Situation Analysis Preparedness and Response*. 2011, WHO.
31. World Health Organization. *Strengthening health security by implementing the International Health Regulations (2005)*, 2020 [cited 2020 27 August 2020]. Available from: https://www.who.int/ihr/procedures/mission-reports-africa/en/.
32. Madu, E., et al., Scoping review and expert reflections: coronavirus disease 2019 – preparedness and response in selected countries of East Africa, West Africa, and Southeast Asia. *Public Health Open Journal*, 2020. **5**(3): pp. 49–57.
33. Coker, R.J., et al., Emerging infectious diseases in southeast Asia: regional challenges to control. *The Lancet*, 2011. **377**(765): pp. 599–609.
34. Horby, P., D. Pfeiffer, and H. Oshitani, Prospects for emerging infections in East and Southeast Asia 10 years after severe acute respiratory syndrome. *Emerging Infectious Diseases*, 2013. **19**(6): pp. 853–860.
35. World Health Organization, *Member countries of WHO South-East Asia region pledge to strengthen emergency preparedness*, 2019 [cited 2020 29 August].
36. Bharali, I., P. Kumar, and S. Selvaraj, How well is India responding to COVID-19?, in *Future Development*. 2020, Brookings.
37. Ghosh, J., A critique of the Indian government's response to the COVID-19 pandemic. *Journal of Industrial and Business Economics*, 2020. **47**: pp. 519–530.

38. MyGov Blogs, Disaster management lessons of COVID-19. 2020, MyGov Blogs.

39. World Health Organization, *WHO supports India's response to COVID-19*, 2020 [cited 2020 29 August 2020]. Available from: https://www.who.int/news-room/feature-stories/detail/who-supports-india-s-response-to-covid-19.

40. World Health Organization, *WHO representative in Mauritius speaking about COVID-19 during the 5th Nelson Mandela Memorial Lecture 2020*, 2020 [cited 2020 30 August 2020].

41. World Population Review, *Countries by density 2020*, 2020 [cited 2020 29 August].

42. Jeeneea, R. and K.S. Sukon, *The Mauritian response to COVID-19: rapid bold actions in the right direction*, 2020 [cited 2020 30 August 2020]. Available from: https://voxeu.org/article/mauritian-response-covid-19.

43. Sun, M.C. and C.B.L.C. Wah, Lessons to be learnt from the COVID-19 public health response in Mauritius. *Public Health in Practice*, 2020. **1**: p. 100023.

44. Mamode, K.N., A.D. Soobhug, and K.M. Heenaye-Mamode, Studying the trend of the novel coronavirus series in Mauritius and its implications. *PLoS One*, 2020. **15**(7): p. e0235730.

45. Ebenso, B. and A. Otu, Can Nigeria contain the COVID-19 outbreak using lessons from recent epidemics? *The Lancet Global Health*, 2020. **8**: e770.

46. World Health Organization. *Nigeria's polio infrastructure bolster COVID-19 response*, 2020 [cited 2020 31 August 2020]. Available from: https://www.afro.who.int/news/nigerias-polio-infrastructure-bolster-covid-19-response.

47. Shuaib, F.M.B., et al., AVADAR (Auto-Visual AFP Detection and Reporting): demonstration of a novel SMS-based smartphone application to improve acute flaccid paralysis (AFP) surveillance in Nigeria. *BMC Public Health*, 2018. **18**: pp. 57–65.

48. Nigeria Centre for Disease Control. *First case of corona virus disease confirmed in Nigeria*, 2020 [cited 2020 31 August 2020]. Available from: https://ncdc.gov.ng/news/227/first-case-of-corona-virus-disease-confirmed-in-nigeria.

49. Nigeria Centre for Disease Control. *COVID-19 outbreak in Nigeria situation report S/N 54*, 2020 [cited 2020 31 August 2020]. Available from: https://ncdc.gov.ng/diseases/sitreps/?cat=14&name=An%20update%20of%20COVID-19%20outbreak%20in%20Nigeria.

50. Onukwue, A., *54gene rolls out mobile laboratory to boost Nigeria's coronavirus test capacity*. 2020, Techcabal.

51. AFP-Agence France Presse, *Nigeria receives COVID-19 test kits from Chinese Billionaire Ma*. 2020, Yahoo News.

52. World Economic Forum. *COVID-19 in Africa: insights from our 23 April WHO media briefing* [Internet]. World Economic Forum. 2020 [cited 2020 9 September]. Available from: https://www.weforum.org/agenda/2020/04/covid19-in-africa-our-media-briefing-with-who/.

53. Grainger, C., *A software for disease surveillance and outbreak response – insights from implementing SORMAS in Nigeria and Ghana*. 2020 [cited 2020 9 September]. Available from: http://health.bmz.de/en/healthportal/ghpc/case-studies/software_disease_surveillance_outbreak_response/index.html.

54. Yan, P. and G. Chowell, *Quantitative Methods for Investigating Infectious Disease Outbreaks*. Vol. 70. 2019, Springer, Cham.

55. Chowell, G., Fitting dynamic models to epidemic outbreaks with quantified uncertainty: a primer for parameter uncertainty, identifiability, and forecasts. *Infectious Disease Modelling*, 2017. **2**(3): pp. 379–398.

56. Meehan, M.T., et al., Modelling insights into the COVID-19 pandemic. *Paediatric Respiratory Reviews*, 2020. **35**: pp. 64–69.

57. Wang, W. and X.-Q. Zhao, Threshold dynamics for compartmental epidemic models in periodic environments. *Journal of Dynamics and Differential Equations*, 2008. **20**(3): pp. 699–717.

58. Silal, S.P., et al., Sensitivity to model structure: a comparison of compartmental models in epidemiology. *Health Systems*, 2016. **5**(3): pp. 178–191.

59. Biswas, M.H.A., L.T. Paiva, and M. De Pinho, A SEIR model for control of infectious diseases with constraints. *Mathematical Biosciences & Engineering*, 2014. **11**(4): p. 761.

60. Kermack, W.O. and McKendrick, A.G., Contribution to the mathematical theory of epidemics. *Proceedings of the Royal Society of London Series A – Containing Papers of a Mathematical and Physical Character*, 1927. **115**(772): pp. 700–721.

61. Di Caro, B., COVID-19 in Africa: insights from our 23 April WHO media briefing, 2020. Available on https://www.weforum.org.

62. World Health Organization, *Calibrating Long-Term Non-Pharmaceutical Interventions for COVID-19: Principles and Facilitation Tools*. 2020, WHO Regional Office for the Western Pacific, Manila.

63. Iboi, E.A., et al., Mathematical modeling and analysis of COVID-19 pandemic in Nigeria. MedRxiv, 2020.

64. Cori, A., et al., A new framework and software to estimate time-varying reproduction numbers during epidemics. *American Journal of Epidemiology*, 2013. **178**(9): pp. 1505–1512.

65. Adekunle, A.I., et al., Is Nigeria really on top of COVID-19? Message from effective reproduction number. *Epidemiology and Infection*, 2020. **148**: p. e166.

66. Chowdhury, R., et al., Dynamic interventions to control COVID-19 pandemic: a multivariate prediction modelling study comparing 16 worldwide countries. *European Journal of Epidemiology*, 2020. **35**(5): pp. 389–399.

67. Manrique-Abril, F., C. Téllez-Piñerez, and M. Pacheco-López, Estimation of time-varying reproduction numbers of COVID-19 in American countries with regards to non-pharmacological interventions. *F1000Research*, 2020. **9**(868): p. 868.

68. Kucharski, A.J., et al., Early dynamics of transmission and control of COVID-19: a mathematical modelling study. *The Lancet Infectious Diseases*, 2020. **20**(5): pp. 553–558.

69. Abbott, S., et al., Estimating the time-varying reproduction number of SARS-CoV-2 using national and subnational case counts. *Wellcome Open Research*, 2020. **5**(112): p. 112.

70. Hridoy, A.-E.E., et al., Estimation of effective reproduction number for COVID-19 in Bangladesh and its districts. medRxiv, 2020.

71. Den Boon, S., et al., Guidelines for multi-model comparisons of the impact of infectious disease interventions. *BMC Medicine*, 2019. **17**(1): p. 163.

72. COVID-19 Multi-Model Comparison Collaboration (CMCC), *Model Fitness-for-Purpose Assessment Report*. 2020, World Health Organization, Geneva.

73. Prem, K., et al., Projecting contact matrices in 177 geographical regions: an update and comparison with empirical data for the COVID-19 era. *PLoS Computational Biology*, 2020. **17**(1): p. e1009098.

74. Walker, P.G., et al., The impact of COVID-19 and strategies for mitigation and suppression in low- and middle-income countries. *Science*, 2020. **369**: 413–422.

75. Hamby, D.M., A review of techniques for parameter sensitivity analysis of environmental models. *Environmental Monitoring and Assessment*, 1994. **32**(2): pp. 135–154.

76. Tobler, W.R., A computer movie simulating urban growth in the Detroit region. *Economic Geography*, 1970. **46**(sup1): p. 234–240.

77. Saffary, T., et al., Analysis of COVID-19 cases' spatial dependence in US counties reveals health inequalities. *Frontiers in Public Health*, 2020. **8**: 579190.

78. Adegboye, O., D. Leung, and Y. Wang, Analysis of spatial data with a nested correlation structure. *Journal of the Royal Statistical Society: Series C (Applied Statistics)*, 2018. **67**(2): pp. 329–354.

79. Elliot, P., et al., *Spatial Epidemiology: Methods and Applications*. 2000, Oxford University Press, Oxford.

80. Franch-Pardo, I., et al., Spatial analysis and GIS in the study of COVID-19. A review. *Science of The Total Environment*, 2020. **739**: p. 140033.

81. Arab-Mazar, Z., et al., Mapping the incidence of the COVID-19 hotspot in Iran – implications for travellers. *Travel Medicine and Infectious Disease*, 2020. **34**: p. 101630.

82. Gayawan, E., et al., The spatio-temporal epidemic dynamics of COVID-19 outbreak in Africa. *Epidemiology and Infection*, 2020. **148**: p. e212.

83. Chalkidou, K., et al., *Introducing the COVID-19 Multi-Model Comparison Collaboration*. 2020, Center for Global Development.

84. Loayza, N.V. and S. Pennings, *Macroeconomic Policy in the Time of COVID-19: A Primer for Developing Countries.* 2020, World Bank, Washington, DC.

85. Bonaccorsi, G., et al., Economic and social consequences of human mobility restrictions under COVID-19. *Proceedings of the National Academy of Sciences,* 2020. **117**(27): pp. 15530–15535.

86. Brodeur, A., et al., A literature review of the economics of COVID-19. IZA Discussion Paper Series No. 13411, 2020.

87. Altig, D., et al., Economic uncertainty before and during the COVID-19 pandemic. *Journal of Public Economics,* 2020. **191**: p. 104274.

88. Coibion, O., Y. Gorodnichenko, and M. Weber, *The Cost of the Covid-19 Crisis: Lockdowns, Macroeconomic Expectations, and Consumer Spending.* 2020, National Bureau of Economic Research, Cambridge, MA.

89. Chetty, R., et al., *The Economic Impacts of COVID-19: Evidence from a New Public Database Built Using Private Sector Data.* 2020, National Bureau of Economic Research, Cambridge, MA.

90. Gopinath, G., Limiting the economic fallout of the coronavirus with large targeted policies, in *Mitigating the COVID Economic Crisis: Act Fast and Do Whatever It Takes,* R. Baldwin and B. Weder di Mauro, Editors. 2020, CEPR Press, London, pp. 41–48.

91. Hepburn, C., et al., Will COVID-19 fiscal recovery packages accelerate or retard progress on climate change? *Oxford Review of Economic Policy,* 2020. **36**: S359–S381.

92. The World Bank Group, *World Bank Group COVID-19 Crisis Response Approach Paper: Saving Lives, Scaling-Up Impact and Getting Back on Track (English).* 2020, The World Bank Group, Washington, DC.

93. International Monetary Fund, *IMF Policy Responses to COVID-19. Policy Tracker.* 2020. IMF, Washington, DC.

94. OECD, The impact of the coronavirus (COVID-19) crisis on development finance, in *OECD Policy Responses to Coronavirus (COVID-19).* 2020.

95. Forsythe, E., et al., Labor demand in the time of COVID-19: evidence from vacancy postings and UI claims. *Journal of Public Economics,* 2020. **189**: p. 104238.

96. International Labour Organization, *Social Protection Spotlight. Social Protection Responses to the COVID-19 Pandemic in Developing Countries.* 2020. International Labour Organization, Geneva.

97. Anner, M., *Abandoned? The Impact of Covid-19 on Workers and Businesses at the Bottom of Global Garment Supply Chains.* 2020, PennState Center for Global Workers' Rights (CGWR), State College, PA.

98. International Labour Organization, *ILO Monitor: COVID-19 and the World of Work. Third Edition. Updated Estimates and Analysis.* 2020. International Labour Organization, Geneva.

99. Rahman, H.Z., et al., Livelihoods, coping and support during Covid-19 crisis, in *PPRC – BIGD Rapid Response Research.* 2020.

100. Hargreaves, S., et al., Targeting COVID-19 interventions towards migrants in humanitarian settings. *The Lancet Infectious Diseases,* 2020. **20**(6): pp. 645–646.

101. Lau, L.S., et al., COVID-19 in humanitarian settings and lessons learned from past epidemics. *Nature Medicine,* 2020. **26**(5): pp. 647–648.

102. Ratha, D.K., et al., *COVID-19 Crisis through a Migration Lens. Migration and Development Brief 32.* 2020, The World Bank Group, Washington, DC.

103. Machasio, I.N. and N.D. Yameogo. *When the well runs dry: finding solutions to COVID-19 remittance disruptions.* 2000 [cited 16 September 2020]. Available from: https://blogs.worldbank.org/africacan/when-well-runs-dry-finding-solutions-covid-19-remittance-disruptions.

104. Acosta, P., et al., What is the impact of international remittances on poverty and inequality in Latin America? *World Development,* 2008. **36**(1): pp. 89–114.

105. Bertoli, S. and F. Marchetta, Migration, remittances and poverty in Ecuador. *The Journal of Development Studies,* 2014. **50**(8): pp. 1067–1089.

106. Yang, D., International migration, remittances and household investment: evidence from Philippine migrants' exchange rate shocks. *The Economic Journal,* 2008. **118**(528): pp. 591–630.

107. International Monetary Fund, *COVID-19 financial assistance and debt service relief.* 2020 [cited 16 September 2020]. Available from: https://www.imf.org/en/Topics/imf-and-covid19/COVID-Lending-Tracker.

108. The World Bank Group, *World Bank Group's operational response to COVID-19 (coronavirus) – projects list*. 2020 [cited 16 September 2020]. Available from: https://www.worldbank.org/en/about/what-we-do/brief/world-bank-group-operational-response-covid-19-coronavirus-projects-list.

109. The World Bank Group, *Debt service suspension and COVID-19*. 2020 [cited 16 September 2020]. Available from: https://www.worldbank.org/en/news/factsheet/2020/05/11/debt-relief-and-covid-19-coronavirus.

110. Brenton, P. and V. Chemutai, *Trade Responses to the COVID-19 Crisis in Africa*. 2020, World Bank, Washington, DC.

111. Gourinchas, P.-O., Flattening the pandemic and recession curves, in *Mitigating the COVID Economic Crisis: Act Fast and Do Whatever*, R. Baldwin and B. Weder di Mauro, Editors. 2020, CEPR Press, London. pp. 31–39.

112. Pak, A. and O.A. Adegboye, The importance of structural factors in COVID-19 response in Western Pacific. *Asia Pacific Journal of Public Health*, 2021. **33**: 977–978.

Bonus Chapter

Systematically Improving Your Collaboration Practice in the 21st Century

Doug Zahn

This chapter explores two processes for improving your practice:

POWER: A guide to successful meetings.
RAPID: A guide to rapidly recover from your breakdowns.

Part 1: The POWER Process

"Power" in this context does not imply controlling, dominating, or manipulating your client; instead, POWER is an acronym for steps to achieve desired results. Let us consider each of the five steps of POWER: *prepare*, *open*, *work*, *end*, and *reflect* and how to apply them effectively.

Prepare

Your preparation for interaction begins during the scheduling of the meeting. Be sure that you are both clear on when and where you are meeting, as well as the estimated length of time allotted for this meeting. Develop an agenda if possible. Ask your client if there is any background information that would be useful for you to review before the meeting.

If you are both willing and able to video this session, begin with asking permission to do so. Explain that videoing this meeting is a part of your continuing effort to systematically improve your services.

Open

Getting prepared for a conversation is one thing. Opening the conversation effectively is another. The open segment is comprised of four distinct conversations: time, wanted, willing, and able.

The first step is to inquire how much time your client has available for this interaction and then make any adjustments considering how much time you have available. If both of you have agreed on a time frame in the preparation step, confirm that nothing has changed. This is the *time conversation*.

Ask what your client wants to receive from this interaction to learn his list of today's requests. For each request, obtain a complete description of the product or service requested, *including* by when it is to be accomplished. This is the *wanted conversation*.

DOI: 10.1201/9781003261148-46

Ask yourself if you are willing to work on this project, based on your moral, ethical, and professional standards and personal preferences. This is the *willing conversation.*

Ask yourself if you can produce the products or services, given your resources. This is the *able conversation.*

Both the willing and able conversations consider time constraints. The number of client requests, the priority of those requests, and the estimated time required for each will directly affect your response.

Work

Address what is wanted from the interaction.

Providing complete, accurate, technical expertise is crucial for serving your client; however, it is not enough. For example, an explanation that is not understood is useless. Periodically assess whether your client is following you. However, if you ask, "Do you understand?" most people will say, "Yes." A better approach is to ask for an explanation of what you have been discussing. It is important that your interaction be cooperative in nature and not an interrogation. A question such as "How do you see this working with your plan?" could provide the confirmation you seek.

Be sure that your explanations address all aspects of information critical to your client's project. Take the time to do this. Be concise. During the preparation step, spend time and effort on the organization and pacing of information you plan to present, as both assist comprehension.

Finally, though stories may be entertaining, facts embedded in a long story have a good chance of being missed. So, attend to your client's stories, listen for facts that relate to what is wanted today, and limit the length of your stories.

End

Develop a workable plan and end the interaction on time.

A plan is workable if both your client and you agree on three questions: who will do what, to what standards, and by when. Assessing this requires reviewing each of the tasks to be completed and verifying your agreement.

Here, at the end, your client and you are trading a series of requests and promises. This concluding step is the last chance in this interaction to ensure that the requests and promises include all the required characteristics of the tasks to be completed as well as the time frame.

Reflect

As soon as possible following the interaction, consider what worked and what did not work, what you have learned, and how this information can be useful for future interactions. Preserve these answers. Later, watch the video and discuss the session with your client, if possible.

As you become familiar with the five steps of the POWER process, you will become more confident and agile in revisiting the steps when required. It is important to address each of the five POWER steps at some point in the meeting.

Rarely does anyone progress linearly through the five steps in POWER. For example, you may have a complete opening conversation. While you are addressing what your

TABLE 1

The **POWER** Process

Step	Essential Activities
Prepare	Identify and handle essential matters to be prepared for this interaction, including getting the okay to video
	Allow enough time for transition from your previous activity to this interaction
Open	Have the four open conversations:
	Agree on a *time* frame for the interaction
	Identify what your client *wants* from the interaction
	Assess whether you are *willing* to produce your client's wants. Assess whether you can produce your client's wants
Work	Address the wants in priority order
	When offering explanations, stop frequently to give your client a chance to rephrase what she has heard. A request to rephrase provides an opportunity to verify that your client understood what you intended to say. The more complex the material, the more frequently you check for comprehension
End	Allow enough time to verify that both of you agree on what was decided today, what to do, to what standards, and by when
	Close the interaction on time
Reflect	Consider what worked and what did not work in the interaction, how this happened, and how this information can be used to improve future interactions. Take the time to do this step in writing as soon as possible. Later, watch the video and, if possible, review the session with your client
	Do not skip this step

client wants from this interaction in the *work* step, you discover that they have additional wants. Go back to the open conversation. Clarify what these wants are, who the stakeholders are, and how the results of achieving these wants will be scrutinized and used. Considering whether you are willing and able to address these additional wants is also essential, as is reassessing the priorities for this interaction in light of the additional wants (Table 1).

Relationship

There are three foundations of the POWER process: relationship, listening, and speaking. We will consider the quality of these three and how each contributes to an effective interaction. Let's begin with relationships.

There are many different degrees of relationships, which you will readily recognize in your daily life. For our purposes, we will examine casual, working and synergistic relationships using examples for each.

A Casual Relationship

Chuck is scheduled to pick up his dry cleaning at 2:00 p.m. As he is walking to the dry cleaners, he passes a stranger on the street, nods, and trades hellos.

> This is an example of a *casual relationship*: casual—without definite or serious intention; relationship—a connection, association, or involvement. This involvement is without serious intention.

A Working Relationship

> In a working relationship, the parties
>
> - have aligned goals,
> - tell each other the truth as they see it, and
> - do what is necessary to clean up incidents in which they have not stated their truth.

Chuck continues around the corner into his dry cleaner's shop and discovers that the stranger he just passed outside is Darrell, the new owner of the shop.

> Now their relationship is no longer casual. It has a definite intention: Darrell provides the service of dry cleaning for Chuck, who compensates Darrell for his service. Chuck and Darrell now have a connection as customer and dry cleaner.

As long as the clothes are Chuck's, they are cleaned to his satisfaction, and he pays what Darrell asks, the interaction is successfully completed.

If, however, problems arise, such as a shirt is missing, a spot on a suit coat was not removed, and Chuck arrived at 2:00 p.m. when Darrell had promised the clothes for 6:00 p.m., their relationship is facing greater challenges than in the smooth-sailing scenario. The next step is for it to *now* evolve into something closer than a casual relationship—namely, *a working relationship*.

At the *heart of telling the truth* as you see it is a *commitment to provide full disclosure of your intentions and keep any promises you make*. Both parties must be committed to staying in the conversation until each issue is sorted out in a mutually satisfactory manner. This can be challenging because many of us prefer to leave a difficult conversation rather than sort out the misunderstandings. A possibility in this scenario is for Chuck to return at 6:00 p.m., giving Darrell time to find the shirt and remove the spot.

While no expectations are present in a causal relationship, specific expectations are at the heart of a working relationship where parties have aligned goals and will keep their word. They will clean up breaks in integrity and their consequences whenever they occur. Meeting these expectations will result in a relationship that consistently produces effective interactions. This is a cooperative way to work together. Trust will grow between both parties with each promise that is kept

A Synergistic Relationship

If Chuck and Darrell resolve this situation to each other's satisfaction, they have begun to build a working relationship that each of them can count on. If Chuck continues to frequent Darrell's shop and they have additional effective interactions, they will grow a stronger, more resilient relationship.

> Steps along this journey will involve addressing and resolving bigger problems—the lost suit, the burned heirloom dress, and so on. If they continue to treat each other with respect, to avoid criticizing each other when mistakes occur, and to stay in the difficult conversations until they are resolved, they will gradually develop a *synergistic relationship*, one in which their combined efforts produce a total effect that is greater than either could have produced alone. Each is trustworthy in the eyes of the other.

In a synergistic relationship, the parties

- have a working relationship;
- treat each other with respect, never abusing each other, overtly or covertly; and
- will do what is necessary to repair incidents in which they have been abusive, overtly or covertly.

Treating each other with respect also produces a relationship that is cooperative rather than adversarial and is not in competition with the other. The synergistic relationship is also collaborative rather than hierarchical. Decisions are made on the basis of conversations that reach a consensus. Neither

POSITIONAL POWER

The power that a person has in an organization due to rank.

tries to dominate the decision-making process by dictating what will be done.

Even in consultations where one person is the other's boss, a collaborative relationship is still possible. What this requires is that the boss does not use her positional power in the organization to resolve disagreements.

A synergistic relationship is stronger than a working relationship by virtue of producing higher-quality results because of the trust developed in each other. When each realizes that he can make a mistake without being put down by the other, he is free to take risks and be more creative. They no longer have to take the safest path through each situation they encounter.

Two of the most important consequences of a synergistic relationship are that both parties are confident that

- they can disagree with the other without being abused in any manner, overtly or covertly, and
- they will disagree when their goals conflict or their integrity is challenged.

A disagreement is resolved in a respectful conversation that both stay in until it is complete to the satisfaction of both. Neither is a "yes man." They value this perhaps more than any other aspect of the relationship, as the more successful each is in their career, the harder it will be for them to find someone they can count on for the truth.

Returning to our example, Chuck is the manager of a marketing firm. He and Darrell have sons on the same soccer team that could really use new uniforms. They live in a small town with plenty of civic spirit but not much money for extra school expenses. As his relationship with Darrell deepens, he begins to see possibilities for working with Darrell to develop a fundraising project for the team: 10% of the money spent during "soccer weeks" at "team-sponsor" stores will be contributed by the stores to the team for new uniforms. Possibilities like this evolve naturally in synergistic relationships as each party is encouraged to look beyond the original reasons they had for working together.

In summary, an effective interaction builds trust between your client and you and strengthens your relationship. Building a strong relationship with your client allows both of you to deal with increasingly difficult challenges and have effective interactions in an ever-widening set of circumstances.

Consultations and Collaborations

As we did with Chuck and Darrell, let's use an anecdote to bolster the point. June takes her daughter Caitlyn to see her pediatrician Dr. Latour. She is seeking advice on how best to treat a cough that has suddenly appeared. The medical aspects of their interaction will always be *consultation* and hierarchical because Dr. Latour has more medical knowledge than does either June or Caitlyn. Typically, fees are paid by June to Dr. Latour for the consultation. However, when they and Caitlyn meet together to produce a treatment plan for the cough, *collaboration* will occur if all three have the opportunity to have a voice in developing the plan, as well as developing a plan for how to resolve any differences that may arise. This process is important to Dr. Latour if the goal is compliance on the part of both June and Caitlyn to apply the treatment successfully. Hopefully, this action will result in curing the cough.

Later that day, Dr. Latour meets with two other pediatricians to explore combining their practices. This meeting is potentially the beginning of a long-term *collaboration* among the three. To successfully birth the collaboration, a prerequisite is that the three of them agree on various questions, including the following:

- How will revenues be distributed among the three?
- In what order will their names appear on the letterhead?
- When there is a disagreement among the partners as to how best to treat a patient, how will the disagreement be resolved?

Conversational Roles: Speaking and Listening

Volumes have been written on "the art of conversation" and for good reason: thoughtful conversation is the foundation of successful communication. The *art* portion of the phrase refers to the relative ease with which we begin and end the exchange, how smoothly we switch between the role of speaker and the role of listener and vice versa, how efficiently we present our message, and how carefully we respect and share the space and time allotted. When done well, a conversation is like a dance; when done poorly, it becomes a duel!

The Role of Speaker

Before you say anything, be clear about what you *intend* to communicate. Concentrate on your message and your choice of words, sequence, and length that will best convey a clear message. Consider your vocabulary and how it will be interpreted. Everyone has certain words that trigger personal reactions, but you will not have this information for each individual you encounter, especially unfamiliar persons, so pay attention to your listener's reactions.

Be attentive to your use of slang—words that are specific to your profession, culture, or generation. Different interpretations of certain words among various groups may surprise you and confuse the other person or worse. My years working in the United Kingdom provided this experience regularly. Being clear and true to your message is your major goal.

After you have spoken, consider what you said and compare it to your intention. If your words don't match your intention, immediately retract your statement and try again.

As a speaker who has delivered a message that reflects your intention, you have one more step to consider: has the listener interpreted the message in a way that reflects your intention? To assume that your listener has understood your message could be a major miscalculation. You can avoid this misstep by saying something similar to, "Would you please rephrase what you just heard so I can check to see if I have omitted anything you will need?" Be certain your listener understands that you are evaluating your delivery and not their comprehension. The more important the message, the more critical it is that you engage in checking for mutual comprehension.

STUMBLING BLOCK

Often what you hear isn't what your client intended to say, and vice versa.

The Role of Listener

In a successful conversation, the roles of speaker and listener are fluid, and exchanges between the conversational partners are relaxed and respectful. If you find yourself in the role of the speaker vastly more than listener, that is called a lecture, and listeners are often offended, so learn to be a great listener. It is a skill that will serve you well.

As a speaker only moments before, you now assume the role of listener, hopefully with as much resolve and attention as when you spoke. Many people find listening more difficult than speaking, often because we are likely to be planning our next comment rather than paying attention to the speaker. Instead, listen as if you have never been presented with any of the material before, especially if you have heard it a hundred times! Your attentiveness often pays off in discovering information that you missed or that was added or revised.

In the role of listener, to assume that you understand the speaker is just as dangerous as it was when you were the speaker. As a listener, you can approach this stumbling block by saying, "This is a complex topic. May I confirm that I have correctly interpreted what you said?"

Listening is a learned behavior. Ask any parent! And careful listening will serve you for a lifetime.

Body Language

Many books have been written about body language. Most people are familiar with the rudiments of what they say. For example, a popular interpretation of body language is to view someone sitting with their arms crossed as resistant to input, perhaps even impervious to outside information. The problem is that sometimes people sit with their arms crossed because they are cold. A good approach to body language is to attend to the hypothesis that this posture suggests and check it out. For example, in the crossed-arms situation, you could ask, "I'm sorry I forgot to ask. Is the room temperature comfortable for you?" There is so much information in what your client says that my usual advice is to attend to the spoken words and sort out any confusion you encounter there.

However, useful information about nonverbal communication that goes well beyond that found in books on body language is presented in a TED talk (Cuddy 2012) entitled "Your Body Language Shapes Who You Are." I encourage you to listen to it.

Speaking

There are four broad categories of statements that are useful in interactions: Requests, Demands, Promises, and Assertions.

These statements differ in their forms. Understanding these differences will enhance your ability to communicate effectively. Each is explored in detail below.

Requests

A complete request has two components: a detailed description of the task requested and a statement of by when the task is to be completed. To omit either component can easily derail an interaction. Sometimes when we are reluctant to make a request, we leave out the detailed description or the "by when."

The three possible responses to a legitimate request are to accept, deny, or deny with a counteroffer. When making a request, be honest with yourself: Are you willing to consider all three of the responses to a legitimate request? Or is "accept" the only tolerable response? If so, you have not made a legitimate request; you have made a demand. People often try to camouflage a demand as a request.

When denying a request, a denial with a counteroffer gives the other person more information and may facilitate finding a mutually satisfactory way forward.

> "What kind of graph will you use in the final report?"
> "I'm not sure. I can give you a simple example of what's possible. Would that work for you for now?"
> "Sure."

Demands

In contrast to a request, the only acceptable response to a demand is yes. Calling a demand a request will damage your relationship with your client. The distinction between a request and a demand is important. Being polite will not transform a demand into a request. Most anyone will see through such an attempt at camouflage.

Rarely does someone say, "I demand that you do this." More frequently, you become aware that a request is in fact a demand when all counteroffers that you make are declined. Eventually you realize that the only acceptable response is yes. You may wish to clarify matters by asking, "Is this really a request, or is this a demand?"

When your boss makes a demand of you, learn what the consequences of denying it are before you do so. If the price of denial is high, you may wish to reconsider your denial.

Promises

A complete promise, like a request, has two components: a detailed description of the task promised and a statement of by when the promise will be fulfilled. Both components are required. Sometimes when we are reluctant to make a promise, we leave out the detailed description or the by-when date.

Whenever you make a promise to your client, you have three options: keep it, break it, or revoke it. The first two are obvious. The third is less so.

As the by-when date draws near and you realize that you don't have the necessary resources to complete the promised task on time, revoke your promise. Likewise, when you realize that any part of your promise is in jeopardy, revoke or replace it with the one you can fulfill.

This action is distinctly superior to breaking your promise. Your client will still be upset that you are not delivering on your original promise and will probably be less upset if you inform them about your situation. The sooner you choose to revoke your promise, the less upset your client will be. Revoking your promise at the last minute will be interpreted as a broken promise.

A friend regularly borrows my lawnmower. His skillful use of requests and promises, coupled with his generosity, makes these interactions effective.

> "It's time for me to mow my yard again. When will your mower be available next week?"
> "Tuesday through Friday."
> "Okay. I will pick it up on Wednesday and return it on Friday with a full gas tank and with the oil checked."
> He picks it up on Wednesday and always returns it by Friday evening, with the gas tank full and additional oil, if necessary.

Over the past 2 years, we have created a synergistic relationship based on these effective interactions. Our conversations have expanded to include discussions of many challenging professional or personal situations, leading to a good neighbor relationship.

Assertions

An assertion is a statement that you can prove. "I have read two books in preparation for this meeting" is a valid assertion if, indeed, you have read two books and can show their application to the topics in this meeting.

Rigorous Listening and Speaking

Clients also make requests, demands, promises, and assertions. An important part of working with your clients is to find out "by when" and "conditions for satisfaction" for each request and promise they make. For critical assertions, it is also important to find out what evidence your client has for them. These conversations can be challenging when your client is your boss or anyone with authority in your world, such as a parent, teacher, doctor, or police officer.

During an interaction, your client will have questions about your statements. You may have noticed that your responses to these questions vary. One response is a willingness to consider any and all questions about your statements. On the other hand, you may get defensive if you discover that your statement under scrutiny has been rejected. Think carefully about your requests, promises, and assertions before making them. Calmly explaining statements when they are questioned reflects your integrity and accountability. Staying present, focusing on your intent while speaking, will improve the quality of any interaction.

RECAP

The POWER process consists of five steps that you can use to produce effective interactions: *prepare, open, work, end,* and *reflect*. The results of using the power process depend on the strength of the relationship you have with your client, how well you listen, and how rigorously you speak.

Part 2: Exploring the RAPID Process

Even while faithfully utilizing the POWER process, unexpected events (breakdowns) will inevitably occur. RAPID examines a plan to recover from any interruption or anomaly that occurs so that you and your client can rapidly return to productive work.

There are five steps in the RAPID process:

Recognize your emotional cloud.

Address your emotional cloud.

Pinpoint the breakdown.

Identify how to get back on track.

Do it.

Dissecting Mistakes and Breakdowns

When you are driving and notice that your speed has crept up to 15 mph over the speed limit, you immediately reduce your speed and continue on your journey without risking a traffic ticket.

This is the gold standard for a rapid recovery from a mistake that could have had adverse consequences if not corrected. You saw the mistake and immediately took action to restore risk-free travel.

The life cycle of a breakdown begins with a process functioning as intended. Then an unintended event occurs.

Any event can be unintended in the eyes of the client or the consultant. An unintended event can occur in at least four ways:

1. **A natural disaster**: a hurricane or flood.
2. **Unmet expectations**: the client expected their consultant to be on time for their meeting today. The consultant wasn't.
3. **Blocked goals**: the client's goal was to complete their entire project in today's interaction, and due to new information, they didn't.
4. **Broken promises and lies**: we expect people to keep their word and tell the truth. When they don't, there is a serious disruption in communication.

The moment you become aware that a breakdown has occurred, you will experience upset. You are now in an "emotional cloud." The larger the breakdown, the darker the emotional cloud. This cloud is a jumble of emotions that obscure your ability to see anything clearly, and it grows with every passing moment. Learning to deal effectively with breakdowns is well worth the cost involved if you are committed to improving the services you provide. Pinpointing exactly what breakdown has occurred, coupled with coaching on how it happened and what alternatives are available, can produce insights strong enough to transform your world.

The Onset of Emotional Clouds

An emotional cloud is triggered by a failure of which you may or may not be aware. Often there is no perceptible amount of time between the actual event and the onset of the associated cloud where your upset has taken residence. Being in an emotional cloud can feel like being overtaken by the black ink cloud that a squid sprays when threatened, the purpose of which is to obscure visibility. For humans, the dark cloud also short-circuits brain function.

Alternatively, this cloud can range from a classic San Francisco pea soup fog to a light morning mist. The thickness of the fog depends on a variety of factors, including how significant the event is in your eyes or your client's eyes, what type of relationship each of you has with the other, and the mental, physical, and emotional condition of each person when the breakdown occurs.

If you are unaware of an emotional cloud or the preceding failure, you are in a precarious position. An unrecognized breakdown can become an invisible force with the potential to destroy a relationship.

An event may be a serious mistake for you and not for your client, and vice versa. People have different expectations, standards, and sensitivities. You may regard a meeting as successful; whereas, your client may view that same meeting as littered with breakdowns. For instance, what you perceive as "good counsel" they interpret as manipulation.

Early Warning Signs

Physical Sensations and Actions

Early warning signs include physical sensations you experience as you realize that an interaction isn't going well: a racing heart, sweaty palms, a tightening of your shoulders. Physical behaviors may also occur at this moment: speaking loudly, squirming in your chair, shuffling papers, doodling. Alternatively, a person may shut down: unable to move, going blank, gazing out the window.

Learning to Recognize Your Emotional Clouds

When you realize that something has gone wrong, what is your immediate reaction? Do you know what your early warning signs are?

I have trouble acknowledging that I have made a mistake. Even the thought that I contributed to a mistake can trigger an emotional cloud. My first reaction under such conditions is to become defensive. I quickly check if anyone has noticed. If not, my knee-jerk reaction is to deny, cover-up, or blame. Here's my historical model: if my mother discovered that half of her newly baked cookies were eaten: I'd plead innocent and point to one of my two younger sisters!

Breakdowns provide signals that the relationship or interaction needs attention, care, or understanding. A session in which all goes well is a worthy goal; however, when mistakes do occur, they are best seen as opportunities to solve problems. A complaint may become an actual blessing. This reminds me of the time when my friend was tested for a minor illness, and a serious condition was discovered. Because the lab technician approached the minor complaint as a possible signal, a rapidly growing tumor was discovered and treated. With an attitude of curiosity rather than blame, a focus on problems can generate a wealth of important information.

When you first begin to study your emotional clouds, you may not become aware of having been in one until well after it has dissipated. A learning strategy is to recall as best you can the events that occurred immediately before the cloud began to form. As you continue to do this with each episode, you will notice some events that occur regularly before your clouds form, like having your advice questioned. These events are ones that you regard as breakdowns.

The Recognize Checklist

1. How rapidly do you recognize the onset of emotional clouds? Are some easier for you to detect than others?
2. What are your early warning signs that indicate you are on your way toward, or already in, an emotional cloud? Physical events? Emotions? Thoughts?
3. Which defensive strategies do you employ: hiding, denying, or blaming? Others?

Mustang

There was a time when I was broke, yet I managed to save enough money to repair the dents and dings in my prized 1969 Mustang and have it painted.

At dinner the first night the car was back from the body shop, one of my sons, then nine, said, "Dad, some of the Mustang letters on the trunk are loose." I immediately reacted with exasperation, "No, they're not! I just paid a lot of money to have the car fixed and painted. It's fine!" The rest of the meal passed in silence.

Instantly an emotional cloud surrounded me, as evidenced by the wave of intense emotions that swept over me. I had invested time, energy, and all my discretionary cash in this project. To have my young son imply that my choice of body shops was mediocre was humbling at best. While in the cloud, I dismissed his input. The result was a bigger breakdown a week later.

This situation helped me to learn how important it is to recognize and manage anger in challenging circumstances. Anger is an early warning sign and part of many emotional clouds. My anger camouflaged my emotional cloud for some weeks until I was discussing this incident with colleagues during a classroom exercise in which we were examining situations in which we had been given good advice and did not take it.

Let's suppose you have identified several of your early warning signs and occasionally *recognize* that you are in an emotional cloud during an interaction. Now what? A key question is whether you think you can *address* and exit this cloud during the interaction without causing damage.

The darker the cloud, the rougher the ride as you seek to exit.

If an emotional cloud has you unnerved, you may find yourself saying words that you know you will regret the moment they leave your mouth. Being distracted by thoughts about how someone else would be handling the problem only results in missing critical information from your client. Part of dealing effectively with breakdowns is operating in a fashion that is consistent with the severity of the cloud you are experiencing. You would not increase your speed in the middle of a fog bank, would you?

Anger is often a reaction to a breakdown in communication and is used as a defense against the experience of vulnerability. No one is perfect, and we do not live in a perfect world. When vulnerable, I find myself using anger—toward myself and others. While anger does distract me from vulnerability, at least in the short term, it does not present a way out of an emotional cloud.

Pride interrupts the examination of our clouds. No one likes to make a mistake, though we all do. Often this threat to pride is handled by blaming the cloud on someone else: "You made me mad at lunch today." Some people develop a habit of blaming others and resist considering the possibility that they may have played a part in creating disagreement, misunderstanding, or mistake.

Fear can also drive both denial and blame.

Intimidation is another barrier to asking important questions.

Denial sometimes shows up when we resist the fact that we are in an emotional cloud.

Overwhelm occurs when a cloud is so intense that you are unable to address it immediately.

Becoming masterful at recognizing emotional clouds during an interaction requires time, commitment, and courage. Is it worth it? Yes, indeed! Consider how long it takes to clean up one nasty comment blurted out in the midst of an emotional cloud. Sometimes our sudden exclamations,

Notice and Act on Early Warning Signs

What to do? First, begin to notice the signs that you are in an emotional cloud and admit it, at least to yourself. We each have our own signature patterns. For example, anger and anxiety form an emotional cloud that regularly envelops me after a breakdown occurs. The progression begins with being only vaguely aware that something is wrong, long before I can pinpoint the actual breakdown. These feelings, thoughts, or hunches are the early warning signs that tell me, "There is a problem here somewhere."

Mustang

A week after my confrontation with my son, I was pumping gas and was surprised to see that I was now the proud owner of a MU TANG. An external force (gravity) intervened; the S dropped off! Now it was impossible to ignore my son's input, no matter how much I disliked hearing it! Time and external factors often lead a person to *recognize* and *address* even the most noxious emotional cloud. I finally checked out my son's assertion. I wiggled the other six letters and discovered that three more were loose! For me, becoming annoyed with my son was an early warning sign that I was about to unconsciously dismiss his input rather than consciously consider whether his observation had merit. From this I learned a strategy to implement when I feel myself becoming irritated: I pause to identify the source of my vulnerability and then address that feeling of dismay. For example, I could have suggested we check out which letters were loose. Implementing this strategy has helped me to stay in the conversation in many difficult situations—a large return for the cost of a single S!

Another issue that hampered my ability to deal with the Mustang cloud was my discomfort with the prospect of confronting mechanics at the body shop. I often am intimidated by experts who know more about my car than I do, so I never talked with the mechanics. This cloud was not addressed.

The Address Checklist

1. Do you recognize that you are in an emotional cloud? How much time passes before you are aware of this cloud?

2. Do you have a partner to help guide you as you *address* your cloud? If not, find one.

3. What are the barriers standing between you and addressing your cloud? Denial, anger, blame, pride, intimidation, guilt, and fear are frequent culprits.

4. What early warning signs between a breakdown and the beginning of your cloud are you beginning to notice?

The Heart of RAPID Recovery

In this chapter, we will explore two steps that are at the heart of RAPID recovery: *pinpoint* the breakdown and *identify* how to get back on track.

Pinpoint the Breakdown

Sometimes you recognize mistakes during an interaction, and sometimes it's a week later. As your awareness increases, you will recognize your errors as they occur, allowing you to address them more quickly. Mastering this process will save both you and your client much time, energy, and emotional distress.

Once you recognize an emotional cloud and decide to face your emotions, move on to *pinpoint* the specific breakdown.

A key step in pinpointing a breakdown is to search for the unintended event that was the source of the breakdown. Sometimes a promise has been broken. This is certain to produce a high level of distress. Surprises and unmet expectations will also produce upset, as will blocked goals. Any one of these can derail an interaction. Sorting out which causes have occurred (broken promise, unmet expectations, or blocked goals) is the challenging work involved in pinpointing a breakdown.

Reluctance to Ask for Assistance Knows No Age Limits

Multiple interactions with seniors in their eighties and nineties demonstrate that being more committed to appearing competent than to getting the job done is a lifelong challenge. Here's a personal story that clarifies the process.

I had a wonderful friend Daniel who was in his mid-80s, having survived a stroke. My wife and I were visiting him and his wife in their retirement community. They invited us to join them for dinner, which was a coat-and-tie event. All of us, except Daniel, were dressed and ready to leave for the dining room. Here's the interaction that followed between Daniel and me:

Daniel:	Just go ahead. I will catch up with you.
Me:	Is there anything I can do that would be of assistance?
Daniel:	No, thank you. That's not necessary. I will be down in a couple minutes.
	We three waited at the table twenty minutes until Daniel appeared, looking distressed. I asked, "Are you okay?"
Daniel:	Yes, it's just this darned tie. It is so hard for me to tie anymore!
	After dinner, I waited until he and I were alone and asked, "How did you get it tied?"
Daniel:	I tried and tried until I was really upset. An aide looked in and asked if she could help. I sheepishly asked her to tie my tie. Ach! This is the darned pride thing again, isn't it? (We had discussed pride on several earlier occasions.)
Me:	I think so.
Daniel:	I thought for sure that I'd be done with vanity by now. Me: I don't know that it ever leaves, Daniel.
Daniel:	It doesn't look to me like it does.

Daniel's expectation that he should be able to tie his necktie was not being met. It would continue not being met until he could accept that this task was something he could no longer do. Accepting the loss of various competencies is tough, as it indicates that he is one step closer to death. Yet, accept it he must if he hopes to move past having frustration accompany every dinner.

Mustang

The first breakdown to pinpoint was that the body shop did not attach the Mustang letters properly. The second was that I rejected input from my son, who had revealed the first one. This judgment was based solely on my assumptions about his powers of observation. Pinpointing the second mistake came to light only when I had conversations with colleagues about this event.

Identify—Getting Back on Track

A complete job of pinpointing the breakdown is an essential first step toward getting back on track.

A key barrier to getting back on track is what you see as your part. Define your part and realize that you and your client may not immediately agree on who is to do what. Stay in the conversation until you both agree on the required actions.

An Effective Apology

When you pinpoint a breakdown as a broken promise, profuse apologies such as "I'm sorry; I am so sorry; you don't know how sorry I am" won't save the day. Two steps are essential for continuing the recovery process.

Step 1. Initiate an *effective apology* by acknowledging that you broke your promise, and then an "I'm sorry" may help. However, stopping here can lead to disaster in the long run.

Step 2. Ask how you can address the consequences of having broken your promise. This step may feel awkward, and it is necessary for the recovery conversation to succeed.

The Pinpoint Checklist

1. How much time passes before you pinpoint your breakdown?
2. Do you have a partner to help guide you as you pinpoint your breakdown? If not, find one.
3. What are the barriers standing between you and pinpointing your cloud? Denial, anger, blame, pride, intimidation, guilt, and fear are frequent culprits.
4. What early warning signs between a breakdown and the pinpointing of your cloud are you beginning to notice?

Identify

Here are some thoughts on identifying how to get back on track in the three cases discussed earlier in Part II.

Mustang

What had to be done to get my relationship with my son back on track was to apologize for snapping at him and discounting his input.

Had I been able to *recognize* and *address* my emotional clouds, I could have said to my son, "Really? Let's go see how bad it is." Then we would have discovered that four letters were loose. I could have returned the car to the shop the next day to get the job done right, provided I could have overcome my fears about the body shop conversation. The longer you wait to address the upset, the harder it is to deal with those fears. With each passing day, they take on a life of their own and negate any "good intentions" that may cross your mind.

In hindsight, what I learned was that embracing and exploring emotional clouds and breakdowns as soon as possible is what produces breakthroughs. Unfortunately, by the time I apologized, enough time had passed so that my son had forgotten the incident. This was a great disappointment and a lesson in humility.

The Identify Checklist

1. Do you make either an effective apology or a complete request when appropriate?
2. Does each of you *identify* what you think has to be done to get back on track?

Do It!

By now, you have gone through the first four steps in the RAPID process. You have *recognized* a breakdown and the attending emotional cloud, *addressed* them both, *pinpointed* the specific difficulty, and *identified* how to get back on track. Now for the last step: *do it*.

After identifying what actions are essential for getting back on track, now is the time for action as you put the cleanup plan into operation. Don't just talk about it. Realize that this concentrated effort will take time, patience, and perseverance, especially if this is a repeater breakdown.

If you are reluctant to spend the time and effort required for this action step, consider the consequences of *not* recovering from this breakdown. Remember that problems like the present one will likely show up again. Practicing the RAPID recovery process at every opportunity will ensure less time spent recovering from future breakdowns. The choice is yours.

Barriers to Taking Action

Another aspect of taking action is staying in the conversation with your client (supervisor, subordinate, spouse, etc.) until the breakdown is resolved. This strategy is a challenge that requires you to manage your anxiety, anger, and impatience so that you can persevere until the recovery is complete.

Staying in the conversation requires looking at your vulnerability when you have been a party to a breakdown. You can ask for forgiveness for your part in creating the breakdown, *and* you cannot force your client to forgive you. This vulnerability is a consequence of creating problems (e.g., breaking promises). Keep in mind that we are all in the same boat as we seek to learn how to recover from the breakdowns that inevitably occur in our lives.

One more point: even if you have broken a promise, recognized, addressed, and pinpointed the breakdown, identified a recovery plan, and executed it, one step remains before recovery is complete: *you must forgive yourself.*

Mustang

It was clear to me shortly after recognizing my mistakes that to restore my relationship with my son, I had to apologize to him for both snapping at him and discounting his input. Unfortunately, I repeated my Brenda behavior and did neither one at the time. By the time an apology was offered, he had forgotten the incident. This outcome was much like the one with my doctor in the Sudafed episode. These experiences have resulted in more timely apologies and reduced the prolonged agony of fuming over an unpleasant breakdown.

The Do It Checklist

1. Review why recovery from this breakdown is important for you.
2. Implement the recovery plan you have created with your client.
3. Verify that this plan has addressed all the fallout from the current breakdown. If not, address the remaining upsets.

RECAP

Review why recovery from a particular breakdown is important to you.

Implement the recovery plan. Verify that the results of the plan have satisfied each of you.

I invite you to consider your next breakdown as an opportunity; use the RAPID process and see what gold you can mine.

This material is extracted from Zahn, D.A. *Stumbling Blocks to Stepping Stones: A Guide to Successful Meetings and Working Relationships.* 2019, iUniverse, Indianapolis.

Index

Note: **Bold** page numbers refer to tables; *italic* page numbers refer to figures.